Geometric Procedures for Civil Engineers

Elias C. Tonias • Constantine N. Tonias

Geometric Procedures
for Civil Engineers

 Springer

Elias C. Tonias
Tonias Engineers
Pittsford, NY, USA

Constantine N. Tonias
The CEDRA Corporation
Pittsford, NY, USA

Additional material to this book can be downloaded from http://extras.springer.com

ISBN 978-3-319-79599-7   ISBN 978-3-319-24295-8   (eBook)
DOI 10.1007/978-3-319-24295-8

Printed on acid-free paper

This Springer imprint is published by Springer Nature
The registered company is Springer International Publishing AG Switzerland

This book is dedicated to all those civil engineers of the 1950s, 1960s, and 1970s who labored endlessly to develop electronic computing software under primitive computer operating systems and hardware for the benefit not only of their business entities but of the civil engineering profession above all.

The authors wish to acknowledge the contributions of their friends and colleagues, too numerous to list, in the civil engineering staff of various civil engineering institutions, the U.S. Army Corps of Engineers, the early members of the Civil Engineering Program Applications (CEPA), the staff of the American Society of Civil Engineers (ASCE), and the International Business Machines (IBM) Corporation.

Preface

It is said that next to the Holy Bible, Euclid's *Elements* and related books are the most published books. Most of these books on geometry concentrate on the theoretical aspects of geometry and various geometric constructions. *Geometric Procedures for Civil Engineers* has been written not only for civil engineers but also for those who are interested in the preparation and use of computer-aided solutions to most common geometric constructions. Computer-aided drafting (CAD) programs have removed the tedious aspects of finding the solution for geometric construction problems, but they also have removed from the user's mind the geometric principles and methodology of the solution of said constructions. The actual product of this book is an extensive set of computer procedures that can be used to solve individual geometric constructions, and even to help create a customized geometric design system or supplement an existing system with specialized add-on procedures. The description of the various computer procedures includes a presentation of the geometric principles of the solution, and a general textual description of the computer steps to accomplish the solution as part of the book. The source code in generic Visual Basic format of the procedure and of a test program is included as part of an accompanying compact disc (CD). Although this book is not intended as an instructional book, its contents do provide solutions to numerous geometric construction problems that could act as a supplement to an instructional book. A basic knowledge of Euclidean Geometry is assumed. The book is comprised of three parts.

Part 1 is composed of the first three chapters and concentrates on certain historic and rudimental aspects of geometry.

Chapter 1 presents an historic synopsis of geometry from its early years through the Hellenistic period, the Middle Ages, the Renaissance years, and up to the current computer revolution. The intent of this chapter is twofold, (a) to familiarize the reader in the evolution of geometry from its infancy to its present state (early twenty-first century), and (b) to acquaint the modern engineer and computer-aided design/drafting (CAD) user of the "good old days" prior to the introduction of and during the early years of not too long ago of the computer revolution. Several photographs of computer hardware have been included. In the early years of the computer there was a distinct appearance among the computers of the various manufacturers, whereas nowadays they all seem to look alike. The reader should note that the names of the various Greek mathematicians and geometers are written in this book in their Greek form and not in their Latinized form such as Apollonios rather than Apollonius. It seems unnecessary to transliterate from one language into a second via a third language.

Chapter 2 presents the definition of the basic geometric features such as points, lines, and curves and classification thereof, as well as of certain terms and conventions as used in the book.

References in this textbook to an accompanying compact disc (CD), which are not published on a CD and there is no disc included with the print edition of this book. The electronic supplements are available without charge at http://extras.springer.com by entering the print ISBN.

Chapter 3 presents various rudimental procedures referred to as support procedures that are used extensively by the construction procedures of the subsequent chapters. These support procedures are assembled in the following groups that:

- Compute basic and inverse trigonometric functions (see the commentary after the last bullet).
- Provide for the conversion of various forms of angles and directions, such as from degrees in decimal form to azimuths in radians, and vice versa, as well as the extraction of degrees, minutes, and seconds from a text string.
- Compute arc lengths, distances, and directions between points.
- Compute angles under a variety of input data conditions.
- Provide for miscellaneous operations including the computation of elevations, Lagrangian interpolation for polynomial curve generation, handling of strings of geometric features, and stationing of distances in the US Customary and SI formats.
- Convert a polyline or polygon comprised of a series of chords into a series of lines and circular arcs if certain chords are small enough to constitute a circular arc defined by a center point, two endpoints, and a radius.
- Convert a distance value into a US Customary or SI station format.

In the infancy years of electronic computing, the trigonometric functions were not a staple item of the programming languages as they are today. Numerous discussions were held regarding methodology, precision and speed until the mid-1960s. Some of the trigonometric procedures presented in this publication are for historic purposes, although some of them are still in use today.

Part 2 is composed of the next five chapters that include various procedures that perform the geometric constructions summarized below. A knowledge of these procedures is considered essential in following the various construction procedures of the subsequent chapters. These procedures may also be used in the programming of entirely new construction procedures that may not be available in this book. In general, in this book, and unless otherwise qualified, the term "curve" in singular or plural form refers to a circular arc. Spirals, parabolas, and ellipses are addressed as such unless it is evident that the term curve or curves pertains to one of them.

The various procedures that are presented in the first two parts of this book may be used to easily develop solutions to problems that may not have been addressed in this book. Similarly, the procedures of the geometric constructions as presented herein may be modified to reflect individual desires, ambitions, and goals.

Chapter 4 contains procedures that create and manipulate points and lines between two points, lines comprised of a series of concatenated two-point lines referred to as polylines and polygons. A polyline comprising a closed figure does not constitute a polygon unless specifically is so made to be. These procedures are assembled in the following groups that:

- Introduce points along a circular arc given a plus arc distance from its start point and an offset distance therefrom, or locating the midpoint of an arc, as well as locating the PC and PT and center point of an arc tangent to two given lines, or finding the PI when knowing the PC, the PT, and the center point of the arc.
- Introduce points along a line or arc given a distance along the line or arc from the start point and a normal offset to the left or right of the line or arc.

- Project points on a two-point line, a polyline or circular arc. Such projections may or may not involve stationing. The procedure projecting a point on a polyline provides a multitude of user options.
- Determine whether a point lies within the extent of a line or circular arc, and rotate and/or scale points about a pivot point.
- Compute the coefficients of the implicit form of a straight line.

Chapter 5 presents procedures that create circular curve features that are tangent to a given line or circular arc, as well as create such features that are not tangent to any other feature. These procedures generate individual curves under a variety of conditions of given curve element composition including length of arc, chord, tangent or middle ordinate, central angle, or passing through a given point. Compound and reverse curves are treated separately in Chap. 8. Also included in this chapter are procedures that provide for the conversion of curves into polyline for graphic presentation or other purposes.

Chapter 6 is dedicated to various geometric intersections of lines and circular arcs under various input requirements. The procedures addressing these intersections are grouped into intersections of (a) lines with lines, (b) lines with arcs, and (c) arcs with arcs. These procedures are followed by a fourth group of five generic intersection procedures that provide for the intersection of two strings of concatenated series of lines and arcs, again under various input and/or desirable results.

Chapter 7 presents procedures that construct tangent lines to circular curves and circular arcs tangent to other circular arcs under various conditions. These procedures are divided into four groups: (a) tangent lines to circular arcs, (b) circular arcs to lines, (c) circular arcs tangent to lines and other circular arcs, and (d) circular arcs tangent to other circular arcs. Lines tangent to spirals are addressed in Chap. 9, and tangent to parabolic curves are addressed in Chap. 11. To construct a circular arc tangent to a spiral or parabolic curve at a specific point, a user needs to locate the point on the pertinent curve and determine the instantaneous radial and therefore the normal direction. Depending on the construction problem and on which parameters have been specified as being known, there could be more than one solution to a problem. The procedures, unless otherwise noted, compute all potential solutions and provide them to the end user for the desirable selection.

Chapter 8 concentrates on the creation of compound and reversed curves. The procedures of this chapter are grouped into three sets or groups depending on what information is to be considered as given, and which features of the curves may vary or float. The first set contains eight procedures that construct pairs of compound curves under different given input requirements, while the second set contains five procedures that construct three compound curves, and the third set contains but one procedure that constructs a multicentric set of compound curves. Generally the combinations of given requirements are the radii, central angles, short and long tangent lengths, and a point through which a curve has to pass through. The back and forward tangents could be fixed or floating as could be the start and endpoint of the curves. The fourth set contains nine procedures that construct pairs of reversed curves. The multicentric procedure of the third group could include one or more reversed curves.

Part 3 is comprised of the remaining four chapters and addresses transition spiral construction problems and composite construction design problems.

Chapter 9 is dedicated to the establishment of highway transition spirals; the intersection of spirals with lines, curves and other spirals; the projection of points on spirals; and the introduction of tangent lines to spirals. A discussion of four other spiral types and a comparison with the highway transition spiral are also provided.

Chapter 10 concentrates on the treatment of dead-end local street alignments such as cul-de-sacs and hammerheads, as well as the introduction of street side bubbles to increase the number of parcels in certain overall land parcel configuration.

Chapter 11 concerns itself with vertical parabolic curves including the location of low and high points; sight and stopping distances and clearances at overpasses; point location along a vertical curve; and parabolic curve fitting through a variety of point conditions, tangencies, and point projections.

Chapter 12 concentrates on ellipses with emphasis on their construction and properties of area and perimeter as they may find use in hydraulic and other civil engineering applications. A simulation of an ellipse by a four centered circular arcs is included.

Appendix A discusses two problems that are encountered in land and marine surveying in locating, determining that is the coordinates of points that can be occupied by an instrument and siting at points of known coordinates that cannot be occupied because they are inaccessible for any reason. These are (a) the three-point otherwise referred to as the Pothenot problem with which a surveyor occupies an unknown point and sights at three known points, and (b) the five-element problem also known as the Hansen problem with which a surveyor occupies each of two unknown points and sights at two known points as well as at the other unknown point.

Appendix B presents three procedures for the fitting of a straight line, circular arc, or polynomial curve through a set of observation points as they may occasionally be encountered in civil engineering. The polynomial curve fitting procedure may be used to fit a polynomial up to the 14th degree with the option to increase said degree if so desired. This procedure may also be used to fit a line (first degree polynomial) or a parabola (second degree polynomial) which can be used instead of the procedure provided in Chap. 11. The first and third of these procedures utilize the method of least squares, but that of fitting a circular arc uses a geometric and statistical approach.

Appendix C concentrates on the algebraic solution to the Apollonian tangency problem of constructing a circle tangent to three given circles or circular arcs. A short history precedes the solution, as is a reference to the correspondence between Rene Descartes and Princess Elisabeth of Bohemia regarding kissing circles.

Geometric Procedures for Civil Engineers is not intended for the theoretical mathematician, nor does it provide new proofs to old lemmata and theorems, and definitely does not introduce any new axioms. It assumes a basic knowledge of Euclidean geometry and concentrates on the actual, production-oriented solution of the construction problem being discussed. For example the orthogonality of the tangent line to a circular arc and of the radial line to the point of tangency is accepted as a fact that does not require proof. Although the primary intent of this book is to present computer-based procedures, it does provide solutions to certain civil engineering and surveying problems.

 Theoretical mathematicians may take issue with the method of solution to certain constructions. While analytic geometry has been used extensively, the use of matrices and calculus-oriented solutions has been avoided as much as possible in favor of the use of rudimental geometric principles. The reason for this dates back to the early days of electronic computing. Matrix operations and high level mathematics involve numerous multiplications and divisions which in civil engineering applications encountered a problem, that of the state plane coordinate systems that treat large tracts of land are treated as being flat and not as ellipsoid areas. State plane coordinates involve numbers of seven and eight digits to the left of the decimal point and a precision requirement of at least four digits to the right of said point in order to ensure three decimal digit precision. For small land-related projects, this precision may be considered by some as too stringent, but for long and complex highway projects it is not. Early computers were not able to meet these requirements, and cumbersome procedures were used to reduce numbers to a workable size and then reconstruct them to their original dimensional basis. Even nowadays this problem surfaces itself when it comes to high demands imposed on geometric information systems (GIS). Surprisingly, low level geometric solutions provide an equal and even faster solution than their high level mathematical solutions.

The computer language used in the production of the procedures is that of Microsoft's Visual Basic. Computer languages come and go, but the concept and structure of Visual Basic is simple enough that it can easily be modified if need be. With very few exceptions use of specialized or proprietary software has been avoided. Regarding the few exceptions, the part of the source code in which they appear can easily be bypassed and replaced with any other appropriate code.

The term procedure pertains to small computer programs that can be called by other procedures, or by a main line or system program to perform a certain operation. Procedures are divided into two classes, *functions* and *subroutines*.

(a) A function procedure accepts one or more input parameters (data or information) and returns a single argument, for example the function procedure for finding:
- The tangent function of a given angle is `Tan` and it is written as
  $A = Tan(alpha)$
  where
  > `alpha` is the given angle in radians and
  > A is the sought tangent function.
- The distance between two known points is `iccomdis` and it is written as
  `D=iccomdis(X1, Y1, X2, Y2)`
  where
  > D is the desired distance between the two points defined by
  > `X1, Y1` the Cartesian coordinates of the first point, and
  > `X2, Y2` the Cartesian coordinates of the second point.

  Each function procedure is described by its function (what it does), its given parameters, and its returned argument. Unless it is quite obvious of what a function procedure does, a description of its geometric logic is presented.

(b) A subroutine procedure accepts as input a number of input parameters and returns a certain number of computed parameters, for example the subroutine procedure `icsmpcrv` for finding the control points of a circular curve tangent to two straight lines when knowing a point on each line (`PTN1, PTE1` and `PTN2, PTE2`), the intersection point of the two lines (`PTN3, PTE3`) and the radius of the circular curve (`R`) is called as follows:

```
Call icsmpcrv (PTN1, PTE1, PTN2, PTE2, PTN3, PTE3, R, _
                CCN, CCE, PCN, PCE, PTN, PTE, IERR, ANGLE, DELTA)
```

where

The underscore (_)	denotes a line continuation (two lines are in reality one),
CCN, CCE	are the returned coordinates of the center point,
PCN, PCE	are the returned coordinates of the PC point,
PTN, PTE	are the returned coordinates of the PT point,
IERR	is a code number denoting a problem that may have been detected, if any,
ANGLE	is the central angle of the arc which could exceed 180°,
DELTA	is the angle of intersection which could be the same ANGLE or its 360° supplement if ANGLE is greater than 180°.

Most all subroutine procedures perform certain input data diagnostics for detecting any possible errors that would cause the computer to "hang up" or return inappropriate results. The parameter `IERR` is used to identify any problem that may have been encountered.

Subroutine procedures are described by their geometric function and geometric logic, a definition of the input and output parameters with any required restrictions, and a verbal description of the general operational (programming) steps. Throughout the book a subroutine and its list of parameters are presented in two parts, the subroutine name and the group of input parameters followed by the group of returned parameters. In the above example the parameter R (the radius) is the last input parameter, and CCN is the first returned parameter starting on a new line. It is possible for the input and returned parameter groups to comprise more than one line as in the following example:

```
Sub icCase3spr1  (xPCN, xPCE, xAZ1, RAD1, RAD2, xROTN, ALPHA1, _
                  THETA, SPRL, STL, STS, xAZ2, _
                  PIN, PIE, PCN, PCE, CSN, CSE, SCN, SCE, PTN, PTE, _
                  xCCN1, xCCE1, xCSN, xCSE,xSCN, xSCE,
                  xCCN2, xCCE2, xPTN, xPTE)
```

in which xAZ2 is the last input parameter and PIN is the first returned parameter. Each input and returned parameter is described individually, and the two sets are presented separately with the returned parameters following the input parameters.

With the available information a person with programming knowledge is capable of following the general logic and creating the necessary source code to develop the procedure. However, no matter how smart and proficient that person may be, the source code and test programs available in the book's accompanying CD will more than compensate for any expended labor. One rule of thumb to remember is that the programming effort of coding the mathematics of a solution is but a small part of the overall programming effort. The handling of the input and output parameter requirements, and logic of handling decision-making processes, requires much more of a programming effort.

All procedures concern themselvnes strictly with the geometric logic and computations of the task at hand. The interactive selection or mass production of the input parameters, and the assessment, interpretation, and presentation of the computed parameters are very much oriented and dependent upon the specific overall computing environment employed and are therefore not considered in this book. For example the above referenced icsmpcrv procedure assumes the coordinates of the three input points as known and does not consider itself on how to obtain their values. Similarly the returned points of the PC and PT and of the center point do not generate or display on the computer monitor the circular curve itself.

The generation of the graphic of a circular curve is a rather simple task. However, spiral, parabolic, and elliptical curves are not as simple. Therefore the procedures that generate for these curves do return an array of coordinates of points along the curve to easily generate the curve graphic. The smoothness of the curve (each curve is represented by a series of chords) is controlled by the spacing of the returned coordinates of the endpoints of said chords, and this is controlled by the user via an input parameter.

Unlike many publications, the numbering system of figures, tables, and plates is not sequential from 1. Instead their number is parted in two with a prefix to the left of a dash line (-) denoting the chapter and a number to the right of it denoting the section of the chapter. The section number is then followed by a letter from a to z, which may be followed with a number from 1 to 9 indicating their sequence within the chapter section. Their number is preceded by the word Fig., Table, or Plate depending on the contents.

It is difficult for current computer programmers to comprehend the difficulties and time consumption required to develop a relatively small program in the early days of electronic computation. Computer memory was extremely small and at a premium monetary value. Thus, the number of hours expended to fit a program into a small computer was numerous. At times it took days to try and eliminate a few words off a program so that it could compile and run. Furthermore, programs could not be developed interactively. A programmer had to write the program source code statements and input data on a special paper form, key punch them on cards or paper tape, load the cards or tape on the computer, run the program and "luck" permitting get and print results to analyze them, identify errors that may have been made, and start all over again. If the programmer was with a larger company, a key punch operator may have been available to relieve the programmer of the key punch operation, but the key punch operator may have to shuffle key punch time between various programmers, and users of the various computer programs. Furthermore, there was also the issue of setting priorities as to which operation or which project had a higher processing priority than another operation or project. Program development and production processing was not as fast as some may have wanted it to be. So program development was not a quick process as one might think today.

The work of this book is the joint effort of a father (Elias C. Tonias, P.E.) and son (Constantine N. Tonias, P.E.) team, and represents the work expended under various electronic computing environments for civil engineering applications from 1958 to the present. It originally began by Elias for the development of a program for the geometric design of curved bridges on an IBM 650 computer operable under the Purdue University X-2 compiler as part of his graduate work, and through various computer adaptations it evolved by Constantine into a full-fledged civil engineering geometric system operating presently within the Environment Systems Research Institute (ESRI) Arc-GIS environment. Special acknowledgements must be made and thanks offered to another son, the now Reverend Father Demetrios E. Tonias, Ph.D., P.E. for his engineering and programming efforts in creating and converting certain of the procedures presented in this book prior to his seeking service in the priesthood of the Almighty in the Holy, Orthodox, Catholic and Apostolic Church.

The authors' appreciation is also offered to the authors and publicists of certain publications for ideas and methods that have been proven helpful in the development of certain procedures in this book. Reference to these authors and publicists is made throughout the text of this book where their contributions have been utilized, and their publications have been listed in the bibliography section of this book.

For a rather extensive and interesting reading on the history of engineering and mathematics as they relate to engineering, reference is made to the book *Engineering in History* by Richard Shelton Kirby, Sidney Withington, Arthur Burr Darling, and Frederick Gridney Kilgour, McGraw Hill Book Company, New York, 1956, and Dover Publications Inc., New York, 1990, and to the books of Sir Thomas Heath such as *A History of Greek Mathematics* (two volumes) and *The Thirteen Books of Euclid's Elements* (three volumes).

Special acknowledgement and thanks must also be extended to Wikipedia and its contributors for various information and particularly for some of the photographs in Chap. 1.

The original version of the preface was revised. An erratum can be found at DOI 10.1007/978-3-319-24295-8_13

Colophon

Geometric Procedures for Civil Engineers was typeset using Adobe® InDesign® CS6 Desktop Publishing Software of Adobe Systems Incorporated, San Jose, California using Minion Pro font of size 10 and leading 15.

With the exception of a few simple equations that were written using said publishing software all other equations were prepared using the ΣMathType™ 6.0c software of Design Science (DSI) of Long Beach, California.

Figures 9.9a, 9.9b1, 9.9b2 and 9.9b3 were created using AVcogo® software of The CEDRA Corporation of Pittsford, New York, operating as an extension of ArcGIS® of Environmental Systems Research Institute (ESRI) of Redlands, California. All other figures were prepared using Corel Designer® 10 and Corel Designer® 12 of Corel Corporation of Ottawa, Ontario, and were then exported into JPEG format at 600 dpi.

All photographs of Chap. 1 were taken with an Olympus Stylus 800 digital camera of Olympus Corporation of the Americas of Center Valley, Pennsylvania, and all internet downloads were slightly enhanced by Adobe® Photoshop® 5.0 LE of Adobe Systems Incorporated, San Jose, California.

Contents

Historic Synopsis

The word "geometry" is derived from the Greek word "γεωμετρία," which is a fusion of the Greek noun "γέα" (also known as "γαία," "γαίη" or just plain "γή," the latter being also the modern Greek word) meaning earth, land, ground, soil, tract, district, nation, native country of a person's origin, or even the entire globe of the earth, and the Greek noun "μέτρον" meaning measure; that is, geometry pertains to the measurement of a piece of land, or even of the earth as a whole, and fairly much to the work of the present-day land surveyors and topographic engineers. From the first time that the ancient Greeks coined the word geometry, geometry has expanded to encompass the measurement, and construction of most of all figures, real and abstract.

Since the formalization of engineering as a university curriculum, and the creation of special engineering schools, surveying has been a staple item of engineering education. Even up to the mid-twentieth century a basic course in surveying was a required subject course included in all engineering curricula in many engineering schools, and has remained a staple branch of civil engineering. Since licensure became a governmental requirement for the practice of engineering in the USA, it was possible for civil engineers to practice land surveying, while non-civil engineers required a special license to do so. During the 1970s, engineers that dealt primarily, if not exclusively, with the survey of lands for the establishment of boundaries and the development of land into finished residential, commercial, and/or industrial parcels that involved title of ownership broke away from civil engineering, and evolved into the distinct practice of land surveying requiring its own licensing process.

However, civil engineers may perform surveying functions for engineering projects as long as title of land and ownership of land is not involved. Thus, there still is a surveying branch within civil engineering practice, and to distinguish it from land surveying, the American Society of Civil Engineers (ASCE) has coined the word Geomatics, which is a composite of the above said word "geo," and "matics" which is an aberration of the Greek form of the

The General Epochs of Geometry	
Babylonian:	3000 BC–1500 BC
Egyptian:	2000 BC–1000 BC
Greek:	600 BC–300 BC
Hellenistic:	300 BC–300 AD
Byzantine:	300 AD–1400 AD
Hindu:	500 AD–1100 AD
Arabic:	800 AD–1200 AD
European:	1500 AD–1900 AD
Modern:	1900 AD–Present

ending of an adjective as used in the word "mathematics." In its true sense, the word "mathematics" refers to a lesson (μάθημα) and not strictly to the mathematics as we know it today.

1.1 The Greek Age

Although the word geometry is of Greek origin, its function and use is not a Greek invention. If we are to consider the literal meaning of the word as the measurement of land, evidence of its use may be traced to more ancient civilizations, such as the Babylonian civilization in ancient Mesopotamia in Asia (present-day Iraq) and the Egyptian civilization in north Africa (present-day Egypt). Babylonians and Egyptians used geometry to mete tracts of land as long ago as the third millennium before Christ. Through the years, this knowledge progressed from land measuring to problem solving, decorative designs, and erecting monuments that survive to the present. Babylonians used geometric terms, length and width, to solve algebraic problems, such as to find a surface area, or to see how far out a ladder leaning against a wall will move if its top is lowered a certain length. Both Babylonians and Egyptians used geometry to create ornamental figures in their floor tiles and wall decorations, and to layout and

© Springer International Publishing Switzerland 2016
E.C. Tonias, C.N. Tonias, *Geometric Procedures for Civil Engineers*, DOI 10.1007/978-3-319-24295-8_1

construct irrigation projects and pyramids. In both of these civilizations, the knowledge of geometry rested within the realm of a few noblemen and particularly of the priests.

US president Abraham Lincoln studied geometry to train himself in deductive logic so as to better prepare himself for his legal arguments before a judge and jury.

According to the Greek historian Herodotos, who had visited Egypt, the need for geometry in that country arose out of the need to measure the extent of impact caused by the floods of the river Nile on the parcels of land. In the British Museum there is an old papyrus scroll dated to about 1700 BC. The papyrus, scribed by a person named Ahmes, contains a treatise on Egyptian geometry, which, according to the writer, is a copy of an older work dating to about 3000 BC.

Due to their thirst for exploration, commerce, and knowledge, the ancient Greeks came into contact with Babylonians and Egyptians, and thus became cognizant of their knowledge in geometry. As said above, it is important to remember that any geometric and related knowledge that the Babylonians and the Egyptians had was the prerogative of, and restricted to the native elite class of priesthood, and was not available to the public. Furthermore, such knowledge did not appear to be based on logic, was rather the result of experience, and was not structured nor well defined by any means. It was the Greeks of that era though that formalized the concepts and the principles of geometry, and raised them to the level of science and even philosophy. They defined terms, discovered propositions, and used deductive logic to prove geometric theorems. Those Greeks valued geometry so much, that knowledge of geometry was considered a prerequisite to the study of philosophy and of the physical sciences. Unlike the Babylonians and the Egyptians, knowledge of geometry was made available to the public in general, or at least to the public that could afford the cost of education.

Plato was so impressed by geometry that he posted over the entrance to his academy the inscription: "μηδείς αγεωμέτρητος εισίτω" "no one ignorant of geometry may enter"

Tomes have been written about the Greek theoreticians of mathematics but relatively little is known by the general public about the application of the geometric theorems and of the solution to geometric construction problems, that is of the Greek engineers and architects. Considerable credit has been given, and justifiably so, to the Romans for the construction of roadways, viaducts, water supply projects, and structural edifices. However, it was the Greeks that drawing upon the construction of public works knowledge from previous civilizations created the foundation upon which the Roman engineers based their construction marvels.

Acropolis The rock of Acropolis of Athens with its Parthenon (shown to the right), Propylaea and other structures, long considered as one of the wonders of the ancient world, was constructed upon a rock formation with marble and other material transported from the neighboring mountain of Pendele. The huge and ultra heavy rocky material had to be excavated, moved down the mountain slopes of Pendele, transported 12 km away to the foot of the said rock on a specially constructed roadway to sustain the loads, and then raised upon the top of the rock of Acropolis. Since the terrain did not provide the opportunity of constructing gently sloping ramps and an excessive labor force was not available as in Egypt or Mesopotamia, mechanical devices such as cranes, winches, and pulleys had to be invented and used for the first time.

Well known for its aesthetics and extensive incorporation of geometric proportionality and use of the golden ratio ($\phi = [1 + 5^{-2}]/2 = 1.618$), its structural stability (withstood numerous earthquakes and canon bombardments throughout its life), installation of anti-seismic connections, and construction in general was an engineering feat for that time in history. The various structures as we know them today replaced older structures during the golden age of Athens during the administration of Perikles (Περικλής) with Pheidias (Φειδίας) as the main sculptor and perhaps the overall planner, Iktinos (Ικτίνος) as the design architect and Kallikrates (Καλλικράτης) as the construction manager.

Diolkos Geographically Greece constitutes the southern tip of the Balkan peninsula. As the Greek landscape progresses southward the land narrows a bit until it reaches just south of Athens. At that point the shape of the land changes drastically by the invasion of two sea gulfs, the Saronicos (Σαρωνικός κόλπος) on the east and the Corinthian (Κορινθιακός κόλπος) on the west. Between these two gulfs

Replica of the Tympanum (Τύμπανον) or Frieze at the Parthenon's Façade

The Rock of Acropolis with the
Parthenon circa May 2012

Model of a Greek crane used in the construction of the Parthenon and
of the Diolkos of Corinth as per professor Theodosios P. Tasios of the
National Metsovian Polytechnic of Athens, Greece

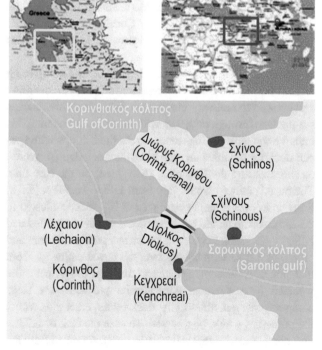

Upper Left: Overall map of Greece. *Upper Right*: Enlargement of the
general vicinity of the isthmos of Corinth southwest of Athens. *Bottom*:
Enlargement about the diolko

there is a short 3.7–5.3 mile (6–8 km) wide (more or less) neck of land at the end of which lies the Peloponnese (Πελοπόννησος). The city of Corinth (Κόρινθος) is located at the north end of the Peloponnese and just south of the said short neck of land which is called isthmos (ισθμός) which means land between two seas. Parenthetically, south of Corinth towards the south end of Peloponnese was the land of the Lacedemonians (Λακεδαιμονίων) commonly referred to as the Spartans (Σπαρτιάτες).

At that time, and even now, Greece has been a sea ferrying nation with commercial interests in the whole of the Mediterranean sea. To sail from Marseilles (a Greek colony at that time) or from Sicily and southern Italy (Magna Graecia as it was called then) to Athens and eastward to Cyprus, Egypt, and the Greek colonies in Asia Minor, ships had to go 450 miles (724 km) around the southern tip of the Peloponnese. This was quite time consuming and kind of dangerous due to the winds and currents in that area. Therefore, a shorter sea ferrying route was of need, and the Corinthians answered the call.

Around 625 BC Periandros (Περίανδρος) the tyrant of Corinth (τύραννος—in those days the word tyrant did not have the real bad connotation of today, it just meant an authoritative ruler) conceived the idea of digging a big ditch from the west to the east across the isthmus. As work began, Pythia (Πυθία), the oracle of Delphi (Δελφοί) warned Periandros that the gods would be mad as his plans, and he stopped digging. Perhaps the real reason was not the fear of the gods of Olympos but the prohibitive cost and the technical problems of the project. One of the technical issues was the fear that the waters of the gulf of Corinth would be higher than that of the Saronic gulf and they would flood the temples on the east side of the isthmos.

Periandros may have stopped digging but did not give up on the idea of a short cut. He conceived and built an overland precursor to the railway, the diolkos (διολκός) which means the passage through or across with the aid of a machine. He built a track with paving stones across the isthmos with special docking facilities on each side, the Lechaion (Λέχαιον) on the west and the Kenchreai (Κεγχρεαί) on the east side. The paved pathway would vary from 11.5 to 16.5 feet (3.5 to 5 m) wide over a length estimated at 3.7 to 5.3 miles (6 to 8.5 km). A ship would reach one dock, it would be lifted and pulled with special cranes and winches over an incline and on top of a base with wheels over the paved pathway. From that point on sheer manpower would pull the ship on the wheeled base overland with special precautionary steps along any steep inclines to the dock on the other side where the processes would be reversed. To ease the task of pulling the ship overland, cargo and sails would be removed from the ship and carried by animal and human power from one port to the other and reassembled on the other side. To help pay for the expenses of construction and maintenance there had

Wastewater clay drain pipe under a street of ancient Athens. Picture at an Athens metro station museum

to be the obligatory toll booth or station. An internet computer animation video prepared under the direction of Theodosios P. Tasios (Θεοδόσιος Π. Τάσιος), professor of Civil Engineering at the Metsovio Polytechnic Institute of Athens, Greece, may be found at: http://vimeo.com/13229017. Although the video audio is in Greek, the animation is quite graphic and self-explanatory.

The diolkos stayed in operation for almost 1500 years before ceasing operation. Although it was intended for commercial purposes, it was of use in various wars, one of them being the 27 year long Peloponnesian War between Athens and Sparta. According to Thucydides' (Θουκυδίδης), an Athenian historian and general, book 3 chapter 15, the Spartan fleet was the first to arrive at the west end of the diolkos, and using machinery moved their fleet on the east side to attack Athens.

Other Greek and Roman rulers such as Demetrios I of Macedon (Δημήτριος 1ος Μακεδονίας), also known as Poliorketes (Πολιορκητής), that is, the besieger, because of his inventions of besieging machines for overwhelming city fortifications, Julius Caesar, Caligula, Nero, and Hadrian (Publius Aelius Traianus Hadrianus Augustus) tried to excavate a canal across the isthmos to replace the diolko but failed to accomplish the task until a French engineering company carried out the canal project between 1880 and 1993 A.D. for the Greek government.

It is interesting to ride the subway (metro) in Athens and see practically at every station a museum of artifacts including structures and wastewater facilities of an era of over two millennia ago that were not available in Europe and other places a mere 500 years ago. It is also interesting to read about the machines and instruments, including computers and land and astronomic measuring instruments, city streets, water works and water ways including an original Suez canal excavated under Ptolemy II (Πτολεμαίος Β' 283–246 BC)

The bell of a wastewater Clay drain pipe under a street of ancient Athens—Not much different from present-day clay pipes. Picture at an Athens metro station museum

and others created and/or constructed by such engineers as Archimedes, Ctecibios, Hero, and others.

Presented below is a short list of certain prominent Greek mathematicians, geometers, and engineers and their contributions to the formalization of geometry and to engineering as we know it today.

Thales—Θαλής ο Μιλήσιος (624–547 BC) from Miletos, a Greek city on the Aegean coast of Asia Minor, is said to be the first to establish geometry as a deductive science depending on general propositions. Not too well known among common geometry users of today, Thales is credited to have discovered the theorems that:

- Opposite angles of two intersecting lines are equal to each other;
- A triangle is defined when a side and the angles at each its ends are given;
- Angles at the base of an isosceles triangle are equal;
- A circle is bisected by its diameter and the angle inscribed in a semicircle is a right angle.
- If two triangles have two angles of one triangle equal to two corresponding angles of the other triangle, and one side of one triangle equal to a corresponding side of the other triangle, the remaining sides and angle of one triangle are equal to the corresponding sides of the other triangle.

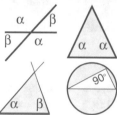

Thales Θαλής ο Μιλήσιος

Trivial as these theorems may seem to someone who may not have forgotten elementary school geometry, they form the basis of most other geometric theorems and geometric construction problems. In addition to being a geometer, Thales being a stargazer was the first recorded Greek astronomer who predicted the solar eclipse of 28 May 585 BC, and also a political statesman. It is said that Thales had traveled to Egypt as a member of a Miletian commerce representation, and as such had the opportunity to meet with Egyptian priests to discuss geometry. It is also purported that during this visit Thales measured the height of the pyramids by measuring the length of the shadow they formed on the ground. Rumor has is that while gazing at the stars near a water well, Thales fell in the well. A young servant girl from Thrace, who happened to be nearby, teased him as being too busy looking at the heavens that he could not see beauty near his feet.

Anaximander—Αναξίμανδρος ο Μιλήσιος (610–546 BC), about a dozen or so years younger than Thales, was a geometer in the literal sense of geometry in that he concerned himself with the shape and size of the earth, and the mapping thereof. While the Egyptians had drawn maps, those maps were of certain districts only and not of the whole earth, and not of the solar system and its composition. Anaximander ventured off the flat shape of the earth, claiming it to be cylindrical with a depth equal to one third of its diameter of the base, upon which the land and seas were situated. Erroneous as he was about the shape of the earth, he definitely was one of the first to question the flatness of the earth. According to the Greek Historian Herodotos, Anaximander learned about the sun dial from the Babylonians, and constructed one in Greece. Upon it he showed the solstices, the times, the seasons and the equinox. It is said that he also constructed a sphere representing the heavens.

Anaximander Αναξίμανδρος ο Μιλήσιος

Pythagoras—Πυθαγόρας ο Σάμιος from the Greek island of Samos (circa 580 BC), is perhaps the most notable of the old geometers. He was a student of Thales, and he proved many of today's geometric theorems, including the most famous that the square of the hypotenuse of a right triangle is equal to the sum of the squares of its two sides. Although Pythagoras is credited with this hypotenuse theorem, which was named the Pythagorean theorem after him, there are some scholars like Apollodoros (Απολλόδορος) according to Plutarch (Πλούταρχος), Athenaeus (Αθηνεύς) and Porphyr (Πορφυρός 233–304 AD) and Proclos (Προκλός 410 AD) that have questioned the fact that Pythagoras was the first to discover it, but not the fact that he did write about it extensively.

During the enlightenment period, students of mathematics were required to derive a new proof of the Pythagorean theorem. To date there are over 400 recorded proofs.

Over the next couple centuries, many other scholars expanded on the work of Thales and Pythagoras. Among these scholars, according to the philosopher **Proclos**—Πρόκλος (circa 410 AD) of Constantinople, are such names as Anaxagoras of Clazomenae—Αναξαγόρας ο Κλαζομένιος (500–428 BC), Oenopides of Chios—Οινοπίδης ο Χίος (500–450 BC), Hippocrates of Chios—Ιπποκράτης ο Χίος (470–410 BC), and Theodoros of Cyrene—Θεόδωρος ο Κυρηναίος (465–398 BC), as well as Antiphon—Αντιφών του δήμου Ραμούντος Αττικής (480–411 BC), Hippias of Elis—Ιππίας ο Ηλείος (circa 470 BC), Eudoxos of Cnidos—Εύδοξος ο Κνίδιος (408–355), Democritos—Δημόκριτος (460–370), Leodamas of Thasos—Λεωδάμας ο Θάσιος (circa 380 BC), Archytas of Taras—Αρχύτας ο Ταραντίνος (428–347 BC), and Theaetetos of Athens—Θεαίτητος ο Αθηναίος (417–369 BC).

Plato—Although most famous for his treatises on philosophy, Plato of Athens—Πλάτων ο Αθηναίος (427–347 BC), the student of Socrates (Σωκράτης), was an enthusiast and proponent of the study of geometry (see inset on page 2). He believed and taught that geometry and mathematics are essential for the development of the mind and knowledge in general, and that they should not be learned strictly for their use in a variety of trades, crafts and works. He argued that astronomers should look at the cosmos beyond gazing at the stars and apply geometric

Pythagoras Πυθαγόρας ο Σάμιος

Statue of Pythagoras in the island of Samos, Greece

Plato Πλάτων ο Αθηναίος

and mathematical principles so as to understand the science of astronomy and learn about the motion of the stars with respect to each other. In writing about the study of geometry he stated that a geometer commences by first promoting a hypothesis and then by the use of geometric diagrams and geometric logic the geometer proves or disproves the original hypothesis. Like his teacher Socrates, Plato delved into the generic and abstract aspect of geometry but did realize the need for a formalization and classification of the various geometric principles. Thus, one may argue that Plato's desire for such formalization may have been the impetus for the work of Euclid a century more or less later on. His real name was Aristocles (Αριστοκλής). Plato in Greek means broad and was given to him as a nickname because of his physic.

Aristotle of Staneira in Northern Greece—Αριστοτέλης ο Στανειρής (384–322 BC) was a student of Plato. Although not a true mathematician per se, he was a student of geometry and employed geometric and mathematical in the formalization of principles of botany and other sciences and even including metaphysics. After the death of Plato, Aristotle returned to northern Greece and he was engaged by Philip II, King of Macedon, to be the teacher and mentor of his son Alexander the Great. When Alexander left northern Greece and proceeded to conquer the eastern then known world, Aristotle relocated to Athens where he established his school of philosophy, which he called the Lyceum, with the aid and support of his student Alexander the Great. Throughout his life's work, Aristotle uses geometric and mathematical arguments to prove his theses, and argues for the formalization of the teaching of geometry, mathematics, and for the use of their logic not only in the study of botany, astronomy, meteorology, anatomy, and other hard sciences, but also in poetics, rhetoric, ethics, politics, and their like.

Euclid—So far, the above scholars, with the exception of Archimedes and Heron (see below) were primarily inter-

Euclid Ευκλείδης ο Αλεξανδρεύς (325–265 BC)

ested in geometry for its own intellectual sake and mental exercise than for its practical use and application. Publication of their work was not extensive, nor easy (try to write on a papyrus, and mass produce such a publication), and whatever there was, it concentrated on distinct axioms, theorems, proofs, and constructions. Their work was not organized in any logical written order until about 300 BC, when Euclid of Alexandria—Ευκλείδης ο Αλεξανδρεύς (325–265 BC) a teacher at the university of Alexandria during the reign of the Ptolemy the First (306–283 BC), published the first known formal text on geometry, and presented its axioms, theorems, and constructions in a logical and systematic order. Euclid's text is known as *The Elements—Στοιχεία Ευκλείδου*, and it is composed of 13 chapters, commonly referred to as books.

The Elements Book I sets forth the essential definitions, postulates, and axioms or common notions as referred to by Aristotle, and contains 48 propositions grouped into three parts:

- Propositions 1–26 deal with lines and intersection thereof, and triangles in general.
- Propositions 27–32 address the theory of parallels, line construction through a point, and orthogonal triangles.
- Propositions 33–48 are concerned with parallelograms, triangles, and squares, and the areas thereof.

Euclid's definition of parallel lines has bothered Euclid himself and many mathematicians of the renaissance era that eventually lead to the development of non-Euclidean geometry.

The Elements Book II is an extension of the third part of Book I with emphasis on rectangles and squares, and the

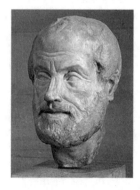

Aristotle Αριστοτέλης ο Στανειρής

transformation of areas. It also establishes in geometric terms what we call today algebraic relationships, such as:

$$ab+ac+ad+ae+\ldots=a\left(b+c+d+e+\ldots\right)$$

The Elements Book III is the subject of the circle, its chords, sectors, and segments, and tangency with a line.

The Elements Book IV contains various problems dealing with circles, triangle inscribed in and circumscribed about circles, rectangles, and regular polygons.

The Elements Book V introduces the theory of ratios and proportions, thus setting the foundation for the next two books, and distinguishes between commensurable and incommensurable quantities. For example, consider the following two cases:

- Two triangles of equal height, say $H=4$, and one having a base $B_1=6$ and the other a base of $B_2=10$. The area of the first triangle is now $A_1=12$ and that of the second triangle is $A_2=20$. The ratio of $B_1/A_1=6/12=0.5$, and the ratio of $B_2/A_2=10/20=0.5$. Now consider any other triangle of height $H=4$ and $B_n=40$, the area of which is $A_n=80$. For this triangle the ratio of $B_n/A_n=40/80=0.5$.

Euclid did not have a word for the radius of a circle. Euclid referred to it as the line from the center to the circle.

When asked by king Ptolemy for an easy way to learn geometry, he responded that there is no royal avenue to geometry.

- Now consider two rectangles of equal height, say $H=4$, and one having a diagonal $C_1=5$ and the other having a diagonal $C_2=8$. The area of the first rectangle is now $A_1=12$ and that of the second rectangle is $A_2=27.7128$. The ratio of $C_1/A_1=5/12=0.4167$, and the ratio of $C_2/A_2=8/27.7128=0.2887$. Now consider any other rectangle of height $H=4$ and diagonal $C_n=10$, the area of which is $A_n=36.6606$. For this rectangle the ratio of $C_n/A_n=10/36.6606=0.2827$. Note that in the first example a triangle's height and base are commensurable, while in the second example, a rectangle's height and diameter are incommensurable.

It is estimated that The Elements has been the most published written work in the world next to the Bible.

The Elements Book VI applies the theorems of Book V to lines, triangles, parallelograms, and circles, and addresses similarities, equalities, and relationships of figures.

The Elements Books VII, VIII, *and* **IX** concentrate on the theory of numbers beginning with definitions and expanding on propositions for:

- A unit and a number;
- Odd and even numbers and multiplication thereof;
- Prime numbers—numbers divisible only by themselves and 1;
- A plane number—the multiplication of two numbers yields a plane figure;
- A solid number—the multiplication of three numbers yields a solid;
- A square number—the multiplication of a number by itself yields a square, or A^2;
- A cubic number—the multiplication of a number by itself twice yields a cube, or A^3;
- Proportionalities and similarities; and
- A perfect number—a number which is equal to the sum of its submultiple parts (divisors), such as:

$28=1+2+4+7+14$	(perfect number—1, 2, 4, 7, and 14 are divisors of 28)
$24<1+2+3+4+6+8+12=36$	(over-perfect number)
$8>1+2+4=7$	(defective number)

The Elements Book X addresses commensurable lines, that is lines that can be measured in total by another line of fixed length (see Book V). Commensurable and incommensurable lines are referred to in modern mathematics as rational and irrational numbers. In addressing the rationality of lines, Euclid employs the method of exhaustion, which may be referred to as the predecessor to integration.

The Elements Books XI, XII, XIII pertain to spatial and solid geometry. Of these books:

- Book XI relates the perpendicularity and parallelism of two dimensions of Books I through VI to three dimensional space, as well as the volumes of parallelepipeds.
- Book XII calculates areas and volumes by using the method of exhaustion for cones, pyramids, cylinders, and the sphere.
- Book XIII relates the golden section, and the five regular Platonic polygons of Book IV to solids inscribed in a sphere.

The Elements Books XIV and XV supplement the preceding 13 books, and they have not been written by Euclid. Book XIV compares a dodecahedron with an icosahedron both inscribed in the same sphere and proving that the ratio of the surface of one to that of the other is the same the ratio of their volumes. This book is purportedly written by Hypsicles (Ὑψικλῆς), a student of Euclid, some time between 200 and 150 BC.

Book XV comprises three parts, and their authors and time of writing are not certain. In the third part there is a reference to "Isidoros (Ισίδορος) our great teacher" who probably is the architect of the Saint Sophia Church in Constantinople about 532 AD. There is another story which maintains that both of these books were drafted by Euclid, and they were found by Hypsicles, who edited them before presenting them to the king.

The *Elements* addressed primarily plane geometry, although some books did advance into solid geometry.

Archimedes of Syracuse—Αρχιμήδης Συρακουσών (287–212 BC) was perhaps the first to transcend theoretical mathematics and science into the practical world of engineering, civil as well as military engineering. We are all familiar with the story of him running naked through the streets of Syracuse to the King's palace yelling "εύρηκα, εύρηκα" (eureka, eureka"—I discovered, I discovered) to announce to King Hieron II (Ιερών 2ος Βασιλεύς Συρακουσών) that he, the King, had been swindled by the craftsman of his gold crown. So buoyancy in a liquid and specific weight of material came to being. He was so good in erecting machinery that would catapult missiles at the attacking Roman ships during the second Punic War, or set them on fire by redirecting and concentrating the sun's rays on the Roman ships with his special concave mirrors, that he was able to hold off the Roman army and navy for 2 years. For this impressed the Roman general Marcellus so that he gave strict instructions to spare him and have him brought to him alive. However, fate had a different desire. Archimedes was about 75 years old when he was slain.

Archimedes' contributions to geometry include the development of propositions and theorems on the construction and mensuration of the sphere, cylinder, conoids, spheroids, the circle, conic curves, and spirals. He applied his knowledge of geometry to the science of optics (prisms, mirrors, and lenses), mechanics (equilibriums, levers, and centers of gravity of various plane figures and solid shapes), hydrostatics, and astronomy.

> When the Romans were finally successful in breaching the defenses of Syracuse, Archimedes was busy in his yard drawing geometric figures, and angrily responded to the Roman soldier's command "Μη μου τους κύκλους τάραττε" ("Do not disturb my circles").
>
> This attitude did not please the soldier, who must have thought of Archimedes as some kind of a weird old sorcerer, so off went his head.

It is from his book the *Sand Reckoner* (Αρχιμήδης Ψαμμίτης—Archimedes, he that measures the grains of sand addressing large numbers) that we learn of the theory regarding the solar system as conceived before him by Aristarchos (Αρίσταρχος), and much later confirmed by Copernicus, that the sun is the center of the universe and that the planets, including earth, revolve about the sun. In this same book, Archimedes conceives of an arithmetic system to handle numbers far beyond the capacity of the then Greek arithmetical notation. According to our present number system, that would require 80,000 million zeros. Several of Archimedes' manuscripts had survived until the ninth century, at which time Leon restored them in the University of Constantinople, and are in existence to date. Many of his books since then have found home in various libraries in western Europe. One of them, *On Mechanical Theorems, Method to Eratosthenes*, or for short *The Method* (Περί Μηχανικών Θεωρημάτων Προς Ερατοσθένη Έφοδος) was discovered in Constantinople as recently as 1906.

> The lever must have been one of the first tools created by man. However, it is Archimedes who formulated its mechanics. He was so impressed by it, that he purportedly said "Δως μοι πα στω και ταν γαν κινάσω" ("Give a place to stand and I will move the earth").

Among his many accomplishments, Archimedes is credited for formalizing the laws of the lever, and the invention of the Archimedes screw, a water pumping device, which has been used in many applications, and particularly in water irrigation and sewage applications, even today. Purportedly Archimedes discovered and used his screw while on a visit to Egypt. Although this type of screw may not be truly an invention of Archimedes (some archaeologist claim that similar type pumping systems have been found in Mesopotamia), he certainly is the first one to formalize and write about its concept and use. Regarding the lever refer to the side bar on the left.

The Archimedes Palimpsest In 1906, Johan Ludvig Heiberg, a Danish scholar and publisher of Greek mathematical treatises, accidentally stumbled on a palimpsest (πάλιμψηστος) containing a prayer in Greek at the Metochion

Archimedes of Syracuse Αρχιμήδης Συρακουσών (287–212 BC)

Plate 1.1 *Top*: The Archimedes palimpsest. *Bottom, from left to right*: William Christens-Barry of the Johns Hopkins University, Roger Easton of The Rochester Institute of Technology entering data and Keith Knox of the Boeing Corporation taking the computer image. Photographs by KenCedeno for The New York Times

(Μετόχοιον) of the Eastern Orthodox Church Patriarchate of Jerusalem in Constantinople (now Istanbul, Turkey). A Metochion is a representation or embassy of one patriarchate or autocephalos church to another, while a palimpsest refers to a used papyrus that has been scrapped or covered and used again for another subject. In those "good old days" it was indeed a "paperless society" and there was a habit of using a papyrus over and over again as it may be needed.

A small, barely visible, circle with a dot in it drew the attention of Heiberg. In tracing the palimpsest with his finger he scrapped some of the ink and revealed the original text, and lo and behold a suspicion arose. The original "erased" or covered text was the original of Archimedes' treaties known as *The Method*. The discovery of this palimpsest made the headlines of The New York Times on the 16th day of July 1907.

Since its discovery the palimpsest, comprising 177 pages in rather very bad condition, went along a complex, and to some degree, adventurous and contested, ownership journey across Europe and America until 1998 when it was sold to an anonymous person for $2,000,000. From 2001 until 2008 the Walters Art Museum and scientists from the Rochester Institute of Technology and the Xerox Corporation both of Rochester, New York, and of the Johns Hopkins University of Baltimore, Maryland were able to preserve the contents of the Palimpsest employing modern photographic and electronic computing techniques (see Plate 1.1). For a further history on this document reference is made to:

- The 17 June 2001 issue of *The Sunday Times Magazine* of the *London Times*,
- The 27 November 2006 issue of *The New York Times*, and to
- The *What's Happening in the Mathematical Sciences* Volume 7, 2009, pp. 84–97.

One point of interest is that the person who overwrote on the original papyrus, most probably a monk who did not care much about geometry and technology, cut the original manuscript pages in half, rotated them 90° not necessarily in the same direction and did not reuse them in sequence. Thus, the scientists that preserved the manuscripts had to involve classical scholars that could translate and place the various pages in order.

Ctecibios of Alexandria—Κτησίβιος ο Αλεξανδρεύς (285–222 BC), a contemporary Archimedes, was the son of barber and perhaps a barber himself, liked to experiment with a variety of contraptions that lead him to the discovery of pneumatic machines. His first endeavor was to hang a mirror in his father's barbershop so that it could be raised and lowered by a hidden chord controlling a leaden ball in a pipe. While playing with his mirror he noticed that as the ball would fall rapidly into the pipe it would compress the air

Hydraulis Restored first century BC hydraulis found at Dion, Greece in 1992

which would escape through a small opening, thus making a sound. Ctecibios was also married to a singer named Lais (Λαΐδο), not the educated courtesan of Corinth, who was interested in a new kind of a musical instrument. To please his wife, Ctecibios built what may be considered to be the precursor to the modern pipe organ and called it hydraulis (ύδραυλις), a composite word of water (hydor—ύδωρ) and pipe (avlos—αυλός). An hydraulis comprised a series of pipes of varied length and piano like keys which when pressed would force water through a pipe which in turn would drive air out of a hole to produce the sound. It is presumed that Ctecibios determined the lengths of the pipes using Pythagoras' mathematical representation of the so-called perfect musical system of the time. While excavating at the feet on the northern side of Mount Olympos in 1992 professor of archaeology Demetrios Pantermalis (Δημήτριος Πνατερμαλής) at the Aristotelian University of Thessaloniki discovered the remnants of an hydraulis presently exhibited at the Archaeological Museum of Dion (Αρχαιολογικό Μουσείο Δίου). Dion (Δίον) was an ancient Greek religious site and now a village near the excavation site.

Heron (aka Hero) of Alexandria—Ήρων ο Αλεξανδρεύς (10–70 AD) is one of those Greek mathematicians that may be considered as the first mechanical engineer in the modern sense of the word. There appears to be a disagreement between historians regarding the years of Hero's life ranging from the first century BC to 250 AD. This is quite a span of years. He was an avid reader and follower of Archimedes' works and put this knowledge in the pursuit of engineering and land surveying. He invented several machines including perhaps the first steam engine for a variety of applications.

Among his contributions to geometry is his formula for the area of a triangle as a function of its three sides, and known as Hero's or Heron's Formula which states that

$$A = \sqrt{s(s-a)(s-b(s-c)}$$

where A is the area of the triangle, a, b and c are the lengths of the triangle's sides, and s is one half the sum of the said sides.

Hero established and ran a polytechnic institute of higher learning in Alexandria in Egypt, and has written several works on geometry, land surveying, mechanics and optics, of which five have survived to date, and they are the *Pneumatica* (Πνευματικά) and the *Automata* (Αυτόματα) that have survived in their original Greek, and the *Mechanics (Μηχανικά)*, the *Metrics* (Μετρικά) and the *Dioptra* (Δίοπτρα) that have survived as translations from Greek into Arabic. The last of the said three books is considered to be an excellent book on land surveying. In addition to his steam engine he is credited for having created various machines for lifting heavy weights with small effort by turning a wheel with a crank, various missile throwing machines for arrows (this is questioned by some historians) and a catapult for military applications of

Heron of Alexandria Ήρων ο Αλεξανδρεύς (10–70 AD)

Heron's Steam Jet Engine Illustration from Heron's book "Pneumatica"

Heron's Odometer National Museum of Athens, Greece

the Roman army, and the first odometer which comprised a set of toothed wheels attached to a wooden screw to record the distance traveled by a wagon. A modification of this device was adapted to the vessels of the Roman ships for measuring the number of Roman miles traveled by a ship.

Other Geometers—There are other prominent mathematicians that have greatly enriched geometry and/or help record it including Apollodoros (Απολλόδορος 260–200 BC), Hipparchos (about 140 BC), Athenaeus (Αθηνεύς), and Porphyr (Πορφυρός 233–304 AD), Proclos (Προκλός 410 AD), Descartes (1596–1650), Saccheri (1667–1733), Gauss (1777–1855), Lobatschewsky (1793–1856), Bolyai (1802–1860), Riemann (1826–1866), and many others. In the above chronology, the reader should notice (a) that all scholars mentioned are European, and (b) that there is a rather large gap in years between Proclos and Descartes.

The Time Gap—Throughout this historic section of this book reference has been made to Byzantium and to the Byzantine Empire. Its name as referred to by its various political and military administrations and its citizenship was not the Byzantine Empire but the Roman Empire. In 285 A.D., the emperor of the original Roman Empire Diocletian split the administration of his empire into eastern and western administrative districts with capitals at Nicomedia in Asia Minor near the Black Sea and Rome, respectively. A few years later in 324 A.D. Emperor Constantine the Great transferred the eastern capital a few miles to the west to the old Greek city of Byzantium on the European side of the Sea of Bosphoros and named it Constantinople (city of

Constantine—present-day Istanbul in Turkey). As the western portion of the Roman Empire disintegrated a century or two later due to invasions by various Germanic and Asiatic tribes, the eastern portion retained the name of the Roman Empire until its final downfall to the Ottoman Turks in 1493. The terms Eastern Roman Empire and Byzantine Empire were terms assigned to the said empire by western European scholars, mostly members of its clergy, during the renaissance years as a distinction of the Roman Empire with its base in Rome.

The Roman Empire of the east gradually began to shed its Latin linguistic and cultural basis in favor of the Greek language and culture with the final official change instituted by Emperor Herakleos (Ηράκλειος) between 610 and 641 A.D. While the citizens of the empire, both Greek and non-Greek referred to themselves as Romans (Ρωμαίοι). The crusaders of the fourth crusade in 1204 when they ransacked Constantinople referred to the citizens of the empire as Greeks.

Because this is not an history book we will suffice to say that time gap for the pursuit of mathematics in general and geometry referred above might have been due to the geopolitical upheaval of the European continent and the middle east caused by (a) the said invasions in the western empire, (b) similar invasions in the eastern empire, (c) the eastern empire's perennial conflicts with the Persians and various Slavic tribes, and (d) the sudden emergence of Islam and Arabic expansionism, all of which distracted the populace from the pursuit of science to self preservation. After the fall of Constantinople to the Ottoman Turks in 1453, it was the Greeks of the fallen Byzantium that escaped to western Europe carrying with them the seeds for the renaissance. In the mean time, it was the time for the Indian and Arabian scholars to carry the torch who translated, reproduced, and preserved many of the Greek documents and introduced advances of their own, including perhaps the most important one the introduction of the number system.

For all great advances that the Greeks made in theoretical geometry, they had a very cumbersome arithmetic notation system. In his second book, Euclid addressed algebraic problems in geometric terms, and carried mathematical equations using that cumbersome notation of letters of the Greek alphabet representing numbers. And the Romans who adopted Greek culture and science did not do any better in numeric notation. The world had to wait for the Arabs at about 600 AD for their number system.

1.2 Trigonometry and Algebra

With geometry, a person is able to construct various figures with a compass and a straight edge, and solve such problems as intersections, tangencies, projections, and the like. With geometry one can also perform certain mensurations and computations such as finding areas, heights of triangles, radii of inscribed and circumscribed circles to mention but a few. However, there are two major drawbacks.

- The first drawback is that we cannot compute the angles of a triangle if we know its sides, and we cannot compute all elements of a triangle if one or more of the given elements is an angle; excepting of course certain special cases as that of equilateral triangles, and right triangles. One may argue that Thales proved that a line of a specified length and an angle at each end of the line define a triangle. Theoretically this is true. However, try to draw a line of a specified length, and then construct an angle of 34°52′44″ at one of the end points, and an angle of 55°32′04″ at the other end point. The only way to do this is with the help of a mechanical device like a protractor, and good luck if you are looking for high precision.
- The second drawback is that of scale and workability. We may be able to design on paper and then lay out in the field a small scale project, but we cannot do the same for a large and complex project because of physical limitations, such as size of paper and potential for errors due to systematic errors (due to the equipment themselves), and due to blunders (due to human imperfection). It is noted that accidental errors have not been forgotten; systematic errors and blunders are enough to prove the point.

As an example, let us consider the layout of a rather simple circular amphitheater having a radius of 50 m (164.04 feet), a portion of which is to be traversed by a chord, beyond which the stage and ancillary support facilities are to be housed (see Figure 1.2). Let us now consider the following:

1. First, let us assume that the length of the above said chord is to be also 50 m (164.04 feet) long. In this case, the geometric solution is simple, and the project could easily be laid out in the field without any computations because we

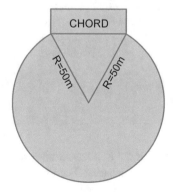

Figure 1.2 Sample Amphitheater

are dealing with an equilateral triangle with each side being 50 m (164.04 feet) long, not a great distance relatively speaking.

2. Now let us suppose that the said chord is not to be 50 m (164.04 feet), but it should be instead 35 m (114.83 feet) long, thus having an isosceles and not an equilateral triangle. For this scenario, we will have to:

   (a) Divide the isosceles triangle into two right triangles (each formed by one of the said radii and one half of the chord length); and then

   (b) Use the Pythagorean theorem to solve for the third side (from the circle's center to the midpoint of the said chord) as being equal to:

   $$\sqrt{50^2 + 17.5^2} = 46.837 \ m \ \left(153.67 \ feet\right)$$

   Alternatively, the surveyor, having located the center point of the circle, could lay out a point on the circumference 50 m away from the said center point along a desired direction, and then intersect two circular arcs, one about the said point having a 35 m (114.83 feet) radius, and the other about the center point of the circle having 50 m (164.04 feet) radius. Two ropes or chains of the same two lengths could do the job.

3. But what do we do if for some reason we want the central angle subtended by the said chord to be 39°43′51.7″. This may sound preposterous, but let us bear with it for a paragraph or two. The only way to construct the project would then be to have an angle measuring instrument. Alternatively, we could use in the office a compass, a scaled ruler, a protractor, and a triangle or two to lay out the project on a rather manageable size paper at a scale of say 1:50 or even 1:100, and then make measurements on the said paper, and use them to lay out the project in the field. Depending on the precision of the equipment, and care taken by the person doing the layout, a decent accuracy within a few decimeters or foot or so might be attained.

Greek theaters were semicircular while Roman amphitheaters were generally oval or circular. Even though Greek theaters had a rather large footprint in outdoor space, they had no need for microphones and for electronic sound amplifiers. Their architectural and acoustic design was such that the actors and the chorus could be heard even by those in the most remote seats.

However, what about a more complex and larger project than that of the last scenario? For this, geometry alone is neither efficient nor adequate. Thus, a set of relations between the sides and the angles of a triangle had to be devised. According to Greek writers of the fourth century AD the three Greek astronomers discussed below were considered to be the pioneers of trigonometry, a word being a fusion of the Greek word "τρίγωνον" (three angled figure) and the word "μέτρον" (measure) yielding "τριγωνομετρία" (trigonometry). According to these writers:

Hipparchos of Rhodes—Ἵππαρχος ο Ρόδιος (190–120 BC) was the originator of trigonometry during the second century BC having generated a table of chords as a function of the half the angle that subtends the chord in a circle, that is C = 2Rsin(½δ) where R denotes the radius of the circle, C denotes the chord, and δ denotes the said angle. Purportedly he had written 12 books with tables relating arc lengths to arc chords, but none of these books have survived to date. We know very little about his personal life except that he was born in Nicea in Bithynia (modern-day Turkey) and that he had made astronomical observations in Bithynia, in the island of Rhodes and in Alexandria in modern-day Egypt. Most of what we know about him is from writings of other mathematicians and particularly from Ptolemy of Alexandria (see below).

Hipparchos of Rhodes Ἵππαρχος ο Ρόδιος (190–120 BC)

Menelaos of Alexandria—Μενέλαος ο Ἀλεξανδρεύς (70–140 AD) is said to be the first to introduce to geodesics spherical triangles composed of arcs of great circles as being analogous to triangles on a plane surface formed by straight lines. Although other geometers has considered spherical surfaces for geodesics, Menelaos is the one that defined their properties and developed a set of trigonometric relations thereof. He too wrote books (purportedly six), of which only one has survived to date, *The Book of Spherical Propositions*. He is also famous for the theorem that bears his name regarding the colinearity of three points that lie one on each side of a triangle, or extension thereof. This theorem was revisited by the Italian geometer Giovanni Ceva in 1678 who offered an alternate proof. Much later on, Albert Einstein presented two different proofs to Menelaos' Theorem. For an extensive discussion on this theorem visit *http://www.mathpages.com/home/kmath442/kmath442.htm*.

Ptolemy of Alexandria (aka Claudius Ptolemaeus)—Κλαύδιος Πτολεμαίος (108–160 AD) was a mathematician, geographer, and astronomer who wrote extensively on each of his studies of interest. Although not a Roman, he was a Roman citizen as his first name Claudius indicates, and is not believed to have any relation to the earlier Greek Ptolemaic dynasty. Several of his books have survived in one form or another, of which the most famous are the Almagest, the Arabic name of the translation from the Greek name *H Μεγάλη Σύνταξις* which means a mathematical treatise. In one of his books there are tables of chords at intervals of 30 min, and accurate to at least five places. The methodology of developing these tables is described, and theorems implicitly referring to what we call today half-angle formulas are introduced.

Ptolemy of Alexandria Μενέλαος ο Αλεξανδρεύς (70–140 AD)

The trigonometric relations that these mathematicians developed were not expressed in the form of what we now call trigonometric functions, laws, and formulas, but in the form of successive written instructions. The sine and cosine functions were first introduced several years later by the Hindus, and books with tables of such functions for plane and spherical triangles were published. None of these books is extant, but prior to their loss these Hindu books had been translated into Arabic about the end of the eighth century. The tangent and cotangent functions and other trigonometric relations were introduced later by the Arab mathematicians.

al-Battani (850–929 AD) are the last two words of his 15 word name was an Arab prince. He is also referred to by his latinized name Albategnius. He was born in modern-day Turkey, but lived most of his life in Damascus Syria delving in mathematics and astronomy, and he also died there. He discovered several trigonometric identities and introduced the law of cosines to spherical triangles. He improved on the work of Ptolemy of Alexandria by introducing trigonometric rather than geometric methods to the solution of right angled

al-Battani (850–929 AD)

triangles. As an astronomer, he made many accurate observations in Syria and he contributed greatly in the astronomical relationship between the sun and the earth.

Abu al-Wafa (aka Buzjani) is an Arab astronomer and mathematician from Baghdad (949–983) who translated into Arabic the work of Euclid and wrote commentaries about this work. He contributed extensively in the development of plane and spherical trigonometry, he introduced more accurate means of computing sines functions, and he introduced the secant and cosecant functions of an angle. He wrote a practical book on arithmetic for the business man, and he developed solutions to various geometric constructions.

Jabir ibn Aflah of Seville, also known by his latinized name Geber, lived about one hundred years after Abu al-Wafa. As an astronomer he commented and supplemented the work of Ptolemy on spherical triangles.

During the same general era, the Persian mathematicians:

al Biruni are the last two names of his much longer multi-word name (973–1048) was born in modern-day Uzbekistan and died in Afghanistan. He studied and wrote (146 books) about astronomy, philosophy, mathematics, geography, and many other subjects. His contribution to trigonometry was

al Biruni (973–1048)

the establishment of the general law of sines to plane triangles. He has been honored with commemorative stamps by both Afghanistan and Russia.

Nasir al-Din al-Tusi (1201–1274) was an astronomer, biologist, chemist, mathematician, philosopher, physician, physicist, scientist, and theologian who formalized the separation of trigonometry from astronomy as an independent discipline of plane and spherical triangles. He has been honored with a commemorative stamp by Iran.

In western Europe, trigonometry was introduced in the fifteenth century. Whether this work was done independently of the Arab and Persian work, or not is in question.

Nasir al-Din Stamp

Johannes Muller von Konigsberg (1436–1476), also known as Regiomontanus, a Latin translation of Konigsberg which means mountain of the king, was a Prussian astronomer. He began to study mathematics at the very early age of eleven in Leipzig, Saxony and three years afterwards in Vienna, Austria. His books *De Triangulis Omnimodis Libri Quinque* was one of the first to introduce the modern state of trigonometry by translating the work of Ptolemy of Alexandria from the Greek rather than from the Arabic,

Johannes Muller von Konigsberg (1436–1476)

while his follow up book on this subject *Epytoma in Almagesti Ptolemei* he critiqued the translation. It should be noted that the first book, comprising five volumes, was completed in 1464 but it was not printed and published until 1533 after his death. The first two volumes dealt with plane trigonometry, while the other three addressed spherical trigonometry.

Excerpt from von Konigsberg's book De Triangulis Omnimodis Libri Quinque: *"You who wish to study great and wonderful things, who wonder about the movement of the stars, must read these theorems about triangles. Knowing these ideas will open the door to all of astronomy and to certain geometric problems."*

In 1475, Pope Sixtus IV appointed Regiomontanus titular bishop of Ratisbon, and invited him to Rome to undertake the reformation of the Julian calendar which had been introduced by Julius Ceasar in 46 BC and had been in effect ever since for both civil and religious (Christian) purposes. The work of Regiomontanus continued after his death, and lead to the development of the so-called *Gregorian calendar*, and its adoption by the *inter gravissimas* papal bull of Pope Gregory XII on 24 February 1582. This change was done in order to correct for the many leap days required by the Julian calendar and move the date of Easter closer to the vernal equinox date of the 21st day of March.

The **Trigonometric Tables** (1968) by Dr. Jean (Johann Theodor) Peters is a 2″ thick 8.5″ × 11″ book (see Plate 1.3) for angles between 0° and 90°:

- 902 pages of eight place tables for sine, cosine, tangent, and cotangent at 1 s interval;
- 50 pages of 21 place tables for sine and cosine at 10 s intervals.

It is a translation of the book *Achtstellige Tafel der Trigonometrischen Funktionen für jede Sexagesimalsekunde des Quadranten* (Berlin 1939), followed by an appendix which is a translation of *Einundzwanzigstellige Werte der Funktionen Sinus und Cosinus* (Berlin 1911), also by Dr. J. Peters. The original work was done in 2 years (1931–1932) by Dr. Peters and Dr. L.J. Comrie for the German Nautical Almanac Office.

	12° 51'				12° 52'				12° 53'				
	473	510	981	107	473	510	978	108	473	510	976	108	
n	sin	tg	ctg	cos	sin	tg	ctg	cos	sin	tg	ctg	cos	
	0,222	0,228	4,38	0,974	0,222	0,228	4,37	0,974	*0,222	0,228	*4,37	0,974	
0	39939	11232	38054	95565	68299	41836	79317	89091	*96656	72445	*20731	82609	60
1	40412	11742	37073	95457	68771	42346	78340	88983	*97129	72955	*19756	82501	59
2	40884	12252	36093	95349	69244	42856	77362	88875	*97601	73465	*18780	82393	58
3	41357	12762	35113	95241	69716	43367	76385	88767	*98074	73975	*17805	82285	57
4	41830	13272	34133	95133	70189	43877	75407	88659	*98546	74485	*16830	82177	56
5	42303	13782	33153	95025	70662	44387	74430	88551	*99019	74996	*15855	82069	55
6	42775	14292	32173	94918	71134	44897	73452	88443	*99492	75506	*14880	81961	54
7	43248	14802	31193	94810	71607	45407	72475	88335	*99964	76016	*13905	81853	53
8	43721	15312	30214	94702	72080	45917	71497	88227	00437	76526	*12930	81745	52
9	44193	15822	29234	94594	72552	46427	70520	88119	00909	77036	*11956	81636	51
10	44666	16332	28254	94486	73025	46937	69543	88011	01382	77547	*10981	81528	50
11	45139	16842	27274	94378	73498	47447	68565	87903	01855	78057	*10006	81420	49
12	45611	17352	26294	94271	73970	47958	67588	87795	02327	78567	*09031	81312	48
13	46084	17862	25315	94163	74443	48468	66611	87687	02800	79077	*08057	81204	47
14	46557	18372	24335	94055	74915	48978	65634	87579	03272	79587	*07082	81096	46
15	47029	18882	23356	93947	75388	49488	64657	87471	03745	80098	*06107	80988	45
16	47502	19393	22376	93839	75861	49998	63680	87363	04218	80608	*05133	80880	44
17	47974	19903	21397	93731	76333	50508	62703	87255	04690	81118	*04158	80771	43
18	48447	20413	20417	93623	76806	51018	61726	87147	05163	81628	*03184	80663	42
19	48920	20923	19438	93516	77279	51528	60749	87039	05635	82138	*02209	80555	41
20	49392	21433	18458	93408	77751	52039	59772	86931	06108	82649	*01235	80447	40
21	49865	21943	17479	93300	78224	52549	58795	86823	06581	83159	*00261	80339	39
22	50338	22453	16500	93192	78696	53059	57818	86715	07053	83669	99288	80231	38
23	50810	22963	15520	93084	79169	53569	56842	86607	07526	84179	98312	80123	37
24	51283	23473	14541	92976	79642	54079	55865	86499	07998	84689	97338	80014	36
25	51756	23983	13562	92868	80114	54589	54888	86391	08471	85200	96364	79906	35
26	52228	24493	12583	92760	80587	55099	53912	86283	08944	85710	95389	79798	34
27	52701	25003	11604	92653	81060	55610	52935	86175	09416	86220	94415	79690	33
28	53174	25513	10625	92545	81532	56120	51958	86067	09889	86730	93441	79582	32
29	53646	26023	09646	92437	82005	56630	50982	85959	10361	87240	92467	79474	31
30	54119	26533	08667	92329	82477	57140	50005	85851	10834	87751	91493	79365	30
31	54592	27044	07688	92221	82950	57650	49029	85743	11307	88261	90519	79257	29
32	55064	27554	06709	92113	83423	58160	48053	85635	11779	88771	89545	79149	28
33	55537	28064	05730	92005	83895	58670	47076	85527	12252	89281	88572	79041	27
34	56010	28574	04751	91897	84368	59181	46100	85419	12724	89791	87598	78933	26
35	56482	29084	03773	91789	84841	59691	45124	85311	13197	90302	86624	78825	25
36	56955	29594	02794	91681	85313	60201	44148	85203	13670	90812	85650	78716	24
37	57428	30104	01815	91574	85786	60711	43171	85095	14142	91322	84676	78608	23
38	57900	30614	00837	91466	86258	61221	42195	84987	14615	91832	83703	78500	22
39	58373	31124	*99858	91358	86731	61731	41219	84879	15087	92343	82729	78392	21
40	58846	31634	*98880	91250	87204	62241	40243	84771	15560	92853	81756	78284	20
41	59318	32144	*97901	91142	87676	62752	39267	84663	16033	93363	80782	78175	19
42	59791	32654	*96923	91034	88149	63262	38291	84555	16505	93873	79809	78067	18
43	60264	33165	*95944	90926	88622	63772	37315	84447	16978	94383	78835	77959	17
44	60736	33675	*94966	90818	89094	64282	36339	84339	17450	94894	77862	77851	16
45	61209	34185	*93988	90710	89567	64792	35363	84231	17923	95404	76888	77743	15
46	61682	34695	*93009	90602	90039	65302	34388	84123	18395	95914	75915	77634	14
47	62154	35205	*92031	90494	90512	65813	33412	84014	18868	96424	74942	77526	13
48	62627	35715	*91053	90386	90985	66323	32436	83906	19341	96935	73969	77418	12
49	63099	36225	*90075	90279	91457	66833	31460	83798	19813	97445	72995	77310	11
50	63572	36735	*89096	90171	91930	67343	30485	83690	20286	97955	72022	77202	10
51	64045	37245	*88118	90063	92403	67853	29509	83582	20758	98465	71049	77093	9
52	64517	37755	*87140	89955	92875	68363	28534	83474	21231	98975	70076	76985	8
53	64990	38265	*86162	89847	93348	68874	27558	83366	21704	99486	69103	76877	7
54	65463	38776	*85184	89739	93820	69384	26583	83258	22176	99996	68130	76769	6
55	65935	39286	*84206	89631	94293	69894	25607	83150	22649	*00506	67157	76661	5
56	66408	39796	*83229	89523	94766	70404	24632	83042	23121	*01016	66184	76552	4
57	66881	40306	*82251	89415	95238	70914	23657	82934	23594	*01527	65211	76444	3
58	67353	40816	*81273	89307	95711	71424	22681	82826	24066	*02037	64238	76336	2
59	67826	41326	*80295	89199	96183	71935	21706	82717	24539	*02547	63266	76228	1
60	68299	41836	*79317	89091	96656	72445	20731	82609	25012	*03057	62293	76119	0
	0,222	0,228	*4,37	0,974	0,222	0,228	4,37	0,974	0,223	*0,229	4,36	0,974	
	cos	ctg	tg	sin	cos	ctg	tg	sin	cos	ctg	tg	sin	n
	472	511	977	108	472	511	975	109	472	511	972	109	
	77° 8'				77° 7'				77° 6'				

14° 75°

16° 73°

18° 71°

20° 69°

Plate 1.3 Excerpt of the 1968 edition of the *Trigonometric Tables* book by Dr. Jean (Johann Theodor) Peters

Francois Viete (1540–1603) was born and raised in France, Viete first studies law following his father's profession, but did not practice law for more than 4 years. He became a private tutor and eventually he was appointed counselor by various royal decrees. Although he never studied mathematics or astronomy formally, he took great interest in them and his mathematical and astronomical contributions were significant. In his book *In Artem Analyticam Isagoge* (*Introduction to Analytic Art*) published at Tours in 1591, Viete introduced the first systematic symbolic notation for algebra by using symbols for operations, and consonant letters for known quantities and vowel letters for unknown quantities. For this he is referred to by some mathematicians as the father of algebra, which might be a bit unfair to mathematicians that preceded him on this subject. In 1593 Viete published two books on geometry in which he examines such problems as doubling the cube, trisecting an angle, constructing the tangent at any point on an Archimedian spiral, and calculating π to ten places and defining it as an infinite product.

Francois Viete (1540–1603)

Georg Joachim von Lauchen (1514–1574) adopted the name Rheticus which was the Roman name of his birthplace, and he was the only student of Copernicus. This Austrian born mathematician and cartographer is best known for developing a full set of trigonometric tables and the half angle formula. His first book *Narratio Prima* (*First Report*) published in 1540 presents the work of Copernicus regarding the celestial revolutionary concept. His student Valentin Otto completed the trigonometric work after his teacher death by publishing the 1500 page volume *Opus Paltinum de Triangulus* accurate enough to be used into the early twentieth century.

1.3   Logarithms

Considering the cumbersome Greek number system, and the equally cumbersome system of the Romans, as well as the absence of zero and negative numbers, it is impressive to comprehend the difficulty that the Greeks encountered in producing the tables of chords. Even with the advent of the Arabic number notation the computation of tables was a horrendous task. The advancements in algebra simplified mathematical expressions, but the task of computation remained a cumbersome operation. Parenthetically, even though the origin of algebra can be traced to **Diophantos of Alexandria**—Διόφαντος ο Αλεξανδρεύς (circa 250 AD), symbolic algebra did not materialize until the sixteenth century by **Francois Viete**, while the proper understanding of negative quantities is attributed to the work of the French mathematician **Albert Girard** in 1629.

Diophantos of Alexandria Διόφαντος ο Αλεξανδρεύς (circa 250 AD)

John Napier or Neper, Baron of Merchistoun Scotland (1550–1617) was the mathematician and astronomer that changed dramatically the computation of complex equations by introducing logarithms in 1614. It was indeed a tremendous invention of the time and years to come in that it was not based on any preceding work. With the use of logarithms trigonometric tables of natural sines, cosines, tangents, and their inverses were published for as many as 21 decimal places (the norm being eight) for each second of an angle. Logarithms have enabled the union of geometry and trigonometry in solving a multitude of mathematical problems, and have contributed in the creation of the various engineering marvels of the industrial age and as recently as the middle of the twentieth century until the introduction of the electronic computer.

John Napier Baron of Merchistoun Scotland (1550–1617)

In the current computer age, the subject of logarithms is but a passage through the general discourse of high school, and even college mathematics, while its use is reserved for special scientific work. However, even up to the late 1970s, and maybe even later in certain engineering and surveying offices, it was a staple computation tool in the design of many projects. Even though computers had invaded several of the private and governmental engineering office in the mid- to late 1960s, one may speculate with confidence that a sizeable portion of the interstate highway system of the USA was designed with the help of logarithms. The senior author of this book remembers countless hours looking up logarithms of distances and trigonometric functions in the geometric design of multilevel interchanges and curved bridges.

Another development that occurred about the same time is the formalization of the spatial coordinate system. The early Greeks had thought of spatial point positioning, but did not pursue it, nor did they do much about it. It was a Frenchman Descartes that brought it to existence.

René Descartes (1596–1650), a French philosopher, politician, and mathematician, is the inventor of the Cartesian coordinate system and of analytic geometry which opened the roadway to the calculus of Newton and Leibniz, and the mathematics of the current era. As was the fashion of those times for philosophers, scientists, and other such elites, Descartes latinized his name to Renatus Cartesius, and hence the name for the said coordinate system. Still in existence to date this system has also been modified to certain other forms, one of them being that of civil engineers and surveyors in the northern hemisphere, and which substitutes north and east for the Y and X axes and advances the quadrants in a clockwise rather than in a counterclockwise rotation.

René Descartes (1596–1650)

An interesting story regarding the correspondence between René Descartes and princess Elisabeth of Bohemia about the solution of the kissing circles is discussed in Appendix C of this book.

Logarithms and the Cartesian coordinate system have found a solid home in the engineering and scientific fields. Since the applications of these fields required the solution of

algebraic equations that involved trigonometric functions of angles, books of logarithmic tables of sines, cosines, and tangents were developed in addition to the books of tables of the natural form of the trigonometric functions.

> Reference is made to Chap. 9 on spirals regarding Pierre de Fermat a contemporary of Descartes

1.4   Slide Rules and Calculators

Although logarithms facilitated greatly the calculation of multiplication, division, and exponentiation, they still left a lot to be desired regarding the speed of calculation. Ever since the dawn of mathematics, man sought means of computation faster than the human mind. Through the ages of computing to date several mechanical devices have been invented, with the first one attributed to a 2000-year-old clock-like mechanism found by divers in 1901 off the coast of the Greek island of Antikythera located just north of the west end of the island of Crete and directly south of Sparta. It is now believed that this device was used to calculate the motion of the stars and planets. For a description of this device and its discovery refer to *http://www.mlahanas.de/Greeks/Kythera.htm*.

William Oughtred (1575–1660) of Eton in Buckinghamshire, England like Rheticus also worked independently on the half-angle formula. His 1631 book *Clavis Mathematicae* (*The Key to Mathematics*) became a standard of its time and was used by such scientists as Isaac Newton. He argued that mathematical publications should be less verbose, and should be more dependent on the use of graphic and symbolic notations as did Francois Viete. Using Napier's logarithms and the logarithmic scales Edmund Gunter, he invented in the mid-1620s the linear slide rule by sliding one logarithmic

William Oughtred of Eton, England

Plate 1.4a Three Views of the replica of the Antikythera mechanism as designed and constructed by the Aristotle University of Thessaloniki, Greece in 2011

scale along another stationary to perform multiplication and division of numbers. Thereafter he introduced the circular slide rule by sliding one circle over another. Oughtred is also credited for having introduced the "x" symbol to denote multiplication, and the terms "sin" and "cos" to denote the sine and cosine trigonometric functions. He is also credited with the invention of the double horizontal sundial.

For those that are interested in mechanical computing devices that existed prior to the advent of electricity, reference is made to the book *Traite de la Construction et Principaux Usage de Instruments de Mathematique* by Nicolas Bion published in France in 1709. Over the years, this book has been published in several editions, and has been translated in several languages. In 1758, it was translated into English and was supplemented by Edmond Stone under the title *The Construction and Principal Uses of Mathematical Instruments*. This translation has been reprinted by *The Astragal Press* of Mendham, New Jersey in 1995. The book contains a multitude of slide rules, or like instruments, that have been created from before Bion's era up to the twentieth century for a variety of purposes. They are fast to use, but not precise; three digits at best. Keuffel & Esser Co. (K+E) slide rule like the one of Plate 1.4a cost about $25 in 1952. Compare with what you can get today sometimes for free.

With the advent of electricity, several electromechanical devices came to existence, the most prominent for engineering work being the *Marchant* (see Plate 1.4c), the *Friden* and the *Monroe*. These calculators looked and operated fairly

much the same way, with addition, subtraction, multiplication, and division requiring the entry of two numbers and the push of one button. As the button was pushed, wheels would begin to rotate and the carriage at the top of the calculator would move noisily to display the result. Exponentiation to a whole integer involved repetitive activation of the multiplication button, while finding the square root of a number required almost a dozen, so to speak, operations and a lot of more noise. The cost of a new Marchant calculator in the early 1970s would cost about $1200 for 45 pounds of steel and of which from time to time a part would break, and the repairman would have to be called to fix replace it. A Hewlett Packard (HP) hand held calculator at that time sold for about $300, weighted a few ounces, and outperformed either calculator many fold.

A Sample Problem Not to insult any reader's knowledge of the use of logarithms but to demonstrate the labor that was required for carry out a simple computing task that would hardly take any time with a computer let us consider a common problem a civil engineer might have encountered in the course of a project design in the good old days of the mid-twentieth century, the intersection of two street center lines (see Plate 1.4d), each defined by a point of known coordinates and a direction expressed in bearings. In solving this problem, the reader should bear in mind that when working with geometric work such as for control survey networks, bridge site location and layout, and the like, a precision of

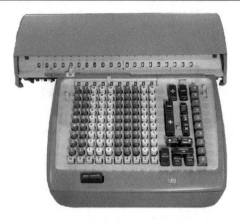

Plate 1.4c Electromechanical Calculator by Marchant (circa late 1960s)

eight significant digits for distances and a decimal of a second of a degree for angles is not uncommon. The required precision of plane state coordinates should be a minimum of three decimal digits to the right of the decimal point and perhaps even more. Needless to say, the use of the slide rule in this work is out of the question due to its associate precision. Thus, the engineer would, as shown in Plate 1.4d:

1. Find the differences in latitude (dx) and departure (dy) between the two points of known point coordinates.
2. Use the Pythagorean theorem to solve for the distance c by:
   - Finding $\log_{10}$ of dx and dy.
   - Multiplying each of dx and dy by two to square them, and then finding the antilogs of each product to find dx2 and dy2.
   - Adding dx2 and dy2, finding the logarithm of the sum, dividing the sum by two to find the square root, and then finding its antilog to yield the distance c.

     Note that in order to find the antilog in Plate 1.4c to obtain the value of c, an interpolation has to be performed. An interpolation was not necessary in finding the values of dx2 and dy2 because dx and dy were assumed to be integers with a small number of significant digits for the sake of simplification.

     Also note that if a mechanical calculator was available, there would not be a need to find the logarithms, but getting the square root would have involved a few more operational steps and clanking machine noise as the calculator would try to give you the square root.
3. Use common trigonometric equations to find angles γ and δ using dx, dy, and c as the known variables.

Plate 1.4b K & E Slide Rule circa 1950

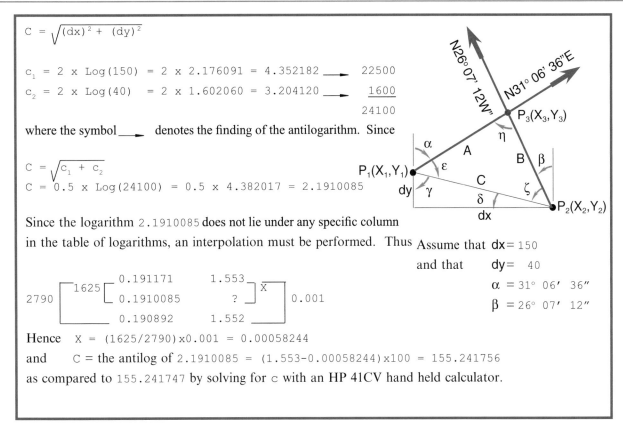

$$C = \sqrt{(dx)^2 + (dy)^2}$$

$c_1 = 2 \times \mathrm{Log}(150) = 2 \times 2.176091 = 4.352182 \longrightarrow 22500$

$c_2 = 2 \times \mathrm{Log}(40) \;= 2 \times 1.602060 = 3.204120 \longrightarrow \underline{\;1600}$

$\qquad\qquad\qquad\qquad\qquad\qquad\qquad\qquad\qquad 24100$

where the symbol $\longrightarrow$ denotes the finding of the antilogarithm. Since

$C = \sqrt{c_1 + c_2}$

$C = 0.5 \times \mathrm{Log}(24100) = 0.5 \times 4.382017 = 2.1910085$

Since the logarithm 2.1910085 does not lie under any specific column
in the table of logarithms, an interpolation must be performed. Thus

Assume that $dx = 150$

and that $\quad dy = \;40$

$\alpha = 31°\;06'\;36''$

$\beta = 26°\;07'\;12''$

$$2790 \left[\begin{array}{c} 1625 \left[\begin{array}{cc} 0.191171 & 1.553 \\ 0.1910085 & ? \end{array}\right] \overline{X} \\ 0.190892 \qquad 1.552 \end{array}\right] 0.001$$

Hence $X = (1625/2790) \times 0.001 = 0.00058244$

and $C =$ the antilog of $2.1910085 = (1.553 - 0.00058244) \times 100 = 155.241756$

as compared to 155.241747 by solving for c with an HP 41CV hand held calculator.

Plate 1.4d Intersection of Two Street Center Lines

4. Solve for angles ε, ζ, and η using the angles γ and δ found in the preceding step and the given bearings α and β.
5. Solve for the lengths A and B using the law of sines with angles ε, ζ, and η and the length C as

$$A = C\sin(\zeta)/\sin(\eta) \quad \text{and} \quad B = C\sin(\varepsilon)/\sin(\eta)$$

6. Compute dx and dy components of the said A and B lengths using the given bearings α and β, and add them to the coordinates of points P1 and P2 respectively to find the coordinates of the intersection point Px.

In perusing Plate 1.4d, note that the logarithmic table used contains only six digit logarithms yielding a length of 155.241756 as compared to a length of 155.241747 obtained with an HP 41CV hand held calculator (see Plate 1.4e) in a matter of seconds.

As seen, the work is rather tedious and time consuming. Since a record of all computations had to be kept, and in order to maintain some sense of clarity and explanation of what had been computed, hand sketches had to drawn, and preprinted table sheets or forms had to be devised in which distances, angles, bearings, trigonometric functions, and logarithms had to be displayed in predesignated cells so that they can "easily" be retraced if the need arose to find an error, or to make a modification in the design.

As an example of a preprinted form reference is made to Plate 1.4f regarding the balancing of a five course traverse. The form portrayed in this plate is a reproduction of an old form created by use of a modern-day spreadsheet program. Note that the associated computations have been done in a bit more "advanced technologic era" than that of the preceding example when a mechanical calculator such as the Marchant of Plate 1.4c might have been used, thus eliminating the need of logarithms.

The senior, in age, author of this book remembers working for several weeks on the geometric layout of a tri-level interchange in Syracuse, New York for which he had been assigned an assistant whose function was to look up the trigonometric functions as he was carrying out the geometric design. This is a far cry from the modern age computerized design in which an engineer could probably do the same in a matter of a few hours.

Plate 1.4e *Top*: The HP 41CV Calculator. *Bottom*: The HP 71B Handheld Basic programmable calculator (circa 1985) The HP 71B was physically 1.3 % larger than the HP 41CV. The pictures as shown are about 43 % and 32 % of actual size

1.5 Enter the Computer Age

1.5.1 The IBM 650

Electronic computers had been around since the mid-1940s in some universities and government installations, but it was 1958 more or less when the friendly sales force of the International Business Machines (IBM) Corporation began knocking on civil engineering doors announcing the first production computer the **IBM 650**. The first delivery was in 1954 to large corporations and universities in general, but it did not hit the public market until 4 years later. It was impressive for its day with a very large footprint, and its need for a lot of air-conditioning at a time when there were not many air-conditioned offices. That was before transistors and the 650 produced more heat then hell. Word has it that in a building in the Rome, New York US Air Force base there were so many 650s and other computers that if the air conditioning system stopped while the computers were operational personnel had but a few minutes to evacuate before they would succumb to the heat. The 650 featured an internal rotating drum memory of 2000 words of ten digits each, and it had punched card input and output with optional magnetic tape. It sold for about $500,000 and rented for close to $10,000 per month for a fully operational installation. User interface with the IBM 650 on a higher level than with machine level or with the assembler language SOAP was with a variety of compiled languages developed at various universities. These compilers hung around until about 1960 when they were all replaced by the newly created computer language FORTRAN (see next section of this chapter).

1.5.2 The IBM 610

In 1957 IBM announced the **IBM 610 Auto-Point Computer** which is the first electronic computer to be considered as a personal computer in the sense that a user could sit behind the typewriter keyboard to program it, and use it for production. It was a lot smaller than the IBM 650 and could be bought for about $55,000 or rented for about $1200 per month. Communication with this computer was done through a series of short sentence like commands that would be more or less conventional manual arithmetic operations. A higher level language like FORTRAN was not available. Some engineering firms began to flirt with this computer, but this computer was not able to find home in a substantial number of firms, if in any at all.

The graduate thesis in 1959 on the geometric design of curved bridges of the senior author of this book used the Purdue University X-2 compiler that required three passes through the IBM 650 before it could be executed for testing and eventual production work.

Sample Traverse Adjustment by the Compass Method without a Computer

Point	Length	Bearing	Latitude Cosine	Latitude North (+) South (-)	Departure Sine	Departure East (+) West (-)	Adjust'nts Lat.	Adjust'nts Dep.	Balanced Lat.	Balanced Dep.	Coordinates North	Coordinates East
S1											10000.00	10000.00
	129.42	N 0 43 55 W	0.99991840	129.41	0.01277449	-1.65	-0.01	0.00	129.40	-1.65		
S2											10129.40	9998.35
	34.55	N 60 2 24 E	0.49939528	17.25	0.86637426	29.93	0.00	0.00	17.25	29.93		
S3											10146.66	10028.28
	132.80	S 29 57 36 E	0.86637426	-115.05	0.49939528	66.32	0.01	0.00	-115.06	66.32		
S4											10031.60	10094.60
	30.38	S 0 43 55 E	0.99991840	-30.38	0.01277449	0.39	0.00	0.00	-30.38	0.39		
S5											10001.22	10094.99
	95.00	S 89 16 5 W	0.01277449	-1.21	0.99991840	-94.99	0.00	0.00	-1.22	-94.99		
S1											10000.00	10000.00
	422.15			dY = 0.018		dX = -0.005			0.000	0.000		

Closure errore

e=(dY^2+dX^2)^0.5 =   dY^2 = 0.000322   dX^2 = 0.000020   e = 0.019

Direction (degr.)= atan(dY/dX) = -75.89

Precision = 1 : 22,812

DATA ENTRY AND COMPUTATIONS

DATA ENTRIES
Point   Point identification.
Length   Course length. Enter the sum at the column bottom.
Bearing   Course bearing.
Coordinates   North and south coordinates of the start point.

COMPUTATIONS
Latitudes and Departures
Cosine   Cosine of the bearing from the book of trigonometric function.
North/South   Product of the Length and Latitude columns noting the $\pm$ sign. Enter the sum at the column bottom.
Sine   Sine of the bearing from the book of trigonometric function.
East/West   Product of the Length and Departure columns noting the $\pm$ sign. Enter the sum at the column bottom.
Course Adjustments
Lat.   Course length times the North/South sum over the traverse length (plus for a north bearing and minus for a south bearing).
Dep.   Course length and the East/West sum over the traverse length (plus for an east bearing and minus for a west bearing).
Course Balancing
Lat.   For a north bearing add the Lat. adjustment to the latitude and subtract it for a south bearing.
Dep.   For an east bearing add the Dep. adjustment to the departure and subtract it for a west bearing.
Coordinates   Add the adjusted latitude to the north coordinate, and the adjusted departure to the east coordinate of the preceding point.
Traverse Closure   The equations for the error of closure and precision are shown in the table.

Plate 1.4f   Non-Computer-Aided Traverse Adjustment by the Compass Rule Method

Plate 1.5a   The IBM 650 Computer Lawrence Livermore National Laboratory

Plate 1.5b   The IBM 610 Auto-Point Computer

Plate 1.5c   The IBM 1620 Model I Level A with Paper Tape (on the *Left*) and Enlargement of the Console and Typewriter Input/Output Devise (on the *Right*)—Circa 1962–1967

1.5.3 The IBM 1620

With the disappointing results of the IBM 610, IBM announced in 1959 the **IBM 1620** with a more impressive console with lots of blinking lights and switches, but the same type of typewriter for man-machine interaction. In addition, a paper tape reader and output device could be added. For those companies that could afford a few more hundreds of dollars per month, a card input–output device could replace the paper tape unit. This computer featured a memory of 20,000 ten digit words, that could be increased to 60,000 words in increments of 20,000 and a substantial cost. The bare bone operable configuration would rent for about $3200 per month.

The paper tape unit was an interesting and relatively inexpensive, for its time, data storage device for saving software source code and for data output for future input. The tape for the IBM 1620, most often made of paper but also of mylar for more durability, was about one inch wide and featured a series of small sprocket holes for moving the tape along the optical reading device of the computer. The tape shown in Plate 1.5d1 is almost three quarters the actual size of such a tape. Notice the three lines of holes over the sprocket holes and the five lines of holes under the sprocket holes. Each column of holes within the tape could have one or up to eight such holes. Depending on the number of holes and on their position within a column, the column would represent a letter or number digit or other permitted character. After some familiarization with the positions of the holes it was not too difficult for a programmer to read such a tape from left to right.

Because of the small amount of memory and cost of hardware certain programs had to be broken up in parts so that they could fit. Thus, it was not unusual to load the first part of a program, then load an empty tape, enter some data in the typewriter, have the program punch on the blank tape intermediate results, unload the tape, load the second program tape, and repeat the process depending on into how many parts a program had been broken up. That was programming

and processing for those that could not afford larger hardware. For those interested for further reading on paper tape media reference is made to LaFarr's introduction on paper tape at http://www.zyvra.org/lafarr/papert.htm and to http://www.science.uva.nl/museum/papertape.php.

The IBM 1620 was the first computer to appeal to several civil engineering firms, and it could be programmed in FORTRAN. As a result of this appeal various generic civil engineering applications began to surface. More about this later on in this book, but next there is another application that emerged about the same time. Most of all civil engineering and all surveying projects involve topographic maps of the land upon which a project is to be constructed. For years photogrammetry had been used but now the computer could really interface the mapping process with the design process, and the tool that could do that was the digitizer. With a digitizer attached to a photogrammetric stereo-compiler, planimetric features, contour lines, and cross sections could be transformed into digital information and stored on some magnetic media such as punched card, paper tape, or magnetic tape and transferred to the computer for processing.

The digitizers of that day were not the light weight toys of today. They were big in size and price. The Wang digitizer of Plate 1.5d3 was about 5 feet (1.5 m) tall and was programmed with an electronic plugboard, a metal box with many, many holes in it with number 14 gauge, more or less, wires with plugs in each and that had to be plugged into the proper holes of the plugboard. A few years after its use the cabinet of this digitizer was gutted and turned into storage bin. The replacement digitizer was a 3 foot (1 m) by 6 foot (2 m) and about 8 in. (200 mm) thick, more or less, flat bed with neon lights that weighted a ton (figuratively speaking) that could be programmed from the computer.

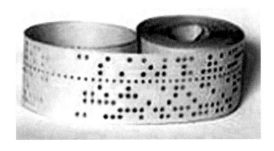

Plate 1.5d1   The IBM 1620 Paper Tape Data Storage Media

Plate 1.5d2 Bausch and Lomb Photogrammetric Stereo-Compiler Attached to a Wang Digitizer (not shown)—Circa 1967

Plate 1.5d3 Partial View of a Wang Digitizer—Circa 1967

1.5.4 The IBM 1620 Competition

During this early time of the computer age The IBM Corporation was not the only one that had entered the computer arena. There were other companies, some of which blossomed later on, but the two that attracted some larger civil engineering firms were the Computer Division of the Bendix Corporation, and the Royal McBee the computer division of Royal Typewriter.

The **Bendix G-15** was introduced in 1956. An operational model cost around $60,000, and could be leased for $1500 per month. FORTRAN was not available. The Bendix G-15 ceased after the computer division of The Bendix Corporation was sold to Control Data Corporation in 1963.

The **LGP-30**, an acronym for Librascope General Purpose, was introduced in the market in 1956 by Royal McBee in partnership with General Precision. Tracing the history of these two companies is a labyrinth of financial intermarriages and well beyond the scope of this book. The LGP-30 was a small, desk size computer that sold for about $47,000.

When the IBM 1620 took hold in the civil engineering computer market, engineering companies with Bendix G-15s and LGP-30s switched to the IBM 1620.

1.5.5 The IBM 1130

All seemed to be progressing gradually and on an even keel, computer-wise use that is, until February 1965 when IBM announced the **IBM 1130 Computing System** and the engineering and scientific communities fell immediately fell in love with it. The bulk of its users held on to their 1130s for at least 10 years, and some even longer. The two authors of this book recall meeting an engineering firm in 1989 that was still using their original IBM 1130. Although announced in 1965, first deliveries were not made until 18 months later. In those early days of the computer it was the norm of IBM to

Plate 1.5e The Bendix G15

Plate 1.5f The LGP-30 with the Front Cover Removed

Plate 1.5g The IBM 1130 Console with Magnetic Disk Storage, High Speed Printer and Card Input/Output Unit and Optional Line Plotter

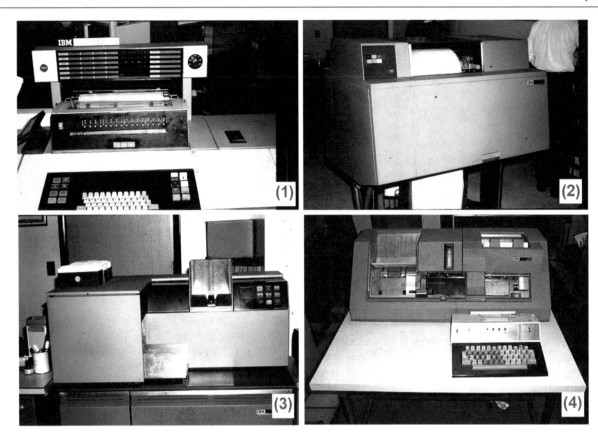

Plate 1.5h   (1) IBM 1130 Console, (2) IBM 1403 Line Printer, (3) 1442 Card Read/Punch, (4) 026 Card Punch

announce a product one and may 2 years ahead of it first working product delivery.

The bare computer could be bought for about $33,000 or rent for about $1000 a month, but could not get the job done. Thus, a basic functional system with 16,384 16-bit word core memory, a 512,000 16-bit word removable 2315 Disk Cartridge, a 1442 high speed card reader and punch, and a 1403 600 lines per minute printer would rent for a little over $4000 per month. The price could go up depending on additional core memory, the use of the 1316 Disk Pack or Packs and printer speed. Also let us not forget the necessary IBM 026 stand alone desk size card punch unit for preparing the data entry stacks of 3.25 by 7.375 in. (82.55 by 187.325 mm) punched cards, which were everyone's nightmare if a card stack happened to be dropped and scattered. Software wise the IBM 1130 could support the FORTRAN, APL, BASIC, COBOL, FORTH, PL/I and RPG programming languages.

With the arrival of the IBM 1130 three other computer peripheral equipment made their debut to the common engineering world, (a) the digitizer, addressed a few pages back, for changing existing hard copy line work into digital format, (b) the line plotters, also referred to as pen plotters and now called large format printers for the production of hard copy engineering drawings, and (c) the graphic display units

or terminals or CRTs for short, now called computer monitors for the digital display of graphic and textual information and interactive processing.

1.5.6   The Plotters

There were two type of plotters, drum plotters and flat-bed plotters. These plotters used ink pens and vector graphic principles to establish points and connect them with a line. Curves were created with a series of concatenated chords (the smaller the chord, the smoother the arc). There were a few companies producing either or both types of plotters, CalComp of which might have been the better known at that time because of its simplicity and relatively low cost. At that time it was the first non-IBM manufactured product used by IBM as the IBM 1627.

The first CalComp drum plotter featured an 11 in. (280 mm) diameter and 30 in. (762 mm) long drum with sprockets at each end rotating clockwise and counterclockwise about a horizontal axle, and a stationary horizontal bar along which a pen unit would slide back and forth. The plotting media, a continuous roll of paper or mylar with sprocket holes at each end, would be mounted on two spindles at the back of the plotter and would lay flat on the surface of the drum near the area of the pen unit's area. As the drum rotated the plotting media would move along with it with the aid of the sprocket holes. The pen within the pen unit could also move up from and down on the plotting media depending on the computer program command and input data, thus creating lines, curves, and letters. Originally, the pen unit would contain a single pen. To create different weight or color lines, the plotter would have to be stopped, the pen type changed and the plotting resumed. Soon units with up to eight different type of pens would be introduced. The plotter was a work horse at a purchase cost of about $8000 and with little if no maintenance except for one "small" issue, that of the $1 or $2 pen. During a production operation the ink could at times plug the pen which had to be cleaned. Many a drawing would be wasted and plotting had to be restarted. This was an issue with all line plotters with pen cleaning a routine maintenance operation.

Another drum plotter that became quite popular in the 1980s was the Zeta plotter which became the Nicolet Zeta plotter and eventually was acquired by Bruning Computer Graphics and became the Bruning Zeta plotter. The Bruning Zeta 836 model proved a good work horse with the 836CS model providing the option of single sheet feed rather than continuous roll media. However, the latter 900 model although had a classier appearance did not measure up to the level of the 836 model.

Plotting software could be developed by use of FORTRAN

The senior author of this book remembers some draftsmen with special arm cloth protectors to hold off the sweat from dripping on the ink-work on linen drafting sheets.

and were use oriented. That is there were specific programs for plotting piers, abutments, cross sections, contours, profiles and the like. During the late 1960s the elder of the two authors of this book was the director of the computing center of Erdman Anthony Associates, a consulting civil engineering firm of Rochester, New York and was producing final design drawings of bridge piers and abutments and reinforcing bar schedules of bridges, and plan and profiles of combined wastewater and storm water sewers with the IBM 1130 and the Calcomp drum plotter.

Like their name implies, flat-bed plotters were flat table like and large enough to accommodate a single sheet of paper or mylar. The size of the table bed varied depending on the desired maximum plotting media that the user might ever use. A single or multi-pen unit would move along a movable axis bar of a large T-square which movable bar would move along a stationary axis of the said T-square. This was the general concept of this type of plotter and there were many

variations thereof. These plotters were bulky, heavy, and expensive.

Ink jet plotters, now called ink jet printers, made their appearance in the computing environment in the late 1970s,

Plate 1.5i IBM 2310 Model 1 The IBM 2310 Model 2 could house two IBM 2315 Disk Cartridges. An IBM 2315 disk of 1 megabyte (MG) cost $250 in 1970. An 8 MG (8000 larger disk space) flash drive cost $18 in 2012. The size of either IBM 2310 model was about $2 \times 2 \times 3$ feet ($0.6 \times 0.6 \times 0.9$ m), while that of a flash drive is smaller than a man's small finger

Plate 1.5j IBM 1627 Calcomp Drum Pen Plotter

Plate 1.5k The Bruning Zeta 836CS Model

Plate 1.5l The Bruning Zeta 900 Model

but their concept dates to the end of the nineteenth century. Their use was not impressive in the production of final design plans due to the coarse resolution. However, as their resolution and ease of use have increased immensely together with the introduction of a multitude of colors have made the pen plotters practically obsolete. Last but not least let us not forget the latest development, that of the laser jet printers. But yet, pen plotters are not quite dead as yet. There are certain graphic artists that do prefer them to ink jet and laser jet printers because of the quality of the vector line work. One must have a very discerning eye for line quality.

1.5.7   The CRTs

The graphic display units were cathode ray tubes and the user interface, besides the interaction dictated by the applicable software, would be with a stylus connected to the tube. The user would hold the stylus and move it with his hand over the tube's surface and would press it on the screen to activate the position of the stylus. Points would be created by a single press and release action of the stylus, while lines would be created by a press, drag and release action of the stylus. These devices were expensive for their time and they did not come into common use until the introduction of the personal computers in the early to mid-1980s. In addition to their large monetary cost, they were also large in size, being more or less 2 foot (600 mm) cubes for a 19 in. screen, and weight (a far cry from the current flat display monitors).

1.5.8   The Microcomputers

The IBM 1130 computer was so successful that several clones, in part or in whole, were created by various compa-

nies that competed with IBM in price. However, the biggest challenge to the IBM 1130 came about 10 years after its introduction by Prime Computer (aka PR1ME Computer) from Natick, Massachusetts, and Digital Equipment Corporation (aka DEC) from Maynard, Massachusetts. Both of these companies were able to wean the majority of the civil engineering market from the IBM 1130, and divide it into two camps, the pro-PR1ME and the pro-DEC.

In the mid-1980s besides the Prime and the DEC computers there were several others such as Sun Microcomputers, Inc. with its SPARC RISC (Reduced Instruction Set Computing) workstation, Hewlett Packard, Apollo, Tektronix, Inc. and others. By the way, Apollo was created in 1980 by William Paduska, the founder of the PR1ME Computer and was acquired by Hewlett-Packard in 1989. IBM tried to counter these competitors in the engineering field in price and performance first with its IBM PS/2, then with the IBM RT and in 1990 with a series of five models, the IBM RISC 6000 POWERstation/POWERserver 220, 340, 350, 520H and the 560 workstations. These were truly excellent computers, but it was too late. The personal computer (PC) had mad its hit in the market place. Incidentally IBM tried to attract the home based computer market with its IBM PCjr using a Charlie Chaplin mimic but failed.

While this type of computers were proliferating, there were major developments on the larger computer monsters. Thus, by the mid-1970s a distinction between the large and small computers came into existence. The large computers were referred to as "mainframes" with the very large ones as "supercomputers" such as the Cray, and the smaller ones referred to as "midrange" or "microcomputers" or "miniframes" or "minicomputers." The latter more or less have disappeared due to the introduction of the so-called work stations and eventually by the personal computers, the PC's, in the mid-1980s.

From 1976 until 1988, **PR1ME Computer** introduced a variety of models ranging from its first, the Prime 400, to the Prime 9955 in 1985, with a variety of in-between models such as the Primes 450, 550, 650, 750 and 850. In addition, there were the 2000 series models such as the Prime 2250 (aka Rabbit), 2450, 2550 and others that emulated their bigger brethren but were intended primarily for the computer aided drafting (CAD) industry. The operating system of these computers was the PRIMOS, and its first CAD software was the MEDUSA system (originally developed in Cambridge, England) which was supplemented by the acquisition of Computervision in 1899. Actually, there were two slightly different versions of MEDUSA, one that was owned by Computervision and was operable on both Prime and DEC VAX computers (see below) and the Prime MEDUSA that ran on Prime computers only. Computervision was one of the original players in the computer aide drafting and manufacturing field since 1969. By July of 1992, Prime Computer could no longer compete price wise with the new

Plate 1.5m (*Left* to *right*). The Prime 2550 Central Processing and Monitor and its Magnetic Tape and External Storage Disks, and the Prime 2250 (aka the Rabbit)

personal computer (PC) market and various proprietary graphics software and closed its doors. The smallest Prime CAD system would sell for $70,000 while that of a PC for under $20,000.

A contemporary to the PR1ME Computer machine series was the **Digital Equipment Corporation (DEC)** VAX series computers. Prior to the VAX series, DEC had entered the computer market with their PDP series, of which the PDP-11 proved a rather respectful computer. Although there were distinct differences in computer architecture and operating systems, there was not much difference between a PR1ME and a DEC VAX computer as far as the final product of civil engineering practitioner's work and cost were concerned. Even in looks, aside from color and insignia, Prime's and DEC's looked fairly much the same. The disagreements between the two camps mentioned earlier lay primarily in philosophical and really in mundane or "computer geek" principles. By this time it was the software that was controlling computer use. DEC was about 15 years old when PR1ME Computer came to being, and lasted six years longer before being acquired by Compaq which in turn was acquired by Hewlett-Packard in 2002.

1.5.9 Project Orpheus

In 1997 Kevin Johnston, a specialist in Prime's marketing group, conceived the idea of utilizing geographic information system (GIS) technology to assist planners, city managers, real estate professionals and industrialists to best manage and develop land to meet the needs of society. For this he assembled a team of specialists in the GIS field, planning, engineering, architecture and related fields to carry out

a pilot case study project with all disciplines utilizing one single computing system, the Prime 9955II™ super-minicomputer. The project under the code name of Project Orpheus was to select a site for the development of a new technology facility and the necessary support facilities including housing and retail store establishments near a major metropolitan area in the USA without adversely impacting the existing land use and infrastructures. To carry out Project Orpheus Prime selected and organized the following technologies and team members:

- ARC/INFO® (now ArcGIS®) a GIS system software of Environmental Systems Research Institute (ESRI) of Redlands, California for the geographic information modeling tasks.
- ERDAS software from ERDAS. Inc. of Atlanta, Georgia to provide remote sensing for creating and updating the databases and image processing.
- Interactive Surface Modeling (IMS) software from Dynamic Graphics of Berkley, California for surface modeling and map generation.
- The CEDRA System™ (CEDRA™), a civil engineering design package of software from The CEDRA Corporation™ of Rochester, New York for the design and drawing preparation of the project's detailed site design including roadways, parcel subdivision, water supply and distribution, wastewater collection and disposal, and storm water management. The two authors of this book and the now Rev. Demetrios E. Tonias represented The CEDRA Corporation.
- The design of the highway to connect the project development with the nearby metropolitan area was provided by highway design software of Edwards and Kelsey of Livingston, New Jersey.
- MEDUSA™ software from Prime were used by Prime architectural staff for the architectural design of the various structures and landscape of the site.

The selected project site was a 36 by 34 miles square parcel of land situated about 50 miles west of Chicago, Illinois. The project work was divided into six phases, some of which were carried out concurrently as follows:

- Phase 1 involved the site selection and the loading of the pertinent databases into the Prime 9955II computer.
- Phase 2 included the development of the models for the ideal land use plans, and the determination of the influx employees and families, and number and type of businesses that would immigrate as a result of the economic development.
- Phase 3 developed preliminary alternate designs conducted by the various independent technologies and data exchanged between them.

Plate 1.5n The IBM RISC 6000 POWERstation/POWERserver Model 650

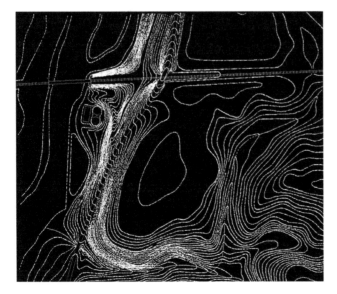

Plate 1.5o Project Orpheus 10 Foot Contours Imported from ARC/INFO into CEDRA Software and Refined to 2 Foot Contours

- Phase 4 involved a team of landscape architects that located building footprints, parking facilities, and a view analysis of the sire from various angles for aesthetic concerns.
- Phase 5 had the teams of engineers and architects and landscape architects resolve project interferences and then finalize the project designs integrating one with the other as needed and updated be base ARC/INFO database.
- Phase 6 involved the overall review of the various designs and drawings and the finalization of the overall project for final presentation.

Project Orpheus proved that the integration of technologies and the coordination of the working teams exceeded the anticipated results with the geographic information technology proving the glue to join the specific technologies associated with the project. Project Orpheus initiated the cooperative efforts of CEDRA and ESRI to bridge the civil engineering avSeries™ software and the geographic information ArcGIS® system.

1.6   Basic Programming Issues

Computer hardware at the time of their introduction to the masses of people were quite fascinating. They were large in size occupying whole rooms, and they had a lot of blinking lights, paper tapes with punched circular holes and paper card with punched rectangular holes, magnetic tapes in glass sealed compartments moving at high speeds, and card readers and punchers making a lot of noise. They all made good props for science fiction movies and tracking aliens from outer space. For the average Joe, though, who tried to communicate with them it was another story.

To put it simply in a rather very rudimental form, computers by "birth" recognize only one thing called a "bit" which could be one of two simple signs, zero (0) or one (1). When eight bits are grouped in a bunch at a time they constitute a byte, and when four bytes are bunched together they constitute one word of 32 bits. Words were used in the early computer era but not nowadays. These byte groups started from being eight at a time, then 16, then 32 and now they have reached 64 bits. To communicate with computers and teach them do what they wanted them to do, humans had to use bits and bytes to create instructions. That is to enter into the computer a lot of zeros and ones. This form of communication was referred to as the machine language, which was an extremely tedious process. The elder author of this book remembers writing a short simple sorting routine and can attest to it.

The next step was for the humans to take these instructions of zeros and ones and create a language that would accept short one word English commands, say "ADD," and

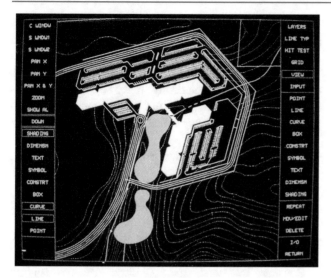

Plate 1.5p CEDRA Software Displaying the Proposed Research and Development Facility, Retention Ponds and Roadways

translate it into a series of zeros and ones so that the computers could understand it. This was referred to as an assembler language. This may sound easy, but it really is not because mathematical instructions are not the only commands one has to give to a computer. You have to tell it to read and write, to translate, to move data up and down, right and left, and do other things that we humans do but never think about them. Did the average Joe we spoke before ever think of every minute detail his brain goes through from the moment the alarm clock rings to the time he opens the door to go to work?

Since different persons created computers, they all wrote assembler languages in different ways, so if one had to change work from one computer to another, a new language had to be learned. So an easier and more common language had to be invented. Such a language was the FORTRAN we spoke of earlier in this chapter, the one developed by **John Warner Backus** (1924–2007) and his team of IBM. It was a great jump up the ladder of communication with, and taming of, the computer beasts. However, humans not being happy with themselves nor with each other invented other similar languages. These were referred to as the compiled languages, among which we have the many offshoots (very loosely speaking) of FORTRAN, such as Pascal, C, Basic, COBOL, ALGOL, APL, FORTH, PL/I, RPG, SNOBOL, and others.

FORTRAN has underwent several upgrades over the years to the present, and it is still in use in many engineering software. Many current computer scientists may scorn and make fun of FORTRAN for its limitations under current computing loads, but it was indeed a workhorse that helped many programmers meet the challenges of that pioneering age. Whereas modern computing techniques simplify pro-

gramming immensely, FORTRAN sharpened a programmer's mind.

Of these languages, C and Basic have become the current standard bearers with FORTRAN still hanging on due to its extensive library of previously developed programs and suitability for engineering and scientific applications. Over the years these three languages have cloned into the interactive processing environment as Visual C++, Visual Basic, and the Microsoft Fortran PowerStation 4.0.

John Warner Backus of IBM (1924–2007)

In addition to this high level programming languages, we have the various proprietary computer operating systems, the ones that control the operation of a computer, of which the Microsoft Corporation's Windows has gained dominance, and with Unix running not very far behind it. Both of these two operating systems are proprietary. Tired of the legal licensing requirements of their use **Linus Benedict Torvalds**, a Finnish software engineer of the University of Helsinki introduced Linux as a free open form computer operating system.

For those young readers of this book that may be involved in the programming of an application, life is easy. A full blown personal computer (PC) may be bought outright for the cost of a 1 month's rental as late as 30 or so years ago. The power of today's PCs is mammoth in comparison to that of the computers, and the programming languages have made giant leaps forward. Two, perhaps the most important, hardware related aspects of programming, computer memory and word size, are no longer an issue.

- **Computer Memory** Whereas nowadays programmers hardly ever think of computer memory size, both due to cost and technology, in the good old days of the 1950s through the 1970s programmers would spend hours, if not days, to save a few words of a program so that it would fit in the computer without breaking it up in parts.
- **Word Size** For many computer applications, and particularly for civil engineering and surveying geometric appli-

cations, an 8 bit word was not enough to yield the required computational precision. This pertained to the size of the coordinates in significant digits, and to the computational procedures or solutions in which distances had to be multiplied or raised to a power. As a result, games had to be played to break up a coordinate in two, carry out the necessary operations and the combine the two parts to display the result of the computation.

Another issue of the early days which was resolved rather quickly was that of certain basic algorithms for the expedient solution of trigonometric functions, the square root, and others.

1.7   Civil Engineering Software—CEPA

The civil engineering profession was one of the first to espouse the computer in the mid- to late 1950s. Before we proceed any further, let us clarify this statement. By Civil Engineering profession we do not imply the profession as a whole, but a few practitioners and a few academicians that foresaw its impact on the profession. In the remaining paragraphs of this chapter we will attempt to present a capsule history of the acceptance of electronic computing by the civil engineering profession.

In the mid- to late 1950s, some civil engineering college departments began to research the computer's potential, but not many. When the elder of the authors of this book suggested to his advisor the development of a computer program in bridge design as the topic of his thesis, he was told to find another advisor. While searching for another advisor, an older professor offered the said author a deal. "I will teach you earthquake engineering" he said, "if you promise to teach me about computers." The deal sealed by a handshake, the said author proceeded to take two computer courses from the mathematics department. When attempting to take a third course, the head of the civil engineering department stepped in and said: "Just because the Ruskies put a Sputnik in the skies, everyone wants to learn Russian. Just because an electrician came up with a box with blinking lights, everyone wants to learn computers. You are a civil engineer, you will never use computers. Two courses are enough." Names have been left out on purpose to save any embarrassment.

Aside from certain universities and colleges, there were some rather large engineering consulting firms that got their electronic computer feet wet by investing in the IBM 650 computer in the mid- and late 1950s, and by 1962 they and several other smaller firms installed an IBM 1620 computer. These companies had no source for the specialized software that they needed for their engineering projects. Therefore, it was up to them to create in-house software development departments and staff to meet their software needs which increased immensely their overhead expenditures.

On 26 April 1965, with the IBM 1130 fresh in the market place, twenty one civil engineers from 19 consulting firms and The IBM Corporation, under the sponsorship of IBM, met at the Palmer House Hotel in Chicago to organize an IBM 1130 user's group for civil engineering. To the best of memory recollection of the elder author of this book, some of the participants included, in company alphabetical order, the following:

- Ammann and Whitney, New York City, Herbert Rothman, P.E.
- Bryozowski, William
- Charles H. Sells, Inc., Bedford Hills, New York, Robert Kenngott, P.E., L.S.
- Charles H. Sells, Inc., Bedford Hills, New York, Max Noach
- Consoer Townsend, Inc., Chicago, Illinois, Rodney G. Dabe, P.E.
- Claude C. Demete, P.E..
- Erdman and Anthony, Rochester, New York, Elias C. Tonias, P.E..
- Erdman and Anthony, Rochester, New York, Paul Treer, P.E.
- Green and Associates, Baltimore, Maryland, Richard V. Reikenis, P.E.
- Greve & O'Rourke Systems, Inc., Los Angeles, California, Norman R. Greve, C.E., S.E.
- IBM Corporation, The, New Rochelle, New York, Mort Sinkoff
- IBM Corporation, The, New Rochelle, New York, Paul Spidell
- McFarland and Johnson, Binghamton, New York, Louis H. Stanley, P.E., L.S.
- Parsons, Brincherhoff, Quade and Douglas, New York City, James Boyd, P.E.
- Sargent and Lundy, Chicago, Illinois, Charles Beck, P.E.
- Skidmore, Owings & Merrill, Chicago, Illinois, Dr. G. Neil Harper, P.E.
- Tippets, Abbett, McCarthy, Stratton, of New York City, Dr. Albert Spalding, P.E.
- Weiskopf and Pickworth, New York City, Morton Aligator

Up until this time, the development of civil engineering software was being carried out individually by the various engineering companies, each of which had instituted to some extent individual company software development departments. Needless to say the company overhead costs began to escalate. Hence the primary objective of the attendees at this Chicago meeting was to foster the cost effective use of electronic computing in the Civil Engineering field. Thus, CEPA, an acronym for Civil Engineering Program Applications came to existence. Several years later, CEPA was reorganized and incorporated as the Society for Computer Applications in Engineering, Planning and Architecture. However, the acronym CEPA was retained. To attain their main goal, CEPA members met twice each year, in the spring and fall, to review and assess the progress of the software development assignments. Two issues faced the newborn CEPA:

- **Development Assignment**, that is which company was going to develop which software to be shared by the group. Towards this end, the various members related the software efforts that their companies had undertaken up to that date, and assignments were made as to which software each company was to produce.
- **Equitable Cost Recovery**, that is how was the sharing of software between member companies to be made equitable for the effort expanded in developing the software. Many a night hours of meetings were expanded by the early CEPA members to design an equitable system of computer program sharing, with the result being the point system. Under this system a company would receive a number of points for each program submitted to the CEPA library, and in exchange for these points a company could receive a computer program that another company had submitted to the said library. In addition, a few companies would pool their efforts and jointly develop a program which could be submitted to this library with participating company receiving an appropriate number of points.

CEPA's membership grew steadily over the years and by the mid-1980s it had over 300 member companies. Not all member companies were actively involved in the development of software. Over the years, the point system was evolved to accommodate them, and points could be purchased.

Another task that was undertaken by CEPA was the accounting and recovery of the computer overhead costs (hardware and software development costs) encountered by the engineering companies that had computer installations, and provided services to governmental agencies. As was the custom of the time, and still is in many cases, governmental agencies would reimburse engineering companies on a cost plus fixed fee, which means wages plus overhead plus some percentage of the first two for profit. Although this may seem equitable to some extent, governmental agencies would place a cap on overhead to about 1.15 of the labor cost, and would limit the fee for profit to roughly 12 %, depending on how overhead was determined. Now this was the sore point for the engineering companies. Many governmental agencies in the early stages of the computer age would accept computer costs as part of overall company overhead while others would not. Since computer hardware costs at that time were very high indeed, and the labor cost for the development of software exceeded the cost of hardware, the overall overhead of an engineering company would far exceed the overhead limits allowed by a governmental agency. Hence CEPA's task was the education of the governmental agencies, and it succeeded to great extent.

Canadian CEPA member companies receiving a deck of cards form the USA would ask the American companies to indicate in the bill of lading that the cards being transported were used and not new cards. For a deck of new and for all intents and purposes worthless cards the Canadian receiver would have to pay a tariff, but for a deck of used cards, those that had valuable software code punched on the cards, there would be no tariff whatsoever. That is government at work.

A few years before CEPA, the American Consulting Engineers Council (ACEC) had undertaken a different path towards the development of computer software and sharing of their costs. This organization established a special software development group referred to as APEC that developed software that was leased by member firms for some monetary value. The software library was rather small and concentrated primarily towards the mechanical engineering aspects (HVAC and plumbing).

As the computer hardware and software industries advanced, and costs were reduced to almost nothing, relatively speaking, the need for in-house software development has for all intents and purposes disappeared, and the need for CEPA ceased to exist. In 1989, or thereabouts, an attempt to revitalize CEPA towards a new direction did not come to fruition. However, during its life CEPA did provide a motivating force in promoting electronic computing in civil engineering.

A reference has been made in the preceding paragraphs regarding governmental agencies and their reaction towards financial accounting and reimbursement for computer use. Since at the time of the computer introduction computer was for the most part in the USA, with Canada, Great Britain, and France not far behind, the governmental agencies referred to above pertain to American agencies. Below are two stories that might be of interest to some readers.

- Although the primary membership of CEPA comprised American engineering companies, there were a few Canadian members. To the representatives of these companies the method of cost plus fixed fee mentioned above and associated methods of determining overhead factors was inconceivable.
- When exchanging programs between CEPA members at that time the means of exchange were to mail a deck of IBM cards which could very in size of a few inches thick to a box or two and maybe more with each box being about $3.25 \times 7.5 \times 18$ inches ($90 \times 190 \times 457$ mm).

1.8 The NICE Pilot Program

One of the highlights of CEPA was the 1971 US National Science Foundation (NSF) grant to assess the need for a pilot program to attempt to define "a national effort to optimize common use of engineering software." The need for this pilot program was a product of the 1971 NSF sponsored *Special Workshop on Engineering Software Coordination* held at the University of Colorado at Boulder under the chairmanship of Dr. Robert L. Schiffman, a professor of the said school's civil engineering department. Conference participants included practitioners and academicians involved in the development of civil engineering software. CEPA was chosen for this grant due to its stature in the profession and experience in the cooperative software development efforts by its membership. The seven member CEPA team was composed of:

- Hugh McGrory, P.E. (Chairman) of Proctor & Redfern, Ltd of Toronto, Canada
- Dr. Albert Spalding, P.E. of Tippets, Abbett, McCarthy, Stratton of New York City
- Robert L. Kenngott, P.E., L.S. of Terra Mettrics of White Plains, New York
- Morton Lipetz, P.E. of Edwards and Kelsey, Inc. of Livingston, New Jersey
- Everett Moone of Comp-Tron, Inc. of Baltimore, Maryland
- Sam H. Yager Jr., P.E., of Bernard Johnson Incorporated of Houston, Texas
- Dr. David Schelling, P.E. of J. E. Greiner Company, Inc. of Baltimore, Maryland

The pilot project was completed in 1975 with a recommendations that "a pilot program towards a National Institute for Computers in Civil Engineering (NICE) be initiated to promote the use of computers and software in civil engineering." Although the recommendation, as supported by the details of the report titled *National Institute for Computers in Engineering*, had certain specific merits, the computer and software market was beginning to change, and the civil engineering profession began to advance on its own. Thus, the recommended pilot program did not materialize.

1.9 The American Society of Civil Engineers—ASCE

In November 1968, at the Fountainebleau Hotel of Miami, Florida CEPA meeting, Elias C. Tonias, P.E., the elder author of this book, recommended to the board of CEPA that CEPA should undertake an effort to promote to the American Society of Civil Engineers (ASCE) to organizationally undertake the full spectrum of computer applications. This was not to compete with, or replace any of such computing efforts undertaken by the various ASCE divisions, but to provide guidance on a cross divisional scope. By the spring of 1969 Tonias, with the great cooperation of Dr. Steven Fenves of Carnegie Mellon University in Pittsburgh, Pennsylvania collected the necessary number of petition signatures from ASCE members to cause the ASCE board to consider the formation of a special council or division within ASCE dedicated to electronic computing in civil engineering. When the petition came before the ASCE board for review, it was rejected.

However, within the ASCE staff there was a young engineer, Donald Taylor, P.E., who thought that the petition had more than some merit, and with the persistence of Taylor and Tonias, a special meeting was arranged with the ASCE research review committee in Washington, DC in the fall of 1970 in which Tonias presented his case. The committee proved to be quite receptive, and recommended to the ASCE board to initiate the *Task Committee on Computer Applications Research* under the auspices of the *ASCE Committee on Research* to assess the need for the formation of a council or division within ASCE. The task committee under the chairmanship of Tonias comprised less than ten volunteers selected by Tonias and Fenves, several of which were CEPA members. Also a member of this task committee was Dr. Kenneth Medearis of KMA Research from Fort Collins, Colorado, who has just completed a study and report for ASCE on civil engineering software development. The committee's organizational meeting was held shortly after its approval at the civil engineering department of the Rose-Hulman Institute of Technology at Terre Haute, Indiana.

This task committee's recommendation to the ASCE board was the formation of an ASCE council on computer applications and practices with the intent that this council would not ever pursue a divisional status. This proposal deviated from the practice of that time within ASCE in which a new civil engineering discipline that was about to emerge would first form a council to get started, and eventually advance to the status of a Division. This proposal was presented to the ASCE board in Boulder, Colorado in the fall of 1971 and was approved with only one objection and with one board member being absent.

Through a fluke, however, for lack of expressing it in any other terms, the board member that was absent, a supporter of the proposal and a dynamic orator, mistook the approval as an approval of a competing proposal, and was able to convince the board to rescind its approval until further study. At its next meeting in the spring of 1972 ASCE approved the formation of the Technical Council on Computer Practices (TCCP), and the termination of TCCP's father the *Task Committee on Computer Applications Research*. The charge of TCCP was the:

- Coordination of civil engineering uses and users across ASCE divisional computing groups and related organization groups outside ASCE.
- Education of future and practicing engineers in the use and implications of electronic computing in civil engineering.
- Promotion of research in civil engineering involving electronic computing to improve current methods and to develop new technologic advances.
- Assess impacts of electronic computing on the professional practice of civil engineering including, but not limited to pricing and cost, facility management and organization, and credibility, ethical and legal impacts of electronic computing upon the civil engineering profession.

The technology of electronic computing has changed drastically since the 1972 birth of TCCP. However, the main charge of TCCP has remained relevant to date. In 1997, TCCP changed its name to the Technical Council on Computer and Information Technology (TCCIT).

1.10   Public Domain Software

Although the USA created the cradle of the computer hardware industry, France and particularly England took a more active governmental interest in the development of software. Much like everything else, software development in the USA was left up to the private sector and the free market. England, France, and Norway may be considered as the first pioneers in computer graphics. In these countries, although their governments provided financial support, the product software were not free, and soon found homes in private enterprises.

In the USA several universities took an interest in electronic computing, but the Civil Engineering Department of the Massachusetts Institute of Technology (MIT) was the first one to actively pursue the development of civil engineering software that were made available to the public practically for free. These software included:

- **COGO** Coordinate Geometry developed by Dr. Charles Miller and his team.
- **STRESS** Structural Engineering System Solver developed by Dr. Steven J. Fenves and his team.
- **STRUDL** Structural Design Language, an outgrowth of STRESS, by Dr. Robert D. Logcher and his team.
- **ICES** Integrated Civil Engineering System comprised several components that included, in addition to the above three, software pertaining to bridge design, roadway design, traffic analysis, and project management.

In addition to MIT, the Civil Engineering Department of the University of Kentucky developed **KYPIPE**, a water distribution analysis and modeling program, under the direction of Dr. Donald Wood. All of these software eventually have found themselves in the private industry, they have been modified and upgraded, and they are not free any more.

In addition to these universities, two USA governmental entities developed software that were made available to the public worldwide for the cost of shipping, and they are:

- The US Environmental Protection Agency (USEPA) which has developed the:
  SWMM Storm Water Management Model which is a dynamic rainfall-runoff simulation model, and the
  EPANET EPA Network which is a dynamic water distribution system simulation model.
- The Hydraulic Engineering Center (HEC) of the US Army Corps of Engineers which has developed a series of hydraulic software modules, all commencing with **HEC**, pertaining to water resources and modeling of river flows and runoff, with **HEC-RAS** being the next generation software for river flow modeling.

Geometry, as used in the civil engineering and surveying professional fields of practice, is actually that branch of mathematics that is referred to as analytic or coordinate geometry. In simple form, analytic geometry employs algebra and trigonometry to solve problems of Euclidean geometry. Although a form of analytic geometry may be traced to Menaechmos—Μέναιχμος (380–320), the Greek mathematician and friend of Plato, it is Rene Descartes (see Chap. 1 of this book) that set the foundation of coordinate geometry. In his honor, coordinate geometry is sometimes referred to as Cartesian Geometry.

2.1 Coordinate Geometry

In order to function, coordinate geometry requires a *medium* on, or in which *geometric figures*, that can be described or defined numerically and worked upon algebraically, are introduced.

The **medium** in the civil engineering and surveying professional fields of practice is a three dimensional coordinate system representing the surface of the earth as an irregular shape which is overlaid on a flat level plane. This flat plane is defined by X and Y coordinates, and the irregularity of the surface is defined by the Z coordinate. These coordinates are also referred to as departure, latitude and elevation, respectively, and they can be both positive and negative with respect to a reference point (see below). The X and Y coordinates are also referred to in the northern hemisphere as the north and east coordinates, indicating the positive direction, while in the southern hemisphere they are referred to as south and east. Many years ago, the south and east system was employed in some states of the USA.

The **geometric figures** that are introduced in the said medium by civil engineers and surveyors include points, lines, curves, and polygons. Since a figure may be a composite of any number, and of any combination of points, lines, curves, and polygons, we will refer to each point, line, curve, or polygon individually as an "element", a "primitive" or a "feature". Thus, these three words may be used in this book

interchangeably to mean the same, a point, a line, a curve, or a polygon.

For the majority of civil engineering and surveying projects a flat earth medium that disregards the earth's curvature is more than adequate. However, when a project is of a magnitude that spans great areas, as the interstate highway system of the USA, in which the earth's curvature becomes an issue, curvature adjustments are made, and the overall project is then broken up into smaller, flat earth coordinate systems. The representation of earth as a flat coordinate system is referred to as a projection. There are several such projections that are employed throughout the world, with the Lambert Conformal Conic, and Transverse Mercator projections being the ones commonly used in the USA.

In the 1930s, the US Coast and Geodetic Survey developed the State Plane Coordinate System (SPCS) to provide a common coordinate medium for engineers and mappers with a maximum scale distortion of the earth's surface of one part in 10,000. Most, but not all States that are longer in a west to east direction, like Tennessee, employ the Transverse Mercator projection, while those that are longer in a south to north direction, like Vermont, employ the Lambert Conformal Conic projection. Also, depending on its shape and size, a State may be divided into two or more zones with different projections. New York State for example is divided into three Transverse Mercator zones, west, central and east, while its Long Island utilizes the Lambert Conformal Conic projection.

With regards to the elevation or Z coordinate, there are numerous local reference datums in all parts of this earth. In the USA there are three reference datums in use:

- The North American Datum of 1927 (**NAD 27**) as used in the USA is based on the Clarke spheroid of 1866, with the assumption that a geoidal height (elevation) of zero at a point in Meades Ranch, Kansas of 39°13′26.686″N (39.22407944) and 98°32′30.506″W (−98.54180722). (*see http://www.ngs.noaa.gov/faq.shtml#WhatDatum*).

© Springer International Publishing Switzerland 2016
E.C. Tonias, C.N. Tonias, *Geometric Procedures for Civil Engineers*, DOI 10.1007/978-3-319-24295-8_2

- The North American Datum of 1983 (**NAD 83**) is based on the Geodetic Reference System of 1980 using 250,000 points and 600 satellite Doppler stations that constrain the system to a geocentric origin. It is currently in use the USA, Canada, Mexico, and Central America.
- The only worldwide referencing system in use today is the **WGS84** datum which stands for World Geodetic System of 1984. It is basically almost the same as the NAD83, and it is the default standard datum for coordinates stored in recreational and commercial GPS units, as well as of the USA Department of Defense (DoD). GPS users must always check the datum of the maps that they are using. This datum must be stored in the user's media together with any GPS coordinates that are observed and recorded. (*see http://www.ngs.noaa.gov/faq.shtml#WGS84*)

With the availability of modern-day computing capability one may argue that the use of flat mediums is passe. In theory this is quite true, but in practice it is practical and convenient.

Before we proceed any further, we need to address the issue of the third coordinate, the elevation. Every civil engineering project and surveying project is three dimensional. However, the introduction of the elevation coordinates for most, if not for all projects is made as a distinct operation of the other two coordinates, with the three coordinates eventually being merged into one three dimensional model. The reasons for this are a historic state of mind, and practicability.

For example, let us consider the design of a new roadway which is to interact with and affect existing and/or new water mains, wastewater and storm water features, and intersecting streets to name but a few. Many years from now the technology may be such that it will possible to visualize and design this roadway in three dimensions at once, but not now. It is still a two concurrent stage design process. It may be argued that this technology exists today, that we do have three dimensional visualization. However, this is done after the preliminary or final design, or as a conceptual design of which the details are yet to be determined.

As another example let us consider the intersection of two street center lines in a rather simplistic manner by assuming that each one is being defined by two three dimensional points and one straight line connecting the said points. Now let us further assume that each of these four points has been assigned an elevation such that the two straight lines intersect at two points, one on top of the other at the same X and Y coordinates of the flat plane medium. Each of these two intersection points has its own elevation as interpolated along its respective line. It is an easy task to have the line intersection computer algorithm determine both elevations, but which one of these two elevations is to be that of the design intersection of the two streets, if either of them.

Among other considerations, a separate drainage design will have to be made in order to properly assign an elevation to the two dimensional solution of the intersection of these two streets.

Thus, this book addresses geometric solutions in the X and Y plane distinctly from the geometric solution in the X and Z, or Y and Z planes.

2.2   The Geometric Features

Points, lines, curves, and polygons may be (a) **actual** or real physical entities in the coordinate geometry medium such as a point denoting a sewer manhole, or a line denoting a wastewater sewer pipe, that is, they are actual features, or (b) **implied**, pseudo or imaginary entities, such as the endpoint of a line at which there is no actual point feature, or an imaginary line such as one between two actual point features used to obtain a direction.

Before proceeding with the definition of the geometric features, it is prudent to address the term "arc". In the 1960s, with the advent of electronic computing, a new science field began to emerge, that of the geographic information systems, or GIS. One of the earlier players in this field, and now the most preeminent is Environmental Systems Research Institute (ESRI) of Redlands, California. This company's ArcGIS computer software system has coined the word "arc" to mean any connectivity between two points. That is, a straight line, a discontinuous, a circular arc, or any other type of curve that connects two points. However, as used in this book, the term "arc" refers only to circular, parabolic, or elliptic arcs.

As used in this book, the term "arc" refers to a circular arc, parabolic, or elliptic and should not be confused with the same term as used in the ArcGIS community.

● This is a point

Points Point features are two dimensional features in the X and Y plane, or in either the X and Z, or Y and Z plane, and defined by their respective coordinates. Points may be created by:

- Direct input of their coordinates,
- Traversing along a straight line, arc, or spiral curve,

- Projecting normally a given point upon a line, arc, spiral curve, or polygon side,
- Intersecting lines, curves, and/or polygons,
- Creating tangent lines to arcs or spiral curves, or creating tangent curves to other spirals.

In addition to the above four basic features, we also have the following feature types:

This is a polyline with three vertices
and real points at its endpoints

Polylines Polyline features are lines that may have none or many vertices between their two endpoints. In essence they are bent and non-straight lines. Regarding the term "line" note the following:

- A line with two endpoints and no in-between vertices is referred to as a *line*, *single line*, or *two-point line*.
- Within the context of this book, the term "*line*" may be used as a generic term for either a two-point line or a polyline, or may pertain to either line type feature when it is obvious as to which type of line is being discussed.
- It is possible for the two endpoints of a polyline to occupy the same space in the medium, and thus form a closed figure. However, even if the polyline forms a closed figure, it is not considered to be a polygon feature.

Line with implied
endpoints

———————————

Line with real
endpoints

●————————●

Lines Line features comprises two actual or implied endpoints, and they are of a finite length as defined by the distance or length between the said two endpoints (see polylines below). Lines may be defined by:

- Specifying a direction and a distance,
- As the result of a point projection on another feature, or
- As a result of a tangency upon a curve.

A polyline that forms a closed figure (same start and end point) does not constitute a polygon.

Curves Curve features may be full circles, circular arcs (arcs), or spiral, parabolic, elliptical, quadratic, or cubic curves. Reference is made in the next section of this chapter regarding curve features. Although the term "*curve*" pertains to any of the said curve types, at times and where obvious, it is used to refer to the curve type under the subject of discussion. A parabolic curve as treated in this book is a special quadratic curve.

Multi-Points A multi-point feature refers to a group of point features that are scattered about a plane, and with the said group being addressed as a single entity. Multi-point features are used in geographic information system (GIS) applications. Multi-point features may be used in the creation of a roadway alignment to contain the key nodes of the alignment.

Multi-Lines A multi-line feature refers to a group of line (two-point line and/or polyline) features that are scattered about a plane, and with the said group being addressed as a single entity. The component line features of a multi-line feature may or may not connect to each other, and they may or may not cross any of the other component features. Multi-line features are used in GIS applications.

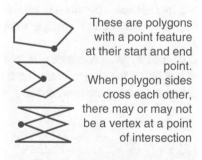

These are polygons
with a point feature
at their start and end
point.
When polygon sides
cross each other,
there may or may not
be a vertex at a point
of intersection

Polygons Polygon features represent closed figures of three or more vertices, one of which is the polygon's start and end point at which a real or pseudo point may or may not exist. Polygons may be convex, concave, or twisted (one side crossing another).

The subsequent chapters of this book that deal with various geometric constructions do not create polylines, but, when pertinent, they operate with polylines. Furthermore, the said chapters do not deal with multi-point or with multi-line features. If such features have to be used with any or the procedures presented in this book, they will have to be decomposed into individual point or line features.

2.3   The Curve Features

The curve types of use in civil engineering and surveying and addressed in this publication include circular arcs, special transition spirals, and parabolic curves. In addition to these curves, ellipses, quadratic curves, and cubic curves are used at times in general drafting for the creation of dimension leaders, and they are addressed below.

As used in civil engineering and surveying, the parabolic curve is a special conic curve of a quadratic form. Circular arcs, or circular curves, and spiral curves are used in the design of roadway, railroad, and other alignments on the X and Y plane, while parabolic curves are used primarily in the design of the vertical profile of an alignment in the X and Z plane. In some municipalities parabolic curves are used to represent the cross-sectional shape of the surface of roadway pavements in the Y and Z plane.

This distinction in the application of these curves to the said two planes stems from years of practice and tradition. Circular curves and spirals offered in the past an easier way of laying them out in the field (stakeout) with deflections from the tangent line, whereas parabolic curves offered an easier way in the determination of differences in elevations. Although one might claim that modern computational means may obviate these difficulties, it is quite doubtful that parabolic curves will replace circular curves and spirals in the transition from one tangent alignment to another tangent alignment.

There are several spiral types, but the one used in roadway alignments and referred to as the highway spiral is of specific form that transitions from a straight line to an arc, or between two arcs so that the instantaneous radius of the spiral varies directly with respect to the distance from the point of the spiral's beginning. A similar form of spiral is used in railways.

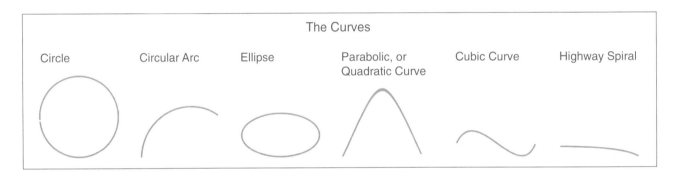

A circular curve may be either a circle or an arc. With the exception of the circle and the ellipse, all other curves are open ended.

When displaying any curve on the computer monitor screen, or drawing it on a plotter or printer by computer means, it is displayed or drawn as a series of chords, that is, as a polyline. However, when a curve is saved in the computer memory or disk, it contains, in addition to the standard polyline attributes, certain special attributes that define the curve. These attributes, as used by the various geometric procedures of this book, are presented in Table 2.3. If the said procedures are to be incorporated in a user specific software system, appropriate provisions for these attributes will probably need to be made in order for the procedure to recognize how the curve has been prepared. In summary, the numbers below are associated with a type of curve **Curve ID #** columns of the said table:

1. Circle defined by the coordinates of three points. The radius and circumference are computed and saved.
2. Circular arc defined by the coordinates of three points. The radius and arc length are computed and saved.
3. Circle defined by the coordinates of its center point and of a point on the circumference, and radius. The circumference is computed and saved.

Circles or arcs thereof and ellipses may be defined clockwise or counterclockwise, but with a few exceptions, they are always drawn in a counterclockwise rotation. All other curve types are drawn as created from their start point towards their end point.

Table 2.3 CURVE ATTRIBUTES BY CURVE TYPE

ATTRIBUTES		CURVE TYPE ID # 1	2	3	4	5	6	7	8	9 (I)	9 (II)
		Circle	Arc	Circle	Arc	Spiral	Spiral	Quadratic curve	Cubic curve	Ellipse	Ellipse
		thru 3 Points	thru 3 Points	Center + Point @ Circle	Center Start Pt. End Pt.	Clockwise	Counterclockwise			Major Axis Pts + Angle of Ellipse	Major Axis Pts + Minor Axis+Mjr.Dir
ID No.	Name										
1	PX1	Pt.1 X	Pt.1 X	CC X	CC X	TS X	TS X	Pt.1 X	Pt.1 X	Pt.1 X	Pt.1 X
2	PY1	Pt.1 Y	Pt.1 Y	CC Y	CC Y	TS Y	TS Y	Pt.1 Y	Pt.1 Y	Pt.1 Y	Pt.1 Y
3	PX2	Pt.2 X	Pt.2 X	Pt. X	Strt X	BckAz	BckAz	Pt.2 X	Pt.2 X	Pt.2 X	Pt.2 X
4	PY2	Pt.2 Y	Pt.2 Y	Pt. Y	Strt Y	DcBck	DcBck	Pt.2 Y	Pt.2 Y	Pt.2 Y	Pt.2 Y
5	PX3	Pt.3 X	Pt.3 X	CC X	End X	DcAhd	DcAhd	Pt.3 X	Pt.3 X	Zero	Zero
6	PY3	Pt.3 Y	Pt.3 Y	CC Y	End Y	Offset	Offset	Pt.3 Y	Pt.3 Y	Zero	Zero
7	PX4	Zero	Zero	Zero	Zero	SL1	SL1	Zero	Pt.4 X	Zero	MnrAx
8	PY4	Zero	Zero	Zero	Zero	SL2	SL2	Zero	Pt.4 Y	Angle	MjrDir
9	RAD	Radius	Radius	Radius	Radius	Radius	Radius	Zero	Zero	Zero	Zero
10	LEN	Arc L	Arc L	Arc L	Arc L	SL	SL	Arc L	Arc L	Arc L	Arc L

Legend

The attribute names in the second column under the heading ATTRIBUTES denote the name of the attribute. The labels below are used to denote the contents of the various attribute fields.

Curve ID#	Identification number assigned to a curve feature to identify its type of curve Pt.N Where N = 1, 2, or 3 denotes point 1, point 2, or point 3, respectively
Pt	Point on a curve
CC	Center point of a circular curve
X	East coordinate
Y	North coordinate
Radius	Radius of a circular curve. For spirals, it is the radius of the circular curve to which the spiral is tangent.
Arc L	Length of a circular curve or ellipse
Strt	Start point of a circular curve
End	Endpoint of a circular curve
TS	The start point of a spiral
BckAz	The back azimuth to a spiral—azimuth of the tangent at the start of the spiral
DcBck	Degree of curve back of the spiral. For a tangent this is infinity represented by a zero value
DcAhd	Degree of curve ahead of the spiral. Exiting spirals are defined as entering spiral is a reverse direction.
Offset	Offset distance of a polyline equidistant from the base or main spiral.
SL1	Distance from the spiral beginning to the start point of a spiral segment.
SL2	Distance from the spiral beginning to the end point of a spiral segment.
SL	Total length of the base or main spiral.
Angle	Ellipse angle when the ellipse is defined by an angle
MnrAx	Length of the minor axis of an ellipse when the ellipse is defined by the major axis and angle.
MjrDir	Cartesian direction of the major axis of an ellipse when the ellipse is defined by the major and minor axes.

4. Circular arc defined by the coordinates of its center point and two endpoints, and radius. The arc length is computed and saved.
5. Clockwise spiral. Refer to Chap. 9.
6.63. Counterclockwise spiral. Refer to Chap. 9.
7. Quadratic curve defined by the coordinates of three points. The arc length is computed and saved.
8. Cubic curve defined by the coordinates of four points. The arc length is computed and saved.
9(I). Ellipse defined by the coordinates of the endpoints of the major axis, and the degree of the ellipse to define the limits of the minor axis. The length of the ellipse is computed and saved.
9(II) Ellipse defined by the center point coordinates, lengths of its major and minor axes, and direction of its major axis. The length of the ellipse is computed and saved.

2.4   Definition of Terms

Presented below are the definitions of certain terms as they are used within the context of this book.

Endpoints Endpoints may be *actual point features* as defined above and located at the endpoints of a line or curve feature, or *implied points* referring to the endpoints of a line or curve feature at the end of which an actual point feature may or may not exist.

Vertex Points Vertex points, or just vertices, like endpoints may be actual point features as defined above, or implied points located at the vertices of a polyline or polygon.

Control points A control point is an implied point used in the creation of a line, curve, or polygon feature. An actual point feature may be introduced at the location of a control point. A feature may have one or more control points depending on the method of its creation. As is to be seen in subsequent chapters, control points enable the user to perform certain feature editing operations. The control points by feature type are identified below.

In addition to control points, curve features require certain additional parameters, again depending on the method of their creation. These parameters are presented in the pertinent curve chapters of this book.

The number of control points by feature type is identified below:

1. A **point feature** has one control feature, itself, and is the only control point that by default is an actual feature.

2. A **two-point line**, or a polyline feature has two control points, one at each endpoint. The vertices of a polyline are not control points.
3. A **Polygon feature** has one control point, its common start and end endpoint.
4. A **curve feature** which could be:
   - A circle defined by its center point and a radius has two control points, its center point and a point on its circumference.
   - A circle defined three point has three control points, the points that defined it.
   - A circular arc defined by its center point, a radius and any other parameter has three control points, its center point and its two endpoints.
   - A circular arc defined three point has three control points, the points that defined it.
   - An ellipse has three control points, the two endpoints of its major axis, and the endpoint on the positive side of the minor axis.
   - A parabolic or quadratic has three control points, the points that defined it.
   - A cubic has four control points, the points that defined it.

As stated previously a curve is saved in the computer database as a polyline with certain attributes such as those of its control points, radius, and the like. At times it is necessary to create an overall polyline comprising a series of concatenated lines and curves as in the case of representing the perimeter of a parcel of land. Also such perimeters as well as center lines may be digitized as single polylines which may require their decomposition into individual lines and curves with their specific attributes. That is, it is necessary to reconstruct the individual components of the polyline. The procedure `Poly2curve` (see Chap. 3) provides this functionality.

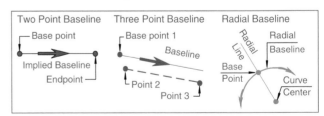

Baselines A baseline is an implied reference line from, along, or about which linear and/or angular measurements are made. There are three baseline types as indicated below. Such a type of a baseline should not be confused with the same term as used in surveying and civil engineering field operations. Reference is also made to the paragraph below regarding directions, and to the subsection regarding conventions.

- **A two-point baseline** is defined by a base point (start point) and an end point, with the direction of the baseline being that from the first towards the second point. These points may be actual, or implied.
- **A three-point baseline** is defined by a base point (point 1) and a direction defined by two other points (point 2 and point 3). It is possible for the base point to be either of the two points that define the direction of the baseline as being from point 2 towards point 3. These points may be actual, or implied points.
- **A radial baseline** is defined as an implied circular arc commencing at a base point located on the said arc, and progressing clockwise or counterclockwise along the said arc.

Pluses and Offsets A **plus distance** is a distance measured along the direction of a linear or circular baseline which may be real or implied from its starting point in order to locate another point, and an **offset** is a normal or radial distance to the baseline at the location of the said last point. Reference is made to the Linear Measures paragraph in the next section of this chapter.

Strings of Features At times, it becomes necessary to address various contiguous two-point lines and/or circular arcs as an entity in the form of a list (collection). An example of this would be the sides of a parcel of land, or the component features of lines and arcs of a street center line. Such a string of features, although it may resemble a polyline, is not a polyline, but a collection of single features.

Directions Certain procedures require as a given argument, or they return as an argument the direction of a line, be it a real or implied line. A direction may be rotational with respect a fixed axis of the coordinate system medium or relative to a predefined line or baseline and they are:

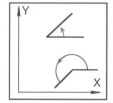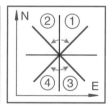

- **Cartesian** measured as a counterclockwise angle about the start point of said line, or from a line parallel to the X (east) axis towards the end of the said line. This direction may be in the range of $0 < \text{direction} < 2\pi$. If it is negative, it implies a clockwise measure.

- **Azimuth** measured as a clockwise angle about the start point of the said line from a line parallel to the due north (positive Y) axis, commonly referred to as a north azimuth, or from the due south (negative Y) axis towards the end of said line, commonly referred to as a south azimuth. This direction may be in the range of $0 < \text{direction} < 2\pi$. If it is negative, it implies a counterclockwise measure. Most of the algorithms (procedures) of this book utilize north azimuths and refer to it as just azimuth.
- **Bearing** measured clockwise or counterclockwise as indicated in the diagram to the left. This direction may be in the range of $0 < \text{direction} < \pi/2$, and carries an identification of the quadrant in the form of NE (northeast), SE (southeast), SW (southwest) or NW (northeast), or their respective numeric equivalents of 1, 2, 3, or 4.

When an algorithm needs to receive from, or to return to an interfacing program a direction, a conversion from a bearing to an azimuth, or the reverse may be required depending on the user's need. Directions when used as input or output arguments in any procedure in this book must be in radians.

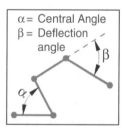

Angles Various geometric operations require the entry or determination of angles, which may be *central angles*, or *deflection angles* in either a clockwise, or counterclockwise rotational direction. Angles when used as input or output arguments in any procedure in this book must be in radians.

2.5 Conventions

In the mathematical world, coordinate geometry utilizes a Cartesian coordinate system. Theoretically, this should not be an issue in civil engineering. However, due to the long standing practice in field work and design, at least in the USA, a special convention system has been employed by engineers and surveyors, the north/east coordinate system, that has survived up to the present, and which convention modifies the Cartesian coordinate system in the following ways:

- The X axis is referred to as the *east axis*, and the Y axis is referred to as the *north axis* with mensurations increasing similarly to the Cartesian system (south to north and west to east).
- Coordinates are expressed as north and east (N, E), and not as X and Y (X, Y) as in the Cartesian system.
- Rotational mensuration is made in a clockwise rotation from the positive direction of the north axis towards the positive direction of the east axis.

The subsequent chapters of this publication present various procedures for defining points, lines, curves, polygons and carrying out various geometric constructions. In describing these procedures the conventions identified below are employed.

Coordinates All geometric work is assumed to be performed in a two dimensional plane, with the occasional provision for the presence of the third dimension of the elevation. Reference to Cartesian coordinates and to north and east coordinates does not imply two different planes, but one, the one upon which a project is being worked. In essence, the

North coordinate = Y coordinate and the
East coordinate = X coordinate

Thus, when the expression

- "**Cartesian coordinates**" is encountered in the text, it emphasizes that the X or east coordinate precedes the Y or north coordinate.
- "**North and east coordinates**" is encountered in the text, it emphasizes that the north or Y coordinate precedes the east or X coordinate.

This is of importance in the various procedures as to which coordinate list argument precedes the other.

All coordinates are expressed in the units of the medium of the coordinate system used which this book assumes them to be in the US Customary System of feet, or in the SI of meters and in certain instances in millimeters. Other types of units such as veras as used in certain US states, or any other type of unit of measure could be used as long as the coordinate system is orthogonal. However, geodetic latitudes and departures may not be used. If such coordinates must be used as input or output, they must be converted to state plane coordinates.

Distances Distances between points and lengths of lines or arcs, may they be given or returned arguments of a geometric procedure presented in this book, are always in the units of the medium of the plane coordinate system that the project is being designed, and they must be linear. Geodetic arcs may not be used. If such arcs must be used as input or output, they must be converted to linear measures.

Linear Measures As a surveyor is laying out points, or observing points along the direction of a straight baseline by measuring a distance along the said direction, and then measuring a normal distance to the left or to the right therefrom commonly referred to as plus and offset distances, respectively. In so doing, a measure:

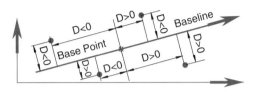

- Along the direction of the baseline (usually a local X axis) is considered positive.
- Opposing the said baseline direction (usually a local X axis) is considered negative.
- To the right of the said baseline direction (usually a local Y axis) is considered positive.
- To the left of the said baseline direction (usually a local Y axis) is considered negative.

A similar analogy may be applied to the case in which the baseline is a circular or spiral curve or even a polyline instead of being a straight line.

Concentric or offset curves When positioning a concentric or offset circular or spiral arc from a corresponding base arc, a positive offset distance positions the offset arc on the outside of the base arc, and a negative offset distance positions the offset arc on the inside of the base arc.

Angular Measures Additionally, in civil engineering and surveying work, when an angle is turned about a line (usually about a baseline) a clockwise rotation is considered positive, and a counterclockwise rotation is considered negative.

Elevations With regards to the Z coordinate (elevation), both, the civil engineering north/east coordinate system and the Cartesian coordinate systems are the same, and elevations can be positive or negative.

2.6 Global Parameters

There are certain parameters referred to as global parameters that have been assigned a value so that they may be used in various procedures without having to be defined each time they are used. These parameters are assigned to a special subroutine of the overall programming system in which the procedures that use the said global parameters reside. The name of this subroutine is user defined.

The global parameters that are called by the procedures presented in this book, the source code of which is included in the accompanying CD, are presented below. There are two types of global parameters, those that are associated with a specific constant such as the value of the value of the geometric π, and those that associated with tolerance values.

Tolerances are used in various geometric operations to determine whether a certain condition has been met, or not. For example if it is necessary to determine if two points are the same, a test is made to determine whether their coordinates are the same within a certain distance that is deemed satisfactory to call the said points as being one and the same point. Similarly a test may be need to be made whether two lines are parallel, or not. The tolerance parameters presented below may be used for testing either linear or angular parameters. The tolerance parameter names and their values shown below for linear measures are in the US Customary System (feet) together with the corresponding the SI (millimeters) value. Also shown are the angular measures in degrees, minutes, and seconds corresponding to the value of the parameter shown.

Preprogrammed tolerance parameters

Name	Feet or radians	Millimeters	Degrees/minutes/seconds
ugTolV1	0.005	1.524	17°11′19.44″
ugTolV2	0.009	2.7432	30°56′23.00″

Name	Feet or radians	Millimeters	Degrees/minutes/seconds
ugTolV3	0.00005	0.01524	00°00′10.31″
ugTolV4	0.00009	0.027432	00°00′18.56″
ugTolV9	0.0009	0.27432	00°03′05.64″

In addition to the above tolerances some procedures use the tolerances indicated below which have been hard coded and have not been assigned to global variables.

Linear tolerances	0.0001	feet	=	0.03048	mm
	0.001	feet	=	0.3048	mm
Angular tolerances	0.0005	radians	=	0°01′43.13″	
	0.0000048	radians	=	0°00′00.99″	
	0.0000001	radians	=	0°00′00.02″	
	0.000000048	radians	=	0°00′00.0099″	

Of the above tolerance values the ones that have been programmed and used in the procedures and test programs of the accompanying CD are those of feet and radians. If the eventual user wishes to have then in millimeters, then their values will have to be changed to those shown above. Alternatively, there can be two sets of the same parameters, one with feet and the other with millimeters, and the global variable **ugunits** (see below) could be used to select the appropriate set of tolerances.

The other global variables and their values that are used are presented below.

Value of the geometric π: **ugpi** = 3.14159265358979'

Units of measure: **ugunits** = 0 for the US or = 1 for the SI

Before proceeding with any of the geometric construction procedures, it is deemed prudent to address certain basic procedures that are called upon by various construction procedures. When developing a computer program of any sizeable magnitude, there is usually a monitor or control program that kind of sets the scope and operation of the application. This program then calls other programs to perform certain dedicated major functions, and which in turn could call upon other smaller programs to perform more dedicated operations. When computers first came of being, there were programs and subprograms. The subprograms then became procedures, which were changed to subroutines or just plain routines, and now they have been renamed procedures. The term "algorithm" has also been applied to them. However, the term algorithm is more generic and pertains to the computational or non-computational steps employed to perform a specific operation which may or may not necessarily be a computer based set of instructions.

Procedures are divided into two classes, subroutines and functions. Each of them is assigned a name and a set of arguments. Both, a subroutine and a function, receive the arguments, and perform certain operations. In performing the operations:

- A subroutine changes the values of some or of all the said arguments. For example, the subroutine

```
Sub icforce (N1, E1, N2, E2, D, AZ)
```
receives the north and east coordinates of points 1 and 2, and returns the distance and azimuth between them. The calling program could then be
Statements of the calling program that set the values of
N1, E1, N2 and E2
```
Call icforce (N1, E1, N2, E2, D, AZ)
```
Statements of the calling program that use the values of
D and AZ

- A function returns a single value represented by the function's name. For example, the function to compute the distance TheDist between two points
```
Function iccomdis (X1, Y1, X2, Y2)
```
receives the X and Y coordinates of points 1 and 2, and returns the distance between the said points in iccomdis. The calling program could then be
Statements of the calling program that set the values of
X1, Y1, X2, and Y2
```
TheDist=iccomdis (X1, Y1, X2, Y2)
```
The procedures that are presented in this chapter are rather simplistic, but are repeatedly used by the various geometric procedures presented in the subsequent chapters.

3.1 Trigonometric Procedures

Most computer operating systems have provided procedures for trigonometric functions such as sine (sin), cosine (cos), and tangent (tan) since the early days of FORTRAN. However, it was not until much later on when the inverse or anti-functions arcsine ($\sin^{-1}$), arccosine ($\cos^{-1}$), and arctangent ($\tan^{-1}$) became available. Prior to that, their solution was up to the individual programmers. From trigonometry, the arcsine, arccosine, and arctangent may be found in terms of each other by use of the following three equations:

$$\sin^{-1}(a) = \tan^{-1}\left(\frac{a}{\sqrt{1-a^2}}\right) \tag{3.1}$$

$$\cos^{-1}(a) = \tan^{-1}\left(\frac{\sqrt{1-a^2}}{a}\right) \tag{3.2}$$

$$\tan^{-1}(a) = \sin^{-1}\left(\frac{a}{\sqrt{1+a^2}}\right) \tag{3.3}$$

E. C. Tonias, C. N. Tonias, *Geometric Procedures for Civil Engineers*, DOI 10.1007/978-3-319-24295-8_3

Thus if one can be determined, it can be used to find the other two. In addition to the said three functions, we also have such other trigonometric functions as the:

Cotangent	$\cot(\alpha)$	$= 1/\tan(\alpha)$
Secant	$\sec(\alpha)$	$= 1/\cos(\alpha)$
External Secant	$\operatorname{exsec}(\alpha)$	$= \operatorname{sen}(\alpha) - 1$
Cosecant	$\operatorname{cosec}(\alpha)$	$= 1/\sin(\alpha)$
Versed Sine	$\operatorname{vers}(\alpha)$	$= 1 - \cos(\alpha)$
Coversed Sine	$\operatorname{covers}(\alpha)$	$= 1 - \sin(\alpha)$

Of these additional functions, the cotangent, external secant, cosecant, and versed sine have found a home in the surveying field for many years prior to the advent of the computer, and tables of their natural and logarithmic values have been created. However, the computer has made their use unnecessary.

Presented below are certain function procedures that compute the arcsine and arccosine of a value in the range of $-1 \leq X \leq 1$, as well as a function procedure that computes the arctangent of a value. With regards to the arcsine function, there are three procedures, with certain distinctions between them. Although these inverse trigonometric functions may be nowadays available in many operating systems they included here from an historic point of view.

The angle results that are returned by each of the trigonometric function procedures presented below are expressed in terms of radians. These angles may be converted into degrees of decimal form, or into a form of degrees, minutes, and seconds by calling the appropriate of the functions presented later on in this chapter.

3.1.1 The RAND Arcsine Function

The `Function icasinRand(X)` determines the arcsine of a value X in the range of $-1 \leq X \leq 1$. This subroutine is based on part of the work of *The RAND Corporation* between 1953 and 1955 for the US Air Force, and published by Cecil Hastings, Jr. under the title *Approximations for Digital Computers*. The approximation equation is

$$\arcsin = \frac{\pi}{2} \ \sqrt{1 \ X} \Psi(X) \qquad (3.4)$$

where

$$\Psi(X) = a_0 + a_1 X + a_2 X^2 + \ldots + a_7 X^7 \qquad (3.5)$$

and

$a_0 = 1.5707,9630,50 = 0.5\pi$	$a_4 = 0.0308,9188,10$
$a_1 = -0.2145,9880,16$	$a_5 = -0.0170,8812,56$
$a_2 = 0.0889,7898,74$	$a_6 = 0.0066,7009,01$
$a_3 = -0.0501,7430,46$	$a_7 = -0.0012,6249,11$

with an anticipated error of $\pm 0.0000,0002$ as per the said RAND publication.

The abovesaid publication contains several other approximations for various mathematical functions, including one for the arctangent function. The arcsine approximation was chosen over that of the arctangent because of slightly lesser error of approximation.

3.1.2 The VB Based Arcsine Function

The `Function icasinan(X)` uses the Visual Basic (VB) arctangent function `Atn (x)` to find the arctangent of X for Equation 3.1. If X is outside the range of $-1 \leq X \leq 1$, a value of $\pi/2$ is returned by the procedure.

3.1.3 The Modified VB Based Arcsine Function

The `Function icasin1(X)` is similar to the above arcsine function `icasinan` with two exceptions. The function `icasin1` checks whether the value of X is:

- Within a tolerance value of 0.0000,0000,1 of being equal to ±1, and if it is, it returns the value of $\pi/2$ as the value of the requested angle.
- Outside the range of $-1 \leq X \leq 1$, and if it is, it returns the value of zero (0.00) as a value of the requested angle.

3.1.4 The Arccosine Functions

The `Function icacosan(X)` and `Function icacosRand(X)` determine the angle whose cosine value is X. To find the said angle, these functions call `icasinan` and `icasinRand`, respectively, to find the angle the sine of which is X. Since in a right triangle the cosine of one of the oblique angles is equal to the sine of the opposite oblique angle, the procedure then subtracts the angle that it has found from $\pi/2$ to find the requested angle. These functions return the values of zero (0.00) if $X \geq 1$, and the value of π if $X \leq 1$.

3.1.5 The Arctangent Function

The `Function icatan(X)` determines the angle whose tangent value is X. To find the tangent of a value, the function `icatan(X)` uses Equation 3.3 in which a=X. Thus `icatan(X) = icasinan(B)` where B=X/(1+X2)1/2.

3.2 Conversion of Angles and Directions

During the solution of various geometric constructions there is a need to compute directions between two points, angles between two directions, and to convert angles or bearings from radians to degrees and vice versa. The procedures below provide this functionality.

3.2.1 Convert Degrees to Radians

The Function icdegrad(X) converts the angle X expressed in degrees in decimal form to radians by multiplying the angle X by $\pi/180$.

3.2.2 Convert Radians to Degrees

The Function icraddeg(X) converts the angle X expressed in radians to degrees of decimal form by multiplying the angle X by $180/\pi$.

3.2.3 Convert Azimuth to Cartesian (Radians)

The Function icaz2car(X) converts a north azimuth (X) expressed in radians to Cartesian angle expressed in radians by subtracting the given angle from $\pi/2$, and then checking if the result is negative. If it is, the procedure adds 2π to make it positive.

3.2.4 Convert Azimuth to Cartesian (Degrees)

The Function icaz2dec(X) converts a north azimuth (X) expressed in decimal degree form into a Cartesian angle also expressed in decimal degree form to six decimal digits.

3.2.5 Convert Azimuth (Radians) to Direction (D/M/S)

The subroutine icaz2dir converts a north azimuth expressed in radians into ten alphanumeric strings nine of which represent the input azimuth into an azimuth, bearing, and Cartesian direction expressed in degrees, minutes, and seconds forms, and the tenth string expressed as a Cartesian rotation in decimal degree form. This procedure is generally used to convert a north azimuth for graphic annotation purposes when preparing a map for printing, such as when annotating metes and bounds of a parcel of land, the direction of a roadway center line, or of any other line.

```
Sub icaz2dir (azim1, _
              string1, string2, string3,
              string4, string5, string6, decS, _
              string7, string8, string9)
```

The input parameters to this procedure include the:

azim1	The given north azimuth in radians to be converted.

The parameters that are returned from this procedure include the:

string1, string4, string7	The north azimuth in three formats as indicated below.
string2, string5, string8	The bearing in three formats as indicated below.
decS	The Cartesian direction in decimal form to six decimal digits.
string3, string6, string9	The Cartesian direction in three formats as indicated below.

For the sample input north azimuth of 4.283,208 radians, the converted directions, with each underscore (_) indicating a blank space, are:

```
String1 =___Azimuth=_245_24_35.1
String4 =245_24_35.1
String7 =245°24′35″

String2 =___SW_65_24_35.1
String5 =SW_65_24_35.1
String8 =SW_65°24′35″

decS =204.590259
String3 = ___Cartesian_204_35_24.9_
              (204.590259)
String6 =204_35_24.9
String9 =204°35′25″
```

In perusing the said example, note that:

- In strings String1, String2, String3, String4, String5, and String6 the seconds component of a direction is expressed to the nearest tenth of a second.
- In strings String7, String8, and String9 the seconds component of a direction is expressed to the nearest second.
- Blanks or zeros are not inserted in lieu of the absent digits if the angle component of:
  - Degrees is less than tree or two digits,
  - Minutes is less than two, and
  - Seconds is less than two.

If the degrees or minutes components is less than one, a zero is introduced.

- The SW denotes the southwest bearing quadrant since the direction of the given azimuth is located in the third quadrant. For any of the other quadrants the annotation would be NE, SE, or NW. Note that numeric beating quadrants are not returned since the are usually not used in the graphic annotation of bearings.
- The angle is converted from radians into degrees, minutes, and seconds by a call to the icdegre procedure.

3.2.6 Convert Azimuth (Radians) to Bearing (D/M/S)

The subroutine icbearg converts a north azimuth expressed in radians into an alphanumeric north bearing quadrant identifier, and into distinct degrees, minutes, and seconds parameters.

```
Sub icbearg  (AZ, _
             IQUAD, ID, IM, C)
```

The input parameters to this procedure include the:

AZ   The given north azimuth in decimal degrees to be converted.

The parameters that are returned from this procedure include the:

IQUAD   The quadrant identifier which could be NE, SE, SW, or NW.
ID   The number of degrees as an integer.
IM   The number of minutes as an integer.
C   The number of seconds in double precision decimal form.

The conversion is attained by comparing the magnitude of the given azimuth with 0°, 90°, 180° and 360° to ascertain the quadrant. Note that:

- If the given azimuth AZ is negative, then 360° are added to it until it becomes positive or zero.
- If the given azimuth AZ is positive and greater than 360°, then 360° are subtracted from it until it becomes positive and less or equal to 360°.
- The angle is converted from radians into degrees, minutes, and seconds by a call to the icdegre procedure.

3.2.7 Convert Bearing (Degrees) to Azimuth (Degrees)

The Function icbrg2az(QUAD, ANGLE) converts a bearing defined by its quadrant (QUAD) and an angle expressed in

decimal degree form (ANGLE) into a north azimuth expressed in decimal degree form.
Note that:

- QUAD should be the integer 1, 2, 3, or 4 denoting the northeast, southeast, southwest, or northwest quadrant, respectively, and that
- ANGLE should be within the range of 0° and 90.00°.

3.2.8 Convert Cartesian Angle (Degrees) to Azimuth (Degrees)

The Function icdec2az(DEG) converts a Cartesian rotation angle expressed in decimal degree form into a north azimuth expressed in decimal degree form. The value of DEG may be in the range between 0° and 720°. This allows for the addition of two or more angles to exceed a full circle in rotation; that is to make a full turn about the X-axis. Should the value of DEG exceed 720°, the resultant azimuth would be negative and 90° out of the proper direction.

3.2.9 Convert Radians to Degrees/Minutes/Seconds

The subroutine icdegre converts an angle expressed in radians into an angle comprising degrees, minutes, and seconds form. The seconds component is not rounded, nor truncated.

```
Sub icdegre   (AZ, _
              ID, IM, C)
```

The input parameters to this procedure include the:

AZ   The given angle in radians. It could represent a north azimuth.

The parameters that are returned from this procedure include the:

ID   The number of degrees as an integer.
IM   The number of minutes as an integer. If the number of minutes is found to be 59.95 or greater, IM is set to be zero and ID is increased by 1°.
C   The number of seconds as a double precision variable.

The given angle AZ may be either positive or negative. If it is negative, it is changed to be positive, the degrees (ID), minutes (IM) and seconds (C) components are determined, and then they are all changed to be negative.

3.2.10 Convert an Azimuth to Be in the Range of 0–360°

The Function icaz0_2pi(AZ) makes certain that an azimuth in radians is within the range of zero and 2π. At times after a series of various operations it is possible for an azimuth to yield proper results even thought it is expressed as a negative direction, and/or as one exceeding 2π in absolute value.

The procedure first queries the input azimuth (in radians) whether it is negative or not. If it is negative, it keeps adding 2π until it becomes positive. Next, if the azimuth is greater then 2π, it keeps subtracting 2π until it becomes less than 2π. If the input azimuth is within the zero to 2π range, it is not altered. The returned azimuth is also in radians.

3.2.11 Extract an Angle or a Bearing from a Text String

The subroutine icInpDms receives a text string containing an angle or direction expressed in degrees, minutes, and seconds form and converts it into five numeric parameters of which (a) the first three contain the number of degrees, minutes, and seconds, (b) the fourth contains the angle expressed as degrees in decimal form, and (c) the fifth contains the angle expressed in terms of radians. There is no distinction between an angle of rotation, azimuth, or Cartesian angle. The method of defining the quadrant of a bearing is addressed later on in this subsection. For a bearing the returned angle is the corresponding north azimuth. This procedure enables the receipt of data from a tabular electronic media and conversion to individual degrees-minutes-seconds format, and the simplification of entering degrees, minutes, and seconds without separating them spaces.

The input text string may be composed of a series of words separated by at least one space (blank character). Leading, trailing and/or embedded spaces are permitted in the string, and if any are found they are disregarded. There are two types of text strings that can be processed, those that represent bearings and those that represent azimuths or just plain angles.

Text strings with bearings: Direction of a bearing must be composed of the angle proper as addressed below for angles, and of the quadrant definition as indicated below.

NE angle	SE angle	SW angle	NW angle
N angle E	S angle E	S angle W	N angle W
angle NE	angle SE	angle SW	angle NW

In the above note that the quadrant designation:

- May be in either lower or upper case letters. The procedure changes them all to be upper case letters.
- May not be defined numerically with 1, 2, 3, or 4 implying NE, SW, SW, or NW, respectively. For the ability to designate the quadrant numerically refer to the icinpbrg procedure presented in the subsequent subsection.

The angle component: An angle, or the angle component of a bearing may be composed of one, two, or three words. More than three words are not acceptable. The angle component is deciphered from right to left word by word. If a string is composed of:

1. **One word**, there can be two interpretations depending on the number of digits in the whole word if there is no decimal points, or in the word to the left of the decimal point if there is a decimal point present. If the said number is:
   (a) Three or less digits, it is assumed that the angle is composed of degrees only expressed in decimal form.
   (b) Six or seven digits, it is assumed that the angle is composed of degrees with two or three digits, and minutes and seconds with each composed of two digits.
   (c) Four or five digits, the input text string is considered to be invalid thus resulting in an error indicator.
      The paragraph (b) above enables the entry of degrees, minutes, and seconds without having to separate them by at least one space. For example:

• 150.50	implies	150° 30′ 00″
• 1503020	implies	150° 30′ 20″
• 000020.54	implies	0° 00′ 20.54″
• 3000020.54	implies	300° 00′ 20.54″
• 403020.1	implies	40° 30′ 20.1″
• −1503020	implies	−150° 30′ 20″
• 0150.50	is incorrect because there are more than three digits and less than six digits to the left of the decimal point. To denote 0° 01′ 50.50″ the entry should be 000150.50.	

2. **Two words**, it is assumed that the angle is composed of degrees and minutes expressed in decimal form, with the decimal being where it is shown. For example:

• 150 50	implies	150° 50′ 00″		
• 0 50.23	implies	0° 50.24′	or	0° 50′ 39.4″

Note that in specifying the minutes part there must be two digits to the left of the decimal point.

3. **Three words**, it is assumed that the angle is composed of degrees, minutes, and seconds expressed in decimal form, with the decimal being where it is shown. For example:

- 150 50 10.05   implies   150° 50′ 10.05″
- 0 05 05.24   implies   0° 05′ 05.24″

Note that in specifying the minutes part there must be two digits, and there should also be two digits to the left of the decimal point for the seconds part.

4. **Negative angles** are indicated by a minus sign in front of the degrees word. A minus sign in front of the minutes and/or seconds is disregarded.

```
Sub icInpDms   (theString, _
                degs, mins, secs, angle, rads, error)
```

The input parameter to this procedure is the:

theString	The text string that contains the given angle in degrees, minutes, and seconds, each separated by at least one space (blank character) and as described above.

The parameters that are returned from this procedure include the:

degs	The number of degrees as an integer.
mins	The number of minutes as an integer.
secs	The number of seconds as a double precision variable.
angle	The angle expressed in terms of degrees in decimal form. If the input parameter theString contains a bearing, then angle contains the corresponding north azimuth, and if not the corresponding value of the input angle.
rads	The parameter angle expressed in terms radians.
error	An error indicator, the value of which if it is:
	0 Denotes that an error has not been detected.
	1 Denotes that an input error has been detected.

Upon invocation the procedure initializes the parameters to be returned and then:

1. Removes from the input string any leading and trailing spaces, determines the number of words in the string, converts any letter characters to upper case, and saves the string in a temporary string.
2. Queries the first and last words to determine whether a bearing direction has been specified.
   - If a bearing is not detected, the procedure skips to the next step.
   - If a proper bearing quadrant is detected, the parameter quad is set to be 1, 2, 3, or 4 to denote the NW, SE, SW, or NW quadrant, respectively, and then proceeds with the next step.
   - If an improper bearing quadrant is detected, error is set to be 1 and the parameters to be returned are set to be zero.
3. Queries the portion of the input text string that contains the word or words representing the angle component, and:
   - If more than three words have been detected error is set to be 1 and the parameters to be returned are set to be zero.
   - If two or three words have been detected the values of the angle's degrees and minutes are extracted, and then continues with the next step to account for the presence of a bearing quadrant.
   - If one word has been detected the procedure assesses whether the special form of input has been used, or not. The special form refers to the specification in the input text string of the minutes and seconds being composed of two digits, with the seconds possibly having a decimal fraction, and the degrees component composed of two or three digits. If the special form:
     - Has been used, the right two digits denote the seconds part with any specified decimal digits, the next pair of digits denotes the seconds part, and the left most two or three digits constituting the degrees part. Thereafter, the second part is divided by 3600 and added to the minutes part which in turn is divided by 60 and added to the degrees part to determine the value of the angle.
     - Has not been used, the one word is assumed to be the angle and it is broken down into minutes and seconds.
4. Queries the value of the angle for exceeding 360°, and if so error is set to be 1 and the parameters to be returned are set to be zero.
5. Computes the value of the angle in terms of radians. The global parameter ugrad2deg is used for this action and it is equal to 180° divided by π.
6. Calls the icbrg2az procedure if a bearing had been detected earlier, to convert it to north azimuth, computes the azimuth to radians, and then calls the icdegree procedure to break it back into degrees, minutes, and seconds.

3.2.12 Extract a Bearing from a Text String

The subroutine icInpBrg receives a text string containing a direction as a bearing, north azimuth or Cartesian angle and converts it into individual components of a bearing composed of the quadrant designator and the angle component in degrees, minutes, and seconds. This procedure is quite similar to the `icInpDms` procedure presented in the preceding subsection with the following differences:

- With `icInpDms` the quadrant designation may be defined only by the letters N, S, E, and W, while with `icInpBrg` the number digits 1, 2, 3, or 4 may be used to imply the NE, SE, SW, or NW quadrant, respectively.
- With `icInpDms` the angle degrees, minutes, and seconds components could constitute one word thus not requiring a space separation, while with `icInpBrg` the degrees, minutes, and seconds parts must comprise individual words separated by a blank space.
- With `icInpBrg` a north azimuth or a Cartesian angle may be converted into a bearing, while with `icInpDms` they cannot.

A bearing must be composed of the angle proper as addressed below for angles, and of the quadrant definition as indicated below.

NE angle	SE angle	SW angle	NW angle
N angle E	S angle E	S angle W	N angle W
angle NE	angle SE	angle SW	angle NW
1 angle	2 angle	3 angle	4 angle

A north azimuth must be comprised as follows:	A angle	or	a angle
A Cartesian angle must be comprised as follows:	C angle	or	c angle

The angle component must be composed of one, two, or three words that contain the following information:

degrees	or	degrees minutes	or	degrees minutes seconds

of which the right most component may have a decimal part.

The input parameter to this procedure is the:

```
Sub icInpBrg   (theString, _
                  quad, degs, mins, secs, angle, _
                                     rads, error)
```

theString	The text string that contains the given direction as indicated above with each word separated by at least one space (blank character).

The parameters that are returned from this procedure include the:

quad	The numeric equivalent 1, 2, 3, or 4 of the quadrant.
degs	The number of degrees as an integer.

mins	The number of minutes as an integer.
secs	The number of seconds as a double precision variable.
angle	The angle expressed in terms of degrees in decimal form. If the input parameter `theString` contains a bearing, then `angle` contains the corresponding north azimuth, and if not the corresponding value of the input angle.
rads	The parameter `angle` expressed in terms radians.
error	An error indicator, the value of which if it is:
	0 Denotes that an error has not been detected.
	1 Denotes that an input error has been detected.

Upon invocation the procedure initializes the parameters to be returned, removes from the input text string any leading, trailing spaces and extraneous embedded spaces, determines the number of words in the string, converts the letter characters to upper case, saves the string in a temporary string, and then parses through the text string to extract the information to be returned in a manner similar to that of the `icInpDms` procedure. The major differences in the code between the two procedures are those identified at the start of this subsection.

3.2.13 Extract a Bearing and Distance from a Text String

The subroutine `icInpBad` receives a text string containing (a) a direction composed of a quadrant identifier and an angle, and (b) a distance composed of a numeric measure and optionally its units of measure, and converts them into a bearing and distance composed of six numeric parameters. The direction may be a bearing, a north azimuth or a Cartesian angle with the quadrant identifier denoting the type of direction. The six numeric parameters that are returned include (a) the numeric identifier of the quadrant, (b) the number of degrees, minutes, and seconds, (c) the angle expressed in decimal degrees form, and (d) the angle expressed in terms of radians. This procedure is quite similar to the `icInpBrg` procedure, which it calls and is presented in the preceding subsection with the following differences:

1. The `icInpBad` procedure requires the presence of a distance, while the `icInpBrg` does not.
2. In the `icInpBrg` procedure the quadrant identifier must be separated from the angle's components by at least a blank space, while in the `icInpBad` procedure the blank space separator between:
   - The quadrant identifier and the degrees part of the bearing *is not required*. For example:
     – **NE33 22 11 55 feet = NE 33 22 11 55 feet.**
     – **N33 22 11 E 55 feet = NE 33 22 11 55 feet.**

- The seconds part and the quadrant identifier *is requited.* For example:
  - **N33 22 11 E 55 feet is proper.**
  - **N33 22 11E 55 feet and 33 22 11NE are not proper.**
3. In the `icInpBrg` procedure the angle's degrees, minutes, and seconds components must be separated from each other by at least a blank space or comma, while in the `icInpBad` procedure they may also be separated by their conventional symbols °, ', and " for degrees, minutes, and seconds, respectively. These symbols are replaced by a blank space.
4. The `icInpBad` procedure, unlike the `icInpBrg` procedure calls the `icAvAsTokens` procedure to process the input text string, thus reference is made to the last said procedure regarding the word separators in the input text string. In the text below the blank space is generally referred to as being the word separator in a text input string. However, any separator as defined in the `icAvAsTokens` procedure could be used instead of the blank space.

Regarding the structure of the input text string note the following:

1. The direction and its angle must precede the distance measure and its optional units of measure as feet, meters, etc.
2. The quadrant identifier may be composed of:
   - A combination of the letters N, S, E and W to imply a bearing.
   - The number 1, 2, 3, or 4 to imply the NE, SE, SW, or NW quadrant of a bearing.
   - The letter A to imply a north azimuth which is converted to a bearing.
   - The letter C to imply a Cartesian angle which is converted to a bearing.
3. The quadrant identifier pairs of letter NE, SE, SW, and NE may precede the angle's degrees, minutes, and seconds, follow them, or straddle them. The letter A and C must precede the said angle components.
4. The numeric quadrant identifier 1, 2, 3, or 4, and the azimuth identifier A or the Cartesian angle identifier C if used they must precede the angle component.
5. The angle may be composed of degrees, minutes, and seconds in one of the following forms:
   - If all three are present, the degrees and minutes must be whole numbers, and the seconds may or may not have a decimal component.
   - If only degrees and minutes are available, then the degrees must be a whole number, and the minutes may or may not have a decimal component.
   - If only degrees are available, then they may or may not have a decimal component.
   - If a numeric quadrant designator is used, than:

- All of degrees, minutes, and seconds of the angle must be present even if they contain zeros.
- The degrees must be separated from the quadrant identifier with a space, unless a period (.) separates the degrees from quadrant identifier and the seconds follow the minutes without a space between them. For example:
  - `403.2846.235 1325 = NE 3 28 46.235 1235`
  - `403.28 46.235 1325` is not proper
- If the azimuth identifier A or the Cartesian angle identifier C is used, then a word separator must separate it from the angle component.
6. A distance component must be present in the input text string. If none is available, than a zero distance must be appended to the string as 0, 0., 0.0, 0.00 or something like that.
7. The units of measure if present must follow the measure itself and they are disregarded.
8. At least one space must separate the angle portion from the distance component of the text string.
9. At least one space must separate the angle portion, or the right letter part of the quadrant identifier if it straddles the angle, from the distance component.

Acomma or a tab character may replace the word separators °, ' and " as well as the blank space. Also refer to the icAvAsTokens procedure.

Sample input text strings include the following:

```
S 3°28'46.235"W 1325 feet
S 3°28'46.235"W 1325
N3°28'46.235"W 1325 meters
N3°28'46.235"W 1325
NE3°28'46.235" 1325 feet
NE3°28'46.235" 1325
N3°28'46.235"W 1325 feet
N3°28'46.235"W 1325
N 3 28 46.235 W 1325 feet
N 3 28 46.235 W 1325
403.2846.235 1325 feet
403.2846.235 1325
```

```
Sub icInpBad    (theString, _
                quad, degs, mins, secs, angle, rads,
                              theDist, error)
```

The input parameter to this procedure is the:

`theString`   The text string that contains the given direction and distance as indicated above.

The parameters that are returned from this procedure include the:

quad	The numeric equivalent 1, 2, 3, or 4 of the quadrant.
degs	The number of degrees as an integer.
mins	The number of minutes as an integer.
secs	The number of seconds as a double precision variable.
angle	The angle expressed in terms of degrees in decimal form in bearing form.
rads	The parameter angle expressed in terms radians.
theDist	The expressed in terms the units of its input.
error	An error indicator, the value of which if it is:
	0 Denotes that an error has not been detected.
	1 Denotes that an input error has been detected.

Upon invocation the procedure initializes the parameters to be returned and then (a) replaces the special symbology of the degrees, minutes, and seconds, if present, with a blank space, (b) calls the avAsTokens procedure to separate into words the input string by checking for comma, blank space or tab separators, (c) deciphers the text string for completeness, (d) separates the distance therefrom, (e) calls the icInpBrg procedure to determine, (f) returns the angle parameters, and (g) terminates.

3.2.14 Extract from a Text String a List of Alphanumeric Words

Two procedures, the avAsTokens and the avAsTokens1 are included in this subsection that extract from an input text string a series of individual words separated from each other by a blank space or any other specified word separator. For example the text string

This#case,#is#a# sentence of# only#letters.

would result in the following seven distinct words (underscores denote allowed blank spaces):

This case, is a _sentence_of _only "letters."

Both of these procedures operate in the same manner with the former being more generic, while the latter is a little simpler and not as comprehensive. There are two differences:

- The characters that may be considered as word delineators, and
- The ability to change all letter characters to upper or lower chase.

With both procedures the following characters are considered to be work delimiters: (a) one or more blank spaces, (b) a comma or (c) a tab character, and all leading, trailing and embedded blank spaces are disregarded except as noted below.

With the avAsTokens procedure the user has the ability to specify as a word delineator, in addition to the characters stated above, any other non-numeric character or string of characters as described below in the description of the input parameters. Also, this procedure enables the user to change all characters to upper or lower case, or leave them as they are. Regarding this refer to the qualifying comments later on in this subsection. This functionality is not available in the avAsTokens1 procedure.

Sub avAsTokens	(theString, delString, UpperLower, _ quad, degs, mins, secs, angle, rads, theDist, error)

The input parameters to this procedure include the:

theString	The input text string comprised numeric and alphanumeric words. The number of words in the input text string is limited only by the computer's operating system as to how many characters can appear in a parameter.
delString	The word delineator that defines what is used in the theString to separate one word from the other. This could be a blank space, a comma, a tab character, or any other non-numeric character or even a string of other non-numeric characters such as XXX, XaX, > or E to mention a few. However, do note that:
	• If the letters N, S, E and W are used they could destroy the quadrant designator if theString represents a bearing.
	• When the word delineator is a single blank character and the first character in any word being extracted is the TAB character, the TAB character is removed from the word as long as UpperLower is not X.
UpperLower	An indicator the value of which if it is:
	U or u all letters in theString are changed to upper case..
	L or l all letters in theString are changed to lower case.
	X, x or any other character or string thereof all letters in theString are not changed and leading, embedded and trailing characters are not disregarded.

The parameters that are returned from this procedure include the:

theList	The collection containing the individual words of the input text string. Each word in the collection could be an alphanumeric text string.
nWords	The number of words that have been extracted.

Upon invocation, the procedure:

1. Queries the value of the UpperLower parameter for being X or x, and if not proceeds to trim (remove) any leading and trailing blank spaces.
2. Counts the number of words in the input string, and initializes a temporary list with the contents of the input string.

3. Processes the temporary string to build the various words that comprised it performing the character conversion to upper or lower case as dictated by the `UpperLower` parameter.
4. Transfers the words to be returned in the `theList` collection, updates the word count and terminates.

```
Sub avAsTokens1(theString, _
                theList, nWords)
```

The input parameters to this procedure include the:

`theString`	The input text string comprised numeric and alphanumeric words. The length of this string is limited by the computer's operating system.

The parameters that are returned from this procedure include the:

`theList`	The collection containing the individual words of the input text string. Each word in the collection could be an alphanumeric text string.
`nWords`	The number of words that have been extracted.

The operation of this procedure similar to that of the `avAsTokens` procedure but without the checking of and associated processing for the additional word delineators and the altering the case of letters in the input text string to upper or lower case.

3.3 Directions and Distances

The procedures below enable the user to determine the distance and direction between two points. Certain of these procedures determine only the direction, or only the distance between two points, while other procedures determine both the direction and the distance between two points. The directions are expressed in radians, while the distances are expressed in the units of measure of the project in which the work is being done.

3.3.1 Length and Central Angle of an Arc

The subroutine `iccomarc` computes the length and central angle in radians of a circular arc that is defined by the Cartesian coordinates of its two endpoints and its radius. The rotational direction of the arc is defined by its radius as indicated below. The specification of a radius with a zero value assumes that the arc has a radius of infinite length (alternatively the arc is really a straight line), and returns the distance between the two input points and a central angle of zero value.

```
Sub iccomarc    (X1, Y1, X2, Y2, R, _
                 arclg, angle)
```

The input parameters to this procedure include the:

`X1, Y1`	The east and north coordinates of the start point of the arc (PC).
`X2, Y2`	The east and north coordinates of the end point of the arc (PT).
`R`	The radius of the arc which is also the indicator of the arc's rotational direction. Note that if: $R > 0$ denotes a clockwise rotation. $R < 0$ denotes a counterclockwise rotation. $R = 0$ denotes a straight line.

The parameters that are returned from this procedure include the:

`arclg`	The computed length of the arc, or of the straight line depending on the value of the radius R.
`angle`	The computed central angle of the arc in radians which has the same sign as the radius R. This value is zero if $R = 0$.

Upon invocation, the procedure initializes the parameters to be returned, and then

1. Computes the differences of the X and Y coordinates of the two endpoints (`X1, Y1`) and (`X2, Y2`), and then queries the differences for being equal to zero within a tolerance of 0.00005 feet (0.01524 mm), and if so sets the length to be zero and terminates; else it continues.
2. Queries the radius R for being zero, and if so computes the length of the straight line using the Pythagorean theorem, and terminates; else it continues.
3. Calls the `icarccnt` procedure to compute the coordinates of the center point of the arc, thereafter calls the `icforce` procedure to compute the direction from the said center point towards each of the said two endpoints, and then calls the `iccenang` procedure to compute the central angle of the arc and terminates.

3.3.2 Distance Between Two Points

The `Function iccomdis(X1, Y1, X2, Y2)` determines the distance between two points being defined by their Cartesian coordinates (`X1, Y1`) and (`X2, Y2`).

The procedure first computes the differences (dX and dY) in the X and Y coordinates between the two given points (`X1, Y1` and `X2, Y2`), and checks if the said differences are equal to zero within a tolerance of 0.00005 feet (0.01524 mm).

- If the said differences are both within the said range, the two points are considered to be the same, and the distance is returned as being zero (0.00).

• Else, the distance between the two given points (d) is computed by use of the Pythagorean theorem as $d=(dX^2+dY^2)^{1/2}$.

The direction between the two input points is not returned. If the direction from point one to point two is also desired the programmer should refer to the `icforce` or to the `icmakdir` procedures presented below.

3.3.3   Distance and Azimuth Between Two Points

The subroutine `icforce` determines the distance and north azimuth from point 1 to point 2, with both points being defined by their north and east coordinates respectively.

```
Sub icforce    (N1, E1, N2, E2, _
               D, AZ)
```

The input parameters to this procedure include the:

`N1, E1`	The north and east coordinates of point 1.
`N2, E2`	The north and east coordinates of point 2.

The parameters that are returned from this procedure include the:

`D`	The distance from point 1 to point 2.
`AZ`	The north azimuth from point 1 to point 2 in radians.

Upon invocation, the procedure initializes the parameters to be returned, and then

1. Computes the differences in the north and east coordinates between the two given points (dN and dE of Figure 3.3a), and checks whether they are within a tolerance of 0.00009 feet (0.0274 mm) of being equal to zero. If the absolute value of:
   • Both of the said differences is zero, the two points are considered to be the same, in which case the distance (`D`) between them is set to be zero (0.00), the azimuth is set to be due west (`AZ`=3/2π), and the procedure terminates.
   • Either of the said two differences is zero within the said range, `AZ` is set to be zero (0.00), and the procedure continues.
2. The procedure then proceeds to check the value of dN whether it is less, equal to, or greater than zero (0.00).
   • If dN<0, the procedure solves for angle α (see Figure 3.3a) by use of the `icatan` procedure, and sets the azimuth between the two points to be $\alpha+\pi$. Note that depending on the sign of dE, the angle α is either positive or negative, thus it is added to or subtracted from π.

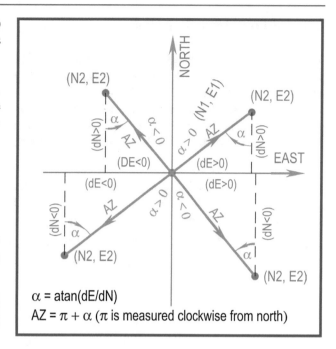

α = atan(dE/dN)
AZ = π + α (π is measured clockwise from north)

Figure 3.3a   Sample Street Center Line Intersection

   • If dN=0, the procedure checks dE whether it is negative or positive, in order to set the north azimuth to be due west (3/2π) or due east (1/2π), respectively.
   • If dN>0, the procedure determines the angle α (see Figure 3.3a) using of the `icatan` procedure in which the arctangent is defined by the ratio of dE/dN. Thereafter, if dE is negative, the procedure adds α to 2π (in essence it subtracts α from 2π since α is negative), and if dE is positive, α is accepted as the azimuth.
3. Computes the distance between the two given points by use of the Pythagorean theorem as $d=(dX^2+dy^2)^{1/2}$ and terminates.

3.3.4   Cartesian Direction (Radians) Between Two Points

The `Function icmakdir(X1, Y1, X2, Y2)` determines the Cartesian direction in radians from point 1 to point 2, with both points being defined by their Cartesian coordinates (`X1, Y1`) and (`X2, Y2`), respectively. To determine the said direction, the procedure calls the above-presented `icforce` procedure to determine the north azimuth, which it then converts into a Cartesian direction. The distance between the two points is not returned. If the distance from point one to point two is also desired, refer to the `icforce` and to `iccomdis` procedures presented above.

3.4   Angles from Directions

The procedures below assist the programmer to (a) reduce an angle to less than 180°, (b) compute the angle formed by two azimuths or differential X and Y coordinates, (c) determine the direction of rotation (clockwise or counterclockwise) of a circular arc, and (d) adjust a computed from two azimuths if the two azimuths straddle the due north direction.

3.4.1   Reduce an Angle to Less than 180° in Absolute Value

The `Function icangl(ANG)` reduces the absolute value of an angle to less than or equal to π. If the absolute value of an angle is less than or equal to π, its value is retained. If it is greater than π, or some multiple of π, the 2π, or multiple thereof, supplement is returned. Note that a tolerance test is not applied in this test. Thus:

If	0	<	ANG	≤	180	then	0	<	icangl	= ANG	≤ 180
If	180	<	ANG	≤	360	then	0	≥	icangl	= ANG − 360	> −180
If	0	>	ANG	≥	−180	then	0	<	icangl	= ANG	≤ −180
If	−180	>	ANG	≥	−360	then	0	≤	icangl	= ANG + 360	≤ 180

For example:

If	ANG	=	+60°	returns	+60°	and if	ANG	=	−60°	returns	−60°
	ANG	=	+120°		+120°		ANG	=	−120°		−120°
	ANG	=	+180°		+180°		ANG	=	−180°		−180°
	ANG	=	+240°		−120°		ANG	=	−240°		+120°
	ANG	=	+300°		−60°		ANG	=	−300°		+60°
	ANG	=	+420°		+60°		ANG	=	−420°		−60°
	ANG	=	+780°		+60°		ANG	=	−780°		−60°

It should also be noted that this procedure operates on an angle without regard to whether the angle is a rotation angle or a direction (azimuth, bearing, or Cartesian rotation) angle. If the angle is a component of a direction, it is up to the programmer to keep track of the type of direction.

3.4.2   Central Angle (Radians) from Azimuths

The subroutine `iccenang` computes the central angle formed between two known north azimuths (`AZ1` and `AZ2`) by rotating from `AZ1` towards `AZ2`. This angle could be greater than 180°. Although this procedure is used primarily to find the central angle and the length of the subtended arc when the directions from the arc's center towards the arc's end points are known, it can be used for any two azimuths with any arbitrary radius length.

```
Sub iccenang    (AZ1, AZ2, DIREC, RAD, opmode, _
                ANG, ARC)
```

The input parameters to this procedure include the:

AZ1	The north and east coordinates of point 1.
AZ2	The north and east coordinates of point 2.

DIREC	The rotational direction of the angle to be computed. Note that if:
	`DIREC = 1` it denotes a counterclockwise rotation.
	`DIREC = 2` it denotes a clockwise rotation.
RAD	The radius, if the angle to be computed is the central angle of a circular arc, or any value if it is the angle between the two given azimuths. Note that if:
	`RAD > 0` it denotes a clockwise rotation.
	`RAD < 0` it denotes a counterclockwise rotation.
opmode	The indicator of whether the procedure is to use the `DIREC` parameter, or the sign of the `RAD` parameter for determining the rotational direction of the angle being computed. Note that if:
	`opmode = 1` use the `DIREC` code.
	`opmode = 2` use the sign of the radius `RAD`.

The parameters that are returned from this procedure include the:

ANG	The computed angle in radians which could be greater than 180°.
ARC	The subtended arc, if `RAD` is a valid radius, in the units of measure of the project.

Figure 3.4a depicts the computational work flow of the procedure in which it is noted that there are four elements that may control the sign of the angle being computed:

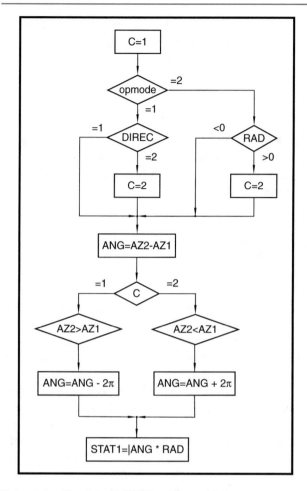

Figure 3.4a Use of the "DIREC" and "opmode" Codes

- The signs of the four input parameters opmode, DIREC and RAD, and
- The magnitude of the azimuth AZ2 as compared to that of the azimuth AZ1.

The angle (ANG) between the two directions is computed by subtracting from the second given direction (AZ2) the first given direction (AZ1) so that

$$ANG = AZ2 - AZ1$$

Note that:

- A positive angle denotes a clockwise angle from AZ1 towards AZ2.
- A negative angle denotes a counterclockwise angle from AZ1 towards AZ2.
- The subroutine accounts for the condition when AZ1 and AZ2 straddle due north direction.

3.4.3 Cartesian Direction (Radians) with dX and dY

The Function icdltoa(dX, dY) computes the Cartesian direction when the departure (dX) and the latitude (dY) between two points are known.

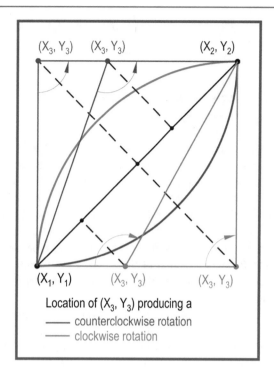

Figure 3.4b Function icgetdir

3.4.4 Rotational Direction of an Arc or Angle

The Function icgetdir (X1, Y1, X2, Y2, X3, Y3) determines the rotational direction of a circular arc from its start point (X1, Y1) towards its endpoint (X2, Y2) when its center point is at (X3, Y3). It returns the index number 1 or 2 denoting the direction of the arc; that is the rotation of the central arc's angle from the first given point towards the second point. A returned index number of:

- 1 denotes a counterclockwise rotation, and
- 2 denotes a clockwise rotation.

Although primarily intended for finding the rotational direction of a circular arc, this procedure may be used to determine the rotational direction of an angle formed by two intersecting lines having a common vertex at (X3, Y3) and rotating from endpoint (X1, Y1) towards endpoint (X2, Y2). With reference to Figure 3.4b, the procedure:

- First inverses from the first point towards the second point to determine the direction of the implied line them, and then
- Projects the third point on the abovesaid line to compute its offset distance.

A positive offset denotes a clockwise rotation (icgetdir = 2), while a negative offset denotes a counterclockwise rotation (icgetdir = 1) provided the angle is less than 180°. If the calling program senses that the angle is greater the 180°, the value of `icgetdir` should be reversed.

3.4.5 Angle Between Two Lines Straddling North

The `Function icstrdln(ORN, ORE, AZ1, AZ2, RAD)` computes and returns the angle in radians formed between two intersecting lines, each of known north azimuth (`AZ1` and `AZ2`) in radians, and having a common vertex point of known north and east coordinates at `ORN` and `ORE`. The angle that is being returned is formed by rotating from `AZ1` towards `AZ2` taking into consideration (a) the specified clockwise or counterclockwise rotation of the angle, and (b) accounting the potential that the two given azimuths might straddle the due north direction.

The procedure assumes that the given north azimuths of the two lines emanate from the said common vertex point, which point could represent the center point of a real or implied circular arc. Although primarily intended for circular alignments when straddling north, it can be used for any condition by assuming an imaginary radius as stated below. With reference to Figure 3.4c, which displays the 32 potential incidences of an angle, it should be noted that:

(a) The procedure first computes the parameter `ANGLE` as indicated in Figure 3.4c for the angle α.
(b) If `AZ1` and `AZ2` are equal to each other within a tolerance of 0.0000001 radians (0.02 s), the value of `ANGLE` is set to be zero radians and the function terminates.
(c) The value of the radius `RAD` (R in Figure 3.4c) could be zero, positive, or negative.
  - If `RAD` is zero, it denotes that there is no circular arc, `ANGLE` is set to be equal to `AZ2 − AZ1` and the procedure terminates without checking for straddling the due north direction.
  - If `RAD` is positive or negative, it denotes that the arc could be real or implied (imaginary) and that the rotation is dependent upon the sign of `RAD` in the test for straddling the due north direction.
  - If `RAD` is positive, it denotes that a clockwise rotation is desired from AZ1 towards AZ2.
  - If `RAD` is negative, it denotes that a counterclockwise rotation is desired from AZ1 towards AZ2.
    The magnitude of the `RAD`, except when it is zero, is of no concern. Only its sign is used when testing for straddling the due north direction.
(d) In testing for straddling the due north direction the procedure queries the signs of `RAD` and of the angle `ANGLE`

(α in Figure 3.4c) to determine the value and sign of the angle θ to be returned as follows. If:

$$RAD > 0 \text{ and } \alpha > 0 \quad \theta = \alpha \quad \text{or} \quad RAD > 0 \text{ and}$$
$$\alpha < 0 \quad \theta = \alpha + 2\pi$$
$$RAD < 0 \text{ and } \alpha > 0 \quad \theta = -\alpha \quad \text{or} \quad RAD < 0 \text{ and}$$
$$\alpha < 0 \quad \theta = -(\alpha + 2\pi)$$

(e) The angle `ANGLE` is now replaced with the value of θ and it is then assigned to the value of the function. That is `icstrdln = Angle` and it is subject to the conditions identified above.

3.4.6 Angle (Radians) Formed by Three Points

The `Function icvecang(XC, YC, X1, Y1, X2, Y2)` computes and returns in radians the Cartesian rotation angle (the angle α in Figure 3.4d) which is formed by two implied lines that emanate from a common vertex at point (`XC, YC`) towards the two endpoints (`X1, Y1`) and (`X2, Y2`) of the said two lines.

The said angle α is always returned as a positive counterclockwise angle that rotates about the said common vertex point from the (`X1, Y1`) endpoint towards the (`X2, Y2`) endpoint. To determine the angle α, the procedure computes the Cartesian directions of the said two implied lines, θ_1 and θ_2 towards the said endpoints, respectively, and then subtracts θ_1 from θ_2 to find the said angle α. If the resultant angle α is negative, the procedure adds 2π to make certain that the angle α is always returned with a positive value. Furthermore, if the angle's value is within a tolerance of 0.0000048 radians (0°00′00.99″) of being equal to 0°, 90°, 180°, 270°, or 360°, it is set to be the nearest of the said angles.

3.5 Elevation Procedures

Presented below are procedures that are used by certain of the procedures presented later on to compute elevations at intersections, projections, and other cases.

3.5.1 Point Elevation on a Line Given Coordinates

The `Function icintel3(X1, Y1, Z1, X2, Y2, Z2, X, Y)` returns the elevation of a point located at the Cartesian coordinates (`X, Y`). This point is assumed to be along a line defined by the points of Cartesian coordinates (`X1, Y1, Z1`) and (`X2, Y2, Z2`). The elevation is linearly interpolated, or extrapolated between the last said two points. Note that:

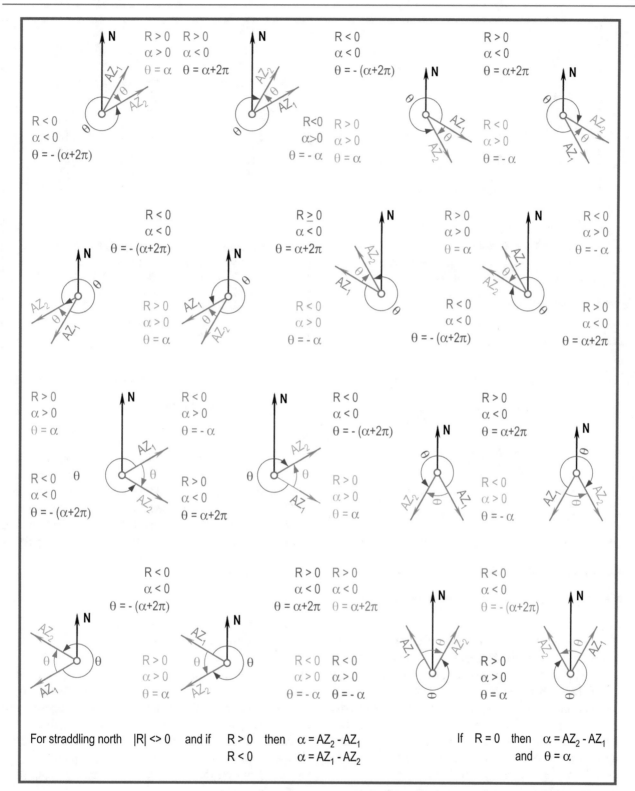

Figure 3.4c Function icstrdln—Angle Between Two Lines that Straddle the Due North Direction

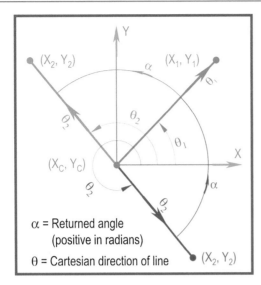

Figure 3.4d Cartesian Angle Formed by two lines

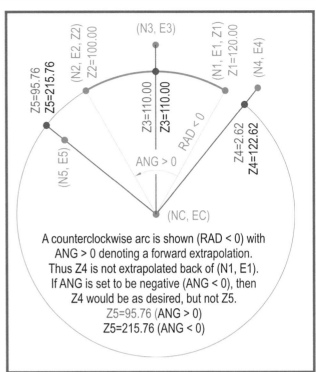

Figure 3.5b Elevations of Points on a Circular Curve

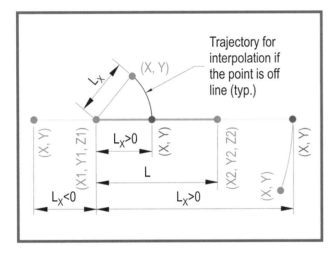

Figure 3.5a Elevations of Point not on Line or not on

- The procedure does not check for colinearity of the three points. It computes the distances of (X, Y) and (X2, Y2) from (X1, Y1), and assumes that (X, Y) is located along a straight line between (X1, Y1) and (X2, Y2).
- The new elevation is computed by dividing the distance LX (see Figure 3.5a) from (X1, Y1) to (X, Y), by total length of the line from (X1, Y1) to (X2, Y2), and multiplying the quotient by the algebraic difference of the elevations of the two endpoints Z2 minus Z1.
- If either of the differences in the X or Y coordinates is within a tolerance of 0.005 feet (1.524 mm), the elevation is set to be zero (0.00).
- If points (X1, Y1) and (X2, Y2) are identical (one is over or under the other—the line is vertical) within the above-said tolerance, the computed elevation is set to be midway Z1 and Z2.

- If (X, Y) is not located along the line implied between point (X1, Y1) and point (X1, Y1) but is offset there-from, the returned elevation may not be the desired one. The distance LX that is used for the point of interpolation or extrapolation is the sloped distance as indicated in Figure 3.5a. In essence the procedure rotates the point so as to be located on line as indicated by the trajectory of the said figure.

3.5.2   Point Elevation on an Arc Given Coordinates

The Function icintelc(N1, E1, N2, E2, Z1, Z2, RAD, ANG, NC, EC, N, E) returns the elevation of a point of north and east coordinates (N, E) located along a circular arc defined by the parameters listed below and as depicted in Figure 3.5b. If the said point is not located on the said arc, the returned elevation is that of the projection of the said point on the said arc. The remaining input para-meters are:

- (N1, E1, Z1) and (N2, E2, Z2) are the north and east coordinates and elevations of the start and end points of the arc.
- (NC, EC) are the north and east coordinates of the arc's center point.

- The arc's radius `RAD` should be positive if the arc is progressing in a clockwise rotation, and negative if it is progressing in a counterclockwise rotation.
- The arc's central angle `ANG` should be positive for a forward extrapolation, and negative for a backward. `ANG` is also used to determine the position of the projection of the given point on the arc.

With reference to Figure 3.5b note the following:

- The displayed given arc is a counterclockwise arc (`RAD < 0`), and any extrapolation is to be made in a forward direction because `ANG > 0`.
- There are three points, (`N3`, `E3`), (`N4`, `E4`), and (`N5`, `E6`), of which the elevations of their projections on the given arc are to be determined and each of which corresponds to the (`N`, `E`) input parameters. Regarding the position of point (`N`, `E`) with respect to the start and end of the arc it is important to note that if the projection of point (`N`, `E`) lies:
  - Within the given arc's limits, its elevation is interpolated between the given two endpoints.
  - If the projection of point (`N`, `E`) lies ahead of the end point of the given arc, its elevation is extrapolated in a forward direction as it should be.
  - If the projection of point (`N`, `E`) lies back of the start point of the given arc, its elevation is extrapolated in a forward direction beyond the endpoint and around the implied extension of the arc. Thus the computed elevation may not be the desired one. In this case this point should be processed with a negative `ANG`. See also comment 5 below.

Regarding the interpolation or extrapolation of the elevation at `N`, `E`, note the following:

1. The procedure assigns `ANG` and `RAD` to temporary parameters for not altering the input data. If the input value of the central angle (`ANG`) is:
   - Negative, the sign of the temporary angle is reversed, and the temporary radius parameter is set to be negative.
   - Positive, both of the said temporary parameters are assigned the input values.
2. If the point at (`N`, `E`) is located off the arc, the elevation to be determined is that of the projection of the said point on the arc (see Figure 3.5b for the interpolated point).
3. The elevation to be determined is computed by dividing the arc length `Lx` from (`N1`, `E1`) to point (`N`, `E`), or to the said projection point, by the total length of the given arc, and multiplying the quotient by the algebraic difference of the elevations of the two arc endpoints (`Z2 − Z1`).

4. The elevation is always computed from the start point of the arc towards its end point in a forward direction. That is, the procedure extrapolates only forwardly.
5. The procedure calls the `icbtwarc` procedure to determine the location of the projection of the point within the limits of the given arc.
6. If (`N`, `E`) lies back of the start point (`N1`, `E1`), the elevation of (`N`, `E`) will be computed as the elevation of a point along an arc ahead of (`N1`, `E1`) times the slope between the arc's endpoints. Thus, the said elevation will be either higher or lower than both of `Z1` and `Z2` depending on whether the said slope is positive or negative. If it is desired to extrapolate back from the start point (`N1`, `E1`), the calling program should change `ANGLE` to be negative, in which case the procedure interchanges the two given endpoints. In so doing the input parameters of the said points are not altered.
7. If the length of the absolute value of the given arc (|`R* Angle`|) is zero (0.00) within a tolerance of 0.001 feet (0.3048 mm), the elevation to be returned is set to be −99999.0 to avoid producing a division by zero.
8. If the absolute value of the arc length from the start point (`N1`, `E1`) to point (`N`, `E`) is, within a tolerance of 0.001 feet (0.3048 mm), equal to:
   - Zero, the elevation to be returned is that of the start point (`N1`, `E1`) of the arc.
   - The length of the given arc, the elevation to be returned is that of the end point (`N2`, `E2`) of the given arc.

3.5.3 Elevation of POL Given Station Value

The `Function icintelv(OF1, EL1, OF2, EL2, OFNEW)` computes the elevation of a point located at station `OFNEW`. The elevation is interpolated between points (`OF1`, `EL1`) and (`OF2`, `EL2`). This new elevation is determined by a straight line interpolation between the said two given points, or by extrapolation if `OFNEW < OF1`, or if `OFNEW > OF2`. It is noted that if `OF1 = OF2` within a tolerance of 0.0001 feet (0.03048 mm), the returned elevation is set to be −99999.0. This procedure may be used for points located along lines, circular arcs, and spirals, as long as the three said stations are known and measured along the path of the line, arc, spiral, or any other path.

3.6 Miscellaneous Support Procedures

The procedures below provide certain support functionality for a variety of programming applications, and which are used by certain of the procedure presented in the subsequent chapters.

Figure 3.6a Curves by
Lagrangian Interpolation of
Given Points

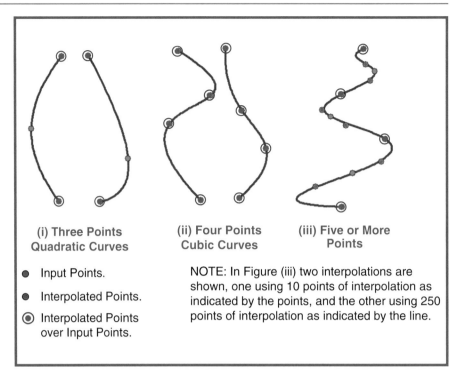

(i) Three Points
Quadratic Curves

(ii) Four Points
Cubic Curves

(iii) Five or More
Points

● Input Points.

● Interpolated Points.

◉ Interpolated Points
over Input Points.

NOTE: In Figure (iii) two interpolations are
shown, one using 10 points of interpolation as
indicated by the points, and the other using 250
points of interpolation as indicated by the line.

3.6.1   Curves by Lagrangian Interpolation

The `iclagr1` procedure applies the Lagrangian interpolation method to generate a polyline that is passing through a
number of user specified (input) points defined by their X
and Y Cartesian coordinates samples of which are shown in
Figure 3.6a. If the number of input points is:

• Two, the generated polyline is a straight line with as many
  vertices as indicated by the number of specified interpolation points (see below).
• Three, the generated polyline is a quadratic curve.
• Four, the generated polyline is a cubic curve.
• Five or greater, the power of the generated curve increases
  accordingly.

The number of points to be interpolated defines the
smoothness of the curve. A rule of thumb used by the authors
of this book when the number of input points exceeds four (4)
is to set the number of interpolated points by the equation

$$A = 50(B-2)$$

where

A = The number of interpolated points, and
B = The number of input points (greater than 4).

The generated curve is represented by a polyline. At each
point of interpolation the procedure introduces a vertex in

the polyline. Thus when there are two input points, the generated polyline would represent a straight line with as many
vertices as the number of the specified points of interpolation minus two. Unless there is some kind of special need,
there is doubt that there is a need to use this procedure to
create a straight line. Reference is made to curve (iii) of
Figure 3.6a and to the note below it at the right side of the
said figure regarding Lagrangian interpolation with more
than five input points.

```
Sub iclagr1    (NPTDAT, XDAT, YDAT, NPTOUT, _
                XOUT, YOUT, IERR)
```

The input parameters to this procedure include the:

NPTDAT	The number of input points through which the new interpolated curve is to pass. This number of points should be two (2) for a straight line, three (3) for a quadratic curve, four (4) for a cubic curve, or more for any other polynomial curve. It cannot be one or less.
XDAT, YDAT	The collections of the X and Y coordinates, respectively, of the points that define the new interpolated curve.
NPTOUT	The number of points to be interpolated and to comprise the polyline curve to be generated. This number should include the input points so that it should be greater than the number of input points (NPTDAT).

The parameters that are returned from this procedure include the:

XOUT, YOUT	The collections of the X and Y coordinates, respectively, of the points which comprise the interpolated curve.
YOUT	The collection of the Y coordinates of the points which comprise the interpolated curve.
IERR	An error indicator the value of which denotes the following: 0 No errors detected. 1 The number of input points is equal to one or less. 2 Then number of points to be interpolated is less then the number of input points.

Upon invocation, the procedure initializes the parameters to be returned, and then

1. Checks the number of input and interpolation points for possible errors, and if any sets in IEER the appropriate value and then aborts; else continues.
2. Resizes the arrays to be returned.
3. Sets the constant interval of interpolation DEL to be performed based on the number of input points and number of desired output points by the equation

$$DEL = (NPTDAT - 1)/(NPTOUT - 1).$$

4. Performs the Lagrangian interpolation over the computed constant interval and computes the coordinates of the points to be returned by the procedure saving them in their respective collections to be returned.

3.6.2 Order of a String of Features

The icordar procedure arranges a string of two-point line and/or circular arc features so that their individual direction of development is the same as that of the sequence of selection of the features that comprise the overall string. That is, if the sequential selection of the various features that comprise a string of features progresses in a clockwise, counterclockwise, forward, or backward direction, then the direction of each individual feature is made to follow the same direction as that of the overall string. So that if the individual features have been selected from left to right in plan view, then each feature will be made to progress from left to right.

For example consider the strings of features shown in Figures 3.6b and 3.6c, in which the direction of development of the individual features of a string of features is indicated with blue arrows, while the sequence of features within the said string is indicated with dark red arrows.

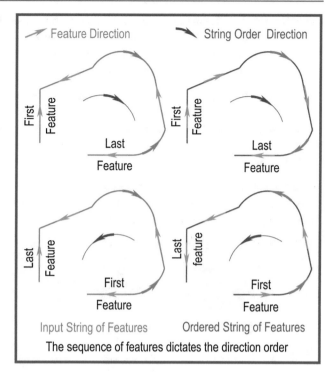

Figure 3.6b Ordering of Open Feature Strings

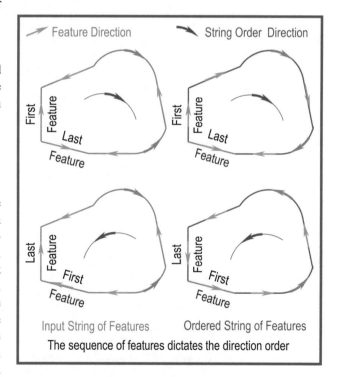

Figure 3.6c Ordering of Closed Feature Strings

The feature strings at the left side of Figure 3.6b and Figure 3.6c indicate the directions of each feature of the original string of features at the time of their selection, while the feature strings at the right side of the said figures indicate the directions of each feature after the completion of the string ordering process. Note the different order of selection between the top and bottom strings of features.

It should be noted that upon ordering a string of features:

- The sequential position of features within the string is not altered. What is altered is the direction of the individual features that oppose the direction of the ordered string.
- The original input strings are not modified. New strings are returned.

This procedure is of use for certain geometric construction operations in which it is required to have all component feature progress in the same direction.

```
Sub icordar    (Y1ListI,   X1ListI,   Y2ListI,
                X2ListI, RdListI, i1, _
                Y1List, X1List, Y2List, X2List,
                            RdList, IERR)
```

The input parameters to this procedure include the:

Y1ListI	The collection of the north coordinates of the start points of each individual feature of the string of features to be ordered.
X1ListI	The collection of the east coordinates of the start points of each individual feature of the string of features to be ordered.
Y2ListI	The collection of the north coordinates of the end points of each individual feature of the string of features to be ordered..
X2ListI	The collection of the east coordinates of the end points of each individual feature of the string of features to be ordered..
RdListI	The collection of the radii of each individual feature of the string of features to be ordered. For lines, RdListI = 0.00.
i1	The number of features in each of the above collections.

The parameters that are returned from this procedure include the:

Y1List	The collection of the north coordinates of the start points of each individual feature of the ordered string of features.
X1List	The collection of the east coordinates of the start points of each individual feature of the ordered string of features.
Y2List	The collection of the north coordinates of the end points of each individual feature of the ordered string of features.
X2List	The collections of the east coordinates of the end points of each individual feature of the ordered string of features.
RdListI	The collection of the radii of each individual feature of the string of features to be ordered, with certain signs possibly reversed. For lines, RdListI = 0.00.

IERR	An error indicator, the value of which if it is:
	0 Denotes that an error has not been encountered.
	1 Denotes that the string is discontinuous (see comments below).

Upon invocation, the procedure initializes the parameters to be returned, and then:

1. Ascertains whether all of the input collections have the same number or elements (should be equal to i1), or not. If all collections:
   - Do not have the same number of elements sets IERR to be 1 and then terminates.
   - Have the same number of elements, continues with the next step.
2. Queries the value of i1. If
   - i1 = 0, sets to be equal to the number of elements in the input collections.
   - i1 = 1, sets the returned lists to be the same as the input lists and then terminates.
   - i1 > 1, continues with the next step.
3. Determines the direction of the string of features by checking the north and east coordinates of the end-points of two successive features as indicated in the flowchart of Figure 3.6d. If a discontinuity is not found, the process is repeated for the remaining features by setting the second feature to be the first one, and the third feature to be the second one, and so on until all of the features in the string have been processed. If a discontinuity is found, the error indicator IERR is set to be 1, and the process terminates.
   It should be noted that:
   - If the direction of a feature is reversed, the coordinates of its start and end points are interchanged. In addition, if the feature is a circular arc, the sign of its radius is changed from plus to minus, and vice versa.
   - If a feature does not have an endpoint, one of the two endpoints is missing, the string is considered discontinuous, and IERR is set to be one (1).
   - A string must be continuous. For a string to be continuous, one endpoint of a feature must be the same as one endpoint of a subsequent feature.
   - The iccompp1 procedure (see later on in this chapter) is used to determine whether two endpoints are the same or not.
   - A string may be open or closed, but it can not have forks or branches. Forks or branches produce a discontinuous string.

Figure 3.6d Order Process
of a String of Features

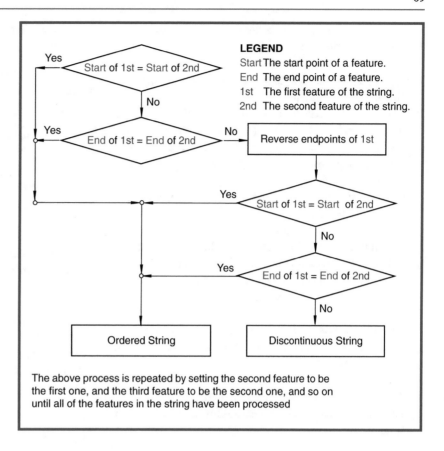

The above process is repeated by setting the second feature to be
the first one, and the third feature to be the second one, and so on
until all of the features in the string have been processed

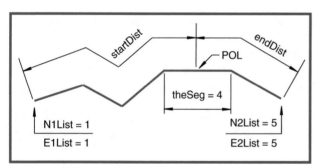

Figure 3.6e POL on a Sample String of Five Features

3.6.3 POL Distances from a String's Endpoints

The icPntOnPoly procedure determines the location of a
point (POL) of given north and east coordinates along a line
defined by a series of two-point lines. The said line could be
any polyline or any curve that has been decomposed into a
series of small chords. Figure 3.6e illustrates a series of such
line segments with each one being defined by the north and
east coordinates of its start and end points. These coordinates
are contained in the collections as identified below. This procedure assumes that:

- The string of features is valid with the endpoints of its
  component segments defined with north and east
  coordinates.

- The end point of a segment is the start point of the next
  segment and the string is not discontinuous. That is there
  are no gaps nor overlaps.
- The POL is located on one component segment of the
  overall string.

```
Sub icPntOnPoly (N1List,    E1List,    N2List,
                 E2List, NPt, EPt, _ theSeg,
                 startDist, endDist)
```

The input parameters to this procedure include the:

N1List, E1List	The collections of the north and east coordinates, respectively, of the start points of each individual line segment.
N2List, E2List	The collections of the north and east coordinates, respectively, of the end points of each individual line segment.
NPt, EPt	The north and east coordinates of the POL.

The parameters that are returned from this procedure
include the:

theSeg	The sequential number of the two-point line, within the said string, upon which the given POL is located.
startDist	The distance from the start point of the string to the given POL.
endDist	The distance from the end point of the string to the given POL.

Upon invocation, the procedure initializes the parameters to be returned, and then:

1. Computes the total length of the string by summing individual segment lengths.
2. Compares the coordinates of the POL with those of the start point of the first line segment and with those of the end point of the last line segment to determine whether the POL happens to coincide with either of these two points.
   - If they do, the returned parameters are set accordingly, and terminates.
   - If they do not, it continues.

   A match is considered as made if both sets of coordinates are equal within a tolerance of 0.005 feet (1.524 mm).
3. Parses through the line segments of the string to identify within which segment the POL is located.

It is noted that if the POL is not located within any component segment of the string, or if it is offset therefrom, the procedure returns the value of

- Zero for the `theSeg` and `endDist` parameters.
- The overall length of the string for the `startDist` parameter.

3.6.4 Distance Measure to Station Format (SI and US Customary)

The `icstatn` procedure converts a distance measure in feet or in meters along an alignment to a station value in the US Customary, or in the SI station format. The selection of the station format is based on the use of the global variable *ugunits* which may have a value of 0 denoting the US Customary station format, and a value of 1 denoting the SI station format. It is assumed that the said global variable has been defined at some time prior to the call for the subject procedure. The examples below display the two formats for the same distance value which may be in feet or meters.

A distance of	yields the station	US	and	SI
15342.413		153+42.413		15+342.413
−15002.413		−150+2.413		−15+002.413

```
Sub icstatn   (OST, _
              IST, IST1, IST2, BST)
```

89.050		0+89.050		0+089.050

The input parameters to this procedure include the:

OST	The given distance measure in feet, or meters to be converted to a station value.

The parameters that are returned from this procedure include the:

IST	The station value to the left of the plus (+) sign.
IST1	The first digit to the right of the plus (+) sign.
IST2	The second digit to the right of the plus (+) sign for the SI format, or the same as IST1 for the US Customary format.
BST	The remaining digits to the right of the plus (+) sign.

It is noted that when using the `icstatn` procedure if the distance measure to be converted `OST` into station format is negative:

- Then the value of the `IST`, and only of the `IST` parameter is displayed as being negative as indicated in the above example.
- And less than one (i.e., −0.54), the minus sign will not be displayed to the left of `IST` because `IST`, `IST1` and `IST2` are treated as individual numeric parameters of zero value (unable to be signed negative) and not as a single text string. Thus, the proper formatting for text display purposes should be handled by the calling program.

3.6.5 Point Coordinate Comparison

The subroutine `iccompp1` compares the Cartesian coordinates (north and east coordinates may be used by interchanging their input sequence) of a given point with those of another point referred to as the base point to determine whether the two points are the same within a tolerance of

```
Sub iccompp1   (XCORD, YCORD, X2, Y2, _
               XCRD2, YCRD2, noFnd)
```

0.005 feet (1.524 mm).

The input parameters to this procedure include the:

XCORD, YCORD	The north and east coordinates of the given point.
X2, Y2	The north and east coordinates of the base point.

The parameters that are returned from this procedure include the:

XCRD2, YCRD2	The coordinates of the: • Base point if a match is made. • Given point if a match is not made.
noFnd	An indicator the value of which denotes the following: 0 the two points are not equal to each other. 1 the two points are equal to each other.

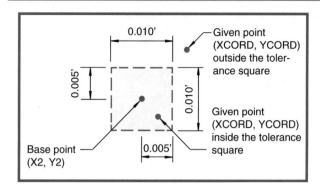

Figure 3.6f   Point Coordinate Comparison

To compare the given point with the base point, the procedure creates a square about the base point having a side equal to twice value of the said tolerance (see Figure 3.6f). If the coordinates of the given point fall:

- Within the said square, the indicator noFnd is set to 1, and the (XCRD2, YCRD2) coordinates are set to be the coordinates of the base point (X2, Y2).
- Outside the said square, the value of noFnd is set to 0, and the coordinates (XCRD2, YCRD2) are set to be those of the given point (XCORD, YCORD).

In either case the coordinates of the input given point and base point remain unaltered.

3.6.6   Decompose a Polyline into Lines and Arcs

The subroutine Polky2curve enables the programmer to decompose a polyline into a series of concatenated lines and circular arcs. That is the procedure reviews the various polyline segments (chords) and replaces any series of cords with a circular arc if certain user input criteria are met. The replacement arc is saved with its start and end points, its central angle that may exceed 180° and its radius. To avoid confusion the text below uses the word "chord" to address the segments of the given polyline and the word "segment" to denote the generated line and curve segments. Use of this procedure arises at times when alignment center lines, right-of-way lines, edges of pavement, parcel sides, and other such lines comprising concatenated lines and circular arcs have been:

- Represented by a single polyline as a series of long and short chords for some use and then there is a need to reconstruct them into their individual line and arc components.
- Digitized as polylines of varied length chords and there is a need to segregate lines from circular arcs for design purposes.
- Surveyed in the field and it is needed to represent lines and arcs as individual entities for geometric processing.

In essence this procedure attempts to reconstruct circular arcs from a polyline of varied length chords that meet certain criteria. In reconstructing the curved portions of the polyline the procedure does not use a best fit curve process but the user specified criteria identified below.

As an example consider the polyline at the upper left of Figure 3.6g defined by the points comprising its vertices as shown at the upper right of the said figure. By applying the abovesaid criteria the polyline is decomposed into four line and two arc segments as indicated in the low left of the said figure. The segments may now be used individually for design purposes, or even reunited into a single but smoother polyline as shown in the low right part of the said figure. In general the procedure accepts two lists of Cartesian coordinates (north and east coordinates are interchanged) of the start point, of the vertices (specified sequentially) and of the end point of the polyline to be processed. A minimum of three points (one vertex) is required to comprise polyline. In addition to these coordinates the procedure requires the specification of the criteria to be applied in identifying polyline portions that could be replaced by a circular arc. These criteria and their operational function are identified below under the description of the input parameters.

```
Sub Poly2curve  (xPts, yPts, _
                 minCurveSeg,        maxChrdDef,
                 chrdLenTol, _ removeColS, rem-
                 oveCTol, compressCC, compcrv-
                 Tol, _ X1List, Y1List, X2List,
                 Y2List, RdList, CAList)
```

The input parameters to this procedure include the:

xPts, yPts	Arrays of the X (east) and Y (north) coordinates of the polyline points with the first being the start and the last being the end point of the polyline.
minCurveSeg	The minimum number of chords for a curve to be recognized as a curve with two being the minimum.
maxChrdDef	The maximum chord deflection in degrees between two adjacent chords. Deflections exceeding this deflection designate the start of a line, or the end of a curve
chrdLenTol	The maximum percentage in decimal form (from 0 to 1.0) tolerance in the difference in length between adjacent chords. Chord differences exceeding this specified tolerance designate the start of a new line segment, or the end of a curve segment being created.
removeColS	A flag denoting whether collinear chords should be compressed into one chord or not as follows: Y: compress collinear chords

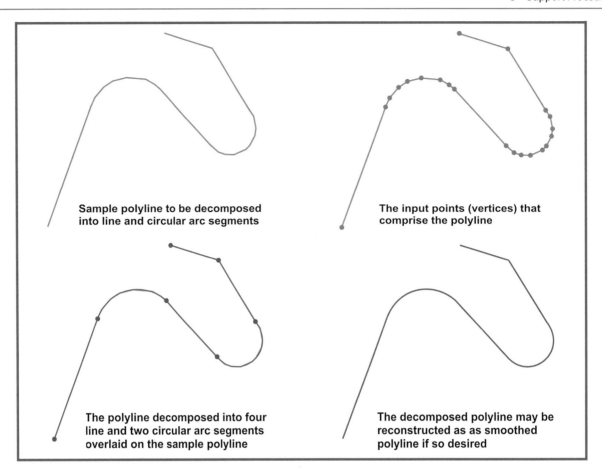

Sample polyline to be decomposed
into line and circular arc segments

The input points (vertices) that
comprise the polyline

The polyline decomposed into four
line and two circular arc segments
overlaid on the sample polyline

The decomposed polyline may be
reconstructed as as smoothed
polyline if so desired

Figure 3.6g   Reconstructing Individual Line and Curve Segments from a Polyline

	N: do not compress collinear chords	`X1List, Y1List`	The list of the X (east) and Y (north) coordinates of the start points of the new decomposed segments.
`removeCTol`	The maximum deflection angle tolerance in degrees for considering two adjacent chords as being collinear.	`X2List, Y2Lis`	The list of the X (east) and Y (north) coordinates of the end points of the new decomposed segments.
`compressCC`	A flag denoting whether compound curves are to compressed into a single curve or not as follows;	`RdList`	The list of radii of the new decomposed segments with their values denoting:
	Y: compress compound curves		Positive: Clockwise curve.
	N: do not compress compound curves into one curve but create two curves.		Negative: Counterclockwise curve.
			Zero: Line.
`compcrvTol`	The maximum tolerance in difference between curve radii for detecting compound curves	`CAList`	The list of central angles of the new decomposed segments (always positive).

The parameters that are returned from this procedure include the:

To decompose the polyline into individual line and curve segments the procedure utilizes the input and output lists (arrays and collections) identified above and a series of internal (local) collections to be identified below. The primary operational work of this procedure is an excessive amount of bookkeeping with a minimal of geometric principle involved. Therefore in general the procedure:

1. First constructs the chords that constitute the given polyline and determines the number of points and chords with the number of points being one higher than that of the chords. This construct process saves in local collections the azimuth, length, radius, delta angle, tangent length, and arc length of each chord with the last four having a value of zero denoting a line.
2. Next parses through these chords and:
   (a) Checks for collinearity of two consecutive chords. If two are found
      (i) to be collinear within the specified tolerance a check is made whether they should be compressed into one or not, and if so the local collections are updated.
      (ii) not to be collinear a check is made whether the current chord could be the end of a curve segment or not, and if so certain flags are set to indicate the action to be taken.
   (b) Detects any series of chords that could constitute a curve segment and sets appropriate flags for later processing. A curve segment is detected if all of the following rules apply:
      (i) A series of contiguous chords are within a tolerance in curve chord length, the exception is when a curve that is drawn based upon alternating deflections is detected (noAcurve = 1)
      (ii) The deflection angles between two equal sides are:
         • Less or equal to a maximum deflection value and not equal to zero,
         • In the same rotation,
         • Equal, excepting the first and last curve chords,
      (iii) The number of such equal contiguous segments satisfies a minimum number of segments to form a curve.
      (iv) The average deflection for the segments comprising the curve is greater than zero. That is a series of chords form a line more or less.
3. Afterwards determines the rotational direction of the polyline at its start, initializes the local collection with first deflection angle and then loops through the chords comprising the polyline to determine which chords may constitute straight line segments and which chords might comprise a curve segment. To detect the start of a curve segment and create the curve the procedure:
   (a) Checks to ascertain that all deflection angles of the chords to comprise the curve rotate either clockwise or counterclockwise. Since it is possible for certain small chords to vacillate left and right with a very small deflection angle such as 0.00005 radians, such exceptions are allowed to occur.
   (b) Checks for compliance with the conditions (i) through (iv) of the above paragraph **2(b)**. When one of these conditions indicates the end of a curve segment the procedure `ParcelFromPoly2` is called to compute the radius and central angle of the curve and remove the intermediate chords comprising the corresponding curve collections that are passed back.

   The procedure `ParcelFromPoly2` is not described in this book but it is included in the accompanying CD. To compute the radius and the central angle of the curve this procedure uses the first and last point of the curve and the middle point. The middle point is the start point of the chord located half way plus one chords within the chords that comprise the curve.
4. Having parsed through all chords the procedure queries the `compressCC` input indicator whether it should compress into one curve two compound curves, and if so queries any compound curves that have radii the difference of which is less the that specified by the input `compcrvTol` tolerance. If any such curves are found they are combined into by adjusting the endpoint of the merged curve.
5. Lastly the procedure updates the collections to be returned and terminates.

3.6.7 Point and Line Feature Display

In testing certain of the procedures presented in this book and included in the accompanying CD a special routine was created to quickly display temporary points, arcs and lines in order to visually confirm the results of the procedures being tested. This procedure is the subroutine `AddLCGraphic` which utilizes certain *Avenue Wraps*© commands since the said testing has been performed within the ArcGIS® environment of the Environmental Research Institute, Inc. (ESRI). *Avenue Wraps*© is a set of specialized commands developed by *The CEDRA Corporation*© that simplify programming under the ArcGIS® environment. These commands are used in the graphic display of points and lines. Two point lines and circular arcs are displayed individually one at a time. Polylines and other type of curves are plotted one polyline segment or arc chord at a time by placing the `AddLCGraphic` subroutine call within a do loop of a number of iterations corresponding to the number of the said segments or chords.

This procedure is not intended as a formal content of the procedures of this book. It is presented herein only because it has been used as a testing tool and is contained in the accompanying CD.

```
Sub AddLCGraphic (PCN,  PCE,  PTN,  PTE,  CCN,
                  CCE, aRAD)
```

The input parameters to this procedure include the:

PCN, PCE	North and east coordinates of the start point of an arc or line.
PTN, PTE	North and east coordinates of the end point of an arc or line.
CCN, CCE	North and east coordinates of the center point of an arc. These parameters are ignored if a line is to be processed.
aRAD	The arc radius with its sign denoting the direction of the arc's rotation. If:
	aRAD <0 Denotes the radius of a counterclockwise curve.
	aRAD =0 Denotes a two point line (no curve).
	aRAD >0 Denotes the radius of a clockwise curve.

Note that if the above three points are the identical (there is no comparison for a tolerance) it is assumed that they all represent a single point to be displayed.

There are no parameters to be returned. The output of this procedure is graphic line or curve.

Upon invocation this procedure:

1. Queries the value of aRAD for being zero or not. If it is:
   (a) Zero, it is assumed that a line is to be displayed, and calls the Avenue Wraps procedure to create a poly-line comprising one segment from (PCN, PCE) to (PTN, PTE).
   (b) Non-zero, it is assumed that a circular arc is to be displayed, and then:
      • Calls the icforce procedure to compute the azimuth from the arc's center point (CCN, CCE) to the arc's start point (PCN, PCE), and repeats the same call to compute the azimuth from the said center point to the arc's end point (PTN, PTE).
      • Calls the iccenang procedure to compute the arc's central angle.
      • Queries the value of aRAD for being positive (clockwise) or negative (counter clockwise) and sets the central angle to have the same sign as the radius.
      • Calls the icarcseg procedure to discretize the arc into polyline chords.
      • Calls the icdrwcrc procedure to create the polyline of the chords.
2. Queries the values of the three input points for being identical and if so calls the appropriate *Avenue Wraps©* procedures to display the point; else calls the appropriate *Avenue Wraps©* procedures to display the polyline of the line or circular arc.

Regarding the icarcseg the icdrwcrc procedures refer to Chap. 5 of this book.

Point and Line Features

The procedures presented in this chapter generate points and lines. In an interactive computing environment, points may be introduced by (a) entering their coordinates, (b) clicking with the computer mouse cursor at a location on the computer monitor screen, (c) importing coordinates from an external file, and (d) constructing points by various geometric methods. Of these four means of creating points, the first three are considered to be highly sensitive to the computing environment at hand, and not subject to any geometric considerations. For this reason, they are to be excluded from this publication. Lines and curves, much like points, they too can be generated by fairly much the same means as points. Thus, the procedures to be presented in this chapter are those that pertain to the points and lines that can be created by various constructions with the aide of coordinate geometry. Curves are to be presented in subsequent chapters.

In using the procedures that are included herein, it is assumed that the coordinate medium, plane or map has been established in the computing environment.

Points and/or lines may be constructed by:

- Mensuration (a) from a given base point, and (b) along, about or offset from a base line.
- Normal projection of a given point upon a line, curve, or polygon, with or without a line connecting the given point with the point of projection.
- Intersection of a line, curve, or polygon with another line, curve, or polygon.
- Tangent projection of a point upon a curve, with or without a line connecting the given point with the point of tangency.

Of these four means, the procedures of this chapter are restricted to the first two, with projections as they pertain to line and circular arcs. Projections on a spiral and all other constructions are to be presented in their distinct chapters of this publication.

4.1 Points Along an Arc

Presented in this section are the procedures that generate points with plus and offset measures along a circular arc. Note that some procedures utilize Cartesian coordinates in their arguments, while others utilize north and east coordinates.

4.1.1 Points with Plus Arc and Offset

The `icarcoff` procedure computes the coordinates of a point by traversing along an arc a specified plus distance along the arc, and then traversing a specified offset distance radially to the right or left of the arc. The arc is defined by the north and east coordinates of its center point, and a north azimuth from the said center point towards the start point from which the plus distance is to commence. Reference is made to Figure 4.1a.

```
Sub icarcoff (CCN, CCE, AZ1, R, ARC, OFF, _
              PTN, PTE, AZZ)
```

The input parameters to this procedure include the:

CCN	The north coordinate of the center point of the arc.
CCE	The east coordinates of the center point of the arc.
AZ1	The north azimuth from the center point towards the start point of the arc from which the plus distance is to be measured.
R	The radius of the arc. A positive radius denotes a clockwise rotation, while a negative radius denotes a counterclockwise rotation.
ARC	The plus arc distance from the start point of the arc to the point to be created. A positive ARC is along the arc direction as defined by the sign of the radius, and a negative ARC opposes the said direction.
OFF	The offset distance of the point to be created from the arc. A positive OFF is to the right of the arc along the arc direction as defined by the sign of the radius, and a negative OFF is to the left of the arc along the said direction. This sign convention should not be confused with certain other procedures in which the outside of the curve positive and the inside is negative.

© Springer International Publishing Switzerland 2016
E.C. Tonias, C.N. Tonias, *Geometric Procedures for Civil Engineers*, DOI 10.1007/978-3-319-24295-8_4

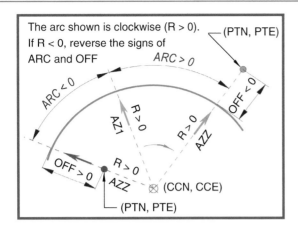

Figure 4.1a   Point with Plus and Offset along an Arc

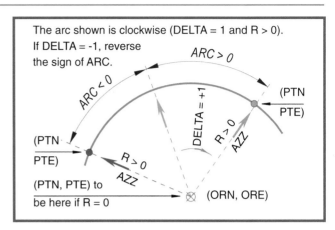

Figure 4.1b   Point an Arc with Arc Length

The parameters that are returned from this procedure include the:

PTN, PTE   The north and east coordinates of the point to be created.
AZZ   The north azimuth from the arc's center point towards PTN, PTE.

Upon invocation, the procedure initializes the parameters to be returned (PTN, PTE to be those of CCN, CCE, and AZZ to be AZ1 in case R is zero), and then:

1. Divides ARC by R to find the subtended angle which is added to AZ1 to find AZZ. Note that depending on the sign of ARC and R the said angle and AZZ could be negative, which should not present any problems in subsequent computations.
2. Calls on the icpt1 procedure (see later on) to compute the (PTN, PTE) coordinates using as plus distance R + OFF.

4.1.2   Points with Plus Arc and Direction

The icceg31 procedure computes the coordinates of a point by traversing along an arc a specified plus distance along the arc in a specified clockwise or counterclockwise direction. Reference is made to Figure 4.1b.

```
Sub icceg31  (R, ARC, AZ, DELTA, ORN, ORE, _
              PTN, PTE, AZZ).
```

R   The radius of the arc, which is intended to be positive. A negative value opposes the direction of DELTA.
ARC   The plus arc distance from the start point to the point to be created. A negative value opposes the direction of DELTA.
DELTA   An indicator the value of which it is:
        +1   Denotes a clockwise rotation, or
        −1   Denotes a counterclockwise rotation.
ORN,   The north and east coordinates of the center point of the
ORE   arc.

The parameters that are returned from this procedure include the:

PTN, PTE   The north and east coordinates of the point to be created.
AZZ   The north azimuth from the arc's center point towards PTN, PTE.

It is noted that this procedure is similar to that of the icarcoff with the following differences:

• If the radius R is zero, a point feature is created at the center of the arc. ORN, ORE and PTN, PTE become equal to each other, respectively.
• The point to be created is always along the arc. There is no provision for an offset.

Also note that if ARC or R are negative, they oppose the direction of DELTA. If they both are negative, the direction of DELTA controls.

4.1.3   Point Midway Along an Arc

The icdmidpc procedure computes the coordinates of the midpoint of a circular arc. The arc is defined by the Cartesian coordinates of its start point, end point, and center point. Reference is made to Figure 4.1c and to the commentary below regarding the endpoints of the arc.

```
Sub icdmidpc  (XS, YS, XE, YE, XC, YC, _
               XP, YP)
```

The input parameters to this procedure include the:

XS, YS   The Cartesian coordinates of the start point of the arc.
XE, YE   The Cartesian coordinates of the end point of the arc.
XC, YC   The Cartesian coordinates of the center point of the arc.

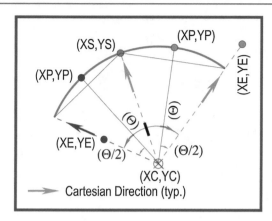

Figure 4.1c Point Midway Along an Arc

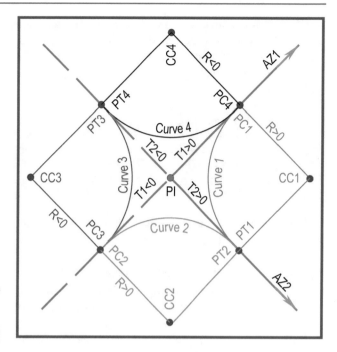

Figure 4.1d Locating the PC, PT, and CC of a Curve from the PI

The parameters that are returned from this procedure include the:

XP, YP	The Cartesian coordinates of the point to be created.

The implied line from (XC, YC) to (XS, YS) defines the radius of the arc and the direction from the center point to the start point of the arc, while the implied line from (XC, YC) to (XE, YE) defines the direction from the center point to the end point of the arc. Thus, the point (XE, YE) need not be located on the arc itself (see Figure 4.1c).

Upon invocation, the procedure initializes to zero the parameters to be returned, and then:

1. Uses the Pythagorean theorem to compute the radius of the arc and the Cartesian direction from (XC, YC) and (XS, YS) using the differences in X and Y coordinates.
2. Inverses from the center point (XC, YC) towards the second point (XE, YE) to find the direction and set the coordinates of the arc's endpoint, in case the said second point is not the actual arc endpoint.
3. Computes the arc's central angle by subtracting the first direction from the second, divides it by two and computes the Cartesian direction from the arc's center point towards its midpoint. Note that it is always determined to be less than 180°.
4. Uses the last found Cartesian direction and radius to compute the coordinates XP, YP of the arc's midpoint.

The procedure icdmidpc *assumes that the arc subtends a central angle of less than 180°.*

4.1.4 Find the PC, PT, and Center Point of a Curve

The icpcptcc procedure computes the coordinates of the PC, PT, and center point of a circular curve of a route alignment. The curve is defined by its PI coordinates, length and direction of its back and forward tangents, and the length of its radius. For this procedure to work properly, the sign convention of Figure 4.1d must be followed.

```
Sub icpcptcc  (PIN, PIE, AZ1, AZ2, T1, T2, R, _
               PCN, PCE, PTN, PTE, CCN, CCE)
```

The input parameters to this procedure include the:

PIN, PIE	The north and east coordinates of the curve's PI.
AZ1	The north azimuth of the back tangent of the curve, which is positive towards the PI.
AZ2	The north azimuth of the forward tangent of the curve, which is positive away from the PI.
T1	The length of the back tangent of the curve, which is positive in the direction of AZ1.
T2	The length of the forward tangent of the curve, which is positive in the direction of AZ1. The absolute value of T2 should equal the absolute value of T1. If not, the procedure makes it so.
R	The length of the radius of the curve, and which should be positive if the center of the arc is to the right of AZ1, and negative if to the left.

NOTE that the directions of AZ1 and AZ2 and the signs of T1 and T2 determine the quadrant in which the curve lies. A negative T1 or T2 opposes its respective direction.

The parameters that are returned from this procedure include the:

PCN, PCE	The north and east coordinates of the PC point to be created.
PTN, PTE	The north and east coordinates of the PT point to be created.
CCN, CCE	The north and east coordinates of the center point to be created.

Upon invocation, the procedure initializes to zero the parameters to be returned, and then:

1. Calls on the icptl procedure to compute the (PCN, PCE) coordinates by traversing from (PIN, PIE) along AZ1 with a plus distance T1 and an offset distance of zero.
2. Calls on the icptl procedure to compute the (PTN, PTE) coordinates by traversing from (PIN, PIE) along AZ2 with a plus distance T2 and an offset distance of zero.
3. Calls on the icptl procedure to compute the (PCC, PCC) coordinates by traversing from (PCN, PCE) along AZ1 with the given plus distance T1 and the given radius R as an offset distance.

For the subject geometric application of this procedure, the back and forwards tangent lengths should be the same. The reason for the entry of T1 and T2 is to provide the programmer the opportunity to oppose the directions AZ1 and AZ2 if need be.

4.1.5　Find the PI of a Curve of Known PC, PT, and Center Point

The iccuvhpi procedure computes the Cartesian coordinates of the point of intersection (PI) of the tangent lines at the start and end points of an arc when the Cartesian coordinates of the start point, end point, and center point of the arc are known as indicated Figure 4.1e).

```
Sub iccuvhpi  (XC, YC, XS, YS, XE, YE, _
               XPI, YPI)
```

The input parameters to this procedure include the Cartesian coordinates of the arc's:

XC, YC	Center point.
XS, YS	Start point.
XE, YE	End point.

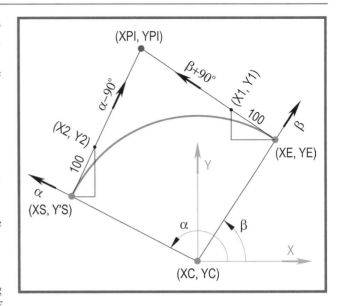

Figure 4.1e PI of a Curve of Known PC, PT, and CC

The parameters that are returned from this procedure include the:

XPI, YPI	North and east coordinates of the PI point of the arc.

Upon invocation, the procedure initializes the parameters to be returned, and then:

1. Computes the Cartesian direction α of the radial line from the arc's center point towards the arc's start point (PC), and then subtracts 90° (π/2) to determine the Cartesian direction of the back tangent line at the PC. This direction would point:
   • Towards the PI for a clockwise arc, and
   • Away from the PI for a counterclockwise arc.
2. Assumes a distance of 100 feet (meters) to locate a temporary point (X1, Y1) along the said tangent line, and computes its Cartesian coordinates as

$$X1 = XS + \left[100.0 * Cos(\alpha)\right]$$
$$and \quad Y1 = YS + \left[100.0 * Sin(\alpha)\right]$$

3. Computes the Cartesian direction β of the radial line from the arc's center point towards the arc's end point (PT), and then adds 90° (π/2) to determine the Cartesian direction of the forward tangent line at the PT. This direction would point:
   • Away from the PI for a clockwise arc, and
   • Towards the PI for a counterclockwise arc.
4. Assumes a distance of 100 feet (meters) to locate a temporary point (X2, Y2) along the said tangent line, and computes its Cartesian coordinates as

$$X2 = XE + \left[100.0 * Cos(\beta)\right]$$
$$and \quad Y2 = YE + \left[100.0 * Sin(\beta)\right]$$

5. Calls the `icln2int` procedure (see Chap. 6) to compute the Cartesian coordinates of the PI (`XPI`, `YPI`) by intersecting the back tangent defined between (`XS`, `YS`) and (X1, Y1) with the forward tangent defined between (`XE`, `YE`) and (X2, Y2), and terminates. In intersecting two lines, the direction of the two line towards or away of the point of intersection is of no consequence.

4.2 Points Along a Line

Presented in this section are the procedures that generate points with plus and offset measures along a two-point line. Note that some procedures utilize Cartesian coordinates in their arguments, while others utilize north and east coordinates.

In order to use these procedures on a polyline, the polyline needs to be first disassembled into its individual two-point line segments. For this, reference is made to the `icprjct2` procedure presented in the next section of this chapter.

4.2.1 Find the Midpoint of a Cartesian Line

The `iccommid` procedure computes the Cartesian coordinates of a point located midway between two given points of known Cartesian coordinates.

```
Sub iccommid  (X1, Y1, X2, Y2, _
               XM, YM)
```

The input parameters to this procedure include the:

X1, Y1	The Cartesian coordinates of the first known point.
X2, Y2	The Cartesian coordinates of the second known point.

The parameters that are returned from this procedure include the:

XM, YM	The Cartesian coordinates of the midpoint between the two known points.

The procedure computes the coordinates of the midpoint by halving the coordinates of the two given points.

4.2.2 Point Along a Cartesian Line

The `icdblpao` procedure computes the Cartesian coordinates of a point by traversing along a given baseline a specified distance and a specified offset therefrom. The baseline is

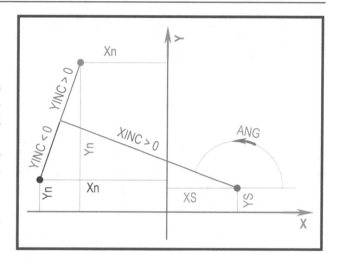

Figure 4.2a Point with Plus and Offset along a Baseline in a Cartesian Plane

defined by the Cartesian coordinates of its start point and a Cartesian direction angle (see Figure 4.2a).

```
Sub icdblpao  (XS, YS, XINC, YINC, ANG, _
               Xn, Yn)
```

The input parameters to this procedure include the:

XS, YS	Cartesian coordinates of the start point of the given baseline. This point may be located in any quadrant.
XINC	Plus distance measured along the baseline. • If it is positive, the measure is along the specified direction of the baseline. • If it is negative, the measure opposes the direction of the baseline.
YINC	Offset distance from the baseline. • If it is positive, the measure is to the right of the baseline. • If it is negative, the measure is to the left of the baseline.
ANG	The Cartesian direction angle, in radians, of the baseline. If the value is: • Positive, the measure is counterclockwise. • Negative, the measure is clockwise.

The parameters that are returned from this procedure include the:

Xn, Yn	The Cartesian coordinates of the new point.

Upon invocation, the procedure initializes to zero the parameters to be returned, and then:

1. Determines the direction of the normal to the baseline by subtracting $\pi/2$ from `ANG`.
2. Computes the X coordinate of the new point by multiplying `XINC` by the cosine of `ANG` and `YINC` by the cosine of the said normal and adding the products to `XS`.

3. Computes the Y coordinate of the new point by multiplying `YINC` by the sine of `ANG` and `YINC` by the sine of the said normal and adding the products to `YS`.

4.2.3   Point Along a Two-Point Line

The `icptl` procedure computes the north and east coordinates of a point by traversing along a baseline a specified distance and offset therefrom. The baseline is defined by the north and east coordinates of its start point and a north azimuth as its direction (see Figure 4.2b).

```
Sub icptl  (PTN1, PTE1, AZ1, D, OFF, _
            PTN3, PTE3)
```

The input parameters to this procedure include the:

PTN1, PTE1	The north and east coordinates of the start point of the given baseline.
AZ1	The north azimuth direction of the baseline.
D	The given plus distance measured along the baseline. If the value is: • Positive, the measure is along the specified direction of the baseline. • Negative, the measure opposes the direction of the baseline.
OFF	The given offset distance from the baseline. If the value is: • Positive, the measure is to the right of the baseline. • Negative, the measure is to the left of the baseline.

The parameters that are returned from this procedure include the:

PTN3, PTE3	The north and east coordinates of the new point.

Upon invocation, the procedure operates in a manner similar to that of the `icdblpao` procedure.

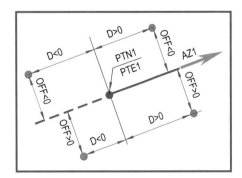

Figure 4.2b Points with Plus and Offset along a Line

4.3   Point Projection on an Arc

Addressed below are the procedures that determine the coordinates of the point of the normal (radial) projection of a given point on a given circular arc. The subroutines in this section utilize north and east coordinates

4.3.1   Point Projection on an Arc with Stations

The `icceg17` procedure projects a given point radially on a given circular arc as indicated in Figure 4.3a.

```
Sub icceg17  (ORN, ORE, AZZ, SCST, N2, E2, R,_
              N1, E1, STAT1, AZ, OFF)
```

The input parameters to these procedures include the:

ORN, ORE	The north and east coordinates of the center point of the given arc.
AZZ	The north azimuth from the said center point towards a reference point on the arc, which point may be a real or implied point.
SCST	The station value of the said reference point. Could be zero or negative (see STAT1).
N2, E2	The north and east coordinates of the given point to be projected on the arc.
R	The arc's radius the value of which should be: • Positive if the arc is to advance in a clockwise rotation. • Negative if the arc is to advance in a counterclockwise rotation.

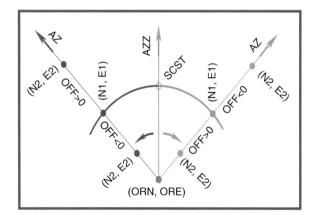

Figure 4.3a Point Projection on a Circular Arc

The parameters that are returned from this procedure include the:

N1, E1	The north and east coordinates of the point of projection.
STAT1	The station value of the point of projection with respect to the reference station (SCST). Note that STAT1 is measured always ahead of SCST in the direction of the rotation (clockwise or counterclockwise). If the point of projection creates a central angle greater than π, and if a backwards station is desired, STAT1 should be subtracted from the circumference of the circle by the calling program.
AZ	The north azimuth from the center point of the arc to the point of projection.
OFF	The radial offset distance of the given point from the arc, the value of which is: • Positive if the given point is to the right of the arc, and • Negative if to the left in the direction of the arc's advance.

Upon invocation, the procedure initialize to zero the parameters to be returned, and then:

1. Calls the icforce procedure to determine the azimuth and the distance from the center point of the arc (ORN, ORE) towards the given point (N2, E2).
2. Compute the offset distance OFF of the given point from the arc taking in account the sign of the radius R.
3. Calls the icptl procedure to determine the north and east coordinates of the point of projection (N1, E1).
4. Calls the iccenang procedure to compute the central angle formed from the reference line toward the azimuth from the center point towards the point to be projected taking into account the sign of the radius R, and then computes the value of the station of the point of projection.

4.3.2 Point Projection on an Arc without Stations

The icprjcrc procedure projects a given point radially on a given circular arc in a manner similar to that of the icceg17 procedure with the difference that the icprjcrc procedure does not require the input of a reference station, nor does it return a station for the point of projection.

```
Sub icprjcrc (CCN, CCE, R, PTKN, PTKE, _
              PTLN, PTLE, AZ, OFFSET)
```

The input parameters to this procedure include the:

CCN, CCE	The north and east coordinates of the center point of the given arc.
R	The arc's radius the value of which should be: • Positive if the arc is to advance in a clockwise rotation. • Negative if the arc is to advance in a counterclockwise rotation.
PTKN, PTKE	The north and east coordinates of the given point to be projected on the arc.

The parameters that are returned from this procedure include the:

PTLN, PTLE	The north and east coordinates of the point of projection.
OFFSET	The radial offset distance of the given point from the arc, the value of which is: • Positive if the given point is to the right of the arc, and • Negative if to the left in the direction of the arc's advance.

Upon invocation, the procedures performs the first three steps of the icceg17 and icceg17B procedures.

4.4 Point Projection on a line

Included below are procedures that compute the coordinates of the point of projection of a given point on a two-point line. Some of these procedures utilize Cartesian coordinates while others utilize north and east coordinates. Of particular interest is the subroutine icprjct2 as its function, presented later on in this section, includes more than the mere projection of a point on line. In addition to projecting a given point on a two-point line, this procedure may be used to project a point radially on a circular arc, as well as on a polyline, and on a polygon.

4.4.1 Point Projection on a Line

The icprjct procedure projects a given point of north and east coordinates on a two point line defined by the north and east coordinates of its start point and its north azimuth direction (see Figure 4.4a).

```
Sub icprjct (PTN1, PTE1, AZ1, PTN2, PTE2, _
             PLUS, OFF, PTN3, PTE3, ANG, D)
```

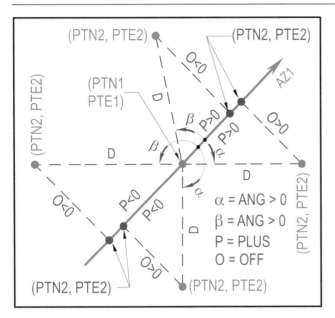

Figure 4.4a   Points Projection on a Line

The input parameters to this procedure include the:

PTN1, PTE1	The north and east coordinates of the start (base) point of the baseline.
AZ1	The north azimuth of the baseline.
PTN2, PTE2	The north and east coordinates of the given point to be projected.

The parameters that are returned from this procedure include the:

PLUS	The plus distance from the point of projection from the base point. • A negative value denotes that the given point lies back of the base point. • A positive value denotes that the given point lies ahead of the base point.
OFF	The offset distance of the given point from the baseline. • A negative value denotes that the given point lies to the left of the baseline. • A positive value denotes that the given point lies to the right of the baseline.
PTN3, PTE3	The north and east coordinates of the point of projection.
ANG	The angle in radians formed by a rotation from AZ1 to the line from the base point (PTN1, PTE1) towards the given point (PTN2, PTE2). • A negative value denotes a counterclockwise angle. • A positive value denotes a clockwise angle.
AZ	The north azimuth in radians from the base point towards the given point.
D	The distance, always positive, from the base point to the given point.

Upon invocation, the procedure initializes to zero the parameters to be returned, and then the procedure:

1. Calls the `icforce` procedure to determine the azimuth AZ and the distance D from the base point (PTN1, PTE1) to the given point (PTN2, PTE2).
2. Checks to determine whether the said distance D is equal to zero within a tolerance of 0.005 feet (1.52 mm), and if so sets the coordinates of the point of projection (PTN3, PTE3) to be those of the base point (PTN1, PTE1), and terminates.
3. Checks to determine whether the said azimuth AZ and the baseline azimuth AZ1 are equal within a tolerance of 0.000001 radians (0°00′00.2″), and if so the angle ANG is set to be zero; otherwise AZ1 is subtracted from AZ to determine ANG.
4. Computes the values of the plus (PLUS) and offset (OFF) distance using the distance D and the cosine and sine functions of ANG, respectively.
5. Recomputes the angle ANG taking into account whether the two azimuths straddle north or not.
6. Calls the `icptl` procedure to compute the coordinates of the point of projection (PTN3, PTE3) by traversing along the baseline with computed plus distance PLUS and an offset distance of zero.

4.4.2   Point Projection on a Feature

The `icprjct2` procedure performs certain operations depending on the value of the input `ipmode` parameter which is composed of 12 unique number codes to be described later on in this subsection. This procedure operates on:

* A group of collections that define a series of concatenated two-point lines and/or circular arcs such as one composed of the sides of a parcel of land. Refer to the input parameters identified below and Figure 4.4b.
* A feature that could be a multipart polyline, or polygon. Since special curves such as spirals, ellipses, quadratics, and cubics are saved as polylines of very small chords, they could be used with this procedure. Note that for a multipart feature, only the first part is processed; all other parts are disregarded.

The icprjct2 procedure provides more than point projection functionality. Read the use of the ipmode parameter.

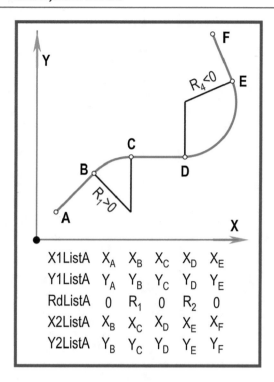

Figure 4.4b Collection Contents of Concatenated Lines and/or Arcs

```
Sub icprjct2   (X1ListA, Y1ListA, RdListA,
                X2ListA, Y2ListA, elmTheme,
                elmRecrd, _
                xPt, yPt, ipmode, _
                XA, YA, RA, XB, YB, cpls,
                off1, xprj, yprj, clen,
                X1List, Y1List, RdList,
                X2List, Y2List, jIdx, Match,
                pl1List, pl2List, pl3List,
                plShap1 As esriGeometry.
                IGeometry, _
                plShap2 As esriGeometry.
                IGeometry, _
                plShap3 As esriGeometry.
                IGeometry, plDist1, plDist2,
                plDist3)
```

The input parameters to this procedure are indicated below. Their contents should be assessed in conjunction with subsequent comments and Table 4.1.

X1ListA, Y1ListA	The collections of the Cartesian X and Y coordinates, respectively, of the start point of a series of concatenated two-point lines and/or circular arcs if ipmode is 1, 5, 6, 7, 15, or 16. For all other ipmode values these collections should contain zeros.
RdListA	The collection of the radii of a series of concatenated two-point lines and/or circular arcs if ipmode is 1, 5, 6, 7, 15, or 16. For all other ipmode values these collections should contain zeros. Note that the radius of a two-point line is represented by a zero value.

X2ListA, Y2ListA	The collections of the Cartesian X and Y coordinates, respectively, of the end point of a series of concatenated two-point lines and/or circular arcs if ipmode is 1, 5, 6, 7, 15, or 16. For all other ipmode values these collections should contain zeros.
elmTheme	The name of the layer of a feature to be used instead of the above identified collections if ipmode is 2, 3, 4, 12, 14, or 22. For all other ipmode values this parameter should contain a blank space.
elmRecrd	The record number in the abovesaid layer of a feature to be used instead of the above identified collections if ipmode is 2, 3, 4, 12, 14, or 22. For all other ipmode values this parameter should contain a zero. Note that record numbering commences at zero (0) and not at one (1).
xPt, yPt	When the value of ipmode is: • 2, 4, 7, 12, 14, or 22 these parameters contain the east and north coordinates of the point to be projected on the string of concatenated lines and/or arcs. • 6 or 16 the value of xPt is used as a plus distance to locate a point along the string of concatenated lines and/or arcs. For this case the value of yPt is not used. It is disregarded. • 1, 5, or 15 the point that is projected is the point that was last selected in the graphic display by the user and resides in the global memory of the control program.
ipmode	The mode of operation or task to be performed by the procedure (refer to the commentary below, as well as to Table 4.1 and Figure 4.4c).

The parameters that are returned from this procedures include the:

XA, YA	The north and east coordinates of the start point of the line segment within which the point of projection lies.
R	The radius of the line segment within which the point of projection is located.
XB, YB	The north and east coordinates of the end point of the line segment within which the point of projection lies.
cpls	The distance along the series of concatenated line segments from the start of the series to the point of projection.
off1	The offset distance of the point to be projected from the point of projection.
xprj, yprj	The north and east coordinates of the point of projection, or those of the point to be created with the xPt distance under ipmode of 6 or 16.
X1List, Y1List	The collections of the Cartesian X and Y coordinates, respectively, of the start point of a series of concatenated two-point lines generated from an input feature when ipmode is 2, 3, 4, 12, 14, or 22. For all other ipmode values these collections should contain −9999.
RdList	The collection of the radii of a series of the concatenated two-point lines generated from an input feature when ipmode is 2, 3, 4, 12, 14, or 22. Since a feature is always treated as a polyline, each of its segments is a two-point line. Hence, this collection is always populated with zeros.

Table 4.1 The Function of ipmode and opmode Codes

| ipmode | opmode | | | |
	1	2	3	4
1	Read List Project Last			
2		Read Shape Project Point No Extrapolation		
3			Read Shape	
4				Read Shape Use xPt as Plus
5	Read List Project Last No Extrapolation			
6				Read List Use xPt as Plus
7	Read List Project Point			
12		Read Shape Project Point		
14				Read Shape Project Point No Progress Bar
15	Read List Project Last No Extrapolation No Progress Bar			
16				Read List Use xPt as Plus No Progress Bar
22		Read Shape Project Point Create 3 Shapes		

Legend

Read List	Use X1ListA, Y1List, RdListA, X2ListA, Y2ListA parameters
Read Shape	Use elmTheme, elmRecrd parameters and force a closed figure to progress in a counterclockwise direction
Project Last	Project the last selected point
Project Point	Project the point defined by xPt, yPt parameters
Use xPt as Plus	Use the xPt parameter as a plus distance to create a point
No Extrapolation	Use projection point within the limits of the line and do not extrapolate
No Progress Bar	Do not display a progress bar during the processing
Create 3 Shapes	Create the 3 shapes defined by the last 9 parameters

For a multipart feature, only the first part is processed. All other parts are disregarded.

Certain ipmode codes have certain commonalities and for this they are grouped under the opmode codes shown on the right.

ipmode	opmode
1	1
2	2
3	3
4	4
5	1
6	4
7	1
12	2
14	4
15	1
16	4
22	2

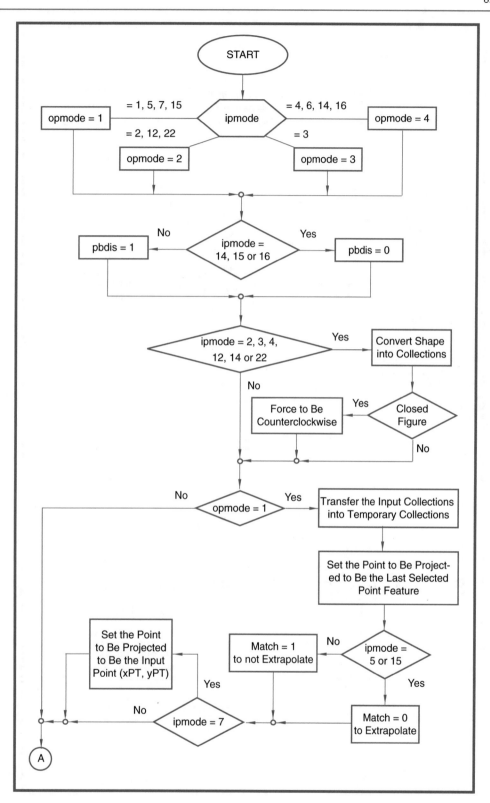

Figure 4.4c   Macro Flowchart for the icprjct2 Procedure

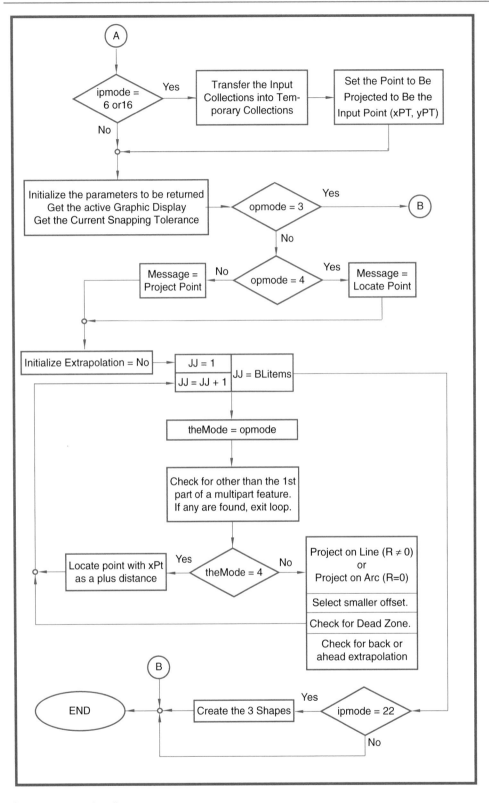

For the ipmode codes that expect to use feature shapes transfer the input collections into temporary collections to preserve the input values, allow for extrapolation, and set the point to be projected.

For the ipmode code of 3 terminate the processing.

Set the progress bar title label.

Process through the various segments of the polyline to select the point of projection yielding the smaller offset, checking for a potential dead zone, or create a point with a plus distance being the value of the xPt input parameter.

Create the three feature shapes if the ipmode code is 22, and terminate the procedure.

Figure 4.4c (continued)

X2List, Y2List	The collections of the Cartesian X and Y coordinates, respectively, of the end point of a series of concatenated two-point lines generated from an input feature when `ipmode` is 2, 3, 4, 12, 14, or 22. For all other `ipmode` values these collections should contain −9999.
jIdx	The segment number of the series of concatenated lines and/or arcs within which the point of projection is located.
Match	An indicator the value of which if it is: 0 Denotes an open shape. 1 Denotes a closed shape. This is pertinent only for opmode codes that read a feature shape and not a collection of coordinates of concatenated lines and/or arcs.
pl1List	The collection of X and Y coordinates of the various points comprising a feature from the start point up to the point of projection when an input feature is operated with the value of `ipmode` being 22.
pl2List	The collection of X and Y coordinates of the various points comprising a feature from the point of projection up to the end point when an input feature is operated with the value of `ipmode` being 22.
pl3List	The collection of X and Y coordinates of all points comprising a feature including the point of projection when an input feature is operated with the value of `ipmode` being 22.
plShap1	The ArcGIS shape of the feature represented by the `pl1List` collection.
plShap2	The ArcGIS shape of the feature represented by the `pl2List` collection.
plShap3	The ArcGIS shape of the feature represented by the `pl3List` collection.
plDist1	The total length of the feature represented by the `pl1List` collection.
plDist2	The total length of the feature represented by the `pl2List` collection.
plDist3	The total length of the feature represented by the `pl3List` collection.

As seen from its calling parameters, this procedure is specifically tailored towards ArcGIS and its ArcObjects component. For any other computing environment, the arguments plShap1, plShap2, plShap3 and the statements of the procedure as included in the accompanying CD will need to be modified. For the same reason, it may be necessary to also modify the procedure statements that used to display a progress bar for certain operations of this procedure.

The writing of a computer program is quite similar to that of writing a technical report. There are three essential aspects of such writing: (a) spelling, (b) syntax, and (c) organization of thought to make the content understandable. Regarding the first two, the program has no choice but to be 100 % correct, because the computer is completely unforgiving for any errors. Many a book has been written with ideas, rules and instructions on how to write a good program, but as many a programmer may attest, at times and particularly for expedience, sometimes it is easier to quickly modify an existing

procedure to add a new "minor" addition than to rethink the overall concept of the procedure and plan for the future. As time goes on, a minor addition becomes two, two become three, and so on. This is the case for this subject procedure. To make it more understandable we provide the following information.

This procedure is governed by the value of `ipmode` as indicated in Table 4.1. Certain `ipmode` code numbers have certain commonalities, and are thus grouped under an `opmode` code number as indicated in the said table. The reader's attention is also directed to the:

- Description of the xPt and yPt input parameters presented above regarding the point to be projected on the specified collection of concatenated lines and/or arcs, or selected feature shape.
- Use of the xPt input parameter to create a point along a polyline rather than projecting a point thereon.
- Use of the ipmode code of 3 which enables a programmer to convert a feature shape into a collection of concatenated lines and/or arcs.

One other issue to be addressed is that of the so called dead zone. Points A, B, C, and D in Figure 4.4d represent any four consecutive vertices of a polyline defined by a collection of concatenated lines and/or arcs or feature shapes, while points E, F, and G represent three points that are to be projected upon the said polyline. Each of the last said three points produces at least two projection points upon the polyline. Since the point of projection to be returned must be located within the physical extent of a segment of the polyline, the point that is returned in the xprj and yprj parameters is:

- The one that produces the first smallest offset distance for point E. Note that if the polyline is extended beyond point D and looped around so that point E could project on any

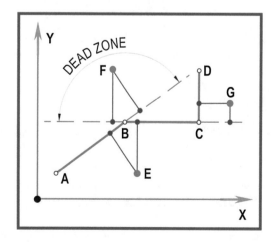

Figure 4.4d The Dead Zone Area

other segment or segments, the returned point would be the one that is located on segment AB or BC, whichever produces the smallest offset, because they encounter the first valid projection.

- The coordinates of point B for point F since both points of projection lie outside the physical extent of both segments AB and BC.
- The one projected on segment CD for point G. In this figure, point G also falls within a dead zone, but it does produce a point of projection that lies within the limits of a segment of the polyline, and thus, its projection point on segment CD is selected.

The logic of the `icprjct2` procedure is strictly that of bookkeeping in interpreting the input and output parameters. Whatever geometric operations are carried out are performed by calling the previously described procedures `iccompp1`, `icforce`, `icprjct`, `icptl`, and `icceg17`. Presented in Figure 4.4c is a macro flowchart that describes the general logic of the procedure.

4.4.3 Point Projection on a Line Implied by Two Points

The `icprojpt` procedure enables the programmer to project a point on a line implied by two pairs of Cartesian coordinates. In essence this is a reduced form of the `icprjct` procedure presented earlier in this chapter

```
Sub icprojpt  (XPT, YPT, XA, YA, XB, YB, _
               Xnew, Ynew)
```

The input parameters to this procedure are indicated below.

XPT, YPT	The Cartesian X and Y coordinates of the point to be projected.
XA, YA	The Cartesian X and Y coordinates of the start point of the implied line.
XB, YB	The Cartesian X and Y coordinates of the end point of the implied line.

The parameters that are returned from this procedures include the:

Xnew, Ynew	The Cartesian X and Y coordinates of the point of projection.

4.5 Other Point and Line Procedures

The procedures presented in this section of this chapter provide certain miscellaneous functionality regarding points and lines.

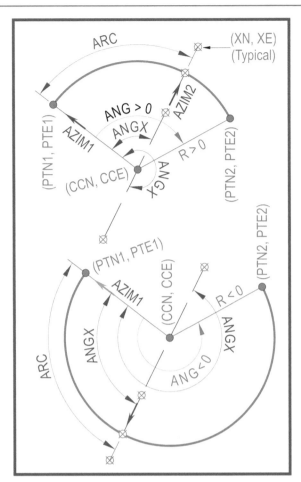

Figure 4.5a Query if a Point Lies on an Arc

4.5.1 Point Lies, or not Within an Arc

The `icbtwarc` procedure enables the programmer to determine whether a given point of north and east coordinates is located along and within the extent of a circular arc. The point in question need not be located on the given arc. It could be located on either side of the arc. The program determines whether the radial line from the arc's center towards the point in question is located within the limits of the arc's central angle (see Figure 4.5a), which could be more than 180°, and could straddle north.

```
Sub icbtwarc  (PTN1, PTE1, XN, XE, PTN2, PTE2,
               R, ANG, CCN, CCE, _
               INTR, XNO, XEO, ARC, AZIM1,
               AZIM2, ANGX)
```

The input parameters to this procedure are indicated below.

PTN1, PTE1	The arc's north and east start coordinates.
XN, XE	The north and east coordinates of the point in question.

PTN2, PTE2	The arc's north and east end coordinates.
R	The radius of the arc which is: • Positive for a clockwise arc rotation, and • Negative for a counterclockwise arc rotation.
ANG	The arc's central angle in radians, which is: • Positive for a clockwise arc rotation, and • Negative for a counterclockwise arc rotation. ANG can exceed 180°. Refer to the operational steps below regarding the case in which the signs of R and ANG oppose each other.
CCN, CCE	The north and east coordinates of the center point of the arc.

The parameters that are returned from this procedures include the:

INTR	An indicator the value of which identifies the location of the point in question with regard to the extent of the arc as follows: **0** The point is the same as one of the arc's endpoints. **1** The point is within the extent of the arc. **2** The point is outside the extent of the arc.
XNO, XEO	The north and east coordinates of the point in question which may be reset as stated later on in this subsection.
ARC	The arc distance (always returned as positive) from the arc's start point (PTN1, PTE1) to the projection point on the arc of the point in question (Xn, XE).
AZIM1	The north azimuth from the arc's center point (CCN, CCE) towards the arc's start point (PTN1, PTE1).
AZIM2	The north azimuth from the arc's center point (CCN, CCE) towards the point in question (Xn, XE).
ANGX	The central angle from AZIM1 towards AZIM2, which is measured in the direction of the arc, is always positive, and may exceed 180°.

Upon invocation, the procedure initializes the parameters to be returned, as well as the values of certain constants, and then:

1. Calls the icforce procedure to determine the azimuth from the center point of the arc (CCN, CCE) to the start point (PTN1, PTE1) of the arc.
2. Queries the sign of the central angle ANG, and forces the signs of ANG and R to be in accord with Table 4.2.
3. Calls the icvecang procedure to determine the central angle ANGX formed from an endpoint of the arc to the point in question (XN, XE). The angle is measured counterclockwise and is positive. If the radius is:
   • Positive, the angle is measured from the end endpoint (PTN1, PTE1) towards the point in question (XN, XE).

Table 4.2 Sign Settings of the Radius and Central Angle

Value	Parameter	Sign Conditions			
Input	ANG	<0	<0	>0	>0
	R	<0	>0	<0	>0
Setting	ANG	>0	>0	>0	>0
	R	<0	<0	<0	>0

• Negative, the angle is measured from the start endpoint (PTN2, PTE2) towards the point in question (XN, XE).
4. Checks the angles below to determine the value of INTR:
   • If ANG=2π within a tolerance of 0.000001 radians (0.2 s), INTR=1.
   • If ANGX of Step 3 is zero within a tolerance of 0.0000048 radians (0.99 seconds), INTR=1, and ANGX=0.0. The value of INTR is reset later to be zero (0).
   • If ANGX=ANG within a tolerance of 0.0000048 radians (0.99 s), INTR=1, and ANGX=ANG.
   • If ANGX<ANG, INTR=1, and if it is greater, INTR=2.
5. Calls the icforce procedure to find the azimuth from the arc's center point to the point in question and compute the coordinates of the point of projection on the arc and the offset distance of the point in question.
6. Checks to see (a) if the point of question has been found in Step 4 to be within the range of the central angle of the arc, and (b) if the offset distance of the said point from the arc is greater than zero, within a tolerance of 0.005 feet (1.524 mm), and if so changes the value of the indicator INTR from 1 to 2. Otherwise, if the point in question (XN, XE) lies within the span of the arc, the coordinates of the point to be returned (XNO, XEO) are set to be those of the said point of projection.
7. Calls the iccomdis procedure to compare the point of question (XN, XE) with each of the start (PTN1, PTE1) and end (PTN2, PTE2) points of the arc, and if a match in both coordinates is made, the coordinates of the point to be returned (XNO, XEO) are set to be those of the end point with which the match has been made.
8. Uses the sign of the radius (R) to force the central angle as computed under Step 3 to be from the start point of the arc towards the point in question.

4.5.2   Point Lies, or not Within a Line or an Arc

The icbtween procedure enables the programmer to determine whether a given point of north and east coordinates is located on and within the extent of a two-point line, or on and within the extent of a circular arc, provided the arc does not exceed 180°.

```
Sub icbtween  (PTN1, PTE1, XN, XE, PTN2,
                       PTE2, R,  _
                       INTR, XNO, XEO)
```

The input parameters to this procedure are indicated below.

PTN1, PTE1	The north and east coordinates of the start point of the line, or arc.
XN, XE	The north and east coordinates of the point in question.
PTN2, PTE2	The north and east coordinates of the end point of the line, or arc.
R	The radius which is equal to zero (0.00) for a line, and positive for a clockwise arc and negative for a counterclockwise arc.

The parameters that are returned from this procedures include the:

INTR	An indicator the value of which identifies the location of the point in question with regard to the extent of the arc as follows:
	0 The point is the same as one of the arc's endpoints.
	1 The point is within the extent of the arc.
	2 The point is outside the extent of the arc.
XNO, XEO	The input north and east coordinates of the point in question which are set to be those of an endpoint if within a tolerance of 0.005 feet (1.524 mm).

Upon invocation, the procedure initializes the variables to be returned, and then:

1. Determines if the point in question (XN, XE) happens to have the same coordinates, within a tolerance of 0.005 feet (1.524 mm), with those of either of the two given endpoints (PTN1, PTE1) and (PTN2, PTE2). If a match:
   - Is made, INTR is set to be zero (0) and the procedure terminates.
   - Is not made, the procedure continues.
2. Queries the value of the input radius (R) to identify whether the point in question is supposed to be located along a line or arc.

In the case of a line (R=0), the procedure:

3a Calls the icforce procedure to determine the azimuths implied between the two given points (PTN1, PTE1) and (PTN2, PTE2), and their reverse, as well as the azimuth between (PTN1, PTE1) and (XN, XE). The reverse azimuth is determined in case the point in question lies back of the first given endpoint. If the three azimuths, within a tolerance of 0.0000048 radians (0°00′00.99″):
   - Are not the same (they are not collinear) the point in question is considered to be off the line, INTR is set to be 2, and the procedure terminates.

   - Are the same (they are collinear) the procedure proceeds to determine whether the point in question lies between the two given endpoints.
4a Creates a rectangle with the two given endpoints being diagonally opposite by determining the maximum and minimum coordinates between the said endpoints.
5a Checks to determine whether each of the north and east coordinates of the point in question is less than the minimum coordinate, or greater than the maximum corresponding coordinate, in both cases as computed above. A positive response to both questions indicates that the point in question is not between the two endpoints and INTR=2, while a negative response indicates the opposite and INTR remains as 1.

Step 5a terminates the procedure.
In the case of an arc (R<>0), the procedure:

3b. Calls the icarccnt procedure (see Chap. 5) to locate the center point of the arc.
4b. Calls the icforce procedure determine the:
   - Azimuths and radius from the said center point towards each of the arc's two endpoints (PTN1, PTE1) and (PTN2, PTE2), and the
   - Azimuth and distance from the said center point towards the point in question.
5b. Checks the last of the said three azimuths for being collinear with either of the other two azimuths within a tolerance of 0.0000048 radians (0.99 s), and if so sets the third azimuth to be that with which a match has been made.
6b. Interchanges the azimuths towards the two endpoints as computed under Step 4b only if the arc is counterclockwise (R<0).
7b. Rotates the three azimuths as computed under Step 4b and as may have been modified under Step 6b, so that the first azimuth is oriented due north. Note that this presumes that the arc subtends an angle of less than 180°. If this angle is greater than 180°, the point in question will be found to be outside the arc.
8b. Checks to see if the azimuth from the arc's center point towards the point in question lies between the other two azimuths.
   - If so, the procedure checks to see if the distance from the arc's center point to the point in question equals the radius, both as computed under Step 4b, within a tolerance of 0.005 feet (1.524 mm), and if so sets INTR to be 1.
   - If not, the procedure sets INTR to be 2.

Step 8b terminates the procedure.

To determine whether the point in question lies on the arc or not, `icbtween` compares the radius computed in Step 4b with the distance computed from the said center point towards the point in question.

4.5.3   Rotate and Scale a Point about a Pivot Point

The `icdrot2s` procedure enables the programmer to rotate and/or scale the X and/or Y Cartesian coordinates of a point with respect to the Cartesian X and Y coordinates of a pivot point as indicated in Figure 4.5b. That is, in the said figure point (X, Y) is rotated about the pivot point (X_S, Y_S) by an angle α to the position at point (X_1, Y_1). Thereafter, the coordinates X_1, and Y_1 are multiplied by the scale factors X_{SC} and YSC, respectively to obtain the coordinates of the point (X_O, Y_O).

The above rotation of the point in question with regard to the given pivot point is made in the `icdrot2s` procedure by use of equations:

$$X_1 = X\cos(\alpha) - Y\sin(\alpha) + \left\{ X_s\left[1 - \cos(\alpha)\right] + \left[Y_s\sin(\alpha)\right] \right\} \quad (4.5a1)$$

and

$$Y_1 = X\sin(\alpha) - Y\cos(\alpha) + \left\{ Y_s\left[1 - \cos(\alpha)\right] + \left[X_s\sin(\alpha)\right] \right\} \quad (4.5a2)$$

where

X = The X coordinate of the point to be rotated
Y = The Y coordinate of the point to be rotated

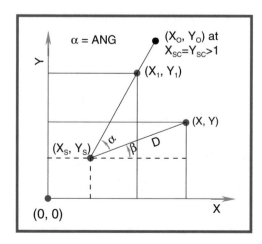

Figure 4.5b   Rotate and Scale a Point about a Pivot Point

X_1 = The X coordinate of the rotated point
Y_1 = The Y coordinate of the rotated point
X_S = The X coordinate of the pivot point
Y_S = The Y coordinate of the pivot point
and α = The angle of rotation

while the scaling of the new point coordinates is attained by use of the equations:

$$X_O = X_1 \cdot X_{SC} \quad (4.5b1)$$

and

$$Y_O = Y_1 \cdot Y_{SC} \quad (4.5b2)$$

where

X_{SC} = The scale factor along the X axis
and Y_{SC} = The scale factor along the Y axis

For the readers that may be questioning the development of Equations 4.5a and 4.5b, reference is made to the prove provided at the end of the discussion of this procedure.

```
Sub icdrot2s  (X, Y, ANG, XSC, YSC, XS, YS,_
               XO, YO)
```

The input parameters to this procedure are indicated below.

X, Y	The Cartesian X and Y coordinates of the point to be rotated and/or scaled.
ANG	The Cartesian angle of rotation in degrees expressed in decimal form with a counterclockwise rotation being positive.
XSC, YSC	The scale factors along the X and Y axes, respectively with a value of: • 1.0 implying no scaling, • Less than 1.0 implying a reduction in the X or Y coordinate, and • Greater than 1.0 implying an enlargement in the X or Y coordinate.
XS, YS	The Cartesian coordinates of the pivot point about which the rotation and/or scaling is to occur.

The parameters that are returned from this procedures include the:

XO, YO	The Cartesian coordinates of the new location of the input point to be rotated and/or scaled (X, Y).

Upon invocation, the procedure:

1. Queries the angle of rotation being equal to zero, and the scaling factors for both being equal to 1. If so, the returned coordinates are set to be those of the input, and the procedure terminates. Otherwise the procedure continues.

2. Applies Equations 4.5a and 4.5b to rotate the input point (X, Y), and then applies Equations 4.5c and 4.5d to scale the coordinates, and then terminates.

An alternate method to solving Equations 4.5a and 4.5 is to replace the statements solving these equations with the following visual basic statements:

```
' Find the distance and direction from
  (XS, YS) to point (X, Y)
  Call icforce (XS, YS, X,Y, D, AZ)
' Add/Subtract the angle of rotation to/
  from the above direction
```

$$Rot = AZ + ANG$$

```
' Compute the rotated coordinates
  Call icptl(XS, YS, Rot, D, 0#, XO, YO)
```

Note that with `icdrot2s` *the rotated coordinates are scaled with respect to the origin point of the coordinate system, and not with respect to the pivot point. If the latter is desired, each scale factor should be 1.0 and the scaling should be done by the calling program.*

If in addition to rotating a point about a pivot pint it is desirable to translate the pivot point (X_S, Y_S) to a new location (X_{NEW}, Y_{NEW}), the calling program of `icdrot2s` could determine the X and Y axis translations as

$$DeltaX = XNEW - XS$$
$$DeltaY = YNEW - YS$$

and apply DeltaX and DeltaY to *X1* and *X2*, respectively. If the calling program were to apply the call to the `icdrot2s` for numerous points at a time, the above two statements should precede the call to the said procedure and the addition of the two deltas should follow it.

Regarding the development of Equations 4.5a and 4.5b, the reader may note from Figure 4.5b that the coordinates of the point to be rotated may be expressed as

$$X = D\cos(\beta) + X_s \qquad (4.5c1)$$

and

$$Y = D\sin(\beta) + Y_s \qquad (4.5c2)$$

Similarly, the coordinates of the rotated point may be expressed as

$$X_1 = D\cos(\alpha + \beta) + X_s \qquad (4.5d1)$$

and

$$Y_1 = D\sin(\alpha + \beta) + Y_s \qquad (4.5d2)$$

By expanding the cosine and sine of the sums of $(\alpha+\beta)$ in the lat two equations we get

$$X = D\left[\cos(\alpha)\cos(\beta) - \sin(\alpha)\sin(\beta)\right] + X_s \qquad (4.5e1)$$

and

$$Y_1 = D\left[\sin(\alpha)\cos(\beta) + \cos(\alpha)\sin(\beta)\right] + Y_s. \qquad (4.5e2)$$

From Equations 4.5e and 4.5f we note that

$$D\cos(\beta) = X - X_s. \qquad (4.5f1)$$

and

$$D\sin(\beta) = Y - Y_s \qquad (4.5f2)$$

With Equations 4.5k1 and 4.5k2 in mind, we can expand Equations 4.5i and 4.j to get

$$X_1 = \left[D\cos(\beta)\right]\cos(\alpha) - [D\sin(\beta)\sin(\alpha) + X_s \qquad (4.5g1)$$

and

$$Y_1 = \left[D\cos(\beta)\right]\sin(\alpha) + \left[D\sin(\beta)\right]\cos(\alpha) + Y_S \qquad (4.5g2)$$

and then we can substitute Equations 4.5k1 and 4.5k2 in the last two equations to get

$$X_1 = (X - X_s)\cos(\alpha) - (Y - Y_s)\sin(\alpha) + X_s \qquad (4.5h1)$$

and

$$Y_1 = (X - X_s)\sin(\alpha) + (Y - Y_s)\cos(\alpha) + Y_s \qquad (4.5h2)$$

By expanding the variables within the parentheses in the last two equations we get

$$X_1 = X\cos(\alpha) - X_S\cos(\alpha) - Y\sin(\alpha) + Y_S\sin(\alpha) + X_S \qquad (4.5i1)$$

and

$$Y_1 = X\sin(\alpha) - X_S\sin(\alpha) + Y\cos(\alpha) - Y_S\cos(\alpha) + Y_S \qquad (4.5i2)$$

By rearranging the terms of the last two equations we get

$$X_1 = X\cos(\alpha) - Y\sin(\alpha) - X_S\cos(\alpha) + X_S + Y_S\sin(\alpha) \qquad (4.5j1)$$

and

$$Y_1 = X\cos(\alpha) + Y\cos(\alpha) + \{Y_S[1-\cos(\alpha)] - X_S\sin(\alpha)\} \qquad (4.5j2)$$

Lastly, by rearranging the three right-most terms of the last two equations we get

$$X_1 = X\cos(\alpha) - Y\sin(\alpha) \\ + \{X_S[1-\cos(\alpha)] + Y_S\sin(\alpha)\} \quad (4.5k1)$$

and

$$Y_1 = X\cos(\alpha) + Y\cos(\alpha) \\ + \{Y_S[1-\cos(\alpha)] - X_S\sin(\alpha)\} \quad (4.5k2)$$

which are the same as Equations 4.5a and 4.5b, respectively. In the last two equations, note that the first two terms pertain to the rotation of the point in question as if the origin of the axes is at the pivot point (X_S, Y_S), and the term within the braces {} pertains to the position of the pivot point with respect to the actual origin of the two axes.

4.5.4   Implicit Form of a Line Between Two Points

The `icimplct` procedure determines the coefficients A, B, and C of the equation of the line between two points represented as

$$Ax + By + C = 0 \quad (4.5.4a)$$

in which the constants A, B, and C can not all be zero for obvious reasons. Two points of coordinates $(X1, Y1)$ and $(X2, Y2)$ are shown in Figure 4.5c. Since each one of these two points lies on the same line, we have the two new equations

$$Ax_1 + By_1 + C = 0 \quad (4.5.4b)$$

and

$$Ax_2 + By_2 + C = 0 \quad (4.5.4c)$$

The above three equations may be solved simultaneously with the determinant

$$\begin{bmatrix} x & y & 1 \\ x_1 & y_1 & 1 \\ x_2 & y_2 & 1 \end{bmatrix} = 0 \quad (4.5.4d)$$

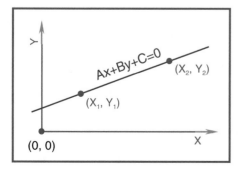

Figure 4.5c Line Equation Between Two Points

In solving this determinant we see that

$$\begin{bmatrix} x & y & 1 \\ x_1 & y_1 & 1 \\ x_2 & y_2 & 1 \end{bmatrix} = 0 \quad (4.5.4e)$$

Once the coefficients A, B, and C have been computed and returned to the calling program, the programmer may use Equation **4.5.4a** in other procedures as may be necessary to represent a line implied by two known coordinates.

This procedure may also be used to determine if a point lies on a line. Once the equation of a line has been determined, the programmer may solve this equation using the (x, y) coordinates of any other point to determine whether this new point is located on the said line or not. Let us rewrite Equation **4.5.4a** as

$$Ax + By + C = v \quad (4.5.4f)$$

in which A, B, and C are the coefficients of the line and x and y are the coordinates of the point in question. The resultant v may now be queried for being zero, positive, or negative. If its value is:

- Negative, then the point in question is located to the right of the given line.
- Positive, then the point in question is located to the left of the implied line.
- Zero, then the point in question is located on the implied line.

Depending on the precision that may be required when testing for the above, the value of v may need to be tested for zero within some dimensionless tolerance, say 0.005.

```
Sub icimplct  (X1, Y1, X2, Y2, _
               A, B, C)
```

The input parameters to this procedure are indicated below.

X1, Y1	The Cartesian X and Y coordinates of the first point on the implied line.
X2, Y2	The Cartesian X and Y coordinates of the second point on the implied line.

The parameters that are returned from this procedures include the:

A, B, C	The coefficients A, B, and C of the line implied from (X1, Y1) towards (X2, Y2).

Upon invocation, the procedure queries the coordinates of the two input points to determine whether the said points form a vertical, horizontal, or inclined line with respect to the X–Y axes, and if so sets the appropriate value of the A, B, and C coefficients as indicated in Figures 4.5d and 4.5e, and then terminates.

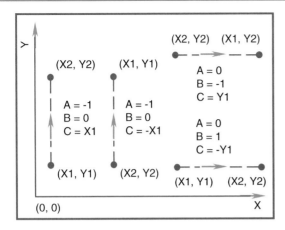

Figure 4.5d Implicit Form of Vertical and Horizontal Lines

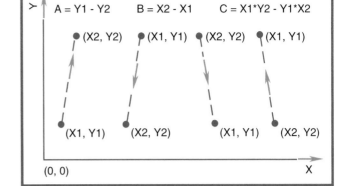

Figure 4.5e Implicit Form of Inclined Lines

Curve Features

5

The procedures addressed in this chapter are those that generate a circle or a circular arc by various construction techniques. Procedures that generate spirals and address their interaction with lines, circular arcs, and other spirals are addressed separately in Chap. 9. Similarly, procedures that address parabolic curves are presented in Chap. 11, while elliptic features are included in Chap. 12. Quadratic or cubic curves may be created by use of the `iclagr1` subroutine addressed in Chap. 3, and in which the argument `NPTDAT` should be set to 3 for a quadratic, and to 4 for a cubic curve. This procedure may also be used to create other curve shapes by increasing the number of points.

Since the predominance of the procedures of this chapter address circular arcs, the terms "curve" and "arc" are used exclusively to imply a circular arc.

The procedures of this chapter address the creation of arcs that may or may not be tangent to a given line, or to another curve. Procedures that address more complicated line–curve and curve–curve tangencies are addressed in Chap. 7. A further discussion on arcs may be found in Chap. 8 which addresses arcs tangent to each other in series, commonly referred to as compound and reversed curves, and in Chap. 10 which addresses the establishment, as well as extraction of the standard parameters of a horizontal alignment composed of tangent lines, a circular curve, and one or two spirals (entering and exiting spirals).

The procedures of this chapter are divided in three sections that contain the following constructions:

1. Procedures that create an arc tangent to one or two lines given certain arc parameters and/or control points (refer to the commentary later on in this introduction).
2. Procedures that create stand alone (non-tangent) arcs given certain arc parameters and/or control points (refer to the commentary later on in this introduction).

3. Decomposition of an arc into a series of chords for graphic display to suit user specific display quality, conversion of arcs into polylines, and extraction of potential arcs from polylines.

There are five component parameters of an arc, and they are its (a) central angle α, (b) radius R, (c) chord length C, (d) length L, and (e) middle ordinate H, all shown in Figure 5.0. Any combination of two of these parameters is able to define an arc. The 27th Edition of the "*Machinery's Handbook*" contains a table of the ten possible combinations of the said parameter pairs together with the related equations to solve for the unknown parameters. Table 5.0 presents the said pairs together with the names of the procedures presented in this chapter. These procedures are divided in two columns, those that create arcs tangent to an existing line or arc, and those that create stand alone arcs.

Circular arcs progress from a start point towards an endpoint in a clockwise or counterclockwise rotational direction.

- This direction is defined by either: (a) the sign of the radius, or (b) the sign of the central angle which is equal to the deflection angle formed by the two tangent lines at the endpoints of the arc (commonly referred to as the delta angle).
- In most procedures, but not in all, a positive radius or delta angle denotes a clockwise rotation, and a negative radius or delta angle denotes a counterclockwise rotation.
- Certain procedures that decompose a curve into a series of chords for drawing (plotting) purposes decompose a curve in a counterclockwise direction, even though if a curve has been defined in a clockwise rotation.

© Springer International Publishing Switzerland 2016
E.C. Tonias, C.N. Tonias, *Geometric Procedures for Civil Engineers*, DOI 10.1007/978-3-319-24295-8_5

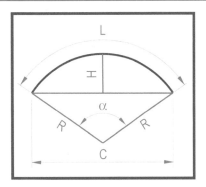

Figure 5.0 Arc Parameters

Table 5.0 Arc Parameters to Define an Arc

Given Parameters			Arc Mode	
			Tangent	Non-tangent
α	and	R	icarcan1	
α		C		icchord1
α		L		icarcAnL
α		H		icarcAnM
R		C	icarccd2	icarccd1
R		L	icarcln1	
R		H		icarcRaM
C		L	a	icarcCdL
C		H		icarcCdM
L		H	a	icarcLnM

aRequires an iterative solution

5.1   Arcs Tangent to a Line or Curve

The procedures of this section assume that the arc to be constructed is to start at a given point defined by its north and east coordinates, and be tangent to a line of known north azimuth that passes through the said point. This point could be a point on a line (POL), or a point on a curve (POC) to which curve the curve to be constructed is tangent. In the case that the POC is the known point, the known north azimuth is that of the common tangent line to both, the existing curve and the curve to be constructed; that is the two curves are to be either compound or reversed (refer to Chap. 8 for additional procedures regarding compound and reversed curves). The procedures of this section include those listed under the Tangent column of Table 5.0 and certain other procedures.

In most procedures, but not necessarily in procedures presented herein, a positive radius or delta angle denotes a clockwise rotation, and a negative radius or delta angle denotes a counterclockwise rotation.

5.1.1   Tangent Arc with Known Radius and Central Angle

The icarcan1 procedure constructs an arc that commences at a known point (POC or POL) of given north and east coordinates, and is tangent to a line of known north azimuth at the said point. The arc is of a known radius, and subtends a known central angle (refer to Figure 5.1a).

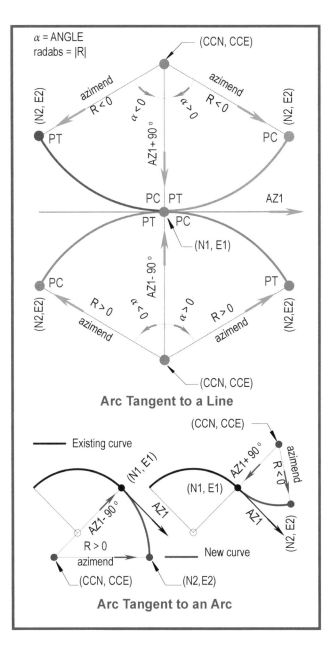

Figure 5.1a Arc of Given Radius and Central Angle Tangent to a Line or Arc

```
Sub icarcan1 (N1, E1, AZ1, R, ANGLE, _
              N2, E2, PCN, PCE, PTN, PTE, CCN,
              CCE, azend, azimPC, azimPT, _
              ARC, angleO, radabs)
```

The input parameters to this procedure include the:

N1, E1	The north and east coordinates of the known point (POC or POL) at which the tangent arc is to start.
AZ1	The north azimuth of the tangent line at the known (N1, E1) point in radians.
R	The radius of the arc to be constructed, which is positive for a clockwise rotation and negative for a counterclockwise rotation.
ANGLE	The central angle in radians of the arc to be constructed and which is to be positive if the arc is to progress in a rotational direction as indicated by the sign of the radius R, and negative if the arc is to oppose the rotational direction of the said radius for display (plotting) purposes. Reference is made to the note below.

The parameters that are returned from this procedure include the:

N2, E2	The north and east coordinates of the end point of the tangent arc to be constructed. Reference is made to the note below.
PCN, PCE	The north and east coordinates of the PC point of the tangent arc to be constructed. Reference is made to the note below.
PTN, PTE	The north and east coordinates of the PT point of the tangent arc to be constructed. Reference is made to the note below.
CCN, CCE	The north and east coordinates of the center point of the tangent arc to be constructed.
azend	The north azimuth from the center point (CCN, CCE) of the constructed arc towards the said arc's PC point (N2, E2). Reference is made to the note below.
azimPC	The north azimuth from the center point (CCN, CCE) of the constructed arc towards the said arc's end point (PCN, PCE). Reference is made to the note below.
azimPT	The north azimuth from the center point (CCN, CCE) of the constructed arc towards the said arc's end point (PTN, PTE). Reference is made to the note below.
ARC	The length of the tangent arc to be constructed.
angleO	The central angle of the constructed arc which is a replica of the input central angle (ANGLE) in radians.
radabs	The absolute value of the input radius (R).

NOTE	With reference to Figure 5.1a, note that the master program that calls the icarcan1 procedure, and for which program this procedure has been written plots all circular curves in a counterclockwise rotation regardless whether the arc has been constructed in a clockwise or counterclockwise direction. Thus: • While points (N1, E1) and (N2, E2) indicate the start and end points of the constructed arc based on the geometric construction of the arc, the returned PC point (PCN, PCE) and PT point (PTN, PTE) indicate the start and end points of the said arc as it is drawn for display (plotting). • The same analogy applies to the returned north azimuths azend, azimPC, and azimPT.

The icarcan1 procedure does not check the start and the end azimuths for straddling north. Thus it is possible for them to be negative rather than a positive 360° supplement. This should not cause any difficulties in subsequent computations.

Upon invocation, the procedure initializes the parameters to be returned, and then:

1. Calls the icptl procedure to compute the north and east coordinates of the center point of the arc by traversing along the given direction AZ1 with a plus distance of zero, and an offset distance equal to the given radius R. Thus, depending on the sign of the said radius, the center point is positioned to the right or to the left of the said north azimuth.
2. Computes the absolute value of the length of the arc to be constructed by multiplying the input central angle (ANGLE) with the input radius (R).
3. Computes the azend, azimPC, and azimPT north azimuths by adding or subtracting the given central angle taking into consideration the sign of the input radius (R) and of the central angle (ANGLE).
4. Calls the icptl procedure to compute the north and east coordinates of the (N2, E2), (PCN, PCE) and (PTN, PTE) points.

5.1.2   Tangent Arc with Known Radius and Chord Length

The `icarccd2` procedure constructs an arc that commences at a POC or point of known north and east coordinates, and is tangent to a line of known north azimuth that passes through the said point. The arc is of a known radius, and known chord length. Reference may be made to Figure 5.1a in which the central angle is not known but the implied chord distance from the PC to the PT points is known. Other than that, the principles of this procedure are the same as those of the preceding one. Refer to the `icarccd1` procedures for similar but non-tangent arcs.

```
Sub icarccd2  (N1, E1, AZ1, R, CHORD, _
               N2, E2, PCN, PCE, PTN, PTE, CCN,
               CCE, azend, azimPC, azimPT, _
               ARC, ANGLE, radabs)
```

The input parameters to these procedures include the:

N1, E1	The north and east coordinates of the known point (POC or POL) at which the tangent arc is to start.
AZ1	The north azimuth in radians of the tangent line at the known point (N1, E1).
R	The radius of the arc to be constructed, and which is positive for a clockwise rotation and negative for a counterclockwise rotation.
CHORD	The chord length of the arc to be constructed and which is to be positive if the arc is to progress in a rotational direction as indicated by the sign of the radius (R), and negative if the arc is to oppose the rotational direction of the said radius for display (plotting) purposes. Reference is made to the note in the preceding subsection.

The programmer is alerted to the argument lists of the `icarccd1` *and* `icarccd2` *procedures regarding the sequence of the chord and radius input parameters.*

The parameters that are returned from this procedure include the:

N2, E2	The north and east coordinates of the end point of the arc to be. Reference is made to the note in the preceding subsection.
PCN, PCE	The north and east coordinates of the PC point of the arc to be. Reference is made to the note in the preceding subsection.
PTN, PTE	The north and east coordinates of the PT point of the arc to be. Reference is made to the note in the preceding subsection.

CCN, CCE	The north and east coordinates of the center point of the arc to be constructed.
azend	The north azimuth from the center point (CCN, CCE) of the arc towards the said arc's PC point (N2, E2). Reference is made to the note in the preceding subsection.
azimPC	The north azimuth from the center point (CCN, CCE) of the arc towards the said arc's end point (PCN, PCE). Reference is made to the note in the preceding subsection.
azimPT	The north azimuth from the center point (CCN, CCE) of the arc towards the said arc's end point (PTN, PTE). Reference is made to the note in the preceding subsection.
ARC	The length of the constructed arc.
ANGLE	The central angle in radians of the constructed arc.
radabs	The absolute value of the input radius (R).

Upon invocation, the procedure:

1. Calls the `icpt1` procedure to compute the north and east coordinates of the center point of the arc (CCN, CCE) by traversing a zero distance along the given azimuth of the tangent line (AZ1) and an offset distance equal to the input radius (R). Depending on the sign of the radius, the center point will be located to the right or left of the tangent line. The procedure then branches to **Step 2b**.
2. Computes the central angle and the length of the arc to be constructed a

$$\theta = 2\left[\sin^{-1}\left(\left| \frac{C}{2R} \right| \right) \right] \tag{5.1a}$$

and

$$L = R\theta \tag{5.1b}$$

where θ = The central angle of the arc
C = The given chord length
R = The given radius
and L = The length of the arc
The procedure then forces the central angle to carry the sign of the chord length.

3. Computes the `azend`, `azimPC`, and `azimPT` north azimuths by adding or subtracting the central angle taking into consideration the signs of the input radius (R) and of the central angle (ANGLE).
4. Calls the `icpt1` procedure to compute the north and east coordinates of the (N2, E2), (PCN, PCE) and (PTN, PTE) points.

5.1.3 Tangent Arc with Known Radius and Arc Length

The `icarcln1` procedure constructs an arc that commences at a given point of north and east coordinates, and is tangent to a line of known north azimuth and which line passes through the said point. This procedure is similar to the preceding two procedures, with the difference being the substitution of the arc length for the central angle or chord length. Therefore, reference is made to the figures and note of the said subsections.

```
Sub icarcln1  (N1, E1, AZ1, R, ARC, _
               N2, E2, PCN, PCE, PTN, PTE, CCN,
               CCE, azend, azimPC, _
               azimPT, CHORD, ANGLE, radabs)
```

The input parameters to these procedures include the:

N1, E1	The north and east coordinates of the known point (POC or POL) at which the tangent arc is to start.
AZ1	The north azimuth (radians) of the tangent line to the arc at (N1, E1).
R	The radius of the arc to be constructed, and which is positive for a clockwise rotation and negative for a counterclockwise rotation.
ARC	The length of the arc to be constructed and which should be positive if the arc is to progress in a rotational direction as indicated by the sign of the radius (R), and negative if the arc is to oppose the rotational direction of the said radius for display (plotting) purposes. Refer to the note in the preceding subsections.

The parameters that are returned from this procedure include the:

N2, E2	The north and east coordinates of the end point of the arc to be. Refer to the note in the preceding subsections.
PCN, PCE	The north and east coordinates of the PC point of the arc to be. Refer to the note in the preceding subsections.
PTN, PTE	The north and east coordinates of the PT point of the arc to be. Refer to the note in the preceding subsections.
CCN, CCE	The north and east coordinates of the center point of the arc to be constructed.
azend	The north azimuth from the center point (CCN, CCE) of the arc towards the said arc's PC point (N2, E2). Refer to the note in the preceding subsections.
azimPC	The north azimuth from the center point (CCN, CCE) of the arc towards the said arc's end point (PCN, PCE). Refer to the note in the preceding subsections.
azimPT	The north azimuth from the center point (CCN, CCE) of the arc towards the said arc's end point (PTN, PTE). Refer to the note in the preceding subsections.

CHORD	The chord length of the constructed arc.
ANGLE	The central angle in radians of the constructed arc.
radabs	The absolute value of the input radius (R).

Upon invocation, the procedure initializes the parameters to be returned, and then:

1. Calls the `icptl` procedure to compute the coordinates of the center point of the arc by traversing along the given direction AZ1 with a plus distance of zero (0.00), and an offset distance equal to the given radius R. Thus depending on the sign of the said radius, the center point is positioned to the right or left of the said north azimuth.
2. Computes the absolute value of the length of the arc to be constructed by dividing the input arc length (ARC) by the input radius (R).
3. Computes the `azend`, `azimPC`, and `azimPT` north azimuths by adding or subtracting the given central angle taking into consideration the sign of the input radius (R) and of the central angle (ANGLE).
4. Calls the `icptl` procedure to compute the north and east coordinates of the (N2, E2), (PCN, PCE), and (PTN, PTE) points.

5.1.4 Arc Tangent to Two Lines Given a Tangent Length

The `iccvt2ls` procedure constructs an arc tangent to two known lines given the arc's tangent length (distance from the point of intersection of the said lines to each of the arc's endpoints). Each of the said lines is defined by the north and east coordinates of their point of intersection (PI), and by their individual north azimuths (see Figures 5.1c and 5.1d). This procedure assumes that the said lines have been intersected, and the values of the (a) coordinates of the said PI point, (b) angle Δ formed by the said lines, (c) one half of the said Δ angle, and the (d) one half of the 180° supplement of Δ have been determined prior to its call.

By specifying the tangent length, a user actually implies the start point (POT) of the arc to be constructed. The POT is assumed to be located on the first line, the azimuth of which is to point from the POT towards the PI, while the arc's end point is assumed to be located on the second line, the azimuth of which is to point from the PI towards the end point of the arc. Note that POT (point on tangent) has been used and not PC because the resultant arc is always created in a counterclockwise rotation causing the PC and PT points not to necessarily correspond with the first and second lines. Figure 5.1d indicates the location of the said points based on the directions of the two lines.

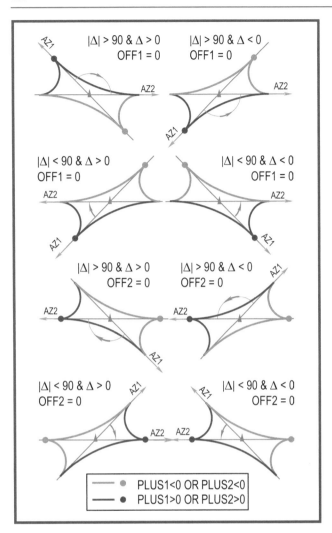

Figure 5.1c Arc Tangent to Two Lines Given the Tangent Length

Since the two intersecting lines create four sectors within which the tangent arc is to be constructed, the procedure requires certain information in order to decide in which sector to position the arc. This is attained by the PLUS1, OFF1, PLUS2, and OFF2 input parameters and by the value of the angle of intersection (Δ or DELTA) of the two lines as described below. Note that DELTA is measured from the first line towards the second line and hence it could be either positive (clockwise) or negative (counterclockwise), and should be less than 180°.

The two given lines and the four sectors that they form, and taking into consideration that each line could have two directions with one opposing the other provide for 32 potential tangent arc conditions as indicated in Figure 5.1c. The position of the POT point with respect to the PI point provides for two potential tangent curves as shown in Figure 5.1c, and which are returned by the procedure. Shown in green color in the said figure are the tangent arcs for a POT located back of the PI, and in red color for a POT located ahead of the PI on the first line.

```
Sub iccvt2ls  (PIN, PIE, AZ1, AZ2, DELTA, Delt1,Delt2,
                PLUS1, OFF1, PLUS2, OFF2, _
                R1, T1, AZ1PC, AZ1PT, DEL1, R2, T2,
                AZ2PC, AZ2PT,DEL2, _
                PCN1, PCE1, PTN1, PTE1, CCN1, CCE1, _
                PCN2, PCE2, PTN2, PTE2, CCN2, CCE2)
```

The input parameters to these procedures include the:

PIN, PIE	The north and east coordinates of the point of intersection of the two known lines to which the arc to be constructed is to be tangent.
AZ1	The north azimuth in radians of the first line to which the said arc is to be tangent, and along which the POT is to be located.
AZ2	The north azimuth in radians of the second line to which the said arc is to be tangent, and along which the arc is to terminate.
DELTA	The angle formed by the two azimuths by turning an angle from AZ1 towards AZ2. A clockwise rotation is considered as positive, and a counterclockwise rotation is considered as negative. Note that this angle although called DELTA is not the conventional deflection angle Δ of a curve.
Delt1	The one half value of the angle DELTA (Delt1=0.5 DELTA).
Delt2	The one half value of the 180° supplement of the absolute value of angle DELTA (Delt2=0.5 [π−|DELTA|]).
PLUS1	The tangent length of the arc to be constructed, which length could be positive or negative, or it could be zero. Reference is made to the note below.
OFF1	An indicator, the value of which could be zero, or any other number. Reference is made to the note below.
PLUS2	The tangent length of the arc to be constructed, which length could be positive or negative, or it could be zero. Reference is made to the note below.
OFF2	An indicator, the value of which could be zero, or any other number. Reference is made to the note below.

The parameters that are returned from this procedure include the:

R1, R2	The radii of the two possible arcs.
T1, T2	The absolute value of the tangent lengths of the two possible arcs.
AZ1PC	The radial north azimuth towards the PC point of the first possible arc.
AZ2PC	The radial north azimuth towards the PC point of the second possible arc.
AZ1PT	The radial north azimuth towards the PT point of the first possible arc.
AZ2PT	The radial north azimuth towards the PT point of the second possible arc.
DEL1, DEL2	The central angles in radians of the two possible arcs, which are always returned with negative values thus denoting a counterclockwise arc rotation for display purposes.

Figure 5.1d Arc Tangent to
Two Lines Given the Tangent
Length

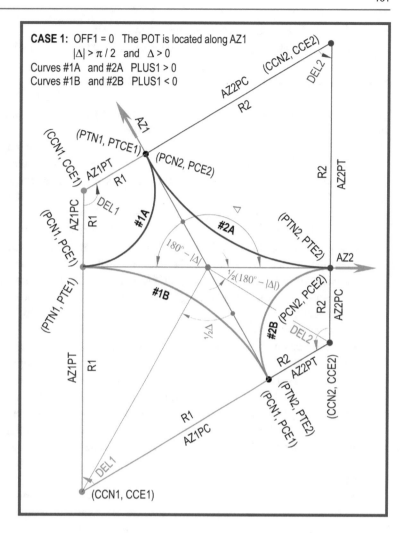

PCN1, PCE1	The north and east coordinates of the start point of the first possible arc.
PTN1, PTE1	The north and east coordinates of the end point of the first possible arc.
PCN2, PCE2	The north and east coordinates of the start point of the second possible arc.
PTN2, PTE2	The north and east coordinates of the end point of the second possible arc.
CCN1, CCE1	The north and east coordinates of the center point of the first possible arc.
CCN2, CCE2	The north and east coordinates of the center point of the second possible arc.

NOTE that the information below, in conjunction with Figure 5.1c, provides an insight about the use of the PLUS1, OFF1, PLUS2, OFF2 and DELTA input parameters:

- A positive PLUS1 or PLUS2 indicates that the POT is to be located ahead of the PI in the direction of the pertinent azimuth AZ1 or AZ2, while a negative PLUS1 or PLUS2 indicates that the POT is to be located back of the PI in the direction of the said pertinent azimuth.

- Either OFF1 or OFF2, but not both, should be zero. If both OFF1 and OFF2 are zero, then the OFF1 condition controls.
- If the POT is to be located on the first line (back tangent), then OFF1 should be 0, in which case OFF2 is disregarded. In this case,
  – PLUS1 should be the tangent length (positive or negative), but not zero.
  – PLUS2 could have any value.
- If the POT is to be located on the second line (ahead tangent), then OFF2 should be 0, and OFF1 should be 1 or any positive number. In this case,
  – PLUS2 should be the tangent length (positive or negative), but not zero.
  – PLUS1 could have any value.
- If OFF1 = 0 and PLUS1 = 0, then control passes on to the OFF2 = 0 condition as stated in the above comment.
- If OFF1 = 0 and PLUS1 <> 0, then the values of OFF2 and PLUS2 are of no concern.
- If PLUS1 and PLUS2 are both zero, all returned coordinates are those of the PI.

- The first of the two curves to be returned is to the left of the line on which the POT is located, and the second curve to the right.
- Both curves are generated in a counterclockwise rotation.

Upon invocation, the procedure computes the tangent functions of the two input half delta angles, and then:

1. Identifies which offset value is equal to zero (0.00), thus identifying the given line AZ1 or AZ2 upon which the POT is to be located.
2. Identifies whether the POT is to be located ahead or back of the PI with respect to the direction of the line defined by the azimuth AZ1 or AZ2. This is done by querying PLUS1 or PLUS2 whether they are positive or negative. Remember that each of these distances represents the input tangent length of the arc to be constructed.
3. Identifies whether the absolute value of DELTA is less or greater than 90°, and whether DELTA is positive or negative.
4. Constructs two tangent arcs, one to the left and one to the right of AZ1 or AZ2 by computing the parameters listed below accounting for the plus or minus sign of the input parameters with respect to the tangent potential arc condition identified above. The numbers 1 and 2 in the parameters refer to the left and right of the said arc.
   - The radius of the arc (R1 or R2) is computed by dividing the tangent line (PLUS1 or PLUS2) by the tangent function of half the DELTA angle or half of the DELTA's 180° supplement (Delt1 or Delt2). The plus distance and angle to be used depends on the particular case in which the arc is to be located. The radius is always returned as positive.
     - The tangent length (T1 or T2) is assigned the value of PLUS1 if OFF1=0, or of PLUS2 if OFF2=0.
     - The north azimuths from the arc's center point towards the arc's PC and PT points (AZ1PC and AZ1PT, or AZ2PC and AZ2PT).
     - The central angles of the said two arcs (DEL1 and DEL2) which are positive if the arc as rotated from the PC towards the PT is clockwise, or negative if the arc as rotated from the PC towards the PT is counterclockwise.

NOTE that the above steps are carried out in a series of IF statements that identify each of the 16 potential tangent arc conditions as depicted in Figure 5.1c.

5. Calls the icpcptcc procedure to compute for each of the said left and right curves the coordinates of the PC (PCN1, PCE1 and PCN2, PCE2), PT (PTN1,

PTE1 and PTN2, PTE2), and center (CCN1, CCE1 and CCN2, CCE2) points.

5.1.5   Arc Tangent to Two Lines Implied by Three Points

The icsmpcrv procedure constructs an arc of a known radius tangent to two lines that are implied by connecting three known points in series as indicated in Figure 5.1e. The first of the said three points should be located on the back tangent, the second points should be the point of intersection of the two implied lines (PI) and the third point should be located on the ahead tangent line.

In constructing the tangent arc, the procedure progresses from the back tangent towards the ahead tangent in the sequence of the three given points in the procedure's argument list. If the given radius is:

- Positive, the arc is placed to the right of the back tangent as represented by the arc displayed in red color in the said figure.
- Negative, the arc is placed to the left of the back tangent as represented by the arc displayed in green color in the said figure. Otherwise said, a negative radius reverses the directions of the given implied lines, or places the arc in opposite sector formed by the said two lines.
- Zero, the coordinates of the start (PC), end (PT) and center points of the tangent arc returned to be those of the second given point, the PI of the two implied lines.

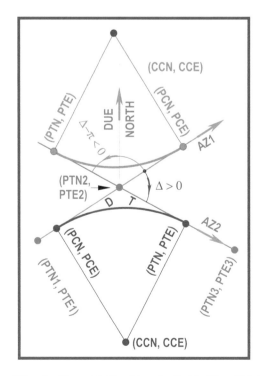

Figure 5.1e Arc Tangent to Two Lines Implied by Three Points

This procedure assumes that Δ is less 180° in absolute value. In roadway alignments in which a Δ exceeds 180°, it should be broken up in two arcs of the same radius.

```
Sub icsmpcrv  (PTN1, PTE1, PTN2, PTE2, PTN3, PTE3,
               R, _
               CCN, CCE, PCN, PCE, PTN, PTE, IERR,
               ANGLE, DELTA)
```

The input parameters to these procedures include the:

PTN1, PTE1	The north and east coordinates of the first of the three known points located on the back tangent back of the PI.
PTN2, PTE2	The north and east coordinates of the second of the said known points which is the intersection (PI) of the two implied lines.
PTN3, PTE3	The north and east coordinates of the third of the said known points located on the ahead tangent ahead of the PI.
R	The known radius of the tangent arc to be constructed which should be: • Positive to place the arc to the right of the back tangent. • Negative to place the arc to the left of the ahead tangent.

The parameters that are returned from this procedure include the:

CCN, CCE	The north and east coordinates of the center point of the tangent arc.
PCN, PCE	The north and east coordinates of the start point (PC) of the tangent arc.
PTN, PTE	The north and east coordinates of the end point (PT) of the tangent arc.
IERR	An error indicator, the value of which if it is 0 Denotes that no error has been encountered, and if it is 6 Denotes that the three point are colinear.
ANGLE	The angle of deflection of the two implied tangent lines (refer to the steps below).
DELTA	The same as ANGLE, but with its absolute value reduced to less than π.

Upon invocation, the procedure initializes to zero the parameters to be returned, and then:

1. Queries the value of the radius for being equal to zero within a tolerance of 0.005 ft (1.524 mm).
   • If so, sets the PC and the PT of the arc to be the same as those of the second given point, and terminates.
   • If not, continues.
2. Calls the icforce procedure to compute the azimuth AZ1 from point 1 to point 2, and the azimuth AZ2 from point 2 to point 3.

3. Computes the values of the ANGLE and of the DELTA parameters by subtracting AZ1 from AZ2. Although theoretically ANGLE = DELTA, it is possible for the absolute value of ANGLE, as computed above, to be greater than 180°, as in the case in which the arc rotates from a northeast direction towards a northwest direction (i.e., if AZ1 = 30° and AZ2 = 330° then 330° − 30° = 300°). To reduce the delta angle to less than 180°, the statement

$$DELTA = icangl(ANGLE)$$

   is used (see Chap. 3 for icangl). Note that the value of DELTA is positive or negative depending on whether the rotation is clockwise or counterclockwise (see Figure 5.1e).

4. Queries the absolute value of DELTA for being equal to zero within a tolerance of 0.0000048 radians (0.99 s).
   • If so, it sets the error indicator IERR to 6, and terminates.
   • If not, it continues.
5. Computes the length of the tangent of the arc to be constructed as

$$T = \frac{R\sin(\alpha)}{|\cos(\alpha)|} \quad\quad (5.1c)$$

   where T = The length of the arc's tangent,
   R = The radius as specified by the input parameter R,
   and α = One half of the deflection angle DELTA.
   Note that the sine over the cosine ratio of the angle is used rather than the tangent function.

6. Calls the icptl procedure (see Chap. 4) to compute the coordinates of the:
   • PC point of the arc by traversing backwards along AZ1 with a plus distance of T and an offset distance of zero.
   • PT point of the arc by traversing forwards along AZ2 with a plus distance of T and an offset distance of zero.
   • Center point of the arc by traversing forwards along AZ2 with a plus distance of T and an offset distance of R.

5.1.6   Arc Tangent to Two Lines and Through a Point

Two procedures are presented in this subsection, the iccrvpoc procedure and the iccvt2lq procedure. Both of these procedures construct an arc that passes through a known point and is tangent to two given lines. However, whereas the iccrvpoc procedure determines and returns only the radius of the arc, the iccvt2lq procedure in addition to the radius, computes and returns the additional parameters that construct the whole arc. In a way, the iccvt2lq

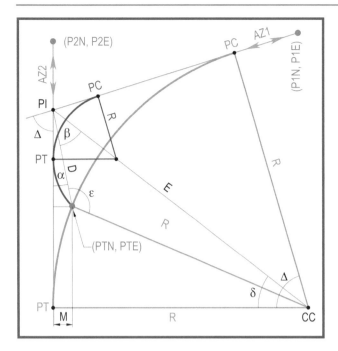

Figure 5.1f Arc Tangent to Two Lines Through a Point

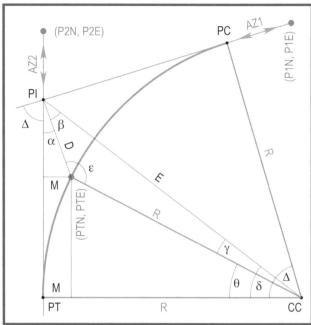

Figure 5.1g Arc Tangent to Two Lines Through a Point (Larger Radius)

procedure operates somewhat in a manner similar to that of the `iccvt2ls` procedure and depends on the calling program to perform some preparatory work for the necessary input parameters as is to be seen later on in this subsection.

As is to be seen below, there is yet another difference between these two procedures. The subject construction of an arc passing through a point and being tangent to two lines has two potential solutions as indicated in Figure 5.1f with one arc having a larger radius than the other. The `iccrvpoc` procedure concerns itself with the tangent arc that has the larger radius, while the `iccvt2lq` procedure addresses both radii. In essence, what it does is that it determines the center point of both tangent arcs that the calling program can display and have the user decide which one is desired and should be used.

5.1.6.1 The Geometric Construction
The given point though which the arc is to pass may be located in any of the four sectors that are formed by the two given lines, but it cannot be located on either of the two given lines. *A test for this condition should be made by the program that calls either of these two procedures.* As for the direction of the azimuths of the said given lines, they may point in either direction with respect to their intersection point. In computing the parameters to be returned, both of the said procedures assume that the tangent arc is to commence on the first given line, and terminate at the second given line.

For any position of the given point through which the tangent arc is to pass within each sector formed by the two

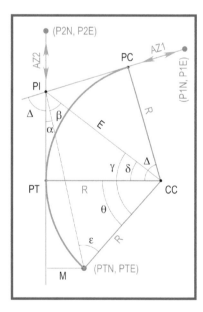

Figure 5.1h Arc Tangent to Two Lines Through a Point (Smaller Radius)

given lines, there are two possible conditions (see Figure 5.1f). The tangent arc with the larger radius is illustrated in Figure 5.1g, while the one with the smaller radius is displayed in Figure 5.1h. Notice that in the former, the point through which the arc is to pass is located between the PC and the PT points, or within the limits, of the tangent arc, while in the latter said given point is located outside the limits of the tangent arc.

As is to be seen below, the solution to these two conditions is very similar with one small difference in one of the equations presented below.

To determine the radius of the tangent arc, it is necessary to determine (see Figure 5.1f):

- The coordinates of the point of intersection (PI) of the two given lines,
- The angle of the said intersection (Δ), and its half value δ,
- The offset distance (M) of the point through which the arc is to pass from the second of the two given lines, and
- The angle α shown in Figures 5.1f and 5.1g.

The first two requirements are attained by calling the `icintrs` procedure (see Chap. 6), and the other two by calling the `icprjct` to procedure.

Since the AZ1 and the AZ2 azimuths of the two given lines may point towards or away from the PI, it is possible for the angle Δ to be greater than 180°. Similarly, and depending on the position of the given points through which the arc is to pass with respect to AZ2, the angle α may also exceed 180°. Thus, the absolute value of both of these two angles is reduced to less than 180° with the `icanlg` function. With the Δ angle known, the angle δ is set to be one half of Δ.

The above preparatory work is done within the `iccrvpoc` procedure; however, for the `iccvt2lq` procedure it should be done by the calling program.

Regarding the Arc with the Larger Radius

The derivation of the equations to be used to determine the value of the radius R is presented below. From Figure 5.1f it is seen that

$$\alpha + \beta = 90 - \delta \tag{5.1d}$$

$$\beta = 90 - (\alpha + \delta) \tag{5.1e}$$

$$\varepsilon = 180 - \beta - \gamma \tag{5.1f1}$$

Substitute Equation 5.1e in Equation 5.1f1 for β, and realizing that the angle $\theta = \delta - \gamma$, we get

$$\varepsilon = 90 + (\alpha + \delta) - \gamma = 90 + (\alpha + \theta) \tag{5.1f2}$$

By applying the law of sines in the triangle formed by the arc's center point CC, the given point through which the arc is to pass and the point of intersection PI we have

$$\frac{\sin(\varepsilon)}{\sin(\beta)} = \frac{E}{R} \tag{5.1g1}$$

Expressing $sin(\beta)$ and $sin(\varepsilon)$ as the sine of the sum of the two angles from Equations 5.1e and 5.1f1, respectively we have

$$\frac{\sin(\varepsilon)}{\sin(\beta)} \frac{\sin[90 + (\alpha + \theta)]}{\sin[90 - (\alpha + \delta)]} \tag{5.1g2}$$

or

$$\frac{\sin(\varepsilon)}{\sin(\beta)} = \frac{\sin(90)\cos(\alpha + \theta) + \sin(\alpha + \theta)\cos(90)}{\sin(90)\cos(\alpha + \delta) + \sin(\alpha + \theta)\cos(90)} \tag{5.1g3}$$

Note that the central angle Δ is the same for both the larger radius and smaller radius tangent arcs.

Since $\sin(90) = 1$ and $\cos(90) = 0$, we can simplify the right side of Equation 5.1g3 and then replace the left side of Equation 5.1g1 with the simplified version of the right side of Equation 5.1g3 to express Equation 5.1g1 as

$$\frac{\cos(\alpha + \theta)}{\cos(\alpha + \delta)} = \frac{E}{R} \tag{5.1h1}$$

Noting from Figure 5.1f that $E = R/\cos(\Delta)$, the above equation may be expressed as

$$\frac{\cos(\alpha + \theta)}{\cos(\alpha + \delta)} = \frac{R}{\cos(\delta)} \frac{1}{R} = \frac{1}{\cos(\delta)} \tag{5.1h2}$$

or

$$\cos(\alpha + \theta) = \frac{\cos(\alpha + \delta)}{\cos(\delta)} \tag{5.1i}$$

The procedures now use the arcsine function `icasinan` to find the 90° complement of the angle$(\alpha + \theta)$ as

$$X = \sin^{-1}(\alpha + \theta) \tag{5.1j1}$$

and with α as a result of the previously described preparatory work of, or for the procedures, the angle θ may now be found as

$$\theta = \frac{\pi}{2} - X - \alpha \tag{5.1j2}$$

With θ and M known, we can now solve for the radius R as follows:

$$\cos(\theta) = \frac{R - M}{R} \tag{5.1k}$$

$$R\cos(\theta) = R - M \tag{5.1m}$$

$$M = R - R\cos(\theta) = R[1 - \cos(\theta)] \tag{5.1n}$$

$$R = \frac{M}{1 - \cos(\theta)} \tag{5.1o}$$

Regarding the Arc with the Smaller Radius

Regarding the condition of Figure 5.1h, the solution for the radius R is the same as that described above with the difference that whereas in Figure 5.1g

$$\theta = \delta - \gamma$$

in Figure 5.1h

$$\theta = \gamma - \delta$$

Thus, Equation 5.1f2 becomes

$$\varepsilon = 90 + (\alpha + \delta) - \gamma = 90 + (\alpha - \theta) \qquad (5.1p)$$

Equation 5.1g2 holds true for the condition of Figure 5.1g, but as a result of the replacement of Equation 5.1f2 with Equation 5.1p, Equation 5.1g2 becomes

$$\frac{\sin(\varepsilon)}{\sin(\beta)} = \frac{\sin\left[90 + (\alpha - \theta)\right]}{\sin\left[90 - (\alpha + \delta)\right]} \qquad (5.1q1)$$

and Equation 5.1g3 becomes

$$\frac{\sin(\varepsilon)}{\sin(\beta)} = \frac{\sin(90)\cos(\alpha - \theta) + \sin(\alpha - \theta)\cos(90)}{\sin(90)\cos(\alpha + \delta) - \sin(\alpha + \delta)\cos(90)} \qquad (5.1q2)$$

Hence Equations 5.1h1 and 5.1h2 become

$$\frac{\cos(\alpha - \theta)}{\cos(\alpha + \delta)} = \frac{E}{R} = \frac{R}{\cos(\delta)R} = \frac{1}{\cos(\delta)} \qquad (5.1r)$$

which causes Equation 5.1i o become

$$\cos(\alpha - \theta) = \frac{\cos(\alpha + \delta)}{\cos(\delta)} \qquad (5.1s)$$

Thereafter, the value of the angle $(\alpha - \theta)$ may be found in a manner similar to that of finding the value of the angle $(\alpha + \theta)$ with Equation 5.1j1, from which the value of angle θ is fond to be

$$\theta = \frac{\pi}{2} - X + \alpha \qquad (5.1t)$$

The value of the radius R may now be found by Equation 5.1o.

5.1.6.2 The `iccrvpoc` Procedure

The `iccrvpoc` procedure introduces an arc which is tangent to two given lines and which passes through a point of known north and east coordinates. Each of the two given lines is defined by the north and east coordinates of a point, and a north azimuth. Each of the said two points may be located anywhere within the extent of its corresponding line.

Regarding the point through which the tangent arc is to pass, it must be located in accord with the statement of the side bar located on the right side of the next page. This procedure computes and returns the radius (R) of the tangent arc.

The point through which the curve is to pass and its corresponding center point must both be located either to the right or to the left of each of the two given lines with respect to their given directions. This must be assured by the calling program of this procedure.

```
Sub iccrvpoc (P1N, P1E, AZ1, P2N, P2E, AZ2,
          PTN, PTE, _
          R)
```

The input parameters to these procedures include the:

P1N, P1E	The north and east coordinates of a point on the first of the two given lines to which the arc is to be tangent.
AZ1	The north azimuth of the first of the two given lines.
P2N, P2E	The north and east coordinates of a point on the second of the two given lines to which the arc is to be tangent.
AZ2	The north azimuth of the second of the two given lines.
PTN, PTE	The north and east coordinates of a point through which the tangent arc is to pass. Reference is made to the side bar of this page regarding the position of this point.

The parameters that are returned from this procedure include the:

R	The radius of the tangent arc.

Upon invocation, the procedure initializes to zero the parameter to be returned, and then:

1. Calls the `icintrs` procedure to intersect the two given lines and determine the coordinates of the PI point (PIN, PIE) and their deflection angle DELTA.
2. Calls the `icangle` function to reduce the DELTA angle to less than 180°, in case it is greater than 180°.
3. Calls the `icprjct` procedure to project the given point (PTN, PTE) through which the arc is to pass on the second of the two given lines, and determine the offset distance M and the angle α addressed in the previous sub-subsection.
4. Computes X, θ and the radius R with Equations 5.1j1, 5.1j2, and 5.1o, respectively.

5.1.6.3 The `iccvt2lq` Procedure

The `iccvt2lq` procedure constructs two arcs tangent to two lines and passing through a known point (POC). Each of the two given lines is defined by the north and east coordinates of the point of their intersection (PI), and their individual north azimuths which may point in either direction with respect to the position of the PI as indicated in Figure 5.1f. The point through which the arcs are to pass must be of known coordinates, but these coordinates are not used as input to this procedure. Instead, this point is introduced to this procedure by its distance from the PI along the first given line and its offset distance from the first given line and from the second given line. This implies that the said POC must be preprocessed by the calling program of this procedure.

The `iccvt2lq` procedure is similar to that of the `iccrvpoc` procedure previously described but with three differences. The `iccvt2lq` procedure:

- Determines the radius of the larger and smaller radius arcs (see Figures 5.1g and 5.1h);
- Determines all ancillary parameters to construct both of the said tangent arcs; and it
- Requires that the previously defined preparatory work be done by the calling program.

```
Sub iccvt2lq (PIN, PIE, AZIM1, AZIM2, DELTA,
              delt1, delt2, PLUS1, off1,
              ang1, _
              PLUS2, off2, ang2, _
              RADIUS1, TANGENT1, azim1pc,
              azim1pt, DEL1, _
              RADIUS2, TANGENT2, azim2pc,
              azim2pt, DEL2, _
              PCN1, PCE1, PTN1, PTE1, CCN1,
              CCE1, _
              PCN2, PCE2, PTN2, PTE2, CCN2, CCE2)
```

The input parameters to these procedures include the:

PIN, PIE	The north and east coordinates of the point of intersection (PI) of the two lines.
AZIM1	The north azimuth of the first of the two given lines.
AZIM2	The north azimuth of the second of the two given lines.
DELTA	The angle (in radians) of intersection of the two given lines computed from the first line towards the second line (DELTA = AZIM2 - AZIM1). A positive angle is considered as being clockwise, and negative as counterclockwise.
delt1	One half the value of the DELTA angle.
delt2	One half of the value of the 180° supplement of the DELTA angle.

PLUS1	The distance of the point of projection of the given POC on the first line from the PI. A positive distance is considered as being ahead of the PI, and negative distance as being back of the PI.
off1	The normal offset distance of the POC from the first line. A positive distance denotes that the POC is to the right of the first given line, and a negative distance to the left of the first given line.
ang1	The angle (in radians) formed from the first given line towards the line joining the PI and the POC. A positive angle is considered as being clockwise, and negative as counterclockwise.
PLUS2	Similar to PLUS1 along the second line. This parameter is not used.
off2	Similar to off1 along the second line.
ang2	Similar to ang1 from the second line. This parameter is not used.

The parameters that are returned from this procedure include the:

RADIUS1	The radius of the larger of the two tangent arcs.
TANGENT1	The tangent length corresponding to the larger of the said arcs.
azim1pc	The radial azimuth to the start of the larger of the said arcs.
azim1pt	The radial azimuth to the end of the larger of the said arcs.
DEL1	The central angle (in radians) of the larger of the said arcs.
RADIUS2	The radius of the smaller of the said arcs.
TANGENT3	The tangent length to the smaller of the said arcs.
azim2pc	The north radial azimuth to the start of the smaller of the said arcs.
azim2pt	The north radial azimuth to the end of the smaller of the said arcs.
DEL2	The central angle (in radians) of the smaller of the said arcs.
PCN1, PCE1	The north and east coordinates of the start point (PC) of the larger of the said arcs.
PTN1, PTE1	The north and east coordinates of the end point (PT) of the larger of the said arcs.
CCN1, CCE1	The north and east coordinates of the center point of the larger of the said arcs.
PCN2, PCE2	The north and east coordinates of the start point (PC) of the smaller of the said arcs.
PTN2, PTE2	The north and east coordinates of the end point (PT) of the smaller of the said arcs.
CCN2, CCE2	The north and east coordinates of the center point of the smaller of the said arcs.

Upon invocation, the procedure:

1. Queries the angle DELTA and the offset distances off1 and off2 whether they are positive or negative, and

accordingly assigns a positive 1 (+1) or negative 1 (−1) value to a code corresponding to the (a) tangent length of each of the curves, (b) radius of each of the curves, (c) central angle, and (d) angle α.

2. Computes the values of the tangent lengths and radii of the said arcs, and the radial directions towards the PC and PT.

3. Calls the `icpcptcc` procedure to compute the coordinates of the PC and PT points of both the PC and PT points of both tangent arcs.

4. Determines the sign of the product of the `DELTA`, `off1` and `off2` parameters, and if positive reverses the coordinates of the PC and PT points to construct the said two arcs in a counterclockwise direction for display (plotting) purposes as previously stated.

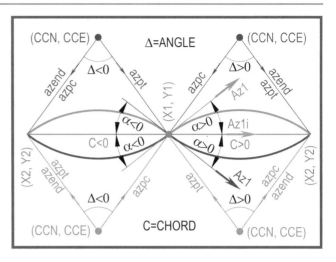

Figure 5.2a   Arc of Given Chord and Radius

5.2   Non-tangent Arcs

The procedures presented in this section are those identified under the non-tangent column of Table 5.0 at the beginning of this chapter. Note that two of these procedures involve the solution of an equation by an iterative method. Also presented in this section is a procedure that creates an arc or circle passing through three known points and one that passes an arc of known radius through two known points.

5.2.1   Arc of Given Chord and Radius

The `icarccd1` procedure constructs an arc given the north and east coordinates of the arc's start point, the length and direction of the arc's chord, and the radius of the arc to be constructed. The `icarccd1` is rather similar to the `icarccd2` procedure of Sect. 5.1.2 with two differences.

- The `icarccd1` procedure requires the direction of the chord from the arc's start point towards its end point whereas `icarccd2` requires the direction of the tangent at the start point of the arc.
- The names of the input parameter are the same in both procedures, but their sequence in the calling argument list is not the same.

Figure 5.2a identifies the various parameters of the procedure which assume that the central angle of the arc to be constructed is less than 180°.

```
Sub icarccd1 (Y1, X1, AZ1i, CHORD, R, _
              Y2, X2, PCN, PCE, PTN, PTE, CCN, _
              CCE, azend, azpc, azpt, ARC, _
              ANGLE, radabs)
```

The input parameters to this procedure include the:

Y1, X1	The north and east coordinates of the start point of the arc to be constructed.
AZ1i	The north azimuth (radians) of the said arc's chord from its start point (Y1, X1) towards its end point.
CHORD	The length of the said chord. A positive value locates the end point along the given chord azimuth (AZ1i) ahead of the start point (Y1, X1), while a negative value locates it along the said direction but back of the said start point.
R	The radius of the said arc. A positive value denotes a clockwise arc rotation, while a negative value denotes a counterclockwise arc rotation.

The parameters that are returned from this procedure include the:

Y2, X2	The north and east coordinates of the end point of the arc to be constructed.
PCN, PCE	The north and east coordinates of the PC point of the said arc (see note below).
PTN, PTE	The north and east coordinates of the PT point of the said arc (see note below).
CCN, CCE	The north and east coordinates of the center point of the said arc.
azend	The north radial azimuth of the said arc from its center point (CCN, CCE) towards its start point (Y1, X1).
azpc	The north radial azimuth of the said arc from its center point (CCN, CCE) towards its PC point (PCN, PCE).
azpt	The north radial azimuth of the said arc from its center point (CCN, CCE) towards its PT point (PTN, PTE).
ANGLE	The central angle (in radians) of the said arc. This angle is assigned the same sign (plus or minus) as that of the input parameter CHORD.
radabs	The absolute value of the input radius (R).

NOTE that since the arc is always displayed (plotted) in a counterclockwise rotation, the (Y1,X1) and (Y2,X2) point coordinates may not necessarily be those of the PC

(PCN,PCE) and PT (PTN,PTE) points, respectively, but rather the reverse.

Upon invocation, the procedure sets azimuth AZ1i in the temporary variable AZ1 to preserve the input value, and then:

1. Computes the arc's central angle as

$$\Delta = 2\left[\sin^{-1}\left(\left|\frac{C}{2R}\right|\right)\right] \tag{5.2a}$$

and the rc length as

$$L = \Delta \cdot R \tag{5.2b}$$

and then causes the sign of the central angle to be that of the chord. Thus, depending on the sign of the input radius, the arc length could be either positive or negative.

2. Computes the deflection angle of the chord as being one half the value of the central angle of the arc, and the azimuth AZ1 of the tangent line to the arc at (Y1, X1) towards the end of the arc (in the general direction of AZ1i) by:
   - Adding to AZ1 the said deflection angle if the input radius is negative, or
   - Subtracting from AZ1 the said deflection angle if the input radius is positive.

3. Calls the icpt1 procedure to compute the coordinates of the center point of the arc (CCN, CCE) by traversing along the azimuth of the tangent line AZ1 a distance of zero (0.00) and an offset distance of R to the right or left depending on the sign of R.

4. Computes the radial north azimuths from the arc's center point towards the end of the arc (azend), the PC of the arc (azpc), and the PT of the arc (azpt). Regarding these azimuths, reference is made to the note preceding these operational steps.

5. Sets the absolute value of the radius, and then calls the icpt1 procedure to compute the coordinates of the arc's PC (PCN, PCE), PT (PTN, PTE), and endpoint (Y2, X2) by traversing along the above defined azimuths with the said absolute value as a plus distance and an offset value of zero.

5.2.2 Arc of Given Radius and Chord or Central Angle

The icchord1 procedure constructs an arc given the north and east coordinates of its start point, its radius, the north azimuth of its chord, and either the chord's length or central angle. This procedure is similar to the previous icarccd1 procedure with the following differences:

- Whereas icarccd1 considers the arc's counterclockwise generation of the graphic in returning the radial azimuths azpc and azpt, icchord1 returns azpc as the azimuth towards the (PCN, PCE) and azpt as the azimuth towards the arc's other end regardless of the arc's clockwise or counterclockwise rotation.
- Whereas icarccd1 is restricted to a central angle of less the 180°, icchord1 can generate an arc with a central angle greater than 180° provided a central angle is entered as input rather than a chord length.

Figure 5.2b identifies the various parameters of this procedure. When the central angle Δ is specified and the given chord length C is zero, or when the central angle Δ is greater than 180° regardless of the value of the chord, the chord length is computed as follows using the symbology of the said figure:

$$\delta = \frac{1}{2}\Delta \tag{5.2c}$$

$$\cos(\delta) = \frac{D}{R} = \frac{R-M}{R} \tag{5.2d}$$

$$M = R - R\cos(\delta) = R\left[1-\cos(\delta)\right] \tag{5.2e}$$

$$D = R - M \tag{5.2f}$$

$$C = \sqrt{R^2 - D^2} \tag{5.2g}$$

With the distances C (one half the computed chord length) and D (the offset distance of the arc center from the chord) known, the procedure uses the icpt1 procedure to tra-

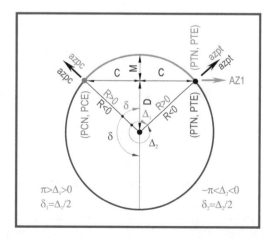

Figure 5.2b Arc of Given Chord and Central Angle

verse from the given PC of the arc along the chord azimuth `AZ1` with the

- *C* and *D* distances to locate the coordinates of the arc's center point.
- The computed chord length (*2C*) and zero (0.00) offset to locate the coordinates of the arc's endpoint PT.

```
Sub icchord1 (PCN, PCE, R, AZ1, CHORD, CAng, _
              CCN, CCE, ANGLE, ARC, azpc, azpt,
              PTN, PTE)
```

The input parameters to this procedure include the:

PCN, PCE	The north and east coordinates of the start point of the arc to be constructed.
R	The radius of the said arc. A positive value denotes a clockwise arc rotation, while a negative value denotes a counterclockwise arc rotation.
AZ1	The north azimuth (in radians) of the arc's chord from its start to its end point.
CHORD	The length of the arc's chord from its start point (Y1, X1) towards its endpoint. • The arc's endpoint is located along AZ1 and ahead of the arc's start point (Y1, X1) if CHORD>0, and back of the said start point if CHORD<0. • If CHORD=0 then CAng is used to compute the chord length.
CAng	The central angle (in radians) of the said arc. • If CAng and R are both positive, a clockwise arc is generated. • If either CAng or R, or both, is negative, a counterclockwise arc is created.

The parameters that are returned from this procedure include the:

CCN, CCE	The north and east coordinates of the center point of the arc to be constructed.
ANGLE	The central angle (in radians) as computed if the chord is used, or the value of the input parameter if the central angle is used.
ARC	The arc length of the arc to be constructed.
azpc	The radial north azimuth towards the PC point of the arc to be constructed.
azpt	The radial north azimuth towards the PT point of the arc to be constructed.
PTN, PTE	The north and east coordinates of the PT point of the arc to be constructed.

As stated above, this procedure is quite similar to that of the

1. Queries the values of the chord and central angle parameters to determine whether the chord or the central angle is to be use to construct the arc. If the:
   - Chord is to be used, the procedure performs the operational steps of the `icarccd1` procedure and terminates.

- Central angle is to be used, the procedure continues as indicated below.
2. Queries the sign of the central angle. If is negative, the procedure makes it positive, and forces the radius to be negative regardless of its input sign, otherwise there is no change to either the central angle or the radius.
3. Computes the length offset distance D of the center point from the chord, and the length of the chord by use of the above presented Equations 5.2c through 5.2g, noting that the chord length is twice the value of C of Equation 5.2g.
4. Calls the `icptl` procedure to determine the north and east coordinates of the center point of the arc by traversing along the chord's azimuth a plus distance of half the chord length, and an offset distance of D.
5. Calls the `icptl` procedure to determine the north and east coordinates of the end point (PT) of the arc by traversing along the chord's azimuth a plus distance equal to the chord length, and an offset distance of zero.
6. Calls the `icforce` procedure to inverse from the center point towards the arc's start point and end point to determine the respective radial north azimuths, and then terminates.

5.2.3 Arc Given the Central Angle and Arc Length

The `icarcAnL` procedure constructs an arc given its central angle (Δ) and the arc length (*L*) as depicted in Figure 5.2c. To position the arc the procedure also requires as input the north and east coordinates of either (**a**) the arc's center point, or (**b**) the arc's start point, and the radial north azimuth from the center point towards the start point of the arc. With Δ expressed in radians and *L* known, the radius

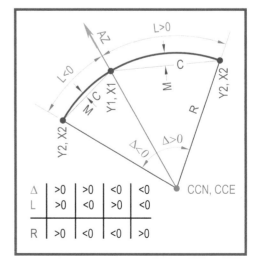

Δ	>0	>0	<0	<0
L	>0	<0	>0	<0
R	>0	<0	<0	>0

Figure 5.2c Arc Given the Central Angle and the Arc Length

(R), chord length (C) and middle ordinate (M) are determined as follows:

$$R = \frac{L}{\Delta} \qquad (5.2h1)$$

$$C2R\sin(-\ ) \qquad (5.2h2)$$

$$M = R\sin[1-\cos]\left(\frac{1}{2}\Delta\right) \qquad (5.2h3)$$

With the above information the coordinates of the PC and PT points and the corresponding azimuths from the center point towards them may now be computed.

The central angle may exceed 180°. However, it should be kept in mind that because of Equation 5.2h1, the arc length must correspond to the central angle. For example for a given angle of say 30° and a radius of 200 ft, the input arc length should be 104.72 ft. If we now introduce in this procedure an angle of 330° (the 360° supplement of the 30° angle), we must know introduce the corresponding arc length of 1151.92 ft and not that of the 30° angle (104.72 ft).

```
Sub icarcAnL(Y, X, thePT, AZ1, CAng, L, _
       Y1, X1, Y2, X2, PCN, PCE, PTN,
       PTE, CCN, CCE, AZ2, azimPC, _
       azimPT, azimCD, R, C, M, IND)
```

The input parameters to this procedure include the:

Y, X	The north and east coordinates of either (a) the center point, or (b) the start point of the arc to be constructed.
thePT	An indicator identifying the point to which the (Y, X) coordinates pertain. • C or c denotes the center point of the arc. • S or s denotes the start point of the arc.
AZ1	The north radial azimuth (in radians) from the center point (CCN, CCE) towards the start point of the arc to be constructed.
CAng	The central angle (in radians), the absolute value of which may exceed 180°, of the arc to be constructed, the value of which if it is: • Positive denotes a clockwise arc rotation. • Negative denotes a counterclockwise arc rotation. • Zero aborts the command.
L	The length of the arc to be constructed, the value of which if it is: • Positive, denotes that the arc progresses along the direction implied by CAng. • Negative, denotes that the arc opposes the direction implied by CAng. • Zero aborts the command.

NOTE that since the arc is always displayed (plotted) in a counterclockwise rotation, the (Y1, X1) and (Y2, X2) point coordinates may not necessarily be those of the PC (PCN, PCE) and PT (PTN, PTE) points, respectively, but rather the reverse.

The parameters that are returned from this procedure include the:

Y1, X1	The north and east coordinates of the start point of the arc to be constructed.
Y2, X2	The north and east coordinates of the end point of the arc to be constructed.
PCN, PCE	The north and east coordinates of the PC point of the arc to be constructed.
PTN, PTE	The north and east coordinates of the PT point of the arc to be constructed.
CCN, CCE	The north and east coordinates of the center point of the arc to be constructed.
AZ2	The north radial azimuth (in radians) from the center point (CCN, CCE) towards the end point of the arc to be constructed.
azimPC	The radial north azimuth towards the PC point of the arc to be constructed.
azimPT	The radial north azimuth towards the PT point of the arc to be constructed.
azimCD	The radial north azimuth towards the midpoint of the chord of the arc to be constructed from the start point towards the end point.
R	The radius of the arc to be constructed.
C	The chord length of the arc to be constructed.
M	The middle ordinate of the arc to be constructed.
IND	An indicator denoting whether an error has been encountered, or not. Its value may be composed of one to three digits, each one denoting the following: 0 No error has been encountered. 1 A zero value has been specified for CAng. 2 A zero value has been specified for ARC. 3 An erroneous code has been assigned to thePT.

Upon invocation, the procedure initializes the parameters to be returned and then:

1. Queries CAng, ARC and thePT for erroneous input information and set the value of IND accordingly. If an error has been detected the procedure aborts; else it continues.
2. Computes the values of the remaining three arc parameters.
3. Computes the coordinates of the center point of the arc to be constructed.
4. Computes the three azimuths to be returned checking for straddling of due north.
5. Computes the coordinates of the end points and those of the PC and PT (refer to the note above regarding the difference between the start and end points and the PC and PT), and then terminates.

5.2.4 Arc Given the Central Angle and Middle Ordinate

The icarcAnM procedure constructs an arc given its central angle (Δ) and the length of the middle ordinate (M) as depicted in Figure 5.2d. To position the arc the procedure also requires as input the north and east coordinates of either

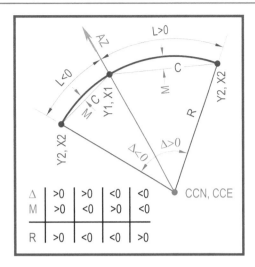

Figure 5.2d Arc Given the Central Angle and the Middle Ordinate

the arc's (a) center point, or (b) start point, and the radial north azimuth towards its start point. With Δ expressed in radians and M known, the radius (R), chord length (C) and arc length (L) are determined as follows:

$$M = R\left[1 - \cos\left(\frac{1}{2}\Delta\right)\right] \qquad (5.2\mathrm{i}1)$$

$$R = \frac{M}{1 - \cos\left(\frac{1}{2}\Delta\right)}$$

$$C = 2R\sin\left(\frac{1}{2}\Delta\right) \qquad (5.2\mathrm{i}2)$$

$$L = R\Delta \qquad (5.2\mathrm{i}3)$$

With this information the coordinates of the PC and PT points and the corresponding radial azimuths towards them may now be computed.

This procedure is very similar to the preceding `icarcAnL` procedure with two basic differences:

- The input specification of the arc length is replaced by the specification of the middle ordinate. However, a positive or negative middle ordinate operates the same as a the arc length in the said preceding procedure (see Figure 5.2d).
- The last three equations replace the preceding three equations in solving for the three unknown parameters of the arc to be constructed.

The central angle may exceed 180°, but if it does, the specified middle ordinate must be that of the 360° supplement.

```
Sub icarcAnM(Y, X, thePT, AZ1, CAng, M, _
            Y1, X1, Y2, X2, PCN, PCE, PTN,
            PTE, CCN, CCE, AZ2, azimPC, _
            azimPT, azimCD, R, C, L, IND)
```

The input parameters to this procedure include the:

Y, X	The north and east coordinates of either (a) the center point, or (b) the start point of the arc to be constructed.
thePT	An indicator identifying the point to which the (Y, X) coordinates pertain. • C or c denotes the center point of the arc. • S or s denotes the start point of the arc.
AZ1	The north radial azimuth (in radians) from the center point (CCN, CCE) towards the start point of the arc to be constructed.
CAng	The central angle (in radians), the absolute value of which may exceed 180°, of the arc to be constructed, the value of which if it is: • Positive denotes a clockwise arc rotation. • Negative denotes a counterclockwise arc rotation. • Zero aborts the command.
M	The middle ordinate of the arc to be constructed, the value of which if it is: • Positive, denotes that the arc progresses along the direction implied by CAng. • Negative, denotes that the arc opposes the direction implied by CAng. • Zero aborts the command.

NOTE that since the arc is always displayed (plotted) in a counterclockwise rotation, the (Y1, X1) and (Y2, X2) point coordinates may not necessarily be those of the PC (PCN, PCE) and PT (PTN, PTE) points, respectively, but rather the reverse.

The parameters that are returned from this procedure include the:

Y1, X1	The north and east coordinates of the start point of the arc to be constructed.
Y2, X2	The north and east coordinates of the end point of the arc to be constructed.
PCN, PCE	The north and east coordinates of the PC point of the arc to be constructed.
PTN, PTE	The north and east coordinates of the PT point of the arc to be constructed.
CCN, CCE	The north and east coordinates of the center point of the arc to be constructed.
AZ2	The north radial azimuth (in radians) from the center point (CCN, CCE) towards the end point of the arc to be constructed.
azimPC	The radial north azimuth towards the PC point of the arc to be constructed.
azimPT	The radial north azimuth towards the PT point of the arc to be constructed.

azimCD	The radial north azimuth towards the midpoint of the chord of the arc to be constructed from the start point towards the end point.
R	The radius of the arc to be constructed.
C	The chord length of the arc to be constructed.
L	The length of the arc to be constructed.
IND	An indicator denoting whether an error has been encountered, or not. Its value may be composed of one to three digits, each one denoting the following: 0 No error has been encountered. 1 A zero value has been specified for CAng. 2 A zero value has been specified for ARC. 3 An erroneous code has been assigned to thePT.

As stated above this procedure is very similar to the preceding `icarcAnL` procedure particularly with regard to its operation. The only difference is the use of the pertinent equation in computing and reporting the three remaining unknown parameters. Thus reference is made to the said `icarcAnL` procedure.

5.2.5 Arc Given the Radius and Middle Ordinate

The `icarcRaM` procedure constructs an arc given its radius (R) and the length of the middle ordinate (M) as depicted in Figure 5.2e. To position the arc the procedure also requires as input the north and east coordinates of either the arc's **(a)** center point, or **(b)** start point, and the radial north azimuth towards the start point of the arc. With R and M known, the central angle (Δ), chord length (C) and arc length (L) are determined as follows:

$$\Delta = 2\cos^{-1}\left(\frac{R-M}{R}\right) \quad\quad (5.2j1)$$

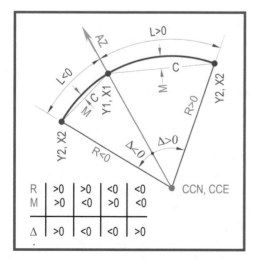

R	>0	>0	<0	<0	CCN, CCE
M	>0	<0	>0	<0	

Δ	>0	<0	<0	>0

Figure 5.2e Arc Given the Radius and the Middle Ordinate

$$L = R\Delta \quad\quad (5.2j2)$$

$$C = 2R\sin\left(\frac{1}{2}\Delta\right) \quad\quad (5.2j3)$$

With the above information the coordinates of the PC and PT points and the corresponding radial azimuths towards them may now be computed. Due to the input parameters, the absolute value of the central angle is restricted to less the 180°. This procedure is similar to the preceding two procedures with the differences being the input and output parameters and associated equations to be solved.

```
Sub icarcRaM(Y, X, thePT, AZ1, R, M, _
             Y1, X1, Y2, X2, PCN, PCE, PTN,
             PTE, CCN, CCE, AZ2, azimPC, _
             azimPT, azimCD, CAng, C, L, IND)
```

The input parameters to this procedure include the:

Y, X	The north and east coordinates of either (a) the center point, or (b) the start point of the arc to be constructed.
thePT	An indicator identifying the point to which the (Y, X) coordinates pertain. • C or c denotes the center point of the arc. • S or s denotes the start point of the arc.
AZ1	The north radial azimuth (in radians) from the center point (CCN, CCE) towards the start point of the arc to be constructed.
R	The radius of the arc to be constructed, the value of which if it is: • Positive denotes a clockwise arc rotation. • Negative denotes a counterclockwise arc rotation. • Zero aborts the command.
M	The middle ordinate of the arc to be constructed, the value of which if it is: • Positive, denotes that the arc progresses along the direction implied by R. • Negative, denotes that the arc opposes the direction implied by R. • Zero aborts the command.

NOTE that since the arc is always displayed (plotted) in a counterclockwise rotation, the (Y1, X1) and (Y2, X2) point coordinates may not necessarily be those of the PC (PCN, PCE) and PT (PTN, PTE) points, respectively, but rather the reverse.

The parameters that are returned from this procedure include the:

Y1, X1	The north and east coordinates of the start point of the arc to be constructed.
Y2, X2	The north and east coordinates of the end point of the arc to be constructed.

PCN, PCE	The north and east coordinates of the PC point of the arc to be constructed.
PTN, PTE	The north and east coordinates of the PT point of the arc to be constructed.
CCN, CCE	The north and east coordinates of the center point of the arc to be constructed.
AZ2	The north radial azimuth (in radians) from the center point (CCN, CCE) towards the end point of the arc to be constructed.
azimPC	The radial north azimuth towards the PC point of the arc to be constructed.
azimPT	The radial north azimuth towards the PT point of the arc to be constructed.
azimCD	The radial north azimuth towards the midpoint of the chord of the arc to be constructed from the start point towards the end point.
CAng	The central angle (in radians) of the arc to be constructed: • Positive for a clockwise arc rotation. • Negative for a counterclockwise arc rotation.
C	The chord length of the arc to be constructed.
L	The length of the arc to be constructed.
IND	An indicator denoting whether an error has been encountered, or not. Its value may be composed of one to three digits, each one denoting the following: 0 No error has been encountered. 1 A zero value has been specified for CAng. 2 A zero value has been specified for ARC. 3 An erroneous code has been assigned to thePT.

Because of the similarity of this procedure with the icar-cAnL procedure their operations are also similar and therefore reference is made to the said icarcAnL procedure.

5.2.6   Arc Given the Arc Length and Middle Ordinate

The icarcLnM procedure constructs an arc given its arc length (L) and the length of the middle ordinate (M) as depicted in Figure 5.2f. To position the arc the procedure also requires as input the north and east coordinates of either the arc's (**a**) center point, or (**b**) start point, and the radial north azimuth towards the start point of the arc. With L and M known, the central angle (Δ), chord length (C), and arc length (L) are found by expressing the middle ordinate (M) and the radius (R) as:

$$M = R - R\cos\left(\frac{1}{2}\Delta\right) = R\left[1 - \cos\left(\frac{1}{2}\Delta\right)\right] \quad (5.2k1)$$

$$R = \frac{L}{\Delta} \quad (5.2k2)$$

Substituting Equation 5.2k2 i Equation 5.2k1 we get

$$M = \frac{L}{\Delta}\left[1 - \cos\left(\frac{\Delta}{2}\right)\right] \quad (5.2k3)$$

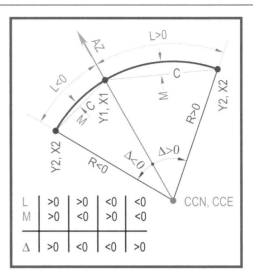

Figure 5.2f Arc Given the Arc Length and the Middle Ordinate

In the last equation we have only one unknown, the angle D, but in two forms, its actual value and the cosine of half of its value. This equation cannot be solved by conventional means. Therefore we will utilize an iterative process. For this we rearrange its terms to express it as

$$\frac{\Delta}{1 - \cos\left(\frac{\Delta}{2}\right)} = \frac{L}{M} \quad (5.2k4)$$

and then let the left and right sides of this equation be expressed as

$$T_1 = \frac{\Delta}{1 - \cos\left(\frac{\Delta}{2}\right)} \quad (5.2k5)$$

and

$$T_2 = \frac{L}{M} \quad (5.2k6)$$

To solve this equation, the procedure assumes a seed central angle (Δ) of 10°, and solves for T_1 and T_2. If they are not equal the angle is increased or decreased by 5°, and the process is repeated. Whenever T_1 and T_2 alternate in size, the process is backed up one step, the increment is halved, and the process is repeated until T_1 and T_2 are equal to each other within a tolerance of 0.0009 ft (0.27432 mm), or until 200 trials have been performed.

With the above information the coordinates of the PC and PT points and the corresponding azimuths from the center point towards them may now be computed. Due to the specified input parameters, the absolute value of the central angle is restricted to less the 180°. Once the remaining three parameters have been determined, this procedure is similar to the preceding three procedures with the differences being the input and output parameters and the equations that solve for the output parameters.

```
Sub icarcLnM (Y, X, thePT, AZ1, ARC, M, _
              Y1, X1, Y2, X2, PCN, PCE, PTN,
              PTE, CCN, CCE, AZ2, azimPC, _
              azimPT, azimCD, CAng, C, R, IND)
```

The input parameters to this procedure include the:

Y, X	The north and east coordinates of either (a) the center point, or (b) the start point of the arc to be constructed.
thePT	An indicator identifying the point to which the (Y, X) coordinates pertain. • C or c denotes the center point of the arc. • S or s denotes the start point of the arc.
AZ1	The north radial azimuth (in radians) from the center point (CCN, CCE) towards the start point of the arc to be constructed.
ARC	The radius of the arc to be constructed, the value of which if it is: • Positive denotes a clockwise arc rotation. • Negative denotes a counterclockwise arc rotation. • Zero aborts the command.
M	The middle ordinate of the arc to be constructed, the value of which if it is: • Positive, denotes an arc progressing along the direction implied by ARC. • Negative, denotes an arc opposing the said implied direction. • Zero aborts the command.

NOTE that since the arc is always displayed (plotted) in a counterclockwise rotation, the (Y1, X1) and (Y2, X2) point coordinates may not necessarily be those of the PC (PCN, PCE) and PT (PTN, PTE) points, respectively, but rather the reverse.

The parameters that are returned from this procedure include the:

Y1, X1	The north and east coordinates of the start point of the arc to be constructed.
Y2, X2	The north and east coordinates of the end point of the arc to be constructed.
PCN, PCE	The north and east coordinates of the PC point of the arc to be constructed.
PTN, PTE	The north and east coordinates of the PT point of the arc to be constructed.
CCN, CCE	The north and east coordinates of the center point of the arc to be constructed.
AZ2	The north radial azimuth (in radians) from the center point (CCN, CCE) towards the end point of the arc to be constructed.
azimPC	The radial north azimuth towards the PC point of the arc to be constructed.
azimPT	The radial north azimuth towards the PT point of the arc to be constructed.
azimCD	The radial north azimuth towards the midpoint of the chord of the arc to be constructed from the start point towards the end point.

CAng	The central angle (in radians) of the arc to be constructed: • Positive for a clockwise arc rotation. • Negative for a counterclockwise arc rotation.
C	The chord length of the arc to be constructed.
R	The radius of the arc to be constructed, the value of which if it is: • Positive denotes a clockwise arc rotation. • Negative denotes a counterclockwise arc rotation.
IND	An indicator denoting whether an error has been encountered, or not. Its value may be composed of one to three digits, each one denoting the following: 0 No error has been encountered. 1 A zero value has been specified for CAng. 2 A zero value has been specified for ARC. 3 An erroneous code has been assigned to thePT.

Because of the similarity of this procedure with the icar-cAnL procedure their operations are also similar and therefore reference is made to the said icarcAnL procedure.

5.2.7 Arc Given the Chord Length and Arc Length

The icarcCdL procedure constructs an arc given its chord length (C) and the arc length (L) as depicted in Figure 5.2g. To position the arc the north and east coordinates of either the arc's (**a**) center point, or (**b**) start point are also required as input. Since when the chord length is a known parameter it is usually the chord's direction that is known, this procedure deviates a bit from the preceding four procedures by requiring (**a**) the radial north azimuth towards the start point of the arc if the coordinates of the center point are specified as being known, or (**b**) the direction of the chord if the coordinates of the start point are specified as being known. With the chord length (C) expressed as

$$C = 2R\sin\left(\frac{\Delta}{2}\right) \tag{5.2m1}$$

and the radius (R) expressed by Equation 5.2k2, where Δ is the central angle, the chord length may now be expressed as

$$C = \frac{2L}{\Delta}\sin\left(\frac{\Delta}{2}\right) \tag{5.2m2}$$

To solve Equation 5.2m2 an iterative process like the one used for Equation 5.2k3 is employed to determine the value of Δ by letting T_2 C, and letting T_1 to be

$$T_1 = \frac{2L}{\Delta}\sin\left(\frac{\Delta}{2}\right) \tag{5.2m3}$$

Once the central angle (Δ) has be determined the remaining parameters, the radius (R) and middle ordinate (M) may be determined by Equations 5.2k2 and 5.2k1, respectively.

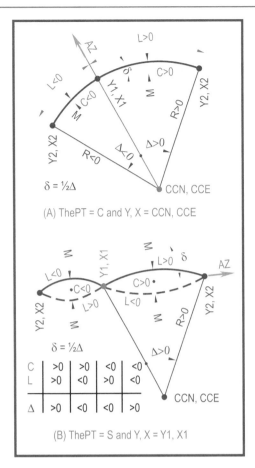

Figure 5.2g Arc Given the Chord and Arc Lengths

In perusing top portion (A) of Figure 5.2g the reader would notice that if the given point is that of the center point, it follows that the chord length and the arc length should both be of the same sign (either both positive or both negative). Should one of these two parameters have an opposite sign, then the resultant curve is flipped 180° about the chord. Thus:

- If C>0 and L<0, the right arc is the one that is so flipped, and
- If C < 0 and L > 0, the left arc is the one that is so flipped.

```
Sub icarcCdL(Y, X, thePT, AZ1, C, ARC, _
            Y1, X1, Y2, X2, PCN, PCE, PTN,
            PTE, CCN, CCE, AZ2, azimPC, _
            azimPT, azimCD, CAng, R, M, IND)
```

The input parameters to this procedure include the:

Y, X	The north and east coordinates of either (a) the center point, or (b) the start point of the arc to be constructed.
thePT	An indicator identifying the point to which the (Y, X) coordinates pertain. • C or c denotes the center point of the arc. • S or s denotes the start point of the arc.
AZ1	The north azimuth (in radians) depending on the value of thePT, and which if it is: C or c AZ1 is from the center point (CCN, CCE) towards the start point of the arc to be constructed. S or s AZ1 is from the start point (Y1, X1) towards the end point (Y2, X2) of the arc to be constructed.
C	The chord length of the arc to be constructed, the value of which if it is: • Positive denotes a clockwise arc rotation if thePT=C or c, or an arc progression along the direction AZ1 if thePT=S or s. • Negative denotes a counterclockwise arc rotation if thePT=C or c, or an arc progression opposing the direction AZ1 if thePT=S or s. • Zero aborts the command.
ARC	The radius of the arc to be constructed, the value of which if it is: • Positive denotes a clockwise arc rotation if thePT=C or c, or an arc progression along the direction implied by C if thePT=S or s. • Negative denotes a counterclockwise arc rotation if thePT=C or c, or an arc progression opposing the direction implied by C if thePT=S or s. • Zero aborts the command.

NOTE that since the arc is always displayed (plotted) in a counterclockwise rotation, the (Y1, X1) and (Y2, X2) point coordinates may not necessarily be those of the PC (PCN, PCE) and PT (PTN, PTE) points, respectively, but rather the reverse. Also refer to the commentary that follows the returned parameters regarding the signs of the chord and arc lengths.

The parameters that are returned from this procedure include the:

Y1, X1	The north and east coordinates of the start point of the arc to be constructed.
Y2, X2	The north and east coordinates of the end point of the arc to be constructed.
PCN, PCE	The north and east coordinates of the PC point of the arc to be constructed.
PTN, PTE	The north and east coordinates of the PT point of the arc to be constructed.
CCN, CCE	The north and east coordinates of the center point of the arc to be constructed.
AZ2	The north radial azimuth (in radians) from the center point (CCN, CCE) towards the end point of the arc to be constructed.
azimPC	The radial north azimuth towards the PC point of the arc to be constructed.

azimPT	The radial north azimuth towards the PT point of the arc to be constructed.
azimCD	The radial north azimuth towards the midpoint of the chord of the arc to be constructed from the start point towards the end point.
CAng	The central angle (in radians) of the arc to be constructed: • Positive for a clockwise arc rotation. • Negative for a counterclockwise arc rotation.
R	The radius of the arc to be constructed, the value of which if it is: • Positive denotes a clockwise arc rotation. • Negative denotes a counterclockwise arc rotation.
M	The middle ordinate of the arc to be constructed, the value of which is positive.
IND	An indicator denoting whether an error has been encountered, or not. Its value may be composed of one to three digits, each one denoting the following: 0 No error has been encountered. 1 A zero value has been specified for CAng. 2 A zero value has been specified for ARC. 3 An erroneous code has been assigned to thePT.

The general of operation of this procedure is similar to its preceding four but with the difference that once all of the parameters to be returned have been initialized, the procedure performs an iterative process similar to that of the preceding procedure to determine the value of the central angle and then computes the values of the remaining parameters. Thereafter, this procedure queries thePT. If its value is:

- S or s   The procedure:
  - Computes the azimuths to be returned using as a start AZ1 by identifying the four cases of Figure 5.2g (refer to the lower part B).
  - Computes the coordinates of the center point, and sets the coordinates of the start point to be those of the input point.
- C or c   The procedure:
  - Computes the five azimuths to be returned using as a start the input azimuth of the radial line towards the start point of the arc by identifying four distinct cases depending on whether the chord length and arc length are positive or negative (refer to the upper part A of Figure 5.2g).
  - Computes the coordinates of the start point, and of the center point in case the arc to be constructed is to be flipped about its chord.

Thereafter the procedure adjusts the azimuths to be returned for straddling north, if need be, and then computes the coordinates of the arc points to be returned, and then terminates.

5.2.8   Arc Given the Chord Length and Middle Ordinate

The icarcCdM procedure constructs an arc given its arc chord length (C) and the length of the middle ordinate (M) as depicted in Figure 5.2h. To position the arc the north and east coordinates of either (**a**) the arc's center point, or (**b**) the arc's start point are also required as input. Since when the chord length is given it is usually its direction that is known, thus this procedure requires (**a**) the radial north azimuth from the arc's center point towards the start point if the coordinates of the center point are specified, or (**b**) the direction of the chord if the coordinates of the start point are specified. With the middle ordinate (M) defined by Equation 5.2k1, the radius (R) may be expressed as

$$R = \frac{M}{1 - \cos(\Delta)} \qquad (5.2n1)$$

By introducing the above value of R in Equation 5.2m1 we can express the chord length (C) as

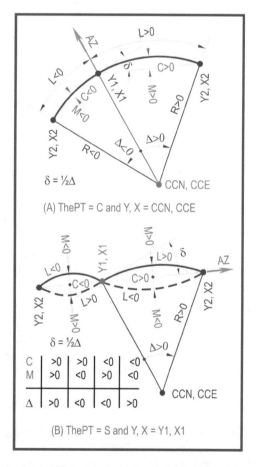

Figure 5.2h Arc Given the Chord Length and the Middle Ordinate

$$C = \frac{2M\sin\left(\dfrac{1}{2}\Delta\right)}{1-\cos\left(\dfrac{1}{2}\Delta\right)} \qquad (5.2n2)$$

If we then rearrange the terms of the above equation with the unknown angle on one side of the equation we get

$$\frac{1-\cos\left(\dfrac{1}{2}\Delta\right)}{\sin\left(\dfrac{1}{2}\Delta\right)} = \frac{2M}{C} \qquad (5.2n3)$$

Next, (**a**) if we remember the half angle tangent trigonometric identity as that of Equation 5.2n4 and we replace the x of the said equation with $\Delta/2$, and (b) if we notice the similarity of the right most term of Equation 5.2n4 with the inverse of the left side of Equation 5.2n3, we can rewrite 5.2n3 as Equation 5.2n5

$$\tan\left(\frac{1}{2}x\right) = \pm\sqrt{\frac{1-\cos(x)}{1+\cos(x)}} = \frac{\sin(x)}{1+\cos(x)} = \frac{1-\cos(x)}{\sin(x)} \qquad (5.2n4)$$

$$\tan\left(\frac{1}{4}\Delta\right) = \frac{2M}{C} \qquad (5.2n5)$$

from which the middle ordinate and the central angle may be expressed as

$$M = \frac{C}{2}\tan\left(\frac{1}{4}\Delta\right) \qquad (5.2n6)$$

and

$$\Delta = 4\tan^{-1}\left(\frac{2M}{C}\right) \qquad (5.2n7)$$

Having found the central angle (Δ), the radius (R) and the arc length (L) may be determined by Equations 5.2n1 and 5.2j2, respectively.

```
Sub icarcCdM(Y, X, thePT, AZ1, C, M, _
        Y1, X1, Y2, X2, PCN, PCE, PTN,
        PTE, CCN, CCE, AZ2, azimPC, _
        azimPT, azimCD, CAng, R, ARC, IND)
```

The input parameters to this procedure include the:

Y, X   The north and east coordinates of either (a) the center point, or (b) the start point of the arc to be constructed.

thePT   An indicator identifying the point to which the (Y, X) coordinates pertain.
- C or c denotes the center point of the arc.
- S or s denotes the start point of the arc.

AZ1   The north azimuth (in radians) depending on the value of thePT, and which if it is:
C or c   AZ1 is from the center point (CCN, CCE) towards the start point of the arc to be constructed.
S or s   AZ1 is from the start point (Y1, X1) towards the end point (Y2, X2) of the arc to be constructed.

C   The chord length of the arc to be constructed, the value of which if it is:
- Positive denotes:
  - A clockwise arc rotation if thePT=C or c, or
  - An arc progression along the direction AZ1 if thePT=S or s.
- Negative denotes:
  - A counterclockwise arc rotation if thePT=C or c, or
  - An arc progression opposing the direction AZ1 if thePT=S or s.
- Zero aborts the command.

M   The middle ordinate of the arc to be constructed, the value of which if it is:
- Positive denotes:
  - A clockwise arc rotation if thePT=C or c, or
  - An arc progression along the direction implied by the C input parameter if thePT=S or s.
- Negative denotes:
  - A counterclockwise arc rotation if thePT=C or c, or
  - An arc progression opposing the direction implied by the C input parameter if thePT=S or s.
- Zero aborts the command.

NOTE that since the arc is always displayed (plotted) in a counterclockwise rotation, the (Y1, X1) and (Y2, X2) point coordinates may not necessarily be those of the PC (PCN, PCE) and PT (PTN, PTE) points, respectively, but rather the reverse.

The parameters that are returned from this procedure include the:

Y1, X1   The north and east coordinates of the start point of the arc to be constructed.

Y2, X2   The north and east coordinates of the end point of the said arc.

PCN, PCE   The north and east coordinates of the PC point of the said arc.

PTN, PTE   The north and east coordinates of the PT point of the said arc.

CCN, CCE   The north and east coordinates of the center point of the said arc.

AZ2   The north radial azimuth (radians) towards the end point of the said arc.

azimPC   The radial north azimuth towards the PC point of the said arc.

azimPT   The radial north azimuth towards the PT point of the said arc.

azimCD   The radial north azimuth towards the midpoint of the chord of the said arc from the start point towards the end point.

CAng   The central angle (in radians) of the said arc:
- Positive for a clockwise arc rotation.
- Negative for a counterclockwise arc rotation.

R	The radius of the arc to be constructed, the value of which if it is: • Positive denotes a clockwise arc rotation. • Negative denotes a counterclockwise arc rotation.
ARC	The radius of the arc to be constructed, the value of which if it is: • Positive denotes a clockwise arc rotation. • Negative denotes a counterclockwise arc rotation.
IND	An indicator denoting whether an error has been encountered, or not. Its value may be composed of one to three digits, each one denoting the following: 0 No error has been encountered. 1 A zero value has been specified for CAng. 2 A zero value has been specified for ARC. 3 An erroneous code has been assigned to thePT.

The operation of this procedure is similar that of `icarcCdL` with the exceptions that (**a**) an iterative process to determine the unknown parameters is not used, and (**b**) when computing the center point coordinates for cases in which the arc is flipped about its chord, the distance from the center point to the midpoint of the chord (tmp2) is computed as:

- tmp2 = 2# * (radabs − Abs(M)) in `icarcCdM` because M is a negative input value, while
- tmp2 = 2# * (radabs − M) in `icarcCdL` because M is computed as a positive value.

5.2.9 Circular Arc Through Three Points

The `icarccn3` procedure constructs an arc defined by the Cartesian coordinates of three points. The procedure computes the north and east coordinates of the arc's center point and the radial north azimuths towards the first and third of the three points that define the arc (see Figure 5.2i). The three given points to define the arc may be specified in either a clockwise or counterclockwise direction.

To determine the center point of the arc the procedure:

- Computes the coordinates of the midpoints of the chords that join the first point with the second point, and the second point with the third point, as well as their respective north azimuths.
- Intersects the normal lines to the said two chords at their midpoints to find the center point's coordinates. However, before this intersection, the procedure checks for a potential collinearity of the three given points. If they are found to be collinear within a tolerance of 0.000001 ft (0.0003048 mm), the procedure assumes a straight line, and returns its midpoint coordinates for the center point parameters and the counterclockwise normal azimuth for each of the azimuth parameters to be returned.

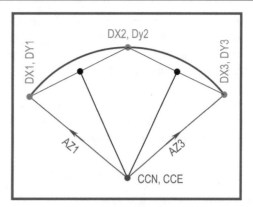

Figure 5.2i Circular Arc Through Three Points

- The radius of the said arc is set to be the average of the distances from the center point to each of the three given points.

```
Sub icarccn3 (DX1, DY1, DX2, DY2, DX3, DY3, _
              CCN, CCE, R, AZ1, AZ3)
```

The input parameters to this procedure include the:

DX1, DY1	The Cartesian coordinates of the start point of the arc to be constructed.
DX2, DY2	The Cartesian coordinates of a point somewhere along the said arc.
DX3, DY3	The Cartesian coordinates of the end point of the said arc.

The parameters that are returned from this procedure include the:

CCN, CCE	The north and east coordinates of the center point of the arc to be constructed.
R	The radius of the said arc.
AZ1, AZ3	The radial north azimuths towards the first and third of the three given points, respectively.

Upon invocation, the procedure:

1. Calls the `icforce` procedure to inverse between (DX1, DY1) and (DX2, DY2), and between (DX2, DY2) and (DX3, DY3) to determine the corresponding chord lengths D1 and D2, and azimuths AZ1 and AZ2.
2. Calls the `icptl` procedure to compute the coordinates of the chord midpoints by traversing along the above chord azimuths a plus distance of half the chord length and an offset distance of zero.
3. Checks for colinearity of the two chords as stated above, and if they:
   - Are colinear computes the midpoint and the counterclockwise normal azimuths.

- Are not colinear computes the normal azimuths to the said chords, and then calls the `icintrs` procedure (see Chap. 6) to intersect the said normal lines and compute the coordinates of the center point.

4. Calls the `icforce` procedure to inverse from the said center point to each of the three given points to find the corresponding lengths and azimuths, and averages the said lengths to determine the radius.

The `icarccn3` *procedure does not treat three collinear points as the diameter of a semicircle. It treats them as implying a straight line.*

5.2.10 Circular Arc Through Two Points and Radius

The `icarccnt` procedure constructs an arc defined by the north and east coordinates of two points and its radius, noting that a positive radius denotes a clockwise curve from the first point towards the second point, while a negative radius denotes a counterclockwise curve. The procedure computes and returns the north and east coordinates of the center point of the arc. With reference to Figure 5.2j note that if the value of the radius is:

- Zero, within a tolerance of 0.0009 ft (0.27432 mm), the returned coordinates are those of the midpoint of the implied line between the two given points.
- One half the length implied between the two points, within a tolerance of 0.005 ft (1.524 mm), the arc is

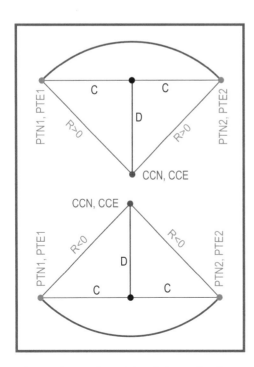

Figure 5.2j Circular Arc Through Two Points and Radius

assumed to be a semicircle with the said implied line being the diameter.

- Not large enough to introduce an arc between the two given points, the returned coordinates of the center point are set to be those of the first given point.

The arc which is created by this procedure subtends a central angle of less than 180°.

```
Sub icarccnt (PTN1, PTE1, PTN2, PTE2, R, _
                CCN, CCE)
```

The input parameters to this procedure include the:

PTN1, PTE1	The north and east coordinates of the start point of the arc to be constructed.
PTN2, PTE2	The north and east coordinates of the arc's end point.
R	The radius of the said arc which may be
	• Positive for a clockwise arc,
	• Negative for a counterclockwise arc, or
	• Zero to imply a straight line between the two given points.

The parameters that are returned from this procedure include the:

CCN, CCE	The north and east coordinates of the center point of the arc to be constructed.

Upon invocation, the procedure:

1. Queries the radius to determine whether its value is equal to zero within the said tolerance, or not, and
   - If so, computes the average of the two north coordinates, and the average of the two east coordinates of the two given point to determine the coordinates of the implied line's midpoint.
   - If not, the procedure continues as indicated below.
2. Calls the `icforce` procedure to inverse from (PTN1, PTE1) to (PTN2, PTE2) to find the chord length and azimuth.
3. Checks to determine whether the chord length is twice the value of the radius within the said tolerance, in which case the chord is the diameter, or not, and
   - If so, computes the coordinates of the center point of the chord in a manner similar to that of computing the coordinates of the implied line under Step 1.
   - If not, continues as indicated below.
4. Applies the Pythagorean theorem to find the square of the offset distance (D) of the arc's center point from the chord as

$$D^2 = R^2 - C^2 = (R+C)(R-C)$$

where C is one half the chord length.

5. Checks to determine whether D^2 is negative, in which case the given radius is too small for the chord length, or not, and
   - If so, and (CCN, CCE) coordinates are set to be those of (PTN1, PTE1).
   - If not, the procedure continues as indicated below.
6. Computes said offset distance D by finding square root of D^2, and then calls the `icptl` procedure to compute the coordinates of the center point (CCN, CCE) by traversing from the first given point towards the other a distance of C and an offset distance of D taking into consideration the sign of the radius R.

5.3 Arcs to Polylines and Vice Versa

Once an arc has been created mathematically, it needs to be displayed on the computer monitor, or plotted on a piece of paper or other media. A point is but a dot, while a straight line is a line between two points, which theoretically is displayed with a staircase effect, except for a Cartesian rotation of 0, 45, or 90° direction. However, an arc cannot be drawn as an arc by rotating a line about a point, the center point. Thus the line has to be decomposed into a series of chords that constitute a polyline. In so doing, the radius and other pertinent arc parameters may be saved with the graphic arc, or they may not so be saved.

In the early years of electronic computing, text was printed and drawings were plotted. Thus, there were printers and plotters. Nowadays text and drawings are being printed, and there are no plotters.

On the other hand, an arc is at times digitized from an existing hard copy. When so digitizing an arc the reverse is taking place. The person that does the digitization recreates the arc in the form of a polyline composed of small chords, in which case the geometric parameters of the arc may or may not be available, or just disregarded by the person that does the digitizing. Furthermore, it is possible for a user to receive an arc that has been created mathematically, but without any of the pertinent arc parameters having been saved.

Thus at times it becomes necessary to transform an arc into a polyline of chords, or to decode whether a graphic arc is indeed an arc, or not. The procedures described in this section address these issues.

5.3.1 Decomposition of an Arc into Chords

The `icarcseg` procedure decomposes a circular arc into a series of chords so that the middle ordinate of each chord does not exceed a user specified value. The intent of this pro-

cedure is to generate a series of chords for drawing a smooth curve depending on the scale of the drawing in which the arc is to be drawn. The smoothness is controlled by the magnitude of the said middle ordinate (the smaller the value of the middle ordinate, the smoother the look of the curve). In this procedure, the arc is defined by its central angle in radians (θ), which is positive for a clockwise arc, and negative for a counterclockwise arc, and its radius (R).

The `icarcseg` procedure is operable in conjunction with two other procedures, the `icdrwic4` procedure that calls the icarcseg procedure, and the `icdrwcrc` that should be called by the main program after the call to the icarcseg procedure.

Ideally, the application of this procedure and its other two associate procedures (refer to the side bar on this page) should be operable interactively with the scale of the instantaneous display of a view on the monitor, or plot without user interruption. However, this is dependent upon the operating system in use and the involved complexity to attain this goal. For the purpose of this text we will disregard this issue. Regarding the scale and middle ordinate, note the following:

1. The scale should be positive, and expressed in terms of 1 plot unit per world units. That is a scale of:
   - $1'' = 200'$ (US Customary system) is specified as 200, and a scale of
   - 1:2400 (SI) is specified as 0.2400.
2. The middle ordinate should be positive, and should be expressed in terms of plot units, that is in terms of inches or millimeters.

```
Sub icarcseg (ANG, R, maxmid, SCALE, _
              SEGS, Cang)
```

The input parameters to this procedure include the:

ANG	The central angle of the arc (positive is clockwise, negative is counterclockwise).
R	The radius of the arc (positive).
maxmid	The maximum desired middle ordinate (positive) in inches (mm).
SCALE	The scale for plotting (display) purposes as per the above comments.

The parameters that are returned from this procedure include the:

SEGS	The number of chords (segments) into which the arc has been decomposed.
Cang	The central angle of each arc segment (positive is clockwise, negative is counterclockwise).

Upon invocation, the procedure preserves the input values of the radius, maximum middle ordinate and scale, and creates temporary variables represented by their absolute values, in case they have been specified with negative values. Thereafter, the procedure:

1. Converts the middle ordinate in terms of world units by multiplying the middle ordinate by the scale. Thus in the above sited example of scales, if the middle ordinate is specified as being:
   • 0.001 in., the middle ordinate becomes 0.001×200, or 0.2 ft, or
   • 0.0254 mm, the middle ordinate becomes 0.0254×0.2400, or 0.006096 m.
2. Queries the middle ordinate as computes above whether it is equal to or greater than the specified radius, or not, and if so, the middle ordinate is set to be one tenth the value of the radius.
3. Computes the central angle θ of the chord as being:

$$q = 2 \left[\frac{\cos^{-1}(R-M)}{R} \right] \qquad (5.3)$$

   where M is the last computed value of the middle ordinate.
4. Queries angle θ whether it is equal to zero within a tolerance of 0.00000009 radians (0.018 s), or not, and if so, it is set to be 0.000001 radians (0.206 s).
5. Computes the number of chords SEGS by dividing the central angle of the arc ANG by the central angle of the chord θ.
6. Queries the number of segments SEGS whether they are less or equal to zero, or not, and if so they are set to be two, and the central angle of the chord is recomputed.
7. Imposes the criterion that the deflection angle of the chord (half angle θ) does not exceed 8°, and recomputes angle θ and the number of chords SEGS.
8. Applies two more conditional checks, which are based on experience, in order to improve the smoothness of the arc's appearance, and yet maintain a minimum number of segment chords.
   (a) If the number of chords SEGS is within the limits indicated below, SEGS is set to the upper limit of the indicated range limits.
       $60 < \text{SEGS} < 80$ $40 < \text{SEGS} < 60$ $20 < \text{SEGS} < 40$ $00 < \text{SEGS} < 20$
   (b) If the absolute value of the central angle of the overall arc is greater than the values indicated below, the number of chords SEGS is multiplied by the indicated value.
       • $|\text{ANG}| > 6.28$ radians (360°) then ANG=ANG * 3.25
       • $|\text{ANG}| > 5.49$ radians (315°) then ANG=ANG * 3.25
       • $|\text{ANG}| > 4.71$ radians (270°) then ANG=ANG * 2.75
       • $|\text{ANG}| > 3.92$ radians (225°) then ANG=ANG * 2.50
       • $|\text{ANG}| > 3.14$ radians (180°) then ANG=ANG * 2.25
       • $|\text{ANG}| > 2.35$ radians (135°) then ANG=ANG * 2.00
       • $|\text{ANG}| > 1.57$ radians (90°) then ANG=ANG * 1.75
       • $|\text{ANG}| > 0.78$ radians (45°) then ANG=ANG * 1.50
9. Recomputes the value of the central angle of the unit chord, and terminates. The calling program of this procedure may now continue by calling the icdrwcrc procedure addressed below, or incorporate the icarcseg procedure into a procedure similar to that of the icdrwic4 procedure also addressed below.

5.3.2 Polyline from Arc (Follows icarcseg)

The icdrwcrc procedure is a follow-up procedure call of the main program to the call for the icarcseg procedure. Whereas icarcseg determines the number chord segments and chord central angle, icdrwcrc creates a polyline from an arc. The source code as included in the accompanying CD and summarized below is oriented for the ArcObjects of ArcGIS environment. For any other environment certain statements will need to be modified.

```
Sub icdrwcrc (XCC, YCC, X1, Y1, AZ1, X2, Y2,
              AZ2, R, Cang, numPnt, _
              pl As esriGeometry.IPolyline)
```

The input parameters to this procedure include the following parameters which refer to the same arc for which the icarcseg procedure had last been called prior to this call:

XCC, YCC	The Cartesian coordinates of the center point of the arc to be converted to a polyline.
X1, Y1	The Cartesian coordinates of the start point of the said arc.
AZ1	The radial north azimuth to the start point (X1, Y1) of the said arc.
X2, Y2	The Cartesian coordinates of the end point of the said arc.
AZ2	The radial north azimuth to the end point (X2, Y2) of the said arc.
R	The radius of the said arc.

Whereas the `Icarcseg` *procedure determines the number of segment chords and the unit chord central angle, the* `icdrwcrc` *procedure is the one that creates the actual polyline.*

`Cang`	The central angle of the unit chord as computed and returned by the `icarcseg` procedure.
`numPnt`	The number of intermediate vertices of the polyline to be created, and which is equal to the value of `SEGS` as returned by the `icarcseg` procedure minus one (the start and end points of the arc are being excepted).

The parameters that are returned from this procedure include the:

`pl`	The actual polyline representation of the input arc in ArcObjects `As esriGeometry.IPolyline` form.

Upon invocation, the procedure:

1. Creates a collection of points.
2. Places in the said collection the coordinates of the starting point of the arc (`X1, Y1`).
3. Adds sequentially to the said collection the coordinates of the intermediate vertex points, each of which is computed by starting with the start radial north azimuth (`AZ1`) and accumulating the central angles of the unit chord `numPnt` times. As each azimuth is accumulated, the procedure calls the `icptl` procedure to traverse from the arc's center point along the last accumulated azimuth a distance equal to the radius `R`, and an offset distance of zero.
4. Places the coordinates of the ending point of the arc (`X2, Y2`).

5.3.3   Polyline from Arc (Calls `icarcseg`)

The `icdrwic4` procedure may be called by the main program to convert an arc into a polyline by combining the calls to the `icarcseg` and to the `icdrwcrc` procedures. The source code of this procedure as included in the accompany-ing CD and summarized below is oriented for the ArcObjects of ArcGIS environment, and should be modified for any other environment.

```
Sub icdrwic4 (XCC, YCC, XS, YS, XE, YE, _
               pl As esriGeometry.IPolyline)
```

The input parameters to this procedure include the:

`XCC, YCC`	The Cartesian coordinates of the center point of the arc to be converted to a polyline.
`XS, YS`	The Cartesian coordinates of the start point of the said arc.

The `icarcseg` *procedure is operable in conjunction with two other procedures, the* `icdrwic4` *procedure that calls the icarcseg procedure, and the icdrwcrc that should be called by the main program after the call to the icarcseg procedure.*

`XE, YE`	The Cartesian coordinates of the end point of the said arc.

The parameters that are returned from this procedure include the:

`pl`	The actual polyline representation of the input arc in ArcObjects `As esriGeometry.IPolyline` form.

Upon invocation, the procedure:

1. Calls the `icforce` procedure to determine the radial north azimuth at the start point (`azimS`) and at the end point (`azimE`) of the arc.
2. Computes the central angle of the arc by subtracting (`azimS`) from (`azimE`).
3. Tests the central angle for straddling north and adjusts it accordingly.
4. Calls the `icarcseg` procedure to determine the number of chord segments, and unit chord central angle.
5. Calls the `icdrwcrc` procedure to create the polyline.

Intersections

6

The procedures addressed in this chapter are those that enable the programmer to intersect lines with lines and lines with circular curves. Procedures that intersect spirals with lines, circular arcs, and other spirals are addressed separately, due to their nature, in Chap. 9.

In perusing this chapter, the reader will notice certain procedures that seem to perform the same operation. However, there are certain distinctions between such similar procedures as to the required input parameters, and the parameters that are returned. Furthermore, there certain procedures that intersection features of a specific type, such as intersecting a line with a line, or a curve with a curve, and there are certain others that are more generic as to what feature type is intersected with what other feature type.

In addition, there are certain procedures, particularly the more generic ones, that their source code as included in the accompanying CD utilizes certain ArcObjects type of statements. Such statements are brought to the attention of the reader for modification to a different programming environment.

The procedures presented below are segregated into two distinct groups, specific intersections and generic intersections. The specific intersection procedures are further divided into three subgroups of intersections, line with line, line with curve, and curve with curve.

6.1   Line with Line Intersections

Two procedures, the `icintrs` and the `icln2int`, that compute the point of intersection of two lines are presented in this section. Neither of them intersect polylines. With the `icintrs` procedure, each of the two lines to be intersected is defined by the north and east coordinates of a point and a north azimuth, while with the `icln2int` procedure, each of the said lines to be intersected is defined by the Cartesian coordinates of its two endpoints. Eventually, `icln2int` calls on `icintrs` to actually compute the coordinates of the point of intersection.

6.1.1   Line with Line Intersection (Basic)

The `icintrs` procedure computes the north and east coordinates of the point of intersection of two lines, each defined by a point and an azimuth, as well as certain other parameters as a result of the intersection as displayed in Figure 6.1a and described below. In computing the point of intersection, the two azimuths are considered to by extended as needed to compute an intersection in either direction from the point through which a line is to pass.

```
Sub icintrs (PTN1, PTE1, AZ1, PTN2, PTE2, AZ2, _
             GANG, DIST, SKEWD, PTN3, PTE3)
```

The input parameters to this procedure include the:

PTN1, PTE1	The north and east coordinates of a known point on the first line.
AZ1	The north azimuth of the first line. It may point in either direction.
PTN2, PTE2	The north and east coordinates of a known point on the second line.
AZ2	The north azimuth of the second line. It may point in either direction.

The parameters that are returned from this procedure include the:

GANG	The angle, in radians, formed by the two intersecting lines. A positive GANG implies a clockwise rotation, while a negative GANG implies a counterclockwise rotation. A value of −9999 denotes that the lines are colinear (GANG=0 or π).
DIST	The distance from the given point on the first line (PTN1, PTE1) to the point of intersection, the sign of which is dependent on the sign and magnitude of PLUS and X, with the sign of X being that of the sine of the angle GANG. Recall how the sine of positive or negative angles varies in the intervals of 0°, 90°, 180°, and 360°.
SKEWD	The distance from the given point on the second line (PTN2, PTE2) to the point of intersection, the sign of which is dependent on the sign and magnitude of OFF.
PTN3, PTE3	The north and east coordinates of the point of intersection.

© Springer International Publishing Switzerland 2016
E.C. Tonias, C.N. Tonias, *Geometric Procedures for Civil Engineers*, DOI 10.1007/978-3-319-24295-8_6

Upon invocation, the procedure initializes the parameters to be returned, and then:

1. Calls the `icprjct` procedure to project (`PTN2`, `PTE2`) on the first line.

   Depending on the magnitude of the two given azimuths, and the orientation of the two given points, whether they lie back or ahead of the intersection point with respect to the point that defines a line, there are eight possible conditions, of which two are shown in Figure 6.1a.

   In Figure 6.1a note the signs of the returned parameters `PLUS` and the `OFF` which for the possible conditions that are not shown in the said figure can be extrapolated depending on the position of the point of projection.

2. Computes the angle (`GANG`) formed by the two given azimuths as indicated above and checks for $|sin(GANG)| < 0.00000048$ radians (0.01 s of a degree), and if
   - So, assumes the lines to be parallel and sets the coordinates of the point of intersection to be midway between the two given points.
   - Not so, continues.

3. Computes the value of `SKEWD` as being $SKEWD = OFF/|sin(GANG)|$.
4. Computes the value of `X` as being $X = SKEWD * |cos(GANG)| * GangSign$

   where $GangSign = 1.0$ with the sign of the product of $sin(GANG) * cos(GANG)$.
5. Computes the value `DIST` as being $DIST = PLUS - X$.
6. Computes the coordinate of the point of intersection (`PTN3`, `PTE3`) by traversing with the `icptl` procedure from the point of the first line along the azimuth of the first line with a distance of `DIST` and an offset distance of zero (0.00).

Note that the angle `GANG` is not checked for straddling north. Thus, its value may be negative or greater than 360°.

6.1.2   Line with Line Intersection (preamble to `icintrs`)

The `icln2int` procedure computes the Cartesian coordinates of the point of intersection of two lines, each defined by the Cartesian coordinates of their endpoints. The point of intersection may lie outside the limits of the lines as defined by their endpoints.

```
Sub icln2int.   (X1, Y1, X2, Y2, X3, Y3, X4,
                 Y4, _
                 XINT, YINT, IFOUND)
```

The input parameters to this procedure include the:

X1, Y1	The X and Y Cartesian coordinates of the start point of the first line.
X2, Y2	The X and Y Cartesian coordinates of the end point of the first line.
X3, Y3	The X and Y Cartesian coordinates of the start point of the second line.
X4, Y4	The X and Y Cartesian coordinates of the end point of second first line.

The parameters that are returned from this procedure include the:

XINT, YINT	The X and Y Cartesian coordinates of the point of intersection.
IFOUND	An indicator, the value of which if it is:
	0 Denotes that an intersection cannot be found (the lines are colinear).
	1 Denotes that an intersection point has been found.

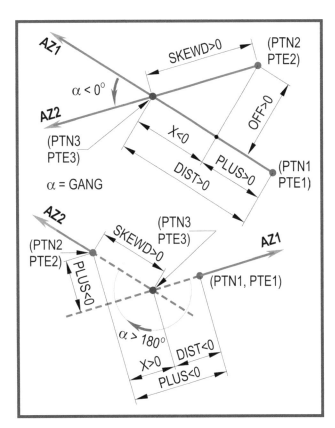

Figure 6.1a   Intersection of Two Lines

Upon invocation, the procedure initializes the parameters to be returned, and then:

1. Calls the `icforce` procedure to determine the azimuth of each of the two given lines.
2. Determines whether the two lines are collinear or parallel in either the same direction or in opposing direction within a tolerance of 0.00005 radians (0° 00′ 10.3″), or not, and:
   - If so, sets the coordinates of the point of intersection are set to be zero (0.00), and `IFOUND` is set to be zero (0), and terminates.
   - If not so, calls the `icintrs` procedure to find the point of intersection and terminates.

6.2   Line with Arc Intersections

Three procedures that compute the point of intersection between a line and a circular arc or with a full circle are presented in this section, and they are the `icceg52`, the `iclincrc` and the `iclincir`. All of these procedures utilize the same basic approach for the determination of the point of intersection as described under the `icceg52` procedure. The specific deviations from `icceg52` of the other two procedures are described under their individual procedures.

It is noted that when intersecting a line with an arc there are two possible points of intersection. The `icceg52` and the `iclincrc` procedures return only one point of intersection as noted below, while the `iclincir` procedure returns both points of intersection. In determining the intersection point or points, each of these three procedures do not limit the solution to the extent of a circular arc as it may actually exist in the work place. If an intersection happens to lie outside the physical extent of the arc, it is returned as an intersection. The procedures do not distinguish between an arc and a full circle.

6.2.1   Line with Arc Intersection (Basic)

The `icceg52` procedure computes the intersection point of a line defined by a point and a north azimuth with a circular arc defined by the north and east coordinates of its center point and its radius (see Figure 6.2a). Of the two possible intersections, this procedure returns the second intersection along the direction of the line, when the said line and arc are extended to intersect twice, if the radius is positive, and the first intersection if the radius is negative.

```
Sub icceg52  (PN, PE, AZ, CCN, CCE, R,_
              PIN, PIE, IERR)
```

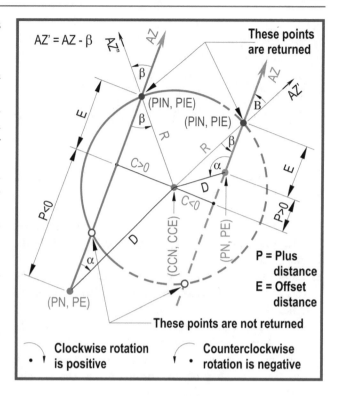

Figure 6.2a   Intersection of Line with Arc

The input parameters to this procedure include the:

PN, PE	The north and east coordinates of a known point on the line.
AZ	The north azimuth of the line. Refer to the commentary below regarding the direction of this azimuth.
CCN, CCE	The north and east coordinates of the center point of the curve.
R	The radius of the curve. • If R>0, the second intersection is chosen. • If R<0, the first intersection is chosen. • If R=0, within a tolerance of 0.005 feet (1.524 mm) the procedure aborts.

The parameters that are returned from this procedure include the:

PIN, PIE	The north and east coordinates of the point of intersection.
IERR	An indicator the value of which if it is: 0 Denotes that an intersection has been found. 1 Denotes that an intersection has not been found. That is the line does not cross the circle, or the arc if extended to be a circle. 2 Denotes that the radius is zero within the abovesaid tolerance. If the line happens to be tangent to the said circle or arc, the point of tangency is returned as the point of intersection.

Of the two possible intersections, the `icceg52` *procedure returns the second intersection along the direction of the line, when the said line and arc are extended to intersect twice, if the radius is positive, and the first intersection if the radius is negative.*

Upon invocation, the procedure initializes the parameters to be returned, and then:

1. Queries the radius to see if it is zero or not within the said tolerance, and if so it beeps, sets `IERR` to 2 and aborts; otherwise it continues.
2. Calls the `icprjct` procedure to project the center point (`CCN`, `CCE`) of the arc on the line to find the plus and offset distances, `P` and `C` respectively (see Figure 6.2a), the angle α, and the distance `D` from (`PN`, `PE`) to (`CCN`, `CCE`). Note from the description of the `icprjct` procedure that the values of `P`, `C`, and α may be positive or negative depending on the position of the center point (`CCN`, `CCE`) with respect to (`PN`, `PE`) and **AZ**, while the value of α is always positive.
3. Queries the absolute value of the radius and the absolute value of the computed offset distance of the center from the given line to determine whether the point of projection lies within, on, or outside the circle (or arc if the arc is considered to be extended) within a tolerance of 0.001 feet (0.3048 mm), and if the point of projection lies:
   - Outside the arc, there is no intersection. The procedure beeps, sets `IERR` to 1, and aborts.
   - On the arc, the line is tangent to the arc. The procedure computes the point of tangency and terminates.
   - Inside the arc, the procedure computes the intersection and terminates.
4. Applies the law of sines to find angle β as bing

$$\beta = \sin^{-1}\left[\sin(\alpha)\frac{D}{R}\right] \qquad (6.2.1\text{a})$$

   Note that β assumes the sign of α since both `D` and `R` are positive.
5. Computes the azimuth from (`CCN`, `CCE`) towards (`PIN`, `PIE`) as

$$AZ' = AZ - \beta \qquad (6.2.1\text{b})$$

6. Computes the coordinates of (`PIN`, `PIE`) by traversing along the above-computed **AZ** a distance of `R` and an offset distance of zero (0.00).

6.2.2 Line with Arc Intersection (Alternate #1)

The `iclincrc` procedure computes the north and east coordinates of the point of intersection of a two-point line, defined by the north and east coordinates of a point thereon and a north azimuth, with a circular arc, defined by the north and east coordinates of its center point and its radius. This procedure is similar to the `icceg52` procedure with the following differences:

Of the two possible intersections, the `iclincrc` *procedure always returns the second intersection along the direction of the line, when the said line and arc are extended to intersect twice regardless of the sign of the radius.*

1. The value of the returned error indicator `INTRSC` is not the same in denoting the type of error that has been encountered.
2. Whereas with the `icceg52` procedure the coordinates of the point of intersection to be returned is dependent upon the sign of the given arc radius, the `iclincrc` procedure always returns the second point of intersection of the given line being considered as extended indefinitely ahead and backwards so as to obtain two intersections regardless of the sign of the radius. Thus with the `iclincrc` procedure it is the direction of the line that specifies the intersection to be returned.
3. Whereas the `icceg52` procedure computes the angle β and azimuth **AZ'** (see Figure 6.2a) to determine the coordinates of the point of intersection, the `iclincrc` procedure computes the value of `E` as

$$E = -\sqrt{R^2 - C^2} = \sqrt{(R+C)(R-C)} \qquad (6.2.2\text{a})$$

   and then calls the `icptl` procedure to traverse from (`PN`, `PE`) along the given azimuth **AZ** a distance of `P + E` and an offset distance of zero (0.00).

```
Sub iclincrc  (PN, PE, AZ, CCN, CCE, R, _
               PIN, PIE, INTRSC)
```

The input parameters to this procedure include the:

`PN, PE`	The north and east coordinates of a known point on the line.
`AZ1`	The north azimuth of the line.

CCN, CCE	The north and east coordinates of the center point of the curve.
R	The radius of the curve (positive or negative—zero aborts the procedure).

The parameters that are returned from this procedure include the:

PIN, PIE	The north and east coordinates of the point of intersection. The same point is returned regardless of the sign of the radius.
AZ	The azimuth from the center point of the arc towards the point of intersection.
INTRSC	An indicator the value of which if it is:
	1 Denotes that an intersection has been found.
	2 Denotes that there is no intersection (the line does not cross the circle, or the arc if it is extended to be a circle).
	3 Denotes that the line happens to be tangent to the said circle, in which case the point of tangency is returned as the point of intersection.

Upon invocation, the procedure operates for all intents and purposes in the same manner as the `icceg52` procedure with the above-stated differences. The testing for the point of tangency is the same as that of the `icceg52` procedure.

6.2.3 Line with Arc Intersection (Alternate #2)

The `iclincir` procedure computes the north and east coordinates of the two points of intersection of a two-point line defined by the north and east coordinates of a point thereon and a north azimuth with a circular arc defined by the north and east coordinates of its center point and its radius which should be positive. This procedure is similar to the preceding one with the difference that `iclincir` returns both intersections by computing D1 and D2 as noted below. In finding these two distances, it should be noted that the distance E as solved by Equation 6.2.2a represents one half of the chord subtended by the arc as intersected by the line when extended across the arc's circle. Thus

$$D1 = P - E \quad \text{and} \quad D2 = P + E$$

Therefore, the computation of the coordinates of the said points of intersection are found by calling the `icptl` procedure and traversing from the point that defines the line along the line's azimuth AZ with plus distances equal to D1 and D2, and offset distance of zero (0.00).

Of the two possible intersections, the `iclincir` procedure always returns both intersection points.

```
Sub iclincir (PN, PE, AZ, CCN, CCE, R, _
              PIN1, PIE1, D1, PIN2, PIE2,
              D2, INTRSC)
```

The input parameters to this procedure include the:

PN, PE	The north and east coordinates of a known point on the line.
AZ	The north azimuth of the line. Refer to the commentary below regarding the direction of this azimuth.
CCN, CCE	The north and east coordinates of the center point of the curve.
R	The radius of the curve (should be positive).

The parameters that are returned from this procedure include the:

PIN1, PIE1	The north and east coordinates of the point of the first (back) intersection with respect to the direction AZ.
D1	The distance from the given point (PN, PE) on the line to the first (back) intersection.
PIN2, PIE2	The north and east coordinates of the point of the second (forward) intersection with respect to the direction AZ.
INTRSC	An indicator the value of which if it is:
	0 Denotes that there is no intersection.
	1 Denotes that the curve is tangent to the line.
	2 Denotes that two intersections have been found.
D2	The distance from the given point (PN, PE) on the line to the second (forward) intersection with respect to the direction AZ.

Note that if the line is tangent to the circle, or to the arc if extended to be a circle, then the returned coordinates PIN2 and PIE2 are the same as those of PIN1 and PIE1, respectively.

Upon invocation, the procedure operates for all intents and purposes in the same manner as the `iclincrc` procedure with the above-stated difference.

6.3 Arc with Arc Intersections

Two procedures that compute the point of intersection between two circular arcs or full circles are presented in this section, and they are the `iccrccrc` and the `icxcvwc`, which use different approaches to find the point of intersection. Note that the former returns only one selected intersection, while the latter returns both points of intersection.

6.3.1   Arc with Arc Intersection (Plus and Offset)

The `iccrccrc` procedure computes the north and east coordinates of the point of intersection between two circular arcs with each arc being defined by the north and east coordinates of its center point and its radius. Of the two potential intersections, the one that is returned is dependent on the sign of the radius of the second arc (R2) which is to be positive or negative as indicated below, while the radius of the first curve (R1) is treated as being be positive. Thus, the value of R2 should be:

- Positive if the desired point of intersection is the one which is to be located to the right of the implied line from the center point of the second arc towards the center point of the first arc, and
- Negative if the desired point of intersection is the one which is to be located to the left of the implied line from center point of the second arc towards the center point of the first arc.

To solve for the point of the desired point of intersection, the methodology implied by the equations below is employed. In Figure 6.3a

$$B^2 = R_2^{\,2} - d_2^{\,2} = R_1^{\,2} - d_2^{\,2} \qquad (6.3.1a)$$

Rearranging the terms to the right of B^2 we have

$$R_2^{\,2} - R_1^{\,2} = d_2^{\,2} - d_1^{\,2} \qquad (6.3.1b)$$

By expanding Equation 6.3.1b to remove the square powers, and realizing that $D = d_2 + d_1$ we have

$$\begin{aligned}(R_2 + R_1)(R_2 - R_1) &= (d_2 + d_1)(d_2 - d_1)\\ (R_2 + R_1)(R_2 - R_1) &= D(d_2 - d_1)\end{aligned} \qquad (6.3.1c)$$

Note that in Figure 6.3a the values of D, d_1, and d_2 are measured from (C2) towards (C1). Therefore, in the case

shown on the right side of the said figure the value of $d_2 < 0$, and thus the equation $D = d_2 + d_1$ holds true.

By rearranging the terms of Equation 6.3.1c, and substituting for d_1 the value of $D - d_2$ we have

$$\frac{(R_2 + R_1)(R_2 - R_1)}{D} = (d_2 - d_1)$$

$$\frac{(R_2 + R_1)(R_2 - R_1)}{D} = d_2 - D + d_2 \qquad (6.3.1d)$$

$$\frac{(R_2 + R_1)(R_2 - R_1)}{D} + D = 2d_2$$

or $\qquad (6.3.1e)$

$$d_2 = \frac{1}{2}\left[\frac{(R_2 + R_1)(R_2 - R_1)}{D} + D\right]$$

With d_2 known, the value of the offset distance of the point of intersection from the line joining the two arc center points may be found with the Pythagorean theorem as

$$B = \sqrt{R_2^{\,2} - d_2^{\,2}} = \sqrt{(R_2 + d_2)(R_2 - d_2)} \qquad (6.3.1f)$$

With values of d2 and B now known, the coordinates of the point of intersection may be computed by traversing from C_2 towards C_1 a distance of d_2 with an offset distance of B. However, before solving for the coordinates of the points of intersection, there is a need to assess the relationship of one curve to the other, for it is possible for the value under the radical in Equation 6.3.1f to be:

- Positive, yielding two intersections,
- Zero, yielding on intersection, that is a point of tangency, or
- Negative, yielding no intersection, that is one curve is completely inside or completely outside the other curve.

Theoretically an assessment of this value for being zero or negative would suffice, but since the values of the radii and of the coordinates of the two center points could be the result of previous computation it is possible not to get a proper solution due to precision of computations. Therefore a determination of which of the relationships of the two curves as depicted in Figure 6.3b needs to be made prior to solving for the points of intersection, or point of tangency.

When the procedure determines the existence of a point of tangency, its coordinates are not computed by the results of Equations 6.3.1e and 6.3.1f, but by traversing from one center points towards the other with the input value of the corresponding radius in order to improve on precision.

```
Sub iccrccrc  (CCN1, CCE1, R1, CCN2, CCE2, R2, _
               XN, XE, INTRSC)
```

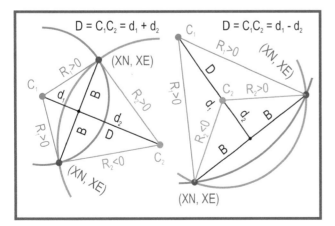

Figure 6.3a Intersection of Arc with Arc (Plus and Offset)

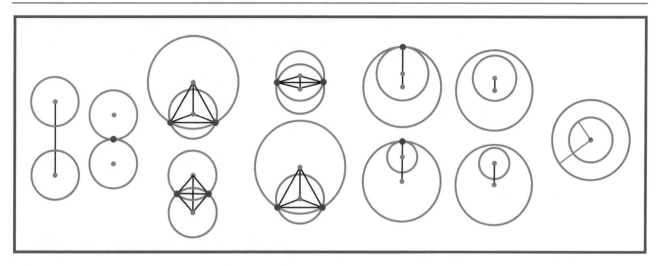

Figure 6.3b Circle to Circle Relationship from Totally Separate to Concentric

The input parameters to this procedure include the:

CCN1, CCE1	The north and east coordinates of the center point of the first curve.
R1	The radius of the first curve (its absolute value is used).
CCN2, CCE2	The north and east coordinates of the center point of the second curve.
R2	The radius of the second curve, the sign of which should be as per the preamble of this procedure to return the desired point of intersection.

The parameters that are returned from this procedure include the:

XN, XE	The north and east coordinates of the desired point of intersection.
INTRSC	An indicator the value of which if it is:
	1 Denotes that an intersection has been found.
	2 Denotes that an intersection has not been found. That is one arc does not cross the other even if either or both arcs are extended to a full circle.
	3 Denotes that one arc is tangent to the other when either or both are extended to a full circle, in which case the returned point is that of the tangency.

Upon invocation, the procedure initializes the parameters to be returned, and then:

1. Calls the icforce procedure to compute the distance D and north azimuth AZ from the center point of the second arc (C2) towards the center point of the first arc (C1) as indicated in Figure 6.3a, and then sets temporary variables with the absolute values of the two radii.
2. Checks the value of D for being equal to zero within a tolerance of 0.005 feet (1.524 mm), and:

- If so, sets the indicator INTRSC to be 2 (concentric arcs), and terminates.
- If not so, continues.
3. Compares the value of D against the sum of the radii to determine whether the arcs are completely each other, and if so whether the arcs are tangent to each other externally.
- If so, calls the icptl procedure to set the coordinates of the point of tangency by traversing from the center point of the second curve, sets INTRSC to be 3 and terminates.
- If not so, continues.
4. Determines whether one curve is located within the other, and if so whether the smaller curve is tangent to the larger curve. This is done by calling the icbtween procedure to see if the center point of the small curve is located within the extent of the radius of the larger curve along the said direction AZ. This is done first for curve one being larger than curve two and then for the reverse. If the curves are found:
- To be tangent, the point of tangency is found similarly to Step 3.
- Not to be tangent, the procedure continues.
5. Uses Equation 6.3.1e to compute the values of d2 and B and the calls the icptl procedure to compute the coordinates of the desired point of intersection.

6.3.2   Arc with Arc Intersection (Three Sides)

The icxcvwcv procedure computes the north and east coordinates of the two points of intersection of two arcs with each arc being defined by the north and east coordinates of

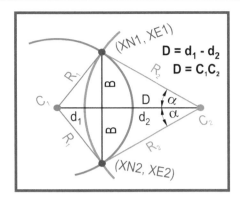

Figure 6.3c Intersection of Two Arcs (Three Sides)

its center point and its radius. The differences between this and the preceding procedure are as follows:

- This procedure returns both points of intersection, instead of one. In the case in which the two curves are tangent to each other, both of the points that are returned have the same coordinates.
- The radii are expected to be positive, but in case they are not their absolute value is used.
- The value of the returned indicator of whether there is or there is not an intersection or tangency point is differs between the said two procedures.

Once the procedure determines that the two curves intersect, it uses the three sides of the triangle, see Figure 6.3c, formed by the two radii from their respective center points of the two arcs towards a point of intersection and the implied line between the said center points to find angle α as being

$$\cos(\alpha) = \frac{R_2^2 + D^2 - R_1^2}{2R_2 D} \qquad (6.3.2a)$$

$$\cos(\alpha) = \frac{(R_2 + R_1)(R_2 - R_1) + D^2}{2R_2 D} \qquad (6.3.2b)$$

$$\cos(\alpha) = \frac{(R_2 + R_1)(R_2 - R_1)}{2R_2 D} + \frac{D}{2R_2} \qquad (6.3.2c)$$

$$\cos(\alpha) = \frac{1}{2R_2}\left[\frac{R_2 + R_1)(R_2 - R_1)}{D} + D \right] \qquad (6.3.2d)$$

With *cos(α)* now known, the value of d_2 may be found as

$$d_2 = R_2 \cos(\alpha) = \frac{1}{2}\left[\frac{(R_2 + R)(R_2 - R_1)}{D} + D \right] \qquad (6.3.2e)$$

The Pythagorean theorem may now be applied to find the offset distance *B* of the point of intersection from the implied line between the center points of the two arcs as

$$B = \sqrt{R^2 - d_2^2} = \sqrt{(R + d_2)(R - d_2)} \qquad (6.3.2f)$$

```
Sub icxcvwcv  (CCN1, CCE1, R1, CCN2, CCE2, R2, _
               XN1, XE1, XN2, XE2, INTRSC)
```

The input parameters to this procedure include the:

CCN1, The north and east coordinates of the center point of the
CCE1   first curve.
R1   The radius of the first curve (should be positive).
CCN2, The north and east coordinates of the center point of the
CCE2   second curve.
R2   The radius of the second curve, the sing of which should
       be as indicated above in order to return the desired point of
       intersection.

The parameters that are returned from this procedure include the:

XN1, XE1 The north and east coordinates of the desired point of
          intersection.
XN2, XE2 The north and east coordinates of the desired point of
          intersection.
INTRSC   An indicator the value of which if it is:
          0 Denotes that an intersection does not exist.
          1 Denotes that the curves are tangent to each other.
          2 Denotes that the two points of intersection have
             been found.

Upon invocation, the procedure initializes the parameters to be returned, and then:

1. Calls the `icforce` procedure to compute the distance D and north azimuth AZ from the center point of the second arc (C2) towards the center point of the first arc (C1) as indicated in Figure 6.3c.
2. Determines and sets the larger (Rmax) and the smaller (Rmin) of the two radii.
3. Queries if the distance D = 0 within a tolerance of 0.0005 feet (1.524 mm), that is the two arcs are concentric, and:
   - If so, sets INTRSC = 0, and terminates.
   - If not so, continues.
4. Queries if (D + Rmin) < Rmax, that is the smaller arc is inside the larger arc, and:
   - If so, sets INTRSC = 0, and terminates.
   - If not so, continues.

5. Queries if the absolute value of the sum $|D - R1 - R2| \leq 0$ within a tolerance of 0.0005 feet (1.524 mm), that is the two arcs are tangent to each other, and:
   - If so, computes the point of tangency by traversing from the center point of the second arc along the azimuth AZ a plus distance of R_2 and an offset distance of zero (0.00), sets INTRSC = 1, and terminates.
   - If not so, continues.
6. Queries if the sum $(D - R1 - R2) > 0$ within a tolerance of 0.0005 feet (1.524 mm), that is the full circles of the arcs are far apart and they do not intersect, and:
   - If so, sets INTRSC = 0, and terminates.
   - If not so, continues.
7. Queries if the sum $(D - R1 - R2) < 0$ within a tolerance of 0.0005 feet (1.524 mm), that is the points of intersection may be found, and:
   - If so, sets INTRSC = 2, and continues with the next step.
   - If not so, terminates.
8. Computes the plus distance d_2 and the offset distance B using Equations 6.3.2e and 6.3.2f, respectively.
9. Calls the icfptl procedure to traverse from the center point of the second curve towards the center point of the first curve a plus distance of $d2$, and an offset distance of B to compute the coordinates of the point of intersection to the right.
10. Calls the icfptl procedure to traverse from the center point of the second curve towards the center point of the first curve a plus distance of $d2$, and an offset distance of $-B$ to compute the coordinates of the point of intersection to the left.

6.4 Generic Intersections

As implied by its title, the procedures described in this section are of a generic nature in that the type of the selected features to be intersected may be that of a two-point line, polyline, circular arc, or polygon. This implies that it is the responsibility of the calling program to determine (preprocess) the type of features that have been selected for intersection. As such, these procedures perform a bookkeeping function that determines the type of feature and then calls upon one of the previously presented geometric procedures to carry out the intersection.

There are five procedures in this section that intersect lines and/or arcs with other lines and/or arcs. Of these five:

- The icelmnx1 and icelmnx4 intersect two features which must be defined as circular curves, with lines having a radius of zero (0.00).
- The icelmnx2 and icelmnx3 intersect lists (collections) each one representing a string of line and/or arc features, with lines having a radius of zero (0.00).

Polylines may be included if they are decomposed by the calling program into individual line segments. Similarly, the perimeter of a polygon may be included, if it is decomposed into a series of line segments. Sides of parcels that may be composed of lines and arcs, may be decomposed and included in the lists.

- The icxExtRetrExtLin is similar to the icelmnx2 procedure with the difference that whereas icelmnx2 returns an intersection within the limits of a segment in a list, the icxExtRetrExtLin extends the ends of the first and last segments of the lists so as to obtain an intersection beyond the limits implied by the lists.

Of the abovesaid procedures, the icelmnx1 procedure is the basic procedure of this group of procedures, in that it deals with only two specific features. The other three provide an in-between preprocessing that establishes certain criteria that specify as to which of all possible intersections should be selected and returned.

6.4.1 Line/Arc Feature Intersection (Radius Only)

The icelmnx1 procedure computes the north and east coordinates of the point of intersection between two features either of which may be a two-point line or a circular arc and is defined as indicated below in the definition of the calling input parameters. This procedure assumes that the central angle of an arc feature is less than 180°. For cases that the central angle is greater than 180° refer to the icelmnx4 procedure.

> The icelmnx1 procedure assumes that a circular arc subtends an arc of less than 180°. For greater arcs the icelmnx4 procedure should be used.

```
Sub icelmnx1 (PTN11, PTE11, PTN12, PTE12, R1,
              PTN21, PTE21, PTN22, PTE22, R2,_
              PTXN, PTXE, _
              XN, XE, INTR1, INTR2, INTRS)
```

The input parameters to this procedure include the:

PTN11, PTE11	The north and east coordinates of the start point of the first feature.
PTN12, PTE12	The north and east coordinates of the end point of the first feature.
R1	The radius of the first feature, the value of which should be:
	• Positive if the feature is a clockwise arc.
	• Negative if the feature is a counterclockwise arc.

- Zero if the feature is a line.

PTN21, PTE21	The north and east coordinates of the start point of the second feature.
PTN22, PTE22	The north and east coordinates of the end point of the second feature.
R2	The radius of the second feature with a value similar to that of R1.
PTXN, PTXE	The north and east coordinates of a point near which the intersection is to be located if there are more than one intersection. This point is used primarily for the intersection of a line with an arc, or for the intersection of an arc with another arc, since in these cases there are two possible intersections. For the intersection of a line with another line, this point could be any point.

The parameters that are returned from this procedure include the:

XN, XE	The north and east coordinates of the point of intersection.
INTR1, INTR2	An indicator regarding the position of the intersection point in relation to the first or second input feature, respectively, the value of which if it is:
	0 Denotes that the said point is at one of the indicated feature's endpoints.
	1 Denotes that the said point is between the indicated feature's endpoints.
	2 Denotes that the said point is outside the indicated feature's limits.
INTRS	An indicator the value of which if it is:
	1 Denotes that the two features intersect each other.
	2 Denotes that the two features do not intersect each other.

Upon invocation, the procedure initializes the parameters to be returned, and then:

1. Checks if the two features share any common endpoints within a tolerance of 0.0005 feet (1.524 mm), by comparing the points in the following sequence:
   (i) Start point of the first feature with the start point of the second feature.
   (ii) Start point of the first feature with the end point of the second feature.
   (iii) End point of the first feature with the start point of the second feature.
   (iv) End point of the first feature with the end point of the second feature.
   If a common point:
   - Exists, the procedure sets the indicators INTR1 and INTR2 to their appropriate value as indicated in the returned parameters above, sets the coordinates (XN, XE) to be those of the common end-

point, and terminates. If the two features share two common endpoints, the common endpoint which is nearer to the input point (PTXN, PTXE) is set to be the intersection point and terminates.
   - Does not exist, the procedure continues.
2. Calls the icforce procedure to compute the distance and azimuth from the start point towards the end point of the first input feature, and if the radius of the feature is not zero, the procedure calls the icarccnt procedure to compute the coordinates of the center point of the arc.
3. Repeats the above step for the second input feature.
4. Queries R1 and R2 to determine if a feature is a line or an arc and accordingly calls
   - The icintrs procedure if both features are lines (R1 and R2 are both zero).
   - The iccrccrc procedure if both features are arcs (R1 and R2 are both nonzero).
   - The iclincrc procedure if one of the features is a line and the other an arc (either R1 or R2 are zero).

At the end of each of the last two calls the procedure selects the intersection that is nearer the input point (PTXN, PTXE).

6.4.2 Line/Arc Feature Intersection (Radius and Angle)

The icelmnx4 procedure computes the north and east coordinates of the point of intersection between two features either of which may be a two-point line or circular arc and is defined as indicated below in the definition of the calling input parameters. With this procedure the central angle of an arc feature may exceed 180°.

```
Sub icelmnx4  (PTN11, PTE11, PTN12, PTE12,
               R1, C1, PTN21, PTE21, PTN22,
               PTE22, _
               R2, C2, PTXN, PTXE, _
               XN, XE, INTR1, INTR2, INTRS)
```

The input parameters to this procedure include the:

PTN11, PTE11	The north and east coordinates of the start point of the first feature.
PTN12, PTE12	The north and east coordinates of the end point of the first feature.
R1	The radius of the first feature, the value of which should be:
	• Positive if the feature is a clockwise arc.
	• Negative if the feature is counterclockwise arc.
	• Zero if the feature is a line.
C1	The central angle in radians of the first feature, the value of which should be:

	• Positive if the feature is an arc.
	• Zero if the feature is a line.
PTN21, PTE21	The north and east coordinates of the start point of the second feature.
PTN22, PTE22	The north and east coordinates of the end point of the second feature.
R2	The radius of the second feature, the value of which should be similar to R1.
C2	The central angle in radians of the second feature, the value of which should be similar to that of C1.
PTXN, PTXE	The north and east coordinates of a point near which the intersection is to be located if there are more than one intersections. This point is used primarily for the intersection of a line with an arc, or for the intersection of an arc with another arc, since in these cases there are two possible intersections. For the intersection of a line with another line, this point could be any point.

The parameters that are returned from this procedure include the:

XN, XE	The north and east coordinates of the point of intersection.
INTR1, INTR2	An indicator regarding the position of the intersection point in relation to the first or second input feature, respectively, the value of which if it is:
	0 Denotes that the said point is at one of the indicated feature's endpoints.
	1 Denotes that the said point is between the indicated feature's endpoints.
	2 Denotes that the said point is outside the indicated feature's limits.
INTRS	An indicator the value of which if it is:
	1 Denotes that the two features intersect each other.
	2 Denotes that the two features do not intersect each other.

Upon invocation, the procedure initializes the parameters to be returned, and then operates in a manner similar to that of the icelmnx1 procedure, but with one difference. When determining, if a common point has not been found, whether a feature is a line or an arc, and finding that a feature is an arc, the procedure determines the coordinates of the center point of the arc by use of the icarccnt procedure by querying the value of C1 or C2. If the central angle C1 or C2 is:

- Less than π, the icarccnt procedure is called by using (PTN11, PTE11) as the start point of the arc, and (PTN12, PTE12) as the end point of the arc.
- Greater than π, the icarccnt procedure is called by using (PTN12, PTE12) as the start point of the arc, and (PTN11, PTE11) as the end point of the arc.

6.4.3 All Intersections of Two Strings

The icelmnx2 procedure computes the Cartesian coordinates of each point of intersection between two strings of features each of which is composed of a series of two-point lines and/or circular arcs, and which are defined as indicated below in the definition of the calling input parameters. This procedure assumes that the central angle of an arc feature is less than 180°. A string of features is intended to be a collection of concatenated two-point line and/or arc features with the end point of one being the start point of the next feature in the collection. However, random features not following this rule may be used.

```
Sub icelmnx2  (X1List, Y1List, X2List, Y2List,
               R1List, _
               X3List, Y3List, X4List, Y4List,
               R2List, opmode, _
               NINTRS, XPList, YPList, ZPList)
```

The input parameters to this procedure include the:

X1List	Collection of the X-coordinates of the start point of the features in the first string.
Y1List	Collection of the Y-coordinates of the start point of the features in the first string.
X2List	Collection of the X-coordinates of the end point of the features in the first string.
Y2List	Collection of the Y-coordinates of the end point of the features in the first string.
R1List	Collection of the radii of the features in the first string, the value of which should be:
	• Positive if the feature is a clockwise arc.
	• Negative if the feature is counterclockwise arc.
	• Zero if the feature is a line.
X3List	Collection of the X-coordinates of the start point of the features in the second string.
Y3List	Collection of the Y-coordinates of the start point of the features in the second string.
X4List	Collection of the X-coordinates of the end point of the features in the second string.
Y4List	Collection of the Y-coordinates of the end point of the features in the second string.
R2List	Collection of the radii of the features in the second string, the value of which should be similar to that of R1List.
opmode	The mode of operation, the value of which dictates which points of intersection are to be selected. If opmode is:
	0 The procedure selects an intersection point if it falls within both intersecting segments.
	1 The procedure selects an intersection point if it falls within the second of the intersecting segments.
	2 Same as opmode of 0 but without displaying a progress bar.
	3 Same as opmode of 1 but without displaying a progress bar.

A progress bar is also not displayed if both strings contain only on feature. It is noted that the source code statements in the accompanying CD is ArcObjects specific. For any other computing environment, it may need to be modified.

The parameters that are returned from this procedure include the:

NINTRS   The number of intersections that have been encountered as per the specified opmode code.

XPList   A collection of the X-coordinates of the points of intersection.

YPList   A collection of the Y-coordinates of the points of intersection.

ZPList   A collection of the Z-coordinates of the points of intersection. These values are all set to be 99999.0 because of the particular application that this procedure has been developed. This collection may be disregarded, as it has no effect on the determination of the points of intersection.

Upon invocation, the procedure initializes the parameters to be returned, and then parses through the various features of the first string of features, and intersects each one of them with the features of the second string of features by calling the icelmnx1 procedure described earlier in this chapter. As each intersection is found, its X and Y coordinates are placed in the XPList and YPList collections, respectively.

6.4.4 All Intersections of Two Strings (Extended)

The icxExtRetrExtLin procedure computes and returns the Cartesian coordinates of each intersection point between two strings composed of two-point line and/or circular arc features that are defined as indicated in the description of the input parameters below. This procedure is very similar to the icelmnx2 procedure with two differences:

(a) It assumes an opmode parameter value of zero, and
(b) It extends the first and last features of each string back and ahead and assumed large distance prior to calling the icelmnx2 procedure to carry out the actual intersection process.

```
Sub icxExtRetrExtLin  (X1List, Y1List, X2List,
                       Y2List, R1List, _
                       X3List, Y3List, X4List,
                       Y4List, R2List, _
                       NINTRS, XPList, YPList,
                       ZPList, _
                       segList1,    Dist1List1,
                       Dist2List1,    segList2,
                       Dist1List2, Dist2List2)
```

The input parameters to this procedure are those of the icelmnx2 procedure, but without the opmode parameter.

The parameters that are returned from this procedure include, in addition to those of the icelmnx2 procedure, the following:

segList1   Collection of the record numbers of the features of the first string in which an intersection exists.

Dist1List1   Collection of the distances of a point of intersection from the start point of the first string.

Dist2List1   Collection of the distances of a point of intersection from the end point of the first string.

segList2   Collection of the record numbers of the features of the second string in which an intersection exists.

Dist1List2   Collection of the distances of a point of intersection from the start point of the second string.

Dist2List2   Collection of the distances of a point of intersection from the end point of the second string.

Upon invocation, the procedure initializes the parameters to be returned, and then:

1. Extends the first feature of each string 10,000 feet or meters back, and the last feature of each string 10,000 feet or meters forward, in order to ensure that a point of intersection beyond the limits of each string could be determined.
2. Calls upon the icelmnx2 procedure with an opmode value of zero (0), and creates a list of the:
   • Coordinates of any points of intersection that lie within both features that are being intersected (note the above extension of the first and last features of the strings).
   • Record numbers of each feature within its respective string at which a point of intersection occurs.
   • Distances of each point of intersection from the start point and end point if the respective string calling the icPntOnPoly procedure.

6.4.5 One Intersection of Two Strings

The icelmnx3 procedure computes and returns the number of intersections that may have been found during the operation, the Cartesian coordinates of only one point of intersection between two strings of two-point line and/or circular arc features that are defined as indicated below, and the feature record number of the first and second string in which the point of intersection is located. It is assumed that arc features subtend an angle of less than 180°. A string of features is intended to be a collection of concatenated two-point line and/or arc features with the end point of one being the start point of the next feature in the collection. However, random features not following this rule may be used. In intersecting the two strings, the procedure computes all possible intersections and returns only one as follows:

(a) If any points of intersection have been found, the one returned is the one that has the lowest record number in the first string, and the highest record number in the second string. That is, the selected intersection is the one closest to the start of the first string, and closest to the end of the second string.
(b) If any points of intersection have not been found, then the program repeats the intersection process, processing the first string in reverse sequence, and returns the first intersection that lies within any of two features of the strings that are being intersected.

```
Sub icelmnx3  (X1List, Y1List, X2List,Y2List,
              R1List, _
              X3List, Y3List, X4List, Y4List,
              R2List,
              NINTRS, theX, theY, theZ,
              the1, the2)
```

The input parameters to this procedure are the same as those of the `icelmnx2` procedure, but without the need for the `opmode` parameter.

The parameters that are returned from this procedure include the:

NINTRS	The number of intersections that have been encountered. It is either one or zero.
theX	The X-coordinate of the selected point of intersection.
theY	The Y-coordinate of the selected point of intersection.
theZ	The Z-coordinates of the selected point of intersection. This value is set to be 99999.0 because of the particular application that this procedure has been developed. This parameter may be disregarded, as it has no effect on the determination of the point of intersection.

Upon invocation, the procedure initializes the parameters to be returned, and then:

1. Parses through the various features of the first string of features, and intersects each one of them with the features of the second string of features by calling the `icelmnx1` procedure described earlier in this chapter. As each intersection is found, its X and Y coordinates are placed in the `XPList` and `YPList` collections, respectively.
2. Queries the number of intersections that have been found.
   • If one or more intersections have been found, the one that meets the first of the abovesaid two criteria is selected, and the procedure terminates.
   • If no intersection has been found, the procedure repeats Step 1 in reverse order, selects the last intersection to be found, and terminates.

Tangencies

The procedures to be presented in this chapter are those that enable the programmer to construct tangent lines to circular curves, or circular arcs tangent to one or more lines and/or other circular arcs. In addition to the procedures of this chapter, there are three other procedures, the `iccvt2ls`, the `icsmpcrv` and the `iccrvpoc`, that have been described in Chap. 5. These three procedures introduce an arc tangent to two lines under three distinct conditions.

The procedures of this chapter are not of generic nature, that is, they do not constitute functions or subroutines that other procedures may use. The procedures of this chapter utilize those procedures that have been presented in previous chapters to build solutions for specific applications, and which are associated with interactive user-computer tools to accomplish the applications. Because of this, certain of these procedures may contain source code statements in the accompanying CD that are specific to the ArcObjects environment. These statements are identified in the descriptions provided below, and they are easily discernible in the said source code.

The subject of tangent constructions is quite extensive, and can range from the very simplistic to some degree of complexity. The procedures addressed in this chapter pertain to tangent constructions that are generally associated with civil engineering and surveying work, and other related applications. On the very simplistic end of the complexity scale, one may consider the following two tangent problems, for which a procedure is not provided. Each of these two problems may be solved with what has been previously presented in this book.

Let us first consider the case in which it is desired to determine the point of tangency of a line of known north azimuth (AZ) to a given circle of a known radius (R) and known north and east coordinates of its center point (CCN, CCE) as indicated in Figure 7.0a. This construction may be accomplished by calling the `icptl` procedure of Chap. 4 as

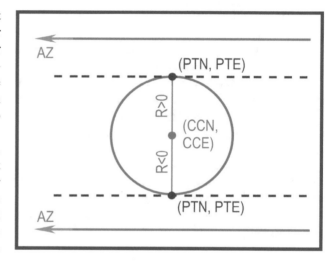

Figure 7.0a Point of Tangency of an Implied Line

```
call icptl(CCN, CCE, AZ, 0#, R, PTN, PTE)
or as call icptl(CCN, CCE, AZ, 0#, -R, PTN, PTE)
```

depending on which side of the circle the point of tangency is desired.

Now, if a line has to be constructed through the point of tangency, `icptl` can be called again as

where D1 and D2 are some distances that define the length of the tangent line to be constructed.

The bookkeeping process of the above construction becomes a bit more cumbersome if the curve is an arc, and not a full circle, and if it is desired to construct the tangent line only if it is to be within the limits of the arc. For this case, the limit points of the arc will need to be made known. The validity check that the point of tangency is within the

```
call  icptl(PTN, PTE, AZ, D1, 0#, PT1N, PT1E)
call  icptl(PTN, PTE, AZ, -D2, 0#, PT2N, PT2E)
```

Circles and arcs may be defined clockwise or counterclockwise, but with a few exceptions they are always drawn in a counterclockwise rotation.

confines of the arc may be made before or after the point of tangency has been found. For the:

- Before case, the azimuths from the center point to each endpoint of the arc will need to be determined and compared with the normal to the given azimuth (`AZ`).
- After case, the `icbtwarc` procedure may be called to determine whether the point of tangency coordinates are within the limits of the arc.

Thus, a simple geometric construction can become a rather cumbersome programming task.

Another similar application is that of constructing a tangent line to a given arc or circle of known radius (`R`) and known north and east coordinates of its center point (`CCN, CCE`) at a specific point (`PTN, PTE`) on the arc or circle as indicated in Figure 7.0a.

This construction may be accomplished by first calling the icforce procedure to determine the azimuth from the said center point towards the point on the curve (`AZR`)

where `D` should be equal to `R`, and then calling the `icptl` procedure as

```
call icforce(CCN, CCE, PTN, PTE, D, AZR)
```

or as

```
call icptl(PTN, PTE, AZR, 0#, D1, PTN, PTE)
```

depending on which side of the circle the point of tan-

```
call icptl(PTN, PTE, AZR, 0#, -D1, PTN, PTE)
```

gency is to be located, with `D1` defining the length of the tangent line.

The procedures to be presented in this chapter are divided into the following groups:

- Those that create lines tangent to arcs or circles.
- Those that create arcs or circles tangent to lines.
- Those that create arcs or circles tangent to lines and other arcs.
- Those that create arcs or circles tangent to other arcs.

When two or more curves are tangent two each another in series, with the first and last being tangent to a line, they are referred to as compound curves if they all progress in the

same rotational direction, and reversed curves if they progress in opposing directions. Such curves are addressed in Chap. 8. The introduction of spirals between lines and curves is addressed separately, due to their nature, in Chap. 9.

7.1 Lines Tangent to Arcs or Circles

Two procedures comprise the contents of this section, the `ictcvept` procedure that constructs a tangent line to an arc from an external point with respect to the arc, and the `iclnt2cv` procedure that constructs a tangent line to two arcs or circles. There are two solutions to the former procedure, and four solutions to the latter procedure.

The curve type code identifies the manner in which a graphic circular feature has been created. The domain of this code is:

1 = circle through 3 points.
2 = arc through 3 points.
3 = Circle with center point and radius.
4 = Arc with center and 2 endpoints.

7.1.1 Line Tangent to an Arc from Point

The `ictcvept` procedure computes and returns the Cartesian coordinates of the point of tangency of a line tangent to an arc or circle from an external point defined by its Cartesian coordinates. Since there are two potential points of tangency, the one that is returned is the one that is nearest to a specific point of which the Cartesian coordinates should be specified. The curve in Figure 7.1a is defined by the Cartesian coordinates of its center point, its radius and a type of curve code number (see side bar on this page). In addition to this information, and if the curve is an arc (curve code of 2 or 4), the Cartesian coordinates of the arc's two endpoints and its central angle, which may exceed 180°, should also be specified. If the external point from which the tangent lines are to be constructed happens to line on the curve, and is within the limits of an arc, it is returned as a point of tangency.

To compute the point of tangency, the procedure computes the distance D and azimuth AZ from the curve's center point towards the specified external point (see Figure 7.1a), and then computes the length of the tangent line T as

$$T = \sqrt{D^2 + R^2} = \sqrt{(D+R)(D-R)} \qquad (7.1.1a)$$

and the central angle α is computed as

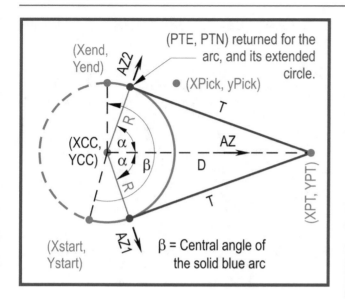

Figure 7.1a   Tangent Lines to Circle from External Point

$$\alpha = \cos^{-1}\left(\frac{R}{D}\right) \qquad (7.1.1b)$$

with which the azimuths from the center point towards the two points of tangency may be found as

$$AZR = AZX + \alpha \qquad (7.1.1c)$$

and

$$AZ2 = AZ - \alpha \qquad (7.1.1d)$$

With these directions known, the coordinates of the points of intersection may be computed.

The proximity point could be set by the calling program to be the user pick on the monitor screen that selects the curve associated with the point of tangency. That is, the user selects the curve by clicking near to where the point of the tangency is expected to be located.

```
Sub ictcvept(XCC,  YCC,  R,  XPT,  YPT,
             xPick,  yPick,  CurveType,
             Angle,  Xstart,  Ystart,  _
             Xend,  Yend,  _
             PTE,  PTN,  IOK)
```

The input parameters to this procedure include the:

XCC, YCC	The Cartesian coordinates of the center point.
R	The radius of the arc (should be positive).
XPT, YPT	The Cartesian coordinates of the external point.
xPick, yPick	The Cartesian coordinates of the proximity point to the tangency point.

CurveType	The curve type code, the value of which denotes the following:

- 1—circle through 3 points
- 2—arc through 3 points
- 3—circle with center and radius
- 4—arc with center, start and end points

NOTE that:

- These codes are due to the database design of the overall system for which this procedure has been developed. They could be reduced to two to distinguish between an arc or a circle.
- The next five parameters require appropriate values for `CurveType` of 2 or 4 curves only; any value could be specified for `CurveType` of 1 or 3.

Angle	The Central angle of the arc which should be:

- Positive for a clockwise rotation, and
- Negative for a counterclockwise rotation.

Xstart, Ystart	The Cartesian coordinates of the arc's start point.
Xend, Yend	The Cartesian coordinates of the arc's end point.

The parameters that are returned from this procedure include the:

PTE, PTN	The Cartesian coordinates of the point of tangency nearest to the proximity point (xPick, yPick).
IOK	An indicator, the value of which if it is:

- 0   denotes that a point of tangency is being returned.
- 1   denotes that the point from which the tangent is to be constructed is not external to the curve, but it is located within the curve.
- 2   denotes that there is no valid point of tangency as per the above sited criteria; the point of tangency does not lie within the limits of the arc.

Upon invocation, the procedure initializes the parameters to be returned, and then:

1. Calls the `icforce` procedure to compute the distance (D) and the azimuth (AZ) from the center point of the arc or circle (XCC, YCC) to the given point (XPT, YPT).
   - D < R, (the external point is not external, but lies within the arc) the indicator IOK is set to be one (1), and the procedure terminates.
   - D ≥ R, the procedure continues
2. Compares D with the given radius R. If it is found, within a tolerance of 0.001 (0.3048 mm) feet, that
3. Computes the length of the tangent line using Equation 7.1.1a, and then the central angle α using Equation 7.1.1b.

4. Computes the azimuths *AZ1* and *AZ2* from the said center point towards each of the two possible points of tangency using Equations 7.1.1c and 7.1.1d.
5. Calls the `icptl` procedure to compute the coordinates of each of the two points of tangency by traversing from (`YCC`, `XCC`) along an azimuth of *AZ1* and *AZ2* a plus distance of `R` and an offset distance of zero (0.00), setting each pair of coordinates in temporary variables.
6. Checks whether the given curve is an arc or a circle, and performs the checks that were identified earlier to return the appropriate point of tangency, if there is one. Thus, if the curve is:
   - An arc, the procedure determines whether either of the two possible solution is within the arc's limits, or not, and accordingly sets the values of the `IOK` indicator, and of the `PTE` and `PTN` coordinates. If both points of tangency fall within the extent of the arc, the one nearer to the point of proximity (`xPick`, `yPick`) is selected.
   - A circle, of the two possible points of tangency, the procedure returns the one nearer to the point of (`xPick`, `yPick`) coordinates.

7.1.2 Line Tangent to Two Circles or Arcs

The `icLnT2cvX` procedure computes and returns the north and east coordinates of the endpoints of all tangent lines to two given circles or circular arcs when considered as been expanded to form a full circle. Also returned are the lengths of the tangent lines and the radial north azimuths to each of the said endpoints. Depending on the position of the two given curves and their size there could be (a) no tangent lines, or (b) up to four tangent lines as indicated by the seven cases of Figure 7.1b. An indicator, the value of which denotes the case encountered, is also returned. Each of the two curves is defined by the north and east coordinates of its center point and its radius. It is up to the calling program to select the appropriate or desired of the potential tangents.

In perusing the general positioning of two curves with respect to each other in Figure 7.1b, and the number of potential tangent lines, it is noted that the starting point of a tangent line is located on the first curve, and the ending point is located on the second curve. In viewing the two curves from the first towards the second curve, there could be:

- A pair of tangent lines with one tangent line on the right side, and one on the left side of both curves, and possible.
- A pair of crisscrossing tangent lines with one tangent line starting on the right side of the first curve and ending on the left side of the second curve, and one starting on the left side of the first curve and ending on the right side of the second curve.

In examining the seven cases of Figure 7.1b, we note that for ase(s):

- **1** and **2**   There are no potential tangent lines, and the procedure aborts.
- **3** and **4**   Although there are no potential tangent lines, two lines are created along the common diameter, each of which connects the circumference of the small curve with that of the larger curve. In Case 4 one of these lines is of zero length, and the common tangent which is normal to the common diameter is not created, but the point of tangency is returned.
- **5** and **6**   There are two tangent lines in these cases, one on each side of both curves. In Case 5 the common tangent which is normal to the line joining the two center points is not created.
- **7**   There are four tangent lines in these cases, one on each side of both curves, and two crossing tangent lines.

```
Sub icLnT2cvX(YCC1, XCC1, R1, YCC2, XCC2,
              R2, _
              Y1R,  X1R,  Y2R,   X2R,   TR,
              AZRT, Y1L, X1L, Y2L, X2L,
              TL, AZLT, _
              YRL1, XRL1, YRL2, XRL2, TRL,
              AZRL1, AZRL2, _ YLR1, XLR1,
              YLR2,  XLR2,   TLR,   AZLR1,
              AZLR2, TheCASE)
```

The input parameters to this procedure include the:

YCC1, XCC1	The north and east coordinates of the center point of the first curve.
R1	The radius of the first curve, which should be positive.
YCC2, XCC2	The north and east coordinates of the center point of the second curve.
R2	The radius of the second curve, which should be positive.

The parameters that are returned from this procedure include the parameters below. Note that a value of zero is returned for any parameters that do not exist as a result of the case 1 through 7 that has been encountered.

Y1R, X1R	The north and east coordinates of the point of tangency of the tangent line to the right of both curves with the first curve.
Y2R, X2R	The north and east coordinates of the point of tangency of the tangent line to the right of both curves with the second curve.
TR	The length of the tangent line to the right of both curves.
AZRT	The north radial azimuth from either center point towards the respective point of tangency [(Y1R, X1R) or (Y2R, X2R)].

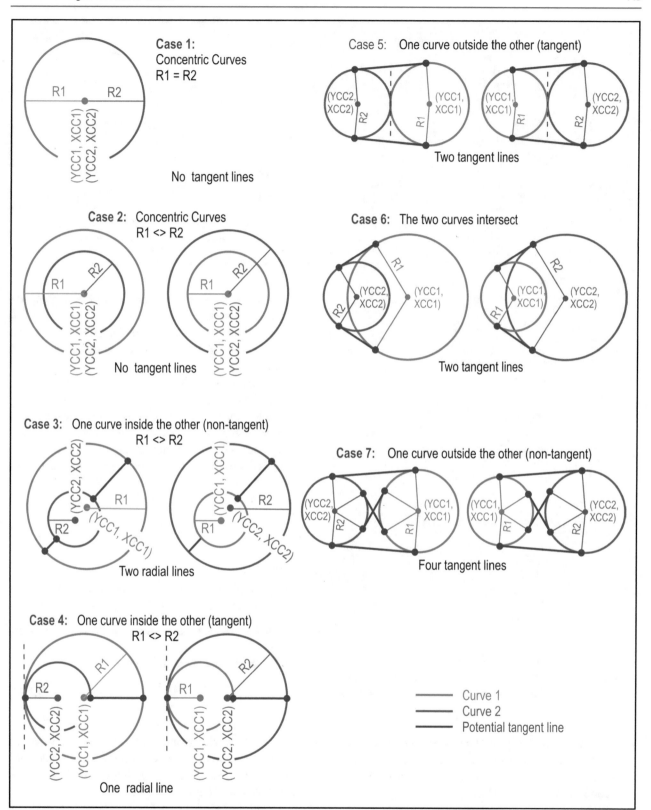

Figure 7.1b   Tangent Lines to Two Circles or Arcs

Y1L, X1L	The north and east coordinates of the point of tangency of the tangent line to the left of both curves with the first curve.
Y2L, X2L	The north and east coordinates of the point of tangency of the tangent line to the left of both curves with the second curve.
TL	The length of the tangent line to the left of both curves (should equal TR).
AZLT	The north radial azimuth from either center point towards the respective point of tangency [(Y1L, X1L) or (Y2L, X2L)].
YRL1, XRL1	The north and east coordinates of the point of tangency of the crossing tangent line to the right of the first curve.
YRL2, XRL2	The north and east coordinates of the point of tangency of the crossing tangent line to the left of the second curve.
TRL	The length of the crossing tangent line to the right of the first curve and to the left of the second curve.
AZRL1	The north radial azimuth from the center point of the first curve towards the point of tangency (YRL1, XRL1).
AZRL2	The north radial azimuth from the center point of the second curve towards the point of tangency (YRL2, XRL2).
YLR1, XLR1	The north and east coordinates of the point of tangency of the crossing tangent line to the left of the first curve.
YLR2, XLR2	The north and east coordinates of the point of tangency of the crossing tangent line to the right of the second curve.
TLR	The length of the crossing tangent line to the left of the first curve and to the right of the second curve.
AZLR1	The north radial azimuth from the center point of the first curve towards the point of tangency (YLR1, XLR1).
AZLR2	The north radial azimuth from the center point of the second curve towards the point of tangency (YLR2, XLR2).
TheCASE	An indicator, the value of which indicates the one of the seven cases of Figure 7.1b that has been encountered.

Upon invocation, the procedure initializes to zero the parameters to be returned, excepting the indicator TheCASE which is initialized to seven, makes certain the radii are positive, and then:

1. Calls the icforce procedure to compute the distance (D) and azimuth (AZx) from the center point of the first curve (YCC1, XCC1) towards the center point of the second curve (YCC2, XCC2).
2. Queries the center point coordinates of the two curves to determine if the curves are concentric, and if so, sets the indicator TheCASE and aborts; else continues.
3. Queries the distance (D) and the two radii to determine which curve lies completely within the other, if any, and if so:

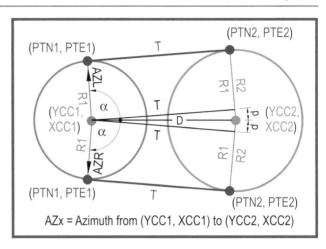

Figure 7.1c Outside Tangent Lines to Two Circles

- Computes the two distances and azimuths along the common diameter that join the two circumferences as stated above, as well as the coordinates of the endpoints.
- Determines whether the two curves are tangent to each other.
- Sets accordingly the indicator TheCASE, and then branches to Step 7 below.

4. Determines the coordinates of the two lines that are tangent to the two circles at their outside (the two tangent lines that do not intersect) by first finding the angle α (see Figure 7.1c) which is formed by the line connecting the two center points and a line parallel to a tangent line and passing through the center point of the first curve (YCC1, XCC1) as

$$\alpha = \sin^{-1}\left(\frac{d}{D}\right) + \frac{\pi}{2} \qquad (7.1.2a)$$

where d = R2 – R1.

With the angle α known, the azimuths from the each of the center points towards the points of tangency may be found as

$$AZR = AZx + \alpha \quad \text{and} \quad AZL = AZX \ \alpha \qquad (7.1.2b)$$

and the tangent length may be found as

$$T = D\sin(\alpha) \qquad (7.1.2c)$$

In Figure 7.1c, R1 is less than R2, and hence α is negative. If R1 would greater than R2, then α would be positive, and thus the above equations would hold true. Furthermore, the said figure portrays Case 7. The same logic would also hold true for Cases 5 and 6.

The north and east coordinates (PTN1, PTE1) and (PTN2, PTE2) as shown in Figure 7.1c of the two tangent lines may now be found by traversing from each of the two center points with the icptl procedure along AZR and AZL a plus distance equal to R1 or R2 and an offset distance of zero (0.00).

5. Computes the sum of the two given radii and compares it with the distance (D) between the center points of the two curves. If this sum is
   - Equal to the said distance within a tolerance of 0.005 feet (1.524 mm), the two curves are considered tangent to each other, TheCASE is set to be 5, and the procedure branches to Step 7 below.
   - Greater than the said distance, the two curves are intersecting, TheCASE is set to be 6, and the procedure branches to Step 7 below.
   - Less than the said distance, the two curves are separated and have two crossing tangent lines, and the procedure continues.
6. Determines the coordinates of the two lines that are tangent to the two circles, and which lines crisscross each other, by finding the angle α shown in Figure 7.1d as

$$\alpha = Cos^{-1}\left(\frac{d}{D}\right) \qquad (7.1.2d)$$

where d = R2 + R1.

With the angle α known, the azimuths from each of the two center points towards the points of tangency may be found as

$$AZR = AZx + \alpha \quad \text{and} \quad AZL = AZR + \pi \qquad (7.1.2e)$$

and T may be found by Equation 7.1.2c.

For this solution, the magnitude of R1 and of R2 is of no consequence. In Figure 7.1d, the solution of only one of the two crisscross tangents is shown. The solution for the other would be similar.

7. Forces each of the azimuths to be returned to be within the range of zero and two π radians, in case any of them is not due to the associated additions and subtractions of the angles α to or from the azimuth AZ or AZx.

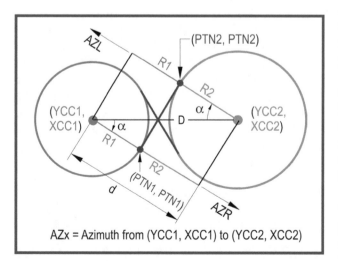

AZx = Azimuth from (YCC1, XCC1) to (YCC2, XCC2)

Figure 7.1d   Crisscross Tangent Lines to Two Circles

7.2   Arcs Tangent to Lines

Six procedures comprise the contents of this section. The first procedure, the icTanCrv2Lin, constructs an arc of a given radius passing through a given point and being tangent to a given line and to another line of which only the direction is considered as being known. The next two, the icTanCrv3Lin and the icCrvInTri procedures, are similar to each other in that each involves three given lines. The former constructs an arc tangent to three lines, while the latter constructs a circle inscribed in a triangle. Two other procedures, the iccvta1p and the iccvta2q, construct an arc or a circle that passes though one or two given points, respectively, and is tangent to a given line. Of these two procedures, the iccvta2q procedure requires the specification of a known radius. Depending on the given data, either procedure may create one or two arcs or circles. Lastly, the iccvtatp procedure is rather simplistic in that it constructs an arc of given radius and arc length tangent to a given line at a given point thereon, which is to be the start point of the arc. This procedure may easily be modified to construct a similar arc given the central angle, or chord length instead of the arc length.

> *Additional procedures of circular arcs tangent to two lines may be found in Chap. 5.*

7.2.1   Arc Through a Point Tangent to a Line and a Direction

The icTanCrv2Lin procedure constructs an arc of given radius tangent to a given line and passing through a point defined by its north and east coordinates (*PN2*, *PE2*) as shown in Figure 7.2a1. The said line is defined by the north and east coordinates of a point (*PN1*, *PE1*) and a north azimuth (*AZ1*). Point (*PN2*, *PE2*) may or may not be located on the said given line, but if it is, it may or may not be the same point as point (*PN1*, *PE1*).

The arc to be constructed commences somewhere along the said given line (*PCN*, *PCE*), and terminates tangent to another line at point (*PTN*, *PTE*) and is defined only by its north azimuth direction (*AZ2*). As a result the procedure returns the north and east coordinates of the center point (*CCN*, *CCE*), the arc endpoints (*PCN*, *PCE*) and (*PTN*, *PTE*), and the intersection point of the two tangent lines (*PIN*, *PIE*).

The icTanCrv2Lin procedure should not be confused with the procedures presented in Sect. 5.1.6 of Chap. 5. This procedure addresses a special case of curve construction that occasionally is encountered in roadway alignments. In Sect. 5.1.6 the two tangent lines are both fixed, the radius is unknown, and there are two potential solutions, while with

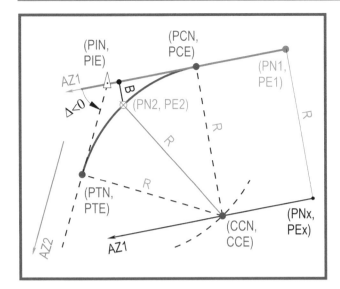

Figure 7.2a1 Arc of Given Radius Passing through a Given Point and Being Tangent to a Given Line and to another Line Implied by a Direction

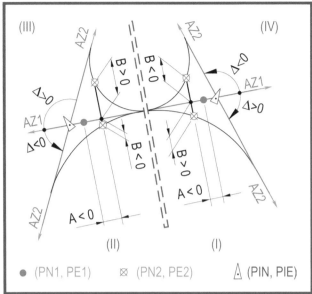

Figure 7.2a2 The Four Cases of an Arc of Given Radius Passing through a Given Point and Being Tangent to a Given Line and to another Line Implied by a Direction

this subject solution the forward tangent is fixed in direction but floating in position, the radius is known, and there is only one solution, the one as defined by the geometric process described below.

To construct the tangent arc, the subject procedure assumes that the azimuth *AZ1* of the given line always points towards the direction of the point of intersection (the PI point of the arc) of the said given line with the implied line of azimuth *AZ2*. The procedure then, with reference to Figure 7.2a2:

- Computes the angle of deflection Δ between the given and the implied line as

$$\Delta = AZ2 - AZ1,$$

- so that a clockwise angle is positive, and then
- Projects the given point (*PN1*, *PE1*) on the given line to find the plus (*A*) and the offset (*B*) distances.

In reviewing Figure 7.2a2, it is noted that there are four different arc condition cases that may arise depending on the two given azimuth directions and the position of the point (*PN2*, *PE2*) through which the arc is to pass. Thus, depending on the computed values of Δ, *A* and *B*, the subject procedure does the following:

- If $\Delta > 0$ and $A > 0$ and $B > 0$, uses *AZ2* as specified.
- If $\Delta > 0$ and $A > 0$ and $B < 0$, reverses the direction of *AZ2* and sets Δ to be the 180° supplement of its originally computed value.
- If $\Delta > 0$ and $A < 0$ and $B > 0$, uses *AZ2* as specified.

- If $\Delta > 0$ and $A < 0$ and $B < 0$, reverses the direction of *AZ2* and sets Δ to be the 180° supplement of its originally computed value.
- If $\Delta < 0$ and $A > 0$ and $B < 0$, uses *AZ2* as specified.
- If $\Delta < 0$ and $A > 0$ and $B > 0$, reverses the direction of *AZ2* and sets Δ to be the 180° supplement of its originally computed value.
- If $\Delta < 0$ and $A < 0$ and $B < 0$, uses *AZ2* as specified.
- If $\Delta < 0$ and $A < 0$ and $B > 0$, reverses the direction of *AZ2* and sets Δ to be the 180° supplement of its originally computed value.

From the above eight possible conditions, it is seen that the plus distance (*A*) has no effect on the construction of the tangent arc. Thus, the said conditions are reduced to four and summarized thusly: *if the deflection angle Δ and the offset distance B are both of the same sign, the azimuth AZ2 is used as specified, otherwise AZ2 is reversed*. The radius of the arc is supposed to be entered as a positive value, but for arc construction purposes whenever *D* is negative, the radius becomes negative as well. However, the input value of the radius is not altered. If the given point is located on the given line, it becomes the start point of the arc and *AZ1* and *AZ2* are used as specified.

```
Sub icTanCrv2Lin(PN1, PE1, AZ1, AZ2, R,
                 PN2, PE2, _
                 CCN, CCE, PCN, PCE, PTN,
                 PTE, PIN, PIE, T, DELTA,
                 IERR)
```

The input parameters to this procedure include the:

PN1, PE1	The north and east coordinates of the point through which the given tangent line (back tangent) is to pass.
AZ1	The north azimuth of the given tangent line (back tangent in radians).
AZ2	The north azimuth of the implied tangent line (forward tangent in radians).
R	The radius of the tangent arc (should be positive).
PN2, PE2	The north and east coordinates of the point through which the arc is to pass.

The parameters that are returned from this procedure include the:

CCN, CCE	The north and east coordinates of the center point of the tangent arc.
PCN, PCE	The north and east coordinates of the start point (PC) of the tangent arc.
PTN, PTE	The north and east coordinates of the end point (PT) of the tangent arc.
PIN, PIE	The north and east coordinates of the point of intersection (PI) of the two lines.
T	The length of the tangent to the arc (from PC or PT to the PI).
L	The length of the arc.
DELTA	The angle of deflection of the two tangent lines.
IERR	An indicator, the value of which if it is:
	0 indicates that a tangent arc has been created.
	1 indicates that a tangent arc cannot be created. This occurs when either the radius or the deflection angle is zero, or the radius is too small with respect to the offset distance of the point (PN2, PE2) from the given line.

Depending on the position of point (PN2, PE2) *with respect to the given line, it is possible for the constructed arc to terminate without passing through the said point. That is, the arc would pass through the said point only if extended beyond the* (PTN, PTE) *point.*

Upon invocation, the procedure initializes the parameters to be returned, and then:

1. Queries the value of R for being zero within a tolerance of 0.005 feet (1.524 mm), and if so sets IERR to be 1, and terminates.
2. Subtracts AZ1 from AZ2 to determine the angle of deflection Δ, taking into consideration the straddling of north. Note that a positive Δ indicates a clockwise rotation, while a negative one indicates a counterclockwise rotation.

3. Queries the value of Δ for being positive, negative or zero within a tolerance of 0.0000048 radians (0.99 seconds). If:
   - Δ is positive, the procedure continues.
   - Δ is negative, the radius R is set to be negative, and the procedure continues.
   - Δ is zero within the said tolerance, IERR is set to be 1, and the procedure terminates.
4. Calls the icprjct procedure to project the given point through which the arc is to pass on the given line and determine the lengths A and B.
5. Queries the value of the offset B for being equal to zero within a tolerance of 0.005 feet (1.524 mm), and depending on the response the procedure encounters one of two case paths:
   (a) **Case 1**—B is equal to zero within the said tolerance: The procedure:
   - Assumes that *AZ1* and *AZ2* are correct,
   - Sets the point of projection under Step 4 to be the point of beginning of the arc (PCN, PCE), and then
   - Continues with the next step.
   (b) **Case 2**—B is not equal to zero: The procedure branches to Step 8.
   Case 1
6. Calls the icptl procedure to establish the center point (CCN, CCE) of the arc by traversing from (PCN, PCE) along AZ1 a plus distance of zero (0.00) and an offset distance of R.
7. Calls the icptl procedure to establish the end point (PTN, PTE) of the arc by traversing from (CCN, CCE) along AZ2 a plus distance of zero (0.00) and an offset distance of −R , and then branches to Step 14.
   Case 2
8. Compares the deflection angle D with the offset distance B.
   - If both are positive or negative, the procedure continues with the next step.
   - If one is positive and the other is negative,
     – *AZ2* is reversed by adding to its specified value 180°,
     – Δ is changed to the 180° supplement of its original value with a reversed sign,
     – The sign of the radius R is changed to coincide with that of D,
   and then the procedure continues with the next step.
9. Calls the icptl procedure to establish a temporary line parallel to the given line of north azimuth AZ1 by establishing the temporary point (PNx, PEx) to the right or left of point (PN1, PE1), depending on the sign of the deflection angle Δ, by traversing along the azimuth AZ1 a plus distance of zero (0.00) and an offset distance of ± (see Figure 7.2a2) as dictated by the sign of Δ.

10. Call the `iclincir` procedure to intersect the temporary parallel line with an imaginary circle having a radius R and its center point at the given point (PN2, PE2).

11. Queries the error indicator of `iclincir` to determine whether an intersection exists or not. If the offset *B* when compared to twice the value of the radius R is:
    - Larger, an intersection does not exist, and the procedure sets IERR to be 1, and terminates.
    - Equal, the parallel line is tangent to the imaginary circle, in which case the point of tangency is returned twice, and the procedure continues.
    - Smaller, two intersections exist, and the procedure continues.

    Of the two possible intersection points that are returned, the first one is selected to become the center point (CCN, CCE) of the tangent arc.

12. Calls the `icpprjct` procedure to project the intersection point that has been selected above on the given line to determine the coordinates of the start point (PCN, PCE) of the tangent arc.

13. Calls the `icptl` procedure to establish the end point (PTN, PTE) of the arc by traversing from (CCN, CCE) along AZ2 a plus distance of zero (0.00) and an offset distance of −R.

Case 1 and 2

14. Computes the length of the arc L by multiplying the absolute value of the radius R by the deflection angle *Δ*, and the length of the tangent T by multiplying the radius R by the tangent of one half of the deflection angle *Δ*.

15. Calls the `icintrs` procedure to intersect the given line with the forward tangent line which is defined by the point (PTN, PTE) and azimuth AZ2 to determine the coordinates of the point of intersection (PIN, PIE) of the two tangent lines.

7.2.2   Arc Tangent to Three Lines

The `icTanCrv3Lin` procedure constructs a circular arc tangent to three lines, each defined by the north and east coordinates of a point and a north azimuth as indicated in the top part (A) of Figure 7.2b. The procedure returns (a) the coordinates of the points of tangency and of the arc's center, (b) the arc's radius, (c) the lengths of the two tangents and arc length components of the overall arc, and (d) the two deflection angles. The `icCrvInTri` procedure presented next in this subsection of this chapter is intended for inscribing a full circle within a triangle. Depending on the selection of the three lines, as is to be shown below, this subject procedure may be used to inscribe an arc within a triangle but not a full circle.

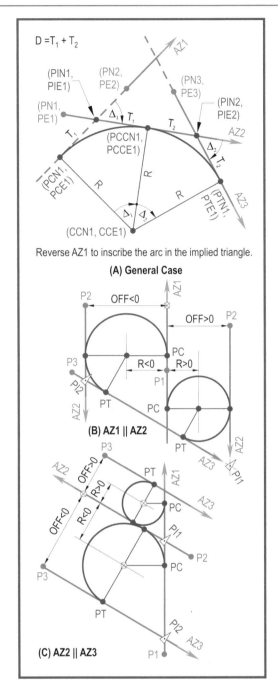

Figure 7.2b Arc Tangent to Three Lines

With reference to Figure 7.2b (A), there are three sectors in which an arc many be constructed so as to be tangent to each line, in addition to being inscribed within the implied triangle. One such arc is shown in the said figure. The other two sectors would be those in which the point (PCCN, PCCE) is located on each of the other two lines.

To construct the tangent arc this procedure requires that the three lines to which the arc is to be tangent should be specified sequentially in the rotational direction that the constructed tangent arc is to follow; that is, the arc should start at the first line, progress to the second, and then terminate at the third line in a clockwise or counterclockwise direction. In introducing these three lines the following should be noted:

• The direction of the first and second line sets the clockwise or counterclockwise rotation of the arc to be constructed.
• The direction of the third line may follow, or oppose the above rotation. If it opposes the rotation it is reversed by the procedure without altering the input value.
• The central angle of the overall arc may exceed 180°.
• The point that defines a line may lie anywhere along the extent of its line.

There are two conditions as indicated in parts (B) and (C) of Figure 7.2b that necessitate a special handling in the course of the procedure and they are addressed below the description of the general case.

The points of intersection between the first and second lines, and between the second and third lines may be found, and the distance D between them may be computed. From Figure 7.2b it is seen that $\delta = T_1 + T_2$, and that if we let $\delta_1 = \frac{1}{2}\Delta_1$ and $\delta_2 = \frac{1}{2}\Delta_2$, the relationships below may be developed.

$$T_1 = R\tan(\delta_1) \tag{7.2.2a}$$

and

$$T_2 = R\tan(\delta_2) \tag{7.2.2b}$$

the length D may be expressed as

$$D = R\cdot\tan(\delta_1) + R\cdot\tan(\delta_2) \tag{7.2.2c}$$

or

$$D = R\left[\tan(\delta_1) + \tan(\delta_2)\right] \tag{7.2.2d}$$

Since the length D has been determined above, the radius may be expressed as

$$R = \frac{D}{\tan(\delta_1) + \tan(\delta_2)} \tag{7.2.2e}$$

The tangent lengths T_1 and T_2 may now be found by use of Equations 7.2.2a and 7.2.2b.

The first of the two special cases referenced above occurs when the first line and the second line are parallel and offset one from the other as indicated in Figure 7.2b Part (B), and the other when the second line and the third line are parallel and offset from the other as indicate in Figure 7.2b Part (C). These are considered to be special cases because the general solution as presented above requires the intersection of these

two pairs of lines to locate their point of intersection, the PI point. Thus, if the lines of such a pair are parallel they would not yield an intersection. The general solution does not encounter this issue when the first line and the second line are not called to intersect each other. The method of solution is addressed under the operational description below.

```
Sub icTanCrv3Lin(PN1, PE1, AZ1, PN2, PE2,
            AZ2, PN3, PE3, AZ3, _
            PCN, PCE, PCCN, PCCE,
            PTN, PTE, PIN1, PIE1, PIN2,
            PIE2, _
            CCN, CCE, R, T1, T2, L1,
            L2, DELTA1, DELTA2)
```

The input parameters to this procedure include the:

PN1, PE1	The north and east coordinates of the point (point 1) through which the first tangent line (back tangent) is to pass.
AZ1	The north azimuth of the first tangent line (back tangent in radians).
PN2, PE2	The north and east coordinates of the point (point 2) through which the second tangent line (middle tangent) is to pass.
AZ2	The north azimuth of the second tangent line (middle tangent in radians).
PN3, PE3	The north and east coordinates of the point (point 3) through which the third tangent line (forward tangent) is to pass.
AZ3	The north azimuth of the third tangent line (forward tangent in radians).

Note that AZ1 and AZ2 should point in the same clockwise or counterclockwise direction.

The parameters that are returned from this procedure include the:

PCN, PCE	The north and east coordinates of the start point (PC) of the tangent arc on the first line.
PCCN, PCCE	The north and east coordinates of the tangency point of the arc with the second tangent line.
PTN, PTE	The north and east coordinates of the end point (PT) of the tangent arc on the third line.
PIN1, PIE1	The north and east coordinates of the point of intersection (PI) between the first and the second line.
PIN2, PIE2	The north and east coordinates of the point of intersection (PI) between the second and the third line.
CCN, CCE	The north and east coordinates of the center point of the tangent arc.
R	The radius of the tangent to the arc.
T1	The length of the tangent to the arc (from PC to the PI1).
T2	The length of the tangent to the arc (from PT to the PI2).
L1	The length of the tangent arc subtended by the first and the second line.

L2   The length of the tangent arc subtended by the second and the third line.

DELTA1   The angle of deflection between the first and the second line.

DELTA2   The angle of deflection between the second and the third line.

IERR   An indicator, the value of which if it is:

0   indicates that a tangent arc has been created.

1   indicates that a tangent arc could not be created. An error occurs when:
- Either AZ1 and AZ2, or AZ2 and AZ3 are colinear.
- When computing the radius a zero divide would occur.

Upon invocation, the procedure initializes the parameters to be returned, and then:

1. Queries the value AZ1 and AZ2, and of AZ2 and AZ3 to determine whether they are equal to each other within a tolerance of 0.0000048 (0.99 radians), and if so, the procedure terminates; else it continues.

2. Queries the absolute values of AZ1 and AZ2 to determine whether they are parallel within a tolerance of 0.0000048 (0.99 radians), and if so continues with the next step (special case of Part B of Figure 7.2b); else it branches to **Step 6**.

Special Case for Parallel Lines 1 and 2

3. Calls the icintrs procedure to intersect the first line with the third line and establish the intersection point (PIN1, PIE1) and the corresponding intersection angle DELTA1, which is adjusted for straddling north.

4. Calls the icprjct procedure to project point 2 on line 1 to get the offset distance OFF which could be positive or negative. Since the first and second line are parallel to each other and tangent to the same arc, the radius of the arc would be one half the value of OFF. Thereafter, the procedure

   (a) Calls the icptl procedure to establish a temporary parallel to AZ1 a distance R away;
   (b) Calls the icptl procedure to establish a temporary point from (PIN1, PIE1) along AZ1;
   (c) Calls the icprjct procedure to project this temporary point on the third line to determine an offset distance to be used as an indicator on which side of the third line a temporary line is to be created;
   (d) Calls the icptl procedure to establish a temporary line parallel to AZ3 and a distance R away in the direction of the last said offset distance, and then
   (e) Calls the icintrs procedure twice, first to intersect the two temporary lines that were constructed above to determine the center point of the arc, and then to intersect the second with the third line to determine point (PIN2, PIE2).

5. Calls the icprjct procedure to project the center point on each of the three given lines to establish the three points of tangency and determine the lengths of the tangent segments, computes the two arc length components of the overall arc and then terminates.

6. Queries the absolute values of AZ2 and AZ3 to determine whether they are parallel within a tolerance of 0.0000048 (0.99 radians), and if so continues with the next step (special case of Part C of Figure 7.2b); else it branches to **Step 10**.

Special Case for Parallel Lines 2 and 3

7. Calls the icintrs procedure to intersect the first line with the second line and establish the intersection point (PIN1, PIE1) and the corresponding intersection angle DELTA1, which is adjusted for straddling north.

8. Calls the icprjct procedure to project point 3 on line 2 to get the offset distance OFF which could be positive or negative. Since the second and third line are parallel to each other and tangent to the same arc, the radius of the arc would be one half the value of OFF. Thereafter, the procedure

   (a) Calls the icprjct procedure to project point (PIN1, PIE1) on the third line to determine an offset distance to be used as an indicator on which side of the third said PI is located;
   (b) Calls the icptl procedure to establish a temporary parallel to AZ3 a distance R away in the direction of the last said offset distance;
   (c) Calls the icptl procedure to establish a temporary parallel to AZ3 a distance R away depending on the sign of last offset distance;
   (d) Calls the icptl procedure to establish a temporary parallel to AZ1 a distance R away depending on the sign of R;
   (e) Calls the icintrs procedure twice, first to intersect the two temporary lines to determine the center point of the arc, and then to intersect the second with the third line to determine point (PIN2, PIE2).

9. Calls the icprjct procedure to project the center point on each of the three given lines to establish the three points of tangency and determine the lengths of the tangent segments, computes the two arc length components of the overall arc and then terminates.

General Case

10. Calls the icintrs procedure to intersects the first line with the second line, and the second line with the third line to determine the points of intersection (PIN1, PIE1) and (PIN2, PIE2), and the deflection angles Δ_1 and Δ_2 formed by the three lines.

11. Queries the sing of Δ_1 and Δ_2 for both being positive or negative. If both are not of the same sign, AZ3 is reversed and DELTA2 assumes the value of the 180° supplement of its original value.

12. Calls the `icforce` procedure to inverse from (PIN1, PIE1) to (PIN2, PIE2) to find the distance D.
13. Assess the denominator of Equation 7.2.2e for being zero, and if so aborts; else it determines the value of the radius of the tangent arc by use of the said equation.
14. Computes the lengths of the two tangents T_1 and T_2 by use of Equations 7.2.2a and 7.2.2b and the arc lengths L_1 and L_2 by multiplying the absolute value of the radius R by the corresponding deflection angle Δ_1 and Δ_2.
15. Calls the `icptl` procedure to determine the center points and points of tangency.

Alternatively to the above method, the construction could also be solved geometrically by bisecting the 180° supplement angles of the two PIs, and intersecting these two bisecting lines to establish the center point of the arc fairly much as is done in the `icCrvInTri` procedure presented next. This approach would also require the detection of the two cases of parallel lines and the taking of the steps to establish the said PIs similarly to the method employed above. The arc parameters, as well as the control checking would have to be as indicated above.

7.2.3 Circle Inscribed in a Triangle

The `icCrvInTri` procedure inscribes a circle within a triangle and tangent to it three sides defined by the north and east coordinates of the three points constituting the three vertices of the triangle (see Figure 7.2c). The procedure returns circle's radius, the north and east coordinates of its center point and of the three tangency points, the north azimuths

and lengths of the triangle sides, as well as the angles of the triangle.

```
Sub icCrvInTri(PN1, PE1, PN2, PE2,
               PN3, PE3, _
               PTN1, PTE1, PTN2, PTE2,
               PTN3, PTE3, CCN, CCE, R, _
               AZ12, AZ23, AZ31, D12, D23,
               D31, A1, A2, A3)
```

The input parameters to this procedure include the:

PN1, PE1	The north and east coordinates of the point at the first vertex of the triangle.
PN2, PE2	The north and east coordinates of the point at the second vertex of the triangle.
PN3, PE3	The north and east coordinates of the point at the third vertex of the triangle.

The parameters that are returned from this procedure include the:

PTN1, PTE1	The north and east coordinates of the point of tangency of the circle with the triangle side between the first and the second vertex.
PTN2, PTE2	The north and east coordinates of the point of tangency of the circle with the triangle side between the second and the third vertex.
PTN3, PTE3	The north and east coordinates of the point of tangency of the circle with the triangle side between the third and the first vertex.
CCN, CCE	The north and east coordinates of the center point of the circle.
R	The radius of the circle.
AZ12	The north azimuth of the triangle side from the first towards the second vertex.
AZ23	The north azimuth of the triangle side from the second towards the third vertex.
AZ31	The north azimuth of the triangle side from the third towards the first vertex.
D12	The length of the triangle side from the first towards the second vertex.
D23	The length of the triangle side from the second towards the third vertex.
D31	The length of the triangle side from the third towards the first vertex.
A12	The triangle angle at the first vertex.
A23	The triangle angle at the second vertex.
A31	The triangle angle at the third vertex.

Upon invocation, the procedure initializes the parameters to be returned, and then:

1. Calls the `icforce` procedure to inverse from:
   - Vertex (PN1, PE1) towards vertex (PN2, PE2) to determine the distance and azimuth AZ12 and distance D12, and

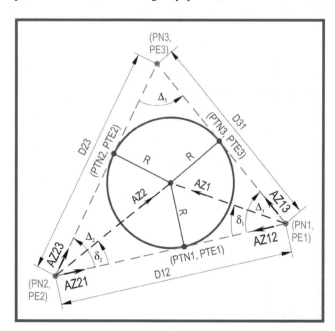

Figure 7.2c Circle Inscribed in a Triangle

- Vertex (PN1, PE1) towards vertex (PTN3, PTE3) to determine the distance and azimuth AZ13 and distance D13.
2. Determines the angle Δ_1 (A1) formed by the said two triangle sides and its half angle value δ, as well as the azimuth of the bisector of Δ_1. The value of the said angles could be positive or negative depending on their orientation and position with respect to each other.
3. Repeats the above two steps to inverse from vertex (PN2, PE2) towards vertex (PN1, PE1) and from vertex (PN2, PE2) towards vertex (PN3, PE3) to determine the angle Δ_2 (A2) formed by the first and second line and its half angle value δ_2 and the azimuth of the bisector of Δ_2.
4. Calls the icintrs procedure to intersect the said two bisecting lines to determine the coordinates of the circle's center point (CCN, CCE), and then calls.
5. Calls the icprjct procedure to project the said center point on each of the three lines to determine the points of tangency (PTN1, PTE1), (PTN2, PTE2), and (PTN3, PTE3).
6. Calls the icforce procedure to inverse from the center point to each of the three points of tangency to determine three distances which are averaged to determine the value of the radius of the circle.

7.2.4 Arc Tangent to a Line and Through Two Points

The iccvta2q procedure constructs a circular arc that passes through two points of known Cartesian coordinates and is tangent to a given line. The line is defined by the Cartesian coordinates of two points. Depending on the coordinate position of the two points that define the line, and of the two points through which the arc is to pass, the number of tangent arcs that can be constructed could be:

- None, if the latter two points straddle the line defined by the former two points,
- None, if both of the latter two points lie on the line defined by the former two points,
- One, if the latter two points are on either side of the said line and equidistant therefrom, or
- Two, if none of the above conditions have been encountered.

If two tangent arcs can be constructed, the procedure returns the Cartesian coordinates of the center points of the two tangent arcs from which the calling program may select the desired one. If only one arc can be constructed, the Cartesian coordinates of its center point is returned twice with one on top of the other.

When the procedure detects that one of the two points through which the arc is to pass is located on the given line

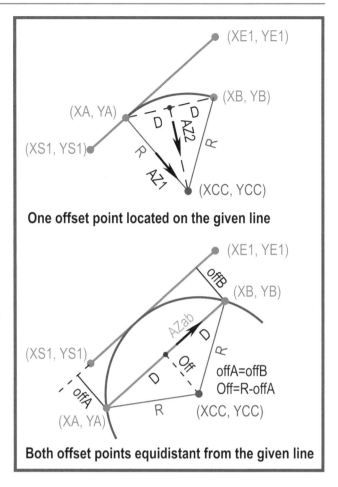

Figure 7.2d Arc Through Two Points Tangent to Line

within a tolerance of 0.005 feet (1.524 mm) the procedure, with reference to the upper part of Figure 7.2d:

- Calls the icforce procedure to find the azimuth of the given line from (YS1, XS1) to (YE1, XE1).
- Calls the icforce procedure to find the length and azimuth of the chord from (YA, XA) to (YB, XB).
- Calls the icptl procedure to locate the coordinated of the midpoint of the said chord.
- Calls the icintrs procedure to intersect the bisector line of the said chord having an azimuth normal to that of the chord with an implied line normal to the given line at the given point that lies on the given line to find the coordinates of the arc's center point (YCC, XCC).

When the procedure detects that both of the two points through which the arc is to pass are equidistant from the given line within a tolerance of 0.005 feet (1.524 mm) and on the same side of the said line the procedure, with reference to the lower part of Figure 7.2d:

- Calls the icforce procedure to find the length and azimuth of the chord from (YA, XA) to (YB, XB).

- Averages the two offset distances *offA* and *offB* in case they are slightly different by the amount of the allowable tolerance.
- Computes the radius *R* of the tangent arc to be constructed as

$$D^2 + \left(R - offA\right)^2 = R^2$$
$$R = \frac{\left(offA\right)^2 + D^2}{2\left(offA\right)} \qquad (7.2.4a)$$

- Computes the distance from the arc's midpoint to the arc's center as:

$$Off = R - offA \qquad (7.2.4b)$$

- Calls the `icptl` procedure to locate the arc's center point by traversing from point *A* towards point *B* a plus distance of *D* and an offset distance of *Off*.

If none of the above conditions are encountered, the solution to this problem as presented in the source code of the accompanying CD is based on the work of Sylvan H. Chasen of the Lockheed-Georgia Company as presented in pages 80–86 of his book *Geometric Principles and Procedures for Computer Graphic Applications* published by Prentice-Hall, Inc. in 1978. This solution, with reference to Figures 7.2e through 7.2h, is based on the premise that each of the center points of the two tangent arcs or circles to be constructed:

- Is located on the line which is normal to the chord from (XA, YA) to (XB, YB) at the chord's midpoint, and
- Is located on a line that is parallel to the given line from (XS1, YS1) to (XE1, YE1) and is offset therefrom a distance equal to the radius of the respective tangent arc or circle, and
- Is equidistant from the chord's endpoints.

The projection of the center point of each arc that is created on the given line is the tangency point of the said arc.

In Figure 7.2e two potential solutions exist when the chord defined by the given points (XA, YA) and (XB, YB) is inclined with respect to the line between points (XS1, YS1) and (XE1, YE1). In this case, the center points of the two circles or arcs are located on the same side of the said chord. Depending on the orientation of the two given points, it is possible for the two center points to be located on the opposite side of the chord as shown on the said figure, or to straddle the chord as shown in Figure 7.2f. Similar conditions may arise if the two given points are located on the opposite side of the said given line. Two additional cases may be encountered if the two given points create a chord

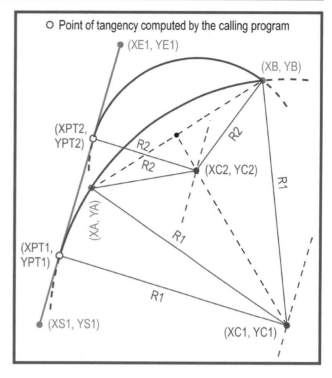

Figure 7.2e Arc Tangent to Given Line and Passing Through Two Given Points (Case 1)

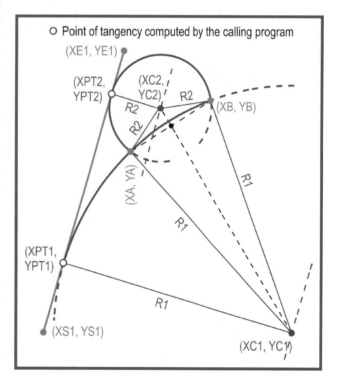

Figure 7.2f Arc Tangent to Given Line and Passing Through Two Given Points (Case 2)

which is normal, or parallel to the said given line as indicated in Figure 7.2g and Figure 7.2h, respectively. Note that in the

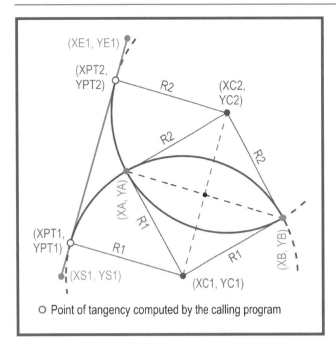

Figure 7.2g Arc Tangent to Given Line and Passing Through Two Given Points (Case 3)

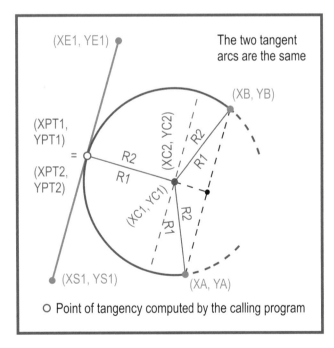

Figure 7.2h Arc Tangent to Given Line and Passing Through Two Given Points (Case 4)

latter case the two solutions to the construction problem are identical, and that this case is handled by the procedure as a special case as indicated earlier in this subsection.

```
Sub iccvta2q(XS1, YS1, XE1, YE1, XA, YA, XB,
             YB, _
             IND, XC1, YC1, XC2, YC2)
```

The input parameters to this procedure include the:

XS1, YS1　The Cartesian coordinates of the start point of the given line.
XE1, YE1　The Cartesian coordinates of the end point of the given line.
XA, YA　The Cartesian coordinates of the first point through which the tangent arc is to pass.
XB, YB　The Cartesian coordinates of the second point through which the tangent arc is to pass.

The parameters that are returned from this procedure include the:

IND　An indicator, the value of which if it is:
　0 denotes that there is no possible solution for the given parameters.
　1 denotes that there is only one solution; both center points coincide.
　2 denotes that there are two solutions, two center points.
XC1, YC1　The Cartesian coordinates of the center point of the first potential arc.
XC2, YC2　The Cartesian coordinates of the center point of the second potential arc.

Upon invocation, the procedure initializes the parameters to be returned, and then:

1. Calls the icforce procedure to construct the given line that is implied from (XS1, YS1) towards (XE1, YE1).
2. Calls the icprjct procedure to project on the given line the points (XC1, YC1) and (XC2, YC2) through which the tangent arc is to pass, and thus determine the offset distances of these two points from the given line.
3. Queries the said offset distances to determine, within a tolerance of 0.005 feet (1.524 mm), whether one or both of the points are located on the given line, and if not, whether they are equidistant therefrom or not, and if not whether they straddle the said line, or not. If:
   • Both points are located on the said line, the procedure sets the indicator to zero, and aborts.
   • If only one point is located on the said line, the center point of the tangent arc is determined by Equations 7.2.4a and 7.2.4b and both sets of the returned coordinates are the same, and the procedure terminates.
   • Both points are not located on the said line, but the straddle the given line, the procedure sets the indicator to zero, and aborts.
   • Both points are not located on the said line, are on same side of the given line, and they are equidistant from the given line, the center point of the tangent arc is determined as described at the start of this procedure (both sets of the returned coordinates are the same), and the procedure terminates.

- Both points are not located on the said line, are on same side of the given line, and they are not equidistant from the given line, the procedure continues as indicated below.
4. Calls the `icimplct` procedure (Chap. 4) to define the given line in its explicit form.
5. Calls the `icimplct` procedure to define chord of the arc formed by the two given points through which the arc is to pass in its explicit form.
6. Determines whether the normal bisector of the chord is parallel to the given line, normal to it, or inclined to it, and accordingly caries out Chasen's solution as described in the in the previously sited reference to compute the Cartesian coordinates of the center points of the two tangent arcs.

Having computed the coordinates of the center points of the two tangent arcs, the calling program of the subject procedure may display one or both of the center points, as the case may be, from which the user or the calling program may select the desired center point. Thereafter, the calling program may call the `icprjct` procedure to project the selected center point on the given line and compute the radius of the selected arc and the point of tangency.

7.2.5 Arc of Given Radius Through a Point Tangent to a Line

The `icCvTa1pX` procedure constructs an arc of known radius that passes through a given point and is tangent to a given line, which line is defined by the north and east coordinates of a point and a north azimuth.

There are two potential solutions to this construction problem as indicated in Figure 7.2i, in which figure the given and returned parameters, as well as the intermediate computational parameters are illustrated. It is noted that:

- If the given point (PN2, PE2) is located at a distance greater than twice the value of the given radius away from the given line (conversely the given radius is too small), then a tangent arc cannot be constructed. In this case the value of each of the returned parameters is a zero.
- In the said figure the point through which the arc is to pass is not located on the line, and thus, both potential tangent arcs are located on the same side of the given line as the given point. However, if the given point (PN2, PE2) is located on the given line, then one of the tangent arcs would be located on one side of the given line and the other tangent arc would be located on the other side. In this case, both ANG1 and ANG2 have a value of zero.

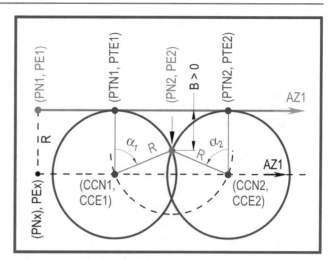

Figure 7.2i Arc of Given Radius and Passing Through a Given Point Tangent to Given Line

```
Sub icCvTa1pX(PN1, PE1, AZ1, PN2, PE2, R, _
              IND, CCN1, CCE1, CCN2, CCE2,
              PTN1, PTE1, PTN2, PTE2, _
              ANG1, ANG2)
```

The input parameters to this procedure include the:

PN1, PE1	The north and east coordinates of a point on the given line.
AZ1	The north azimuth, in radians, of the given line.
PN2, PE2	The north and east coordinates of the point through which the arc is to pass.
R	The radius of the tangent arc to be created which should be positive.

The parameters that are returned from this procedure include the:

IND	An indicator, the value of which if it is:
	0 denotes that a tangent arc cannot be constructed because the radius is too small for the position of the point with respect to the line.
	1 denotes that there is only one solution; the point through which the arc is to pass is located on the given line.
	2 denotes that there are two solutions, two center points.
CCN1, CCE1	The north and east coordinates of the center point of one tangent arc.
CCN2, CCE2	The north and east coordinates of the center point of the other tangent arc.
PTN1, PTE1	The north and east coordinates of one point of tangency.
PTN2, PTE2	The north and east coordinates of the other point of tangency.
ANG1, ANG2	The angles α_1 and α_2, in radians, as shown in Figure 7.2i, and noting that a positive angle denotes a clockwise rotation.

Upon invocation, the procedure initializes the parameters to be returned, and then:

1. Calls the `icprjct` procedure to project the given point (`PN2`, `PE2`) on the given line to determine the point's offset distance **B** (see Figure 7.2i) from the given line.
2. Queries whether the absolute value of **B** is equal to zero, within a tolerance of 0.005 feet (1.524 mm), or not:
   - If so, the given point is the start point of the arc, and the procedure continues with the next step.
   - If not so, the procedure branches to Step 6.
3. Sets the coordinates of points (`PTN1`, `PTE1`) and (`PTN2`, `PTE2`) to be those of the point of projection of the given point (`PN2`, `PE2`) on the given line.
4. Calls the `icptl` procedure to compute the coordinates of the center points (`CCN1`, `CCE1`) and (`CCN2`, `CCE2`) by traversing from (`PTN1`, `PTE1`) along `AZ1` a plus distance of zero (0.00) and an offset distance of +R and −R, respectively.
5. Sets the value of `IND` to be one (1) and terminates.
6. Calls the `icptl` procedure to compute the coordinates of the temporary point (`PNx`, `PEx`) by traversing from (`PTN1`, `PTE1`) along `AZ1` a plus distance of zero (0.00) and an offset distance of
   - +R, if **B** is positive, or
   - −R, if **B** is negative,
        to create a temporary line parallel to the given line.
7. Calls the `iclincir` procedure, to intersect the temporary parallel line with an imaginary circle having its origin at the given point (`PN2`, `PE2`), and a radius of R, to locate the center points (`CCN1`, `CCE1`) and (`CCN2`, `CCE2`) of the two tangent circles.
8. Queries the indicator `INTRSC` which is returned in **step 7**. If its value is:
   - Zero, the procedure sets ind to be zero and terminates because this is an indication that the radius is too small to construct a tangent arc.
   - 1 or 2, the procedure accepts the coordinates of the above step and continues. A value of one denotes that the temporary parallel line is tangent to the imaginary circle, in which case two identical arcs are created.
9. Calls the `icprjct` procedure to project points (`CCN1`, `CCE1`) and (`CCN2`, `CCE2`) on the given line to locate the points of tangency (`PTN1`, `PTE1`) and (`PTN2`, `PTE2`).
10. Calls the `icforce` procedure to determine the azimuth from each center point towards the given point (`PN2`, `PE2`), determines the normal azimuth from each center point normal to the azimuth of the given line, computes the values of the inclosed angles `ANG1` and `ANG2`, and then terminates.

7.2.6　Arc of Given Radius and Start Point Tangent to a Direction

The procedure `icCvTaTpX` constructs a circular arc of known radius that starts at a given point (the PC) and is tangent to a real line of given azimuth through the said point, or implied (imaginary) line parallel to said azimuth as indicated in Figure 7.2j. That is, if the given point is not located on the given line, it is assumed that the arc is to be tangent to a line parallel to the given line that passes through the given start point.

The start point is defined by its north and east coordinates, and the given line is defined by a north azimuth and a point of known north and east coordinates that position the given line in the plane. In addition to the tangent arc's start point, the arc is defined by its radius, and one of the following: (a) arc length, (b) central angle, or (c) chord length. The methodology of this construction is rather simplistic and it is described in the operational steps below. In essence this construction is quite similar to that described for the procedures of Chap. 5 such as the `icarcan1`, `icarccd2`, and `icarcln1` procedures.

```
Sub icCvTaTpX(PN1, PE1, AZ1, PN, PE, R,
              L, A, C, _
              CCN, CCE, PCN, PCE, PTN,
              PTE, IND)
```

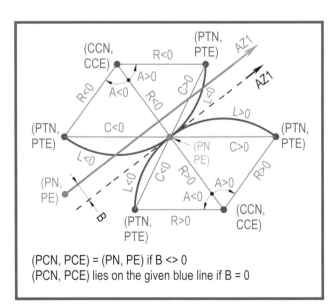

(PCN, PCE) = (PN, PE) if B <> 0
(PCN, PCE) lies on the given blue line if B = 0

Figure 7.2j Arc of Given Radius and Start Point Tangent to a Given Direction

The input parameters to this procedure include the:

PN1, PE1	The north and east coordinates of a point on the given line.
AZ1	The north azimuth of the given line in radians.
PN, PE	The north and east coordinates of the start point of the arc to be constructed.
R	The radius of the tangent arc to be constructed, and which should be: • Positive for a clockwise arc (to the right of the given line), or • Negative for a counterclockwise arc (to the left of the given line) with respect to the direction of the given line as indicated in Figure 7.2j.
L	The length of the arc, or zero (refer to the note below).
A	The central angle (in radians) of the arc, or zero (refer to the note below).
C	The chord length of the arc, or zero (refer to the note below).
NOTE	Of the last three input parameters, only one must have a nonzero value, and the other two should have a value of zero (0.00). If more than one parameter has a nonzero value, the first one to be encountered with a nonzero value controls. It is the calling program's responsibility to pass the appropriate of these last said three parameter values. If the nonzero value is: • Positive, the arc follows the direction implied by the radius. • Negative, the arc opposes the direction implied by the radius.

The parameters that are returned from this procedure include the:

CCN, CCE	The north and east coordinates of the center point of the tangent arc.
PCN, PCE	The north and east coordinates of the start point of the arc (see Step 2 below).
PTN, PTE	The north and east coordinates of the end point of the arc.
IND	An indicator, the value of which if it is: 1 denotes that an error has not been encountered. 2 denotes that an error has been encountered (insufficient input parameters).

Upon invocation, the procedure initializes the parameters to be returned, and then:

1. Checks the input value of the radius for being equal to zero within a tolerance of 0.005 feet (1.524 mm), and if so sets IND to be 2 and aborts.
2. Checks the values of the arc and chord length for being equal to zero within a tolerance of 0.005 feet (1.524 mm), and the central angle for being equal to zero within a

tolerance of 0.0000048 radians (0°00′00.99″) and if so sets IND to be 2 and aborts.
3. Calls the icprjct procedure to project the start point (PN, PE) on the given line to determine the point's offset distance **B** from the given line.
4. Queries whether the offset distance **B** is equal to zero, within a tolerance of 0.005 feet (1.524 mm), or not:
   • If so, point (PCN, PCE) is set to the coordinates of the projection point on the given line.
   • If not so, point (PCN, PCE) is set to the coordinates of the given point (PN, PE).
5. Calls the icptl procedure to compute the coordinates of the arc by traversing from (PCN, PCE) along AZ1 a plus distance of zero and an offset distance of R, noting that a positive R positions the center point to the right of the given line, and a negative R to the left.
6. Computes the radial azimuth from the center point of the arc towards the arc's point of beginning (PCN, PCE) by adding or subtracting $\pi/2$ from the given azimuth AZ, depending on the sign of R.
7. Queries the three input parameters, L, A, or C, for a nonzero value in the indicated input sequence, and selects the first one that has a nonzero value. If the first nonzero parameter is the:
   • L, the procedure computes the central angle as A=L/R.
   • A, the procedure:
     – Accepts A as the value of the central angle as is, if R>0.
     – Assigns −A as the value of the central angle, if R<0.
   • C, the procedure computes the central angle as A=2sin-1(C/2R).
8. Computes the radial azimuth from the arc's center point towards the arc's end point (PTN, PTE) by adding or subtracting the value of A, depending on the value of A.
9. Calls the icptl procedure to compute the coordinates of the arc's endpoint (PTN, PTE) along the last computed radial azimuth with a plus distance equal to the absolute value of the radius R and with an offset distance of zero (0.00).

7.3   Arcs Tangent to Lines and Arcs

The two procedures, the icctlc01 and the icctlc02, that constitute this section construct an arc, or circle that is tangent to a line and to another circular arc or circle. Whereas the former assumes a given radius, the latter requires a point through which the constructed arc is to pass.

7.3.1 Arc with Radius Tangent to Line and Curve

The `icctlc01` procedure constructs an arc of given radius tangent (a) to a given line defined by the north and east coordinates of two points thereon and (b) to a given arc or circle defined by the north and east coordinates of its center point and its radius.

Depending on the location of the given line and of the given arc or circle with respect to each other, and depending on the magnitudes of the radii of the given arc or circle and of the curve to be constructed, there can be no possible tangent curve, or there can be from one up to a maximum of eight possible tangent curve solutions. Figure 7.3a identifies the potential tangent curves that can exist depending on whether the given curve is tangent to, intersects or is apart from the given line.

Not shown in this figure is the condition which yields no possible tangent curve construction. This condition is similar to that displayed at the bottom of the said figure, but with the difference that the given curve is situated away from the given line a distance exceeding the diameter of the curve to be constructed.

When the given curve is situated away from the given line a distance equal to the diameter of the curve to be constructed, there can be but one possible tangent curve construction. This condition is shown in the top portion of the said figure.

Since there can be more than one possible tangent curve, the procedure computes and returns in six collections the north and east coordinates of the center points and of the points of tangency of the potential tangent curves with the given line and with the given curve. The calling program may then display all potential tangent curve from which the user may select the desired one (refer to the side bar on this page).

Although the `icctlc01` procedure returns the coordinates of the center point and of each of the points of tangency of each curve, under an interactive user environment it may be wise to display only the center points, and upon selection of the desired one to recompute the points of tangency. This will relieve the task of keeping track of which curve is in which collection.

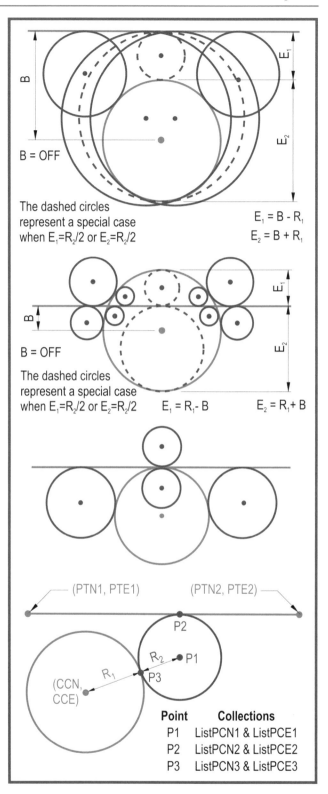

Figure 7.3a Potential Curves of Given Radius Tangent to a Line and a Curve

To accomplish this construction, the procedure solves for the coordinates of the center points of the potential tangent curves by intersecting an imaginary line which is parallel to the given line and is located at a distance away therefrom equal to the radius of the tangent curve to be constructed with an imaginary circle having its center point at the center point of the given curve and a radius equal to the sum of the radii of the given curve and of the tangent curve to be constructed. The solution process is outlined in Figure 7.3b, and it is described in detail in the operational steps below. It is noted that the procedure utilizes a local coordinate system

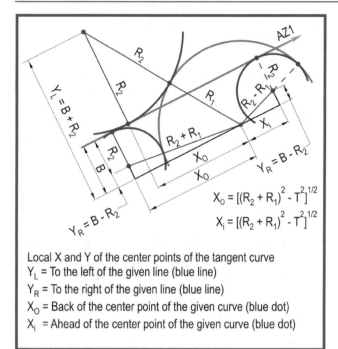

Local X and Y of the center points of the tangent curve
Y_L = To the left of the given line (blue line)
Y_R = To the right of the given line (blue line)
X_0 = Back of the center point of the given curve (blue dot)
X_i = Ahead of the center point of the given curve (blue dot)

Figure 7.3b Computation of the Center Points of Curves of Given Radius Tangent to a Line and a Curve

that has its origin at the center point of the given curve, and its X axis as being along the direction of the given line and the Y axis normal thereto. This is only for the internal relationship of the potential points of tangency with respect to the given line and given curve. The coordinates that are returned are the actual north and east coordinates.

```
Sub icctlc01(PTN1, PTE1, PTN2, PTE2, CCN,
             CCE, R1, R2, _
             ListPCN1, ListPCE1, ListPCN2,
             ListPCE2, ListPCN3, ListPCE3,
             iErr)
```

The input parameters to this procedure include the:

PTN1, PTE1	The north and east coordinates of the start point of the given line.
PTN2, PTE2	The north and east coordinates of the end point of the given line.
CCN, CCE	The north and east coordinates of the center point of the given curve.
R1	The radius of the given curve (positive).
R2	The radius of the tangent curve to be constructed (positive).

The parameters that are returned from this procedure include the:

ListPCN1	The collection of the north coordinates of the center points of a tangent arc.
ListPCE1	The collection of the east coordinates of the center points of a tangent arc.
ListPCN2	The collection of the north coordinates of the points of tangency of a tangent arc with the given line.
ListPCE2	The collection of the east coordinates of the points of tangency of a tangent arc with the given line.
ListPCN3	The collection of the north coordinates of the points of tangency of a tangent arc with the given curve.
ListPCE3	The collection of the east coordinates of the points of tangency of a tangent arc with the given curve.
iErr	An indicator, the value of which if it is:
	1 denotes that at least one tangent curve has been created.
	2 denotes that a tangent curve cannot be created.

Upon invocation, the procedure initializes to 9999999999.0 the contents of the collections to be returned, assumes that a solution will not be encountered, and then:

1. Calls the icforce procedure to determine the azimuth AZ1 of the given line from (PTN1, PTE1) to (PTN2, PTE2).
2. Calls the icprjct procedure to project the center point of the given curve on the given line and determine the offset distance B (see Figure 7.3b). If this offset distance B is negative, the procedure reverses the azimuth of the given line AZ1 by 180°, and changes the sign of this offset. This forces the center point of the given curve to be to the right of the given line.
3. Checks the offset B against R1, and if they are equal to each other within a tolerance of 0.005 feet (1.524 mm), sets R1 to be equal to B.
4. Queries whether B is less than or equal to R1, and:
   - If so, continues with the next step (the given line is tangent to, or intersects the given curve).
   - If not so, branches to Step 8 (the given line and curve are apart from each other).
5. Computes the values of E1 and E2 as indicated in Figure 7.3a.
6. Queries whether E1 is equal R2 within a tolerance of 0.005 feet (1.524 mm), and if so, sets R2 equal to E1.
7. Compares the value of R2 with E1 and E2 for special conditions, and then computes the so called local coordinates of the center points of the potential tangent curves to the left of the given line (see Figure 7.3b) These coordinates are saved in the temporary lists LISTX and LISTY, in a sequential order along AZ1 from right to left.
8. Performs similar computations to those of above for the curves on the right side of the given line from right to left.
9. Calls the icptl procedure to traverse from (CCN, CCE) along AZ1 using the local X coordinates as the plus distance and the local Y coordinates as the offset distance to establish the points of tangency, taking in consideration the location of the tangent curve with respect to the:
   - Given line and its azimuth AZ1 (left or right therefrom) and
   - Center point (CCN, CCE) of the given curve.

7.3.2 Arc Tangent to Line and Arc Through a Point

The `icctlc02` procedure introduces a arc that is tangent to a given line and to a given arc or circle and passes through a given point. The given line is defined by the north and east coordinates of two points, and the given curve is defined by the north and east coordinates of its center point and its radius.

Displayed in Figure 7.3c1 is a generic arrangement of a given line, a given arc, a given point and the possible arc that passes through the said point and is tangent to both the given line and arc. In this figure note the following conventions that are employed by the procedure:

1. The direction of the given line is assumed to be such that the given curve is located to the right of the given line.

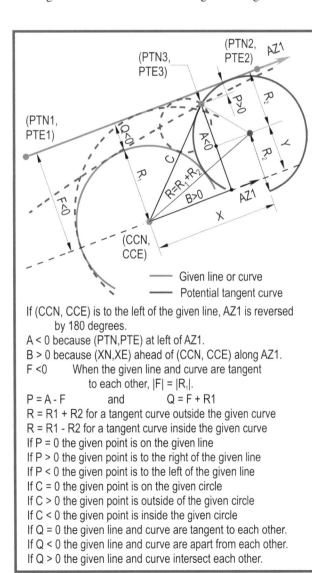

If (CCN, CCE) is to the left of the given line, AZ1 is reversed
    by 180 degrees.
A < 0 because (PTN,PTE) at left of AZ1.
B > 0 because (XN,XE) ahead of (CCN, CCE) along AZ1.
F <0   When the given line and curve are tangent
        to each other, |F| = |R₁|.
P = A - F   and   Q = F + R1
R = R1 + R2 for a tangent curve outside the given curve
R = R1 - R2 for a tangent curve inside the given curve
If P = 0 the given point is on the given line
If P > 0 the given point is to the right of the given line
If P < 0 the given point is to the left of the given line
If C = 0 the given point is on the given circle
If C > 0 the given point is outside of the given circle
If C < 0 the given point is inside the given circle
If Q = 0 the given line and curve are tangent to each other.
If Q < 0 the given line and curve are apart from each other.
If Q > 0 the given line and curve intersect each other.

Figure 7.3c1 Arc Through Point Tangent to Line and Arc

2. The radius is assumed to be positive.
3. The offset distance (F) of the center point of the given curve is assumed to be negative, and when the given curve is tangent to the given line, the absolute value of this offset distance is set to be the absolute value of the given radius (F = −R1).
4. The offset distance (P) of the given point, through which the arc to be constructed is to pass, from the given line is assumed to be positive.
5. The distance (C) of the abovesaid point from the center point of the given arc is assumed to be positive, and when the said point is on the given arc the said distance is set to be the value of the given radius.
6. A local coordinate system is assumed in the solution of this construction problem with its origin at the center point of the given curve, its X axis along the azimuth of the given line as said above, and its Y axis normal to the said X axis and to the left thereof.
7. The distances A and B are signed as indicated in Figure 7.3c1.
8. A tolerance test of 0.005 (1.524 mm) is made to determine whether the given curve and line are tangent to each other, and whether the given point through which the arc to be constructed is located on the given curve or given line.

The positioning of the given line, curve and point with respect to each other controls (a) whether a tangent arc can be constructed or not, and (b) the number of possible tangent arcs that can be constructed, if any. This positioning is dependent on the values of the Q, P, and C parameters as defined in Figure 7.3c1. Thus, these three parameters yield the 27 different conditions that are displayed in Figure 7.3c2, and in which figure the number of potential tangent arcs is shown as the number of curves.

In this figure, the blue dot denotes the given point through which the tangent arc is to pass. Note that:

- For conditions 8, 9, 13, and 14 the positioning of the three given features is such that a tangent arc cannot be constructed.
- Conditions 15, 17, 18, 21, 23, 24, 26, and 27 cannot be encountered. For example, under condition 15 in which the given curve is offset from the given line, the given point cannot be located on the given curve (C = R1) and at the same time be located on the given line (P = 0).
- Condition 20 can generate only one tangent arc only if the given point is located midway between the center point of the given arc and its projection on the said line.
- For condition 25 the said figure displays two pairs of tangent arcs, of which only one pair can be constructed depending on whether the given point is located to the left or to the right of the said line.

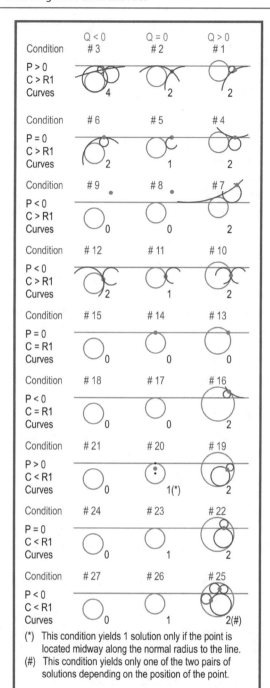

Figure 7.3c2   Arc Through Point Tangent to Line and Arc

(*) This condition yields 1 solution only if the point is located midway along the normal radius to the line.
(#) This condition yields only one of the two pairs of solutions depending on the position of the point.

The procedure computes and returns the coordinates of the center points and of both points of tangency for each tangent arc that can be constructed. The maximum number of tangent arcs that can be constructed is four. The north and east coordinates of these three points are placed and returned in the six collections identified later on.

To compute the coordinates of the center point of the arc to be constructed, the procedure lets R denote the distance

between the center point of the given curve and the center point of the arc to be constructed. Thus, depending on whether the given point is located within or without the given curve the said distance can be expressed as

$$R = R_1 \pm R_2 \tag{7.3.2a}$$

where

R1 denotes the radius of the given curve and
R2 denotes the radius of the tangent arc to be constructed.

Based on the local coordinate system referenced above

$$R^2 = X^2 + Y^2 \tag{7.3.2b}$$

and

$$Y = F \pm R_2 \tag{7.3.2c}$$

depending on whether the tangent arc to be constructed is to be located to the right or left of the given line.

The square of the radius of the new tangent arc may also be expressed as

$$R_2^2 = (A \pm Y)^2 + (B \pm X)^2 \tag{7.3.2d}$$

If we substitute Equation 7.3.2a for R and Equation 7.3.2c for Y in Equation 7.3.2b we get:

$$(R_1 \pm R_2)^2 = X^2 + (F \pm R_2)^2 \tag{7.3.2e}$$

Next, if we substitute Equation 7.3.2c for Y in Equation 7.3.2d we get

$$R_2^2 = (A \pm F \pm R_2)^2 + (B \pm X)^2 \tag{7.3.2f}$$

In Equations 7.3.2e and 7.3.2f we notice that we have two unknowns, the radius of the arc to be constructed R_2 and the local coordinate X, which can be determined by the simultaneous solution of Equations 7.3.2e and 7.3.2f. With X and R_2 now known, Equation 7.3.2c may be used to determine the local coordinate Y, from which the actual coordinates of the center point and of the points of tangency may be computed as indicated in the operational steps below.

The procedure, in order to account for potential solutions, first determines which of the 27 conditions of Figure 7.3c2 has been encountered, and accordingly calls the appropriate of the iccase01, iccase02, iccase03, and iccase04 procedures which solve Equation 7.3.2e for R_2 as

$$R_2 = \frac{X^2 + F^2 - R_1^2}{2(R_1 - F)} \tag{7.3.2g}$$

and then Equation 7.3.2f for X as

$$X = \sqrt{\left(\frac{R_1 - F}{A - R_1}\right)^2 B^2 + \left(\frac{R_1 - F}{A - R_1}\right)\left[C^2 + AR_1 - F(A + R_1)\right]}$$
$$- \left(\frac{R1 - F}{A - R_1}\right)B \qquad (7.3.2h)$$

Each of these procedures customizes the appropriate plus (+) or minus (−) sign of these two equations depending on which of the 27 conditions has been encountered.

```
Sub icctlc02(PTN1, PTE1, PTN2, PTE2, PTN3,
             PTE3, CCN, CCE, R1, _
             R2, PCN1, PCE1, PCN2, PCE2,
             PCN3, PCE3, iErr)
```

The input parameters to this procedure include the:

PTN1, PTE1	The north and east coordinates of the start point of the given line.
PTN2, PTE2	The north and east coordinates of the end point of the given line.
PTN3, PTE3	The north and east coordinates of the point through which the arc is to pass.
CCN, CCE	The north and east coordinates of the center point of the given curve.
R1	The radius of the given curve (may be positive or negative). If a negative radius is entered, the procedure uses its absolute value.

The parameters that are returned from this procedure include the:

R2	The collection of the radii of the potential tangent arcs.
PCN1	The collection of the north coordinates of the center point of the potential tangent arcs.
PCE1	The collection of the east coordinates of the center point of the potential tangent arcs.
PCN2	The collection of the north coordinates of the PC point of the potential tangent arcs.
PCE2	The collection of the east coordinates of the PC point of the potential tangent arcs.
PCN3	The collection of the north coordinates of the PT point of the potential tangent arcs.
PCE3	The collection of the east coordinates of the PT point of the potential tangent arcs.
iErr	An indicator, the value of which if it is:
1 denotes that at least one tangent curve has been created.
2 denotes that a tangent curve cannot be created.

Although the icctlc02 *procedure returns the coordinates of the center point and of each of the points of tangency of each curve, under an interactive user environment it may be wise to display only the center points, and upon selection of the desired one to recompute the points of tangency. This will relieve the task of keeping track of which curve is in which collection.*

Upon invocation, the procedure initializes to 9999999999.0 the contents of the collections to be returned, assumes that a solution will be encountered, and then:

1. Calls the icforce procedure to determine the azimuth AZ1 of the given line from (PTN1, PTE1) to (PTN2, PTE2).
2. Calls the icprjct procedure to project the center point of the given curve on the given line and determine the offset distance F. If the offset is negative, the procedure reverses the direction of AZ1 by 180°. By changing the sign of the offset distance F to be negative, the procedure assumes a local coordinate system that has its origin at the center point of the given curve (CCN, CCE), its positive X axis direction along the direction of AZ1, and its Y axis at 90° therefrom in a counterclockwise rotation.
3. Changes the sign of the offset so that it will be negative.
4. Calls the icprjct procedure to project point (PTN3, PTE3) on the X axis of the local coordinate system to determine the plus and offset distances (A and B), and the distance C from the center point to the given point as indicated in Figure 7.3c1.
5. Computes the parameters P and Q so that:

   P = A − F   and   Q = F + R1   (keep in mind that F < 0).

   With reference to Figure 7.3c1, note that if:

   - Q = 0   the given curve is tangent to the given line.
   - Q < 0   the given curve is apart from the given line.
   - Q > 0   the given curve and the given line intersect.

   Also note that the determination of which of the 27 possible conditions has been encountered is dependent on the position of the given point (PTN3, PTE3):
   - With respect to the given line (to its right or left with respect to AZ1),
   - With respect to the center point of the given curve (back or ahead of it in the direction of AZ1), and with respect to the
   - Sign and value of P and Q as computed above.

6. Determines (a) whether the given point (PTN3, PTE3) is located on the given line or curve, and (b) whether the given curve is tangent to the given line by comparing the distance C with the radius of the given curve R1, and the distance P and Q with zero, all within a tolerance of 0.005 feet (1.524 mm), and if so sets C equal to R1 and P and Q equal to zero.

7. Determines which of the 27 possible conditions of Figure 7.3c2 has been encountered by testing C versus R1, P versus 0.00 and Q versus 0.00 as follows:

   IF ((C > R1) and (R1 > 0.00) and (R > 0.00)) Then

   . . . . . .

   ElseIf ((C > R1) and (R1 > 0.00) and (R = 0.00)) Then

   . . . . . .

   ElseIf ((C > R1) and (R1 = 0.00) and (R > 0.00)) Then

   . . . . . .

ElseIf ((C<R1) and (R1<0.00) and (R>0.00)) Then

. . . . . .

and so on.

8. Once the condition has been determined, the procedure:
   • Sets the `iErr` parameter equal to 2 and terminates if
     there is no solution.
   • Calls the appropriate of the `iccase01`, `iccase02`,
     `iccase03` and/or `iccase04` procedures to compute
     the so called local coordinates of the center points and
     tangency points of the potential tangent curves, and
     place them in the appropriate collections. Note that the:
     – Local coordinates are placed in temporary
       collections.
     – Aforesaid four procedures are used only by the sub-
       ject procedure.
9. Parses through the various solutions and converts the
   local coordinates to the final global coordinates, and
   terminates.

7.4   Arcs Tangent to Other Arcs

The four procedures that constitute this section construct a
circular arc, or circle tangent to another circular arc or circle.
The first pair of procedures, the `iccvtcrp` and the
`iccvcrl`, constructs an arc of known radius tangent to
another arc. In addition to the known radius, the `iccvtcrp`
procedure requires two known points through which the arc
is to pass and at which it is to commence and terminate,
while the `iccvcrl` procedure requires the specification of
either the central angle or the length of the chord of the arc.

The other procedures, the `iccvt2cr` and the `iccvt2cp`,
construct an arc or circle tangent to two other arcs, with the
former requiring the radius of the arc to be constructed, and
with the latter requiring a known point through which the
tangent arc is to pass.

7.4.1   Arc Tangent to Arc Given Radius and Endpoints

The `icCvtCrpX` procedure constructs an arc of given
radius tangent to a given arc or circle starting at a given point
(PN1, PE1) on the given arc or circle and ending at another
given point (PN2, PE2) as indicated in Figure 7.4a. The
given arc or circle is defined by the north and east coordinates of its center point and its radius. The new tangent arc to
be constructed is to:

• Commence at the point of projection of (PN1, PE1) on the
  given arc or circle, and it is to
• Terminate at the point of projection of (PN2, PE2) on the
  new tangent arc.

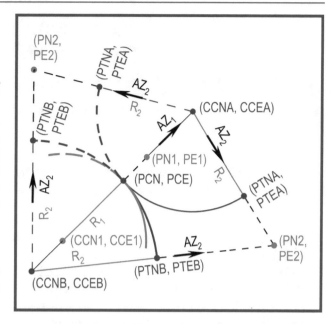

Figure 7.4a   Arc of Given Radius Tangent to a Given Arc at a Point on
the Said Arc and Ending at a Given Point

As seen in the said figure, there are two potential solutions to this construction problem depending on the position
of (PN2, PE2). Therefore, the procedure computes and
returns both tangent arcs which the calling program may display and have the user select which of the two arcs is the
desirable one. At this time the calling program may also
query if the 380° supplement arc is desired.

```
Sub icCvtCrpX(CCN1, CCE1, R1, PN1, PE1,
              PN2, PE2, R2, _
              PCN, PCE, CCN2A, CCE2A, PTNA,
              PTEA, CCN2B, CCE2B, PTNB, PTEB)
```

The input parameters to this procedure include the:

CCN1, CCE1	The north and east coordinates of the center point of the given curve.
R1	The radius of the given curve (either positive or negative).
PN1, PE1	The north and east coordinates of the point to set the start of the new arc.
PN2, PE2	The north and east coordinates of the point to set the end of the new arc.
R2	The radius of the new tangent arc (either positive or negative).

The parameters that are returned from this procedure
include the:

PCNA, PCEA	The north and east coordinates of the start point of the tangent arc 1.
CCNA, CCNA	The north and east coordinates of the center point of the tangent arc 1.

PTNA, PTEA	The north and east coordinates of the end point of the tangent arc 1.
PCNB, PCEB	The north and east coordinates of the start point of the tangent arc 2.
CCNB, CCNB	The north and east coordinates of the center points of the tangent arc 2.
PTNB, PTEB	The north and east coordinates of the end point of the tangent arc 2.

Upon invocation, the procedure initializes to zero the parameters to be returned, forces the two given radii to be positive, and then:

1. Calls the `icforce` procedure to determine the azimuth `AZ1` from (`CCN1`, `CCE1`) towards (`PN1`, `PE1`).
2. Calls the `icptl` procedure to determine the coordinates of the start point of the new arc (`PCN`, `PCE`), the point of tangency, by traversing from (`CCN1`, `CCE1`) along `AZ1` a plus distance of `R1` and an offset distance of zero.
3. Sets `D` to be the value of `R1` plus `R2`, and then calls the `icptl` procedure to determine the coordinates of the center point of the new arc (`CCN2A`, `CCE2A`) by traversing from (`CCN1`, `CCE1`) along `AZ1` a plus distance of `D` and an offset distance of zero.
4. Sets `D` to be the value of `R1` minus `R2`, and then calls the `icptl` procedure to determine the coordinates of the center point of the new arc (`CCN2B`, `CCN2B`) by traversing from (`CCN1`, `CCE1`) along `AZ1` a plus distance of `D` and an offset distance of zero.
5. Calls the `icforce` procedure to determine the azimuth `AZ2` from the center point (`CCN2A`, `CCE2A`) computed in Step 3 towards (`PN2`, `PE2`), and then calls the `icptl` procedure to determine the coordinates of the end point of the new arc (`PTN2A`, `PTE2A`) by traversing from the said center point (`CCN2A`, `CCE2A`) along `AZ2` a plus distance of `R2` and an offset distance of zero.
6. Calls the `icforce` procedure to determine the azimuth `AZ2` from the center point (`CCN2B`, `CCE2B`) computed in Step 4 towards (`PN2`, `PE2`), and then calls the `icptl` procedure to determine the coordinates of the end point of the new arc (`PTN2B`, `PTE2B`) by traversing from the said center point (`CCN2B`, `CCE2B`) along `AZ2` a plus distance of `R2` and an offset distance of zero, and then terminates.

7.4.2 Arc Tangent to Arc Given Radius and L, A, or C

The `iccvtcrL` procedure constructs an arc of given radius tangent to a given arc or circle defined by the north and east coordinates of its center point and radius, and (a) commencing at a point on the given arc located along the line joining the center point of the given arc and a given point, and (b)

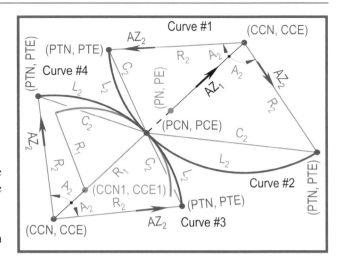

Figure 7.4b Arc of Known Radius and Arc Length, Central Angle, or Chord Length Tangent to Given Arc at a Point on the Given Arc

terminating at a point defined by the length of the new arc, central angle of the new arc, or chord length of the new arc as indicated in Figure 7.4b.

As seen in the said figure, there are four potential solutions to this construction problem. Therefore, the procedure generates all four curves which the calling program can display the center points of each one from which the user may select the desired one. The arcs to be created may exceed 180° if the arc length or central angle are specified rather than the chord length. If the chord is specified, the new arc cannot exceed 180°.

Regarding the new arcs to be constructed, note the following:

- Each of the four new arcs shares the same start point. The center points and the end points of these arcs are each saved in arrays.
- The records of these arrays contain the following information in the indicated order:
  - 1st The curve the center of which is located outside of the given curve, and which new curve is developed in a clockwise direction.
  - 2nd The curve the center of which is located outside of the given curve, and which new curve is developed in a counterclockwise direction.
  - 3rd The curve the center of which is located in the opposite direction of that stated for the first two records, and which new curve is developed in a clockwise direction.
  - 4th The curve the center of which is located in the opposite direction of that stated for the first two records, and which new curve is developed in a counterclockwise direction.
  - Center points of the first pair of records and of the second pair of records are the same.

```
Sub icCvtCrLX(CCN1, CCE1, R1, PN, PE, R2,
              L2, A2, C2, _
              PCN, PCE, CCN, CCN, PTN, PTE)
```

The input parameters to this procedure include the:

CCN1, CCE1	The north and east coordinates of the center point of the given curve.
R1	The radius of the given curve (positive).
PN1, PE1	The north and east coordinates of the given point to set the start of the arc.
R2	The radius of the new tangent arc (positive).
L2	The length of the new tangent arc (positive), or zero.
A2	The central angle in radians of the new tangent arc (positive), or zero.
C2	The length of the chord of the new tangent arc (positive), or zero.

Note that only one of the L2, A2 and C2 parameters should have a nonzero value, and the other two should have a value of zero (0.00). If more than one has a nonzero value, the first one to be encountered in the L2, A2 and C2 sequence with a nonzero value governs.

The parameters that are returned from this procedure include the:

PCN, PCE	The north and east coordinates of the start point of the two potential tangent arcs (same point for each of the arcs located on the given curve).
CCN, CCN	The collection of the north and east coordinates of the center points of the four potential tangent arcs.
PTN, PTE	The collection of the north and east coordinates of the end points of the four potential tangent arcs.

Upon invocation, the procedure initializes to zero the parameters to be returned, forces the two given radii to be positive, and then:

1. Queries the values of L2, A2 and C2 to determine the first one specified with a nonzero value, and if the:
   * Arc length L2 is so specified, the procedure computes the central angle of the new tangent arc as A2 = R2/L2.
   * Central angle A2 is so specified, the procedure uses the absolute value of A2.
   * Chord length C2 is so specified, the procedure computes the central angle of the new tangent arc as A2 = 2 sin-1(|C2|/2/R2).
2. Calls the icforce procedure to determine the azimuth AZ1 from (CCN1, CCE1) towards (PN1, PE1).
3. Calls the icptl procedure to determine the coordinates of the start point of the new arc (PCN, PCE), the point of tangency, by traversing from (CCN1, CCE1) along AZ1 a plus distance of R1 and an offset distance of zero.

The procedure next proceeds to compute the coordinates of the center point and of the end point of each of the four potential new arcs as stated in the preamble to this construction problem.

4. Sets D to be the value of R1 plus R2, and then calls the icptl procedure to determine the coordinates of the center point of the first and second new arcs in the array (CCN, CCN) by traversing from (CCN1, CCE1) along AZ1 a plus distance of D and an offset distance of zero.
5. Sets D to be the value of R1 minus R2, and then calls the icptl procedure to determine the coordinates of the center point of the third and fourth new arcs in the array (CCN, CCN) by traversing from (CCN1, CCE1) along AZ1 a plus distance of D and an offset distance of zero.
6. Computes the azimuth AZ as the reverses the azimuth AZ1 by adding 180° (2π) in order to preserve the value of AZ1.
7. Computes the azimuth AZ2 by adding ANG to AZ, and then calls the icptl procedure to determine the coordinates of the end point of the first new arc in the array (PTN, PTE) by traversing from the first record of the collection (CCN, CCE) computed in Step 4 along AZ2 a plus distance of R2 and an offset distance of zero.
8. Repeats Step 7 above but this time by subtracting ANG from AZ1.
9. Repeats Steps 7 and 8, but this time by using AZ1 and working from the center points of the third and fourth new arcs, which are the same, in order to compute the end point of the third and fourth new arcs, and terminates.

7.4.3 Arc of Given Radius Tangent to Two Arcs

The icCvt2crX procedure constructs an arc of known radius tangent to two given arcs or circles each defined by the north and east coordinates of its center point and its radius. Depending (a) on the position of the two given curves with respect to each other, (b) on the size of the radii (R_1 and R_2) of the given curves, and (c) on the size of the radius (R_3) of the tangent arc to be constructed, the number and location of the potential tangent curves is indicated by the six cases in Figures 7.4c through 7.4g. As seen in these figures, due to the above stated dependencies there could be no solution, up to eight potential solutions, or an infinite number of solutions to this construction program. The procedure having determined the potential curve solutions saves the north and east coordinates of their center points in two corresponding arrays, and the calling program may then display them so that the user may select the desired tangent curve.

Table 7.1 Potential Common Arcs of Known Radius Tangent to Two Given Arcs

Intersect	RA = R1 + R3	with	RB = R2 + R3	and	Potential curve #1
	RA = R1 + R3	with	RB = – (R2 + R3)	and	Potential curve #2
Intersect	RA = R1 – R3	with	RB = R2 + R3	and	Potential curve #3
	RA = R1 – R3	with	RB = – (R2 + R3)	and	Potential curve #4
Intersect	RA = R1 + R3	with	RB = R2 – R3	and	Potential curve #5
	RA = R1 + R3	with	RB = – (R2 – R3)	and	Potential curve #6
Intersect	RA = R1 – R3	with	RB = R2 – R3	and	Potential curve #7
	RA = R1 – R3	with	RB = – (R2 – R3)		Potential curve #8

In the cases of Figures 7.4c through 7.4g:
R_1 = *radius of the smaller given curve.*
R_2 = *radius of the larger given curve.*
R_3 = *radius of the new tangent curve.*

The six cases of Figures 7.4c through 7.4g display the various conditions that may be encountered by the subject procedure. It should be noted that in these figures R_1 denotes the radius of the smaller of the two given curves, R_2 denotes the larger of the said two curves, and R_3 denotes the radius of the new tangent curve to be constructed. The two given radii may be given in any sequence, but the procedure changes them to meet the said sequence.

To compute the coordinates of the center points of the potential new tangent curve, the procedure calls the `iccrc-crc` procedure described in Chap. 6 two intersect two implied circles. In intersecting these two implied circles, the procedure:

• Causes the first implied circle to have its center point at the center point of the smaller given curve, and the second implied circle to have its center point at the center point of the larger given curve.
• Assigns to the first implied circle at first a radius equal to the sum of the radius of the said smaller curve and that of the new tangent curve as $R_A = R_1 + R_3$, and then a radius equal to the difference of the radius of the said smaller curve and that of the new tangent curve as $R_A = R_1 - R_3$.
• Assigns to the second implied circle at first a radius equal to the sum of the radius of the said larger curve and that of the new tangent curve as $R_A = R_2 + R_3$, and then a radius equal to the difference of the radius of the said larger curve and that of the new tangent curve as $R_A = R_2 - R_3$.

These four radii produce four intersection processes, with each one generating two intersection points. However, the user is reminded that the `iccrccrc` procedure returns (a) the coordinates of only one intersection point, and (b) the indicator INTRS which denotes whether an intersection or point of tangency has been found or not. Since the intersec-

tion of two circles produces two points of intersection, this procedure is called twice as indicated in Table 7.1. Note that in this table the sign of R_B produces once the first intersection point and the other time it produces the second intersection point (refer to the side bar on this page).

When intersecting two curves with the `iccrccrc` procedure the sign of the radius of the second curve dictates which of the two potential intersections is to be returned.

It has been stated above that the `iccrccrc` procedure returns only one intersection point or the point of tangency if the two circles are tangent to each other. If an intersection point has been encountered the indicator INTRS is returned with a value of 1, and if a point of tangency has been encountered INTRS is returned with a value of 3. If a the two intersecting circles are tangent to each other, then the tangency point will be returned twice, thus creating duplicate points. These duplicate points are weeded out by the procedure. Thus, in Figure 7.4c there are two sets of potential tangent

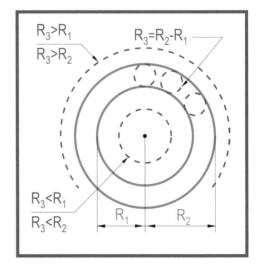

Figure 7.4c The Two Given Curves Are Concentric

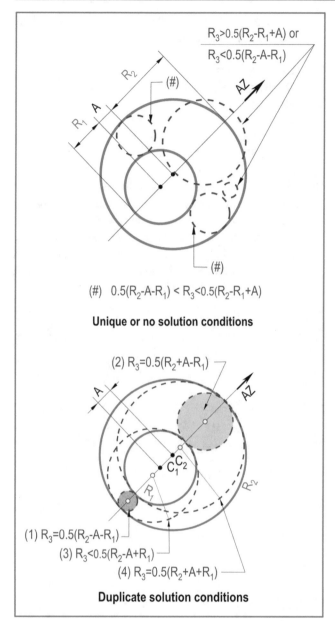

(#) $0.5(R_2-A-R_1) < R_3 < 0.5(R_2-R_1+A)$

Unique or no solution conditions

(2) $R_3=0.5(R_2+A-R_1)$

(1) $R_3=0.5(R_2-A-R_1)$
(3) $R_3<0.5(R_2-A+R_1)$
(4) $R_3=0.5(R_2+A+R_1)$

Duplicate solution conditions

Figure 7.4d One Given Curve Inside and Non-tangent to the Other

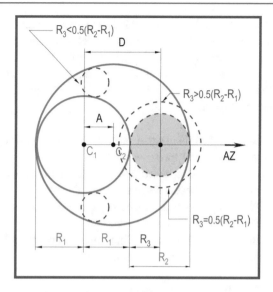

Figure 7.4e One Given Curve Inside and Tangent to the Other

If the two given curves are found to be concentric within a tolerance of 0.005 feet (1.524 mm), the procedure aborts.

Case 2—One curve inside and non-tangent to the other If the smaller of the two given curves is located inside the larger curve as indicated in Figure 7.4d, then depending on the size of the new tangent curve to be constructed there could be:

- No solution if:
  $R_3>0.5(R_2-R_1+A)$
  or
  $R_3<0.5(R_2-R_1-A)$
- Two unique solutions and one pair of duplicated solutions if:
  $R_3<0.5(R_2-R_1+A)$
  and
  $R_3>0.5(R_2-R_1-A)$
- One pair of duplicated solutions if:
  $R_3=0.5(R_2-R_1+A)$
  or
  $R_3=0.5(R_2-R_1-A)$

Case 3—One curve inside and tangent to the other If the smaller of the two given curves is located inside and tangent to the larger curve as indicated in Figure 7.4e, then the conditions of the preceding case apply with one difference. Whereas in the preceding case there could be four duplicate curve conditions, in this case there is only one as indicated by the sole solid fill circle in Figure 7.4e. This condition arises when the radius $R_3=R_2-R_1$.

Case 4—One curve outside and tangent to the other If one of the two given curves is located outside the other

curves, one with and without the removal of the duplicate points. It should be noted that the occurrence of duplicate points is dependent on the size of the radius R_3 of the new tangent curve to be constructed

Case 1—Concentric Curves If the two given curves are concentric as indicated in Figure 7.4c, there can be:

- No common tangent curve solution if the radius of the new arc R_3 is greater or smaller than radii R_1 and R_2 of both of the two given curves.
- An infinite number of common tangent curves if the radius of the new arc R_3 is equal to one half the difference of the radii R_1 and R_2 the two given curves.

given curve and is tangent to it as indicated in Figure 7.4f, then there can be six or eight potential solutions depending on the magnitude of the radius R_3 of the new curve to be constructed. Shown in the said figure are the conditions under which the said tangent curve conditions may occur. Thus, if:

- $R_3 \leq R_1$, then there can be two unique plus two pairs of duplicated tangent curve solutions.
- $A > R_3 > R_2$, then there can be two unique plus two pairs of duplicated tangent curve solutions.
- $R_3 = A$, then there can be two unique plus three pairs of duplicated tangent curve solutions.
- $R_3 > A$, then there can be four unique plus two pairs of duplicated tangent curve solutions.

Case 5—The two given curves are separated from each other If the two given curves are separated from each other as indicated in Figure 7.4g, then depending on the size of the radius R_3 of the new tangent curve to be constructed and the separation distance B between the two given curves there could be two, four, six or eight potential solutions. Thus, if:

- $R_3 \leq 0.5B$, then there can be no tangent curve solutions.
- $B \geq R_3 > 0.5B$, then there can be two unique tangent curve solutions.
- $B + R_1 \geq R_3 > B$, then there can be four unique tangent curve solutions.

- $B + R_2 \geq R_3 > B + R_1 + R_2$, then there can be six unique tangent curve solutions.
- $R_3 > B + R_2$, then there can be eight unique tangent curve solutions.
- Duplicate tangent curve solutions occur when:
  $R_3 = 0.5B$
  $R_3 = (0.5B) + R_1$
  $R_3 = (0.5B) + R_2$
  $R_3 = (0.5B) + R_1 + R_2$

Case 6—The two given curves intersect each other If the two given curves intersect each other as indicated in Figure 7.4h, then depending on the size of the radius R_3 of the new tangent arc to be constructed there could be two, four, six or eight potential solutions. The number and location of the unique and duplicate tangent curves that can be created is indicated in the said figure.

```
Sub icCvt2CrX (CCN1,   CCE1,   R1,   CCN2,
               CCN2,   R2,   R3,   _  CCN3,
               CCE3,  NPT)
```

Throughout the text of this procedure reference to the number of tangent curves includes duplicate curves. However, at the end of the procedure duplicate center points are removed. Thus, the parameter NPT does not include any duplicate curves.

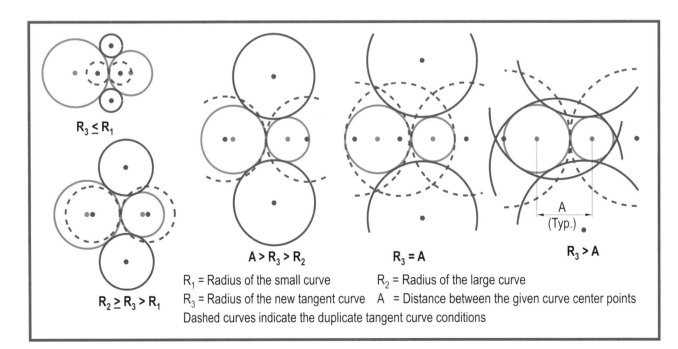

R_1 = Radius of the small curve R_2 = Radius of the large curve
R_3 = Radius of the new tangent curve   A = Distance between the given curve center points
Dashed curves indicate the duplicate tangent curve conditions

Figure 7.4f One Given Curve Outside and Tangent to the Other

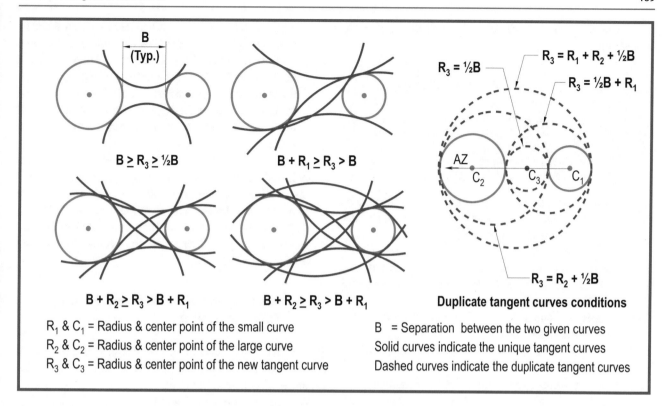

Figure 7.4g Tangent Curves to Two Separated Given Curves

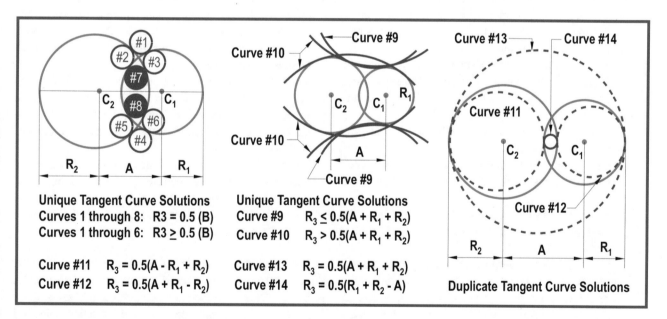

Figure 7.4h Intersecting Given Curves

The input parameters to this procedure include the:

CCN1, CCE1	The north and east coordinates of the center point of the first given curve.
R1	The radius of the first given curve (positive).
CCN2, CCE2	The north and east coordinates of the center point of the second given curve.
R2	The radius of the second given curve (positive).
R3	The radius of the tangent curve to be constructed (positive).

The parameters that are returned from this procedure include the:

CCN3, CCE3   The arrays of the north and east coordinates of the center points of the tangent curve to be constructed.

NPT   An indicator, the value of which if it is:
             • Zero denotes that the two given curves are concentric, or that there is no solution.
             • One or greater denotes the number of tangent arcs that have been constructed (see side bar on this page).

Upon invocation, the procedure initializes the contents of the coordinate arrays to be returned to −9999999.0 and the number of tangent curves to zero, and then:

1. Makes certain that the values of the three input radii are positive by setting RR1 and RR2 to be the absolute value of R1 and R2 respectively.
2. Compares RR1 with RR2, and if RR1 is greater than RR2 reverses the radii and the coordinates of the center points of the two given curves so that the first one is that of the smaller radius.
3. Compares the coordinates and the radii of the two given curves against each other to determine whether the said curves are the same within a tolerance of 0.005 feet (1.524 mm), and if so the procedure terminates.
4. Inverses between the center points of the smaller curve towards the center point of the larger curve to compute the distance (A) and azimuth (AZ).
5. Determines whether the two given arcs are concentric by comparing the above computed distance to zero within a tolerance of 0.005 feet (1.524 mm), and if so the procedure terminates.
6. Calls the iccrccrc procedure eight times to intersect two implied circles with:
   • One having its center at the center point of the smaller given curve, and having a radius RA as indicated in Table 7.1, and the
   • Other having its center point at the center point of the larger given curve, and having a radius equal to RB as indicated in Table 7.1,
      to define the center points of the potential tangent arcs. Since the iccrccrc procedure returns only one intersection point, it is called twice, once with RB being positive and once with RB being negative.
      Each time an intersection has been performed with the iccrccrc procedure, the subject procedure queries INTRS, the indicator returned by the iccrccrc procedure, to determine whether an intersection or tangency point has been encountered, or not. If an intersection (INTRS = 1) or tan-

gency (INTRS = 3) point has been encountered the indicator NPT is increased by one, and the north and east coordinates of the intersection point are saved in the CCN3 and CCE3 arrays, respectively. Since the iccrccrc procedure is called twice with the same radii, if a tangency point is encountered it will be returned twice, thus producing duplicate center points.

7. Parses through the CCN3 and CCE3 arrays, once all eight intersections have been performed, to determine whether any duplicated center points exist, or not, and if so the duplicate coordinates are replaced with −9999999.0. Duplicate points could have been eliminated in the preceding step by not repeating the second intersection whenever INTRS = 3.

7.4.4 Arc Tangent to Two Arcs and Passing Through a Point

The icCvt2CpX procedure introduces an arc that is tangent to two given arcs and passes through a given point with some examples as indicated in Figure 7.4i.

The number of potential solutions is dependent on:

• The relationship of the two given arcs with respect to each other, such as (a) one curve lies within the other, (b) the two curves are separated from each other, (c) the curves intersect each other, and (d) whether the two curves are tangent to each other.
• The location of the given point through which the new tangent curve is to pass with respect to the two given curves, such as the point is located (a) within one or both curves, (b) on the circumference of a curve, and (c) outside one or both curves.
• The location of the given point through which the new tangent curve is to pass with respect to the common tangent line to the two given curves (when one curve is not within the other), such as indicated in Figure 7.4i with the two hollow points.

The methodology employed by this procedure to solve this construction problem could produce up to eight potential solutions. Thus, the procedure computes and saves in three arrays the north and east coordinates of the center point and the radius of each potential new tangent curve. However, depending on the abovesaid conditions there could be duplicate solutions, and possibly a solution could be one of the two given curves. The latter case could occur when the given point is located on the circumference of one of the given curves. Since in most practical solutions such duplicate curves are not useful nor desirable, they are removed by the

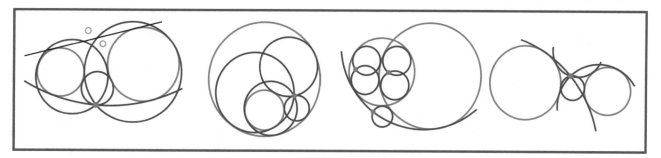

Figure 7.4i Arc Tangent to Two Given Arcs and Passing Through a Given Point

procedure. The returned arrays can be used by the calling program of this procedure to display them, and have the user select the desired one.

The examples of Figure 7.4i depict certain of the potential cases when one of the two given curves is located within the other, and when the said two curves are separated from, or intersect each other. Variations to these conditions could be when the two given curves are also tangent to each other, as well as when the given point through which the new arc is to pass is located on the circumference of one of the two given curves.

To describe the methodology used by this procedure in constructing the new tangent curve reference is made to Figure 7.4j. Under this methodology it is immaterial whether the smaller or the larger curve is introduced first or second, and both radii are assumed to be positive. Furthermore, if the given point through which the new tangent arc is to pass is located within a tolerance of 0.005 feet (1.524 mm) of the circumference of one of the two given curves, it is forced to be on the said circumference. In this figure four cases are indicated. Of these four, Case 3 and Case 4 are basically the same with the difference depending on which curve (the smaller or the larger) is introduced first and which second.

It is noted that the solution for Case 1 presented below is quite similar to that of Sylvan H. Chasen on page 69 of his book titled *Geometric Principles and Procedures for Computer Graphic Applications* by Prentice-Hall, Inc. However, the subtle changes imposed by Cases 2, 3, and 4 are not presented in the said book.

Case 1   This case is encountered when the two known curves are located one outside the other. Let (X_1, Y_1) represent the coordinates of the center point of the first curve, (X_2, Y_2) represent the coordinates of the center point of the second curve, (X_3, Y_3) represent the coordinates of the given point through which the new tangent curve is to pass, and (X, Y) represent the coordinates of the center point of the said new

curve to be constructed. For the sake of the construction, let us translated the given coordinates to a local coordinate system with its origin at point (X_3, Y_3) as indicated in the said figure. Correspondingly, let R_1, R_2, and R_3 represent the radii of the first and second given curves and of the new tangent curve, respectively. In this case there are two solutions, one above and one below the local X axis. The Equations 7.4.1, 7.4.2, and 7.4.3 present the relationships of the three center points and three radii. In these equations, the values of X_1, Y_1, X_2, and Y_2 may be computed by inversion from the given point to the center points of the respective given curves, and thus may be considered as being known. Thus, the three unknowns are the values of X, Y, and R_3 which may be obtained by solving the said three equations simultaneously.

$$\left(X - X_1\right)^2 + \left(Y - Y_1\right)^2 = \left(R_1 + R_3\right)^2 \qquad (7.4.1)$$

$$\left(X - X_2\right)^2 + \left(Y - Y_2\right)^2 = \left(R_2 + R_3\right)^2 \qquad (7.4.2)$$

$$X^2 + Y^2 = R_3^2 \quad or \quad R_3 = \sqrt{Xs^2 + Y^2} \qquad (7.4.3)$$

If we know substitute the value of R_3 from Equation 7.4.3 into the right sides of Equations 7.4.1 and 7.4.2, we get

$$\left(X - X_1\right)^2 + \left(Y - Y_1\right)^2 = \left(R_1 + \sqrt{X^2 + Y^2}\right)^2 \qquad (7.4.3a)$$

$$\left(X - X_2\right)^2 + \left(Y - Y_2\right)^2 = \left(R_2 + \sqrt{X^2 + Y^2}\right)^2 \qquad (7.4.3b)$$

If we next expand the squares of the right sides of the last two equations, and remove the X^2 and Y^2 that appear on the left and on the right sides of the resultant equations, we get

$$X_1^2 + Y_1^2 - 2X_1X - 2Y_1Y = R_1^2 + 2R_1\sqrt{X^2 + Y^2} \qquad (7.4.4)$$

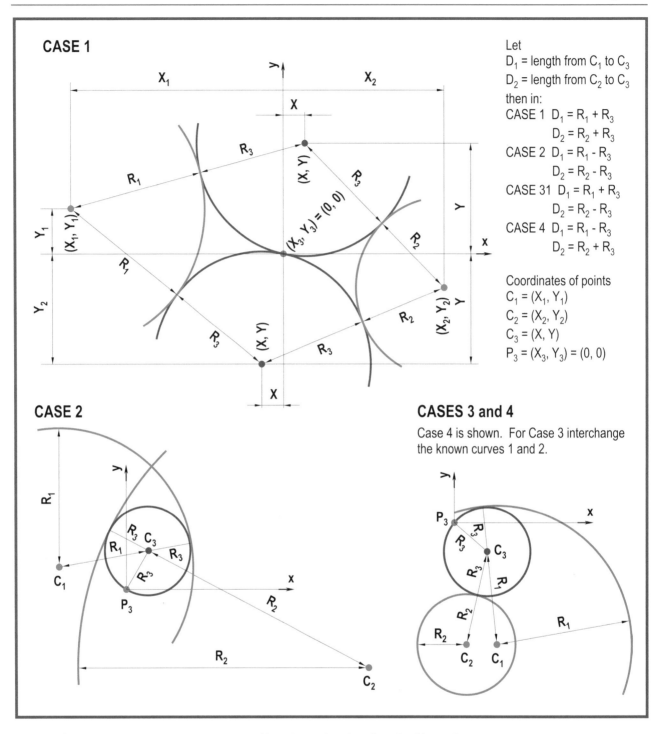

CASE 1

Let
D_1 = length from C_1 to C_3
D_2 = length from C_2 to C_3
then in:
CASE 1 $D_1 = R_1 + R_3$
         $D_2 = R_2 + R_3$
CASE 2 $D_1 = R_1 - R_3$
         $D_2 = R_2 - R_3$
CASE 31 $D_1 = R_1 + R_3$
          $D_2 = R_2 - R_3$
CASE 4 $D_1 = R_1 - R_3$
         $D_2 = R_2 + R_3$

Coordinates of points
$C_1 = (X_1, Y_1)$
$C_2 = (X_2, Y_2)$
$C_3 = (X, Y)$
$P_3 = (X_3, Y_3) = (0, 0)$

CASE 2

CASES 3 and 4

Case 4 is shown. For Case 3 interchange
the known curves 1 and 2.

Figure 7.4j The Four Cases of an Arc Tangent to Two Given Arcs and Passing Through a Given Point

$$X_2^2 + Y_2^2 - 2X_2X - 2Y_2Y = R_2^2 + 2R_2\sqrt{X^2 + Y^2} \qquad (7.4.5)$$

If we next multiply both sides of Equation 7.4.4 by R_2, and both sides of Equation 7.4.5 by R_1, we get

$$R_2X_1^2 + R_2Y_1^2 - 2R_2X_1X - 2R_2Y_1Y$$
$$= R_2R_1^2 + 2R_2R_1\sqrt{X^2 + Y^2} \qquad (7.4.5a)$$

$$R_2X_2^2 + R_2Y_2^2 - 2R_1X_2X - 2R_1Y_2Y$$
$$= R_1R_2^2 + 2R_2R_1\sqrt{X^2 + Y^2} \qquad (7.4.5b)$$

If we then subtract Equation 7.4.5b from Equation 7.4.5a, the terms on the right side of the said equation with the two radicals will be eliminated, and if we then move the right side of the resultant equation to the left side of the equation (thus reversing the sign of the two terms), we get

$$R_2 X_1^2 - R_1 X_2^2 + R_2 Y_1^2 - R_1 Y_2^2 - 2R_2 X_1 X + 2R_1 X_2 X$$
$$- 2R_2 Y_1 Y + 2R_1 Y_2 Y - R_2 R_1^2 + R_1 R_2^2 = 0 \qquad (7.4.5c)$$

By combining terms with X and Y, an rearranging terms, we get

$$\left(2R_1 X_2 - 2R_2 X_1\right)X + \left(2R_1 Y_2 - 2R_2 Y_1\right)Y$$
$$+ R_2\left(X_1^2 + Y_1^2 - R_1^2\right) - R\left(X_2^2 + Y_2^2 - R_2^2\right) = 0 \quad (7.4.6)$$

So that we may simplify a bit this and the subsequent equations, let us define the following constants the values of which may be considered as being known.

$$K_1 = 2\left(R_1 X_2 - R_2 X_1\right)$$
$$K_2 = 2\left(R_1 X_2 - R_2 X_1\right) \qquad (7.4.6a)$$
$$K_3 = R_2\left(X_1^2 + Y_1^2 - R_1^2\right) - R_1\left(X_2^2 + Y_2^2 - R_2^2\right)$$

Thus, Equation 7.4.6 may now be expressed as

$$K_1 X + K_2 Y + K_3 Y \qquad (7.4.7)$$

If we now express Equation 7.4.7 in terms of X and define the constants K and K_5 as indicated below

$$X = -\left[\frac{K_2}{K_1}\right]Y - \left[\frac{K_3}{K_1}\right]$$
$$K_4 = -\left(K_4 / K_1\right) \qquad (7.4.7a)$$
$$K_5 = -\left(K_3 / K_1\right)$$

the value of X may be expressed as

$$X = K_4 Y + K_5 \qquad (7.4.8)$$

We can now substitute the value of X from Equation 7.4.8 in either Equation 7.4.4 or 7.4.5 so that we may solve for Y. Let us arbitrarily select Equation 7.4.4 which may now take the form of

$$X_1^2 - 2X_1\left(K_4 Y + K_5\right) + Y_1^2 - 2Y_1 Y \qquad (7.4.8a)$$
$$= R_1^2 + 2R_1\sqrt{\left(K_4 Y + K_5\right)^2 + Y^2}$$

or

$$\frac{-2X_1 K_4 Y - 2X_1 K_5 + X_1^2 + Y_1^2 - 2Y_1 Y - R_1^2}{2R_1}$$
$$= \sqrt{K_4^2 Y^2 + K_5^2 + 2K_4 K_5 Y + Y^2} \qquad (7.4.8b)$$

or

$$\frac{\left(-X_1 K_4 + Y_1\right)Y}{R_1} + \frac{\left(-2X_1 K_5 + X_1^2 + Y_1^2 - R_1^2\right)}{2R_1}$$
$$= \sqrt{K_4^2 Y^2 + K_5^2 + 2K_4 K_5 Y + Y^2} \qquad (7.4.9)$$

To simplify a bit the above equations, let us define two new constants, and rearrange certain terms under the radical t express Equation 7.4.9 as

$$K_6 = \frac{-X_1 K_4 + Y_1}{R_1}$$
$$K_7 = \frac{-2X_1 K5 + X_1^2 + y_1^2 - R_1^2}{2R_1} \qquad (7.4.9a)$$

$$K_6 Y + K_7 = \sqrt{\left(K_4^2 + 1\right)Y^2 + 2K_4 K_5 Y + K_5^2} \qquad (7.4.9b)$$

If we now square both sides of Equation 7.4.9b to remove the radical, and we then transfer the terms of the right side of the resultant equation to the left we get

$$K_6^2 Y^2 + K_7^2 + 2K_6 K_7 Y - K_4^2 Y^2 - Y^2$$
$$- 2K_4 K_5 Y - K_5^2 = 0 \qquad (7.4.9c)$$

or

$$\left(K_6^2 - K_4^2 - 1\right)Y^2 + \left(2K_6 K_7 - 2K_4 K_5\right)Y + \left(K_7^2 - K_5^2\right) \qquad (7.4.10)$$

By introducing three more constants

$$K_8 = K_6^2 - K_4^2 - 1$$
$$K_9 = 2K_6 K_7 - 2K_4 K_5 \qquad (7.4.10a)$$
$$K_{10} = K_7^2 - K_5^2$$

Equation 7.4.10 may be expressed as

$$K_8 Y^2 + K_9 Y + K_{10} = 0 \qquad (7.4.11)$$

from which the value of Y may be determined by use of the quadratic formula as follows

$$Y = \frac{-K_9 \pm \sqrt{K_9^2 - 4K_8 K_{10}}}{2K_8} \qquad (7.4.12)$$

With Y now known we cab substitute its value in Equation 7.4.8 to find the value of X, and then we can introduce these values of X and Y in Equation 7.4.3 to find the value of R_3.

There are certain issues that need be discussed about the above methodology and particularly about Equation 7.4.12, but we will do so after we discuss Cases 2, 3, and 4.

Case 2 This case is encountered when the two known curves intersect each other. In dealing with this case we use the same terminology to define the various known and unknown features, and the same methodology as used for Case 1. Thus, the fundamental difference between Case 2 and Case 1 is the expression of Equations 7.4.1 and 7.4.2. Since the number and form of equations to be used in Case 2 is quite similar to those of Case 1, and the number of equations is the same, rather than carrying the equation number system sequentially, we increase the numbers by the value of 20 in the right most component of the equation number. Thus, the basic three equations become

$$\left(X - X_1\right)^2 + \left(Y - Y_1\right)^2 = \left(R_1 - R_3\right)^2 \qquad (7.4.21)$$

$$\left(X - X_2\right)^2 + \left(Y - Y_2\right)^2 = \left(R_2 - R_3\right)^2 \qquad (7.4.22)$$

$$X^2 + Y^2 = R_3^{\ 2} \quad \text{or} \quad R_3 = \sqrt{X^2 + Y^2} \qquad (7.4.23)$$

Notice the difference in the sign of the R_3 term in the first two equations. By executing the same steps of Case 1 Equations 7.4.4, 7.4.5, 7.4.5a, and 7.4.5b become as indicated below

$$X_1^{\ 2} + Y_1^{\ 2} - 2X_1 X - 2Y_1 Y = R_1^{\ 2} - 2R_1 \sqrt{X^2 + Y^2} \qquad (7.4.24)$$

$$X_2^{\ 2} + Y_2^{\ 2} - 2X_2 X - 2Y_2 Y = R_2^{\ 2} - 2R_2 \sqrt{X^2 + Y^2} \qquad (7.4.25)$$

$$\begin{aligned} R_2 X_1^{\ 2} + R_2 Y_1^{\ 2} - 2R_2 X_1 X - 2R_2 Y_1 Y \\ = R_2 R_1^{\ 2} - 2R_2 R_1 \sqrt{X^2 + Y^2} \end{aligned} \qquad (7.4.25a)$$

$$\begin{aligned} R_1 X_2^{\ 2} + R_1 Y_2^{\ 2} - 2R_1 X_2 X - 2R_1 Y_2 Y \\ = R_1 R_2^{\ 2} - 2R_2 R_1 \sqrt{X^2 + Y^2} \end{aligned} \qquad (7.4.25b)$$

The subsequent equations of Case 1 remain the same until we come to the constants K_6 and K_7 of Equation 7.4.9a which now become

$$\begin{aligned} K_6 &= \frac{X_1 K_4 + Y_1}{R_1} \\ K_7 &= \frac{2X_1 K_5 - X_1^{\ 2} - Y_1^{\ 2} + R_1^{\ 2}}{2R_1} \end{aligned} \qquad (7.4.29a)$$

All remaining equations of Case 1 remain the same.

   Case 3   This case is encountered when one of the two known curves is located within the other. Similarly as with the preceding case we use the same terminology to define the various known and unknown features, and the same methodology as used for Case 1. Thus, the fundamental difference between Case 3 and Case 1 is the expression of Equation 7.4.2. Since the number and form of equations to be used in Case 3 is quite similar to those of Case 1, and the number of equations is the same, rather than carrying the equation number system sequentially, we increase the numbers by the value of 40 in the right most component of the equation number. Thus, the basic three equations become

$$\left(X - X_1\right)^2 + \left(Y - Y_1\right)^2 = \left(R_1 + R_3\right)^2 \qquad (7.4.41)$$

$$\left(X - X_2\right)^2 + \left(Y - Y_2\right)^2 = \left(R_2 - R_3\right)^2 \qquad (7.4.42)$$

$$X^2 + Y^2 = R_3^{\ 2} \quad \text{or} \quad R_3 = \sqrt{X^2 + Y^2} \qquad (7.4.43)$$

Notice the difference in the sign of the R_3 term in the second equation. By executing the same steps of Case 1 Equations 7.4.5, 7.4.5b, 7.4.5c, 7.4.6, and 7.4.6a become as indicated below.

$$X_2^{\ 2} + Y_2^{\ 2} - 2X_2 X - 2Y_2 Y = R_2^{\ 2} - 2R_2 \sqrt{X^2 + Y^2} \quad (7.4.45a)$$

$$\begin{aligned} R_1 X_2^{\ 2} + R_1 Y_2^{\ 2} - 2R_1 X_2 X - 2R_1 Y_2 Y \\ = R_1 R_2^{\ 2} - 2R_2 R_1 \sqrt{X^2 + Y^2} \end{aligned} \qquad (7.4.45b)$$

$$\begin{aligned} R_2 X_1^{\ 2} + R_1 X_2^{\ 2} + R_2 Y_1^{\ 2} + R_1 Y_2^{\ 2} - 2R_2 X_1 X - 2R_1 X_2 X \\ -2R_2 Y_1 Y - 2R_1 Y_2 Y - R_2 R_1^{\ 2} + R_1 R_2^{\ 2} = 0 \end{aligned} \qquad (7.4.45c)$$

$$\begin{aligned} -\left(2R_1 X_2 - 2R_2 X_1\right)X - \left(2R_1 Y_2 + 2R_2 Y_1\right)Y \\ +R_2\left(X_1^{\ 2} + Y_1^{\ 2} - R_1^{\ 2}\right) + R_1\left(X_2^{\ 2} + Y_2^{\ 2} - R_2^{\ 2}\right) = 0 \end{aligned} \qquad (7.4.46)$$

$$\begin{aligned} K_1 &= -2\left(R_1 X_2 - R_2 Y_1\right) \\ K_2 &= -2\left(R_1 Y_2 - R_2 Y_1\right) \\ K_3 &= R_2\left(X_1^{\ 2} + Y_1^{\ 2} - R_1^{\ 2}\right) + R_1\left(X_2^{\ 2} + Y_2^{\ 2} - R_2^{\ 2}\right) \end{aligned} \qquad (7.4.46a)$$

All remaining equations of Case 1 remain the same.

   Case 4   This case, like Case 3, is also encountered when one of the two known curves is located within the other. The difference between Case 3 and Case 4 is dependent upon which known curve is the first and which is the second. Similarly as with the preceding cases we use the same terminology to define the various known and unknown features, and the same methodology as used for Case 1. Thus, the fundamental difference between Case 4 and Case 1 is the expression of Equation 7.4.1. Since the number and form of equations to be used in Case 3 is quite similar to those of Case 1, and the number of equations is the same, rather than carrying the equation number system sequentially, we increase the numbers by the value of 60 in the right most component of the equation number. Thus, the basic three equations become

$$\left(X - X_1\right)^2 + \left(Y - Y_1\right)^2 = \left(R_1 - R_3\right)^2 \qquad (7.4.61)$$

$$\left(X - X_2\right)^2 + \left(Y - Y_2\right)^2 = \left(R_2 + R_3\right)^2 \qquad (7.4.62)$$

$$X^2 + Y^2 = R_3^{\ 2} \quad \text{or} \quad R_3 = \sqrt{X^2 + Y^2} \qquad (7.4.63)$$

Notice the difference in the sign of the R_3 term in the first equation. By executing the same steps of Case 1 equations there will be an impact on the various equations identified in Case 2 and Case 3. As seen from the preceding two cases, the basic impact is on the values of certain K constants. Thus, for Case 4 the values of K_1, K_2, and K_3 will be those of Case 3, and the values of K_6 and K_7 will be those of Case 2.

Having presented the methodology for solving the subject construction problem, it is important to identify and handle following inherent issues with the said methodology.

- It is noted that both constants K_4 and K_5 use the constant K_3 as a denominator. So if the value of K_4 is zero the procedure will encounter a division by zero and will issue a system error and will abort. Thus, the subject procedure checks for such a zero value of K_4 and if encountered computes the value of Y in terms of known values as indicated below and then introduces this known Y value into Equation 7.4.4 to solve for X. Thus, if K_1 is zero, Equation 7.4.7 becomes

$$K_2Y + K_3 = 0 \qquad (7.4.77)$$

and

$$Y = -\frac{K_3}{K_2} = Y_C \qquad (7.4.77a)$$

Like before, if we now substitute Y_C for Y into Equation 7.4.4 we get

$$X_1^2 - 2X_1X + Y_1^2 - 2Y_1Y_C = R_1^2 + 2R_1\sqrt{X^2 + Y_C^2} \qquad (7.4.78a)$$

or

$$\frac{X_1^2 - 2X_1X + Y_1^2 - 2Y_1Y_C - R_1^2}{2R_1} = \sqrt{X^2 + Y_C^2} \qquad (7.4.78b)$$

or

$$\left[\frac{-X_1}{R_1}\right]X + \frac{X_1^2 + Y_1^2 - 2Y_1Y_C - R_1^2}{2R_1} = \sqrt{X^2 + Y_C^2} \qquad (7.4.78c)$$

We can now redefine he constants K_6 and K_7 as

$$K_6 = \frac{-X_1}{R_1}$$
$$K_7 = \frac{-2Y_1Y_C + X_1^2 + Y_1^2 - R_1^2}{2R_1} \qquad (7.4.79a)$$

so that Equation 7.4.9b becomes

$$K_6X + K_7 = \sqrt{X^2 + Y_C^2} \qquad (7.4.79b)$$

If we now square both sides of Equation 7.4.79b to remove the radical, and we then transfer the terms of the right side of the resultant equation to the left we get

$$K_6^2X + K_7^2 + 2K_6K_7 = X^2 + Y_C^2 \qquad (7.4.79c)$$

or

$$\left(K_6^2 - 1\right)X^2 + \left(2K_6K_7\right)X + \left(K_7^2 - Y_C^2\right) = 0 \qquad (7.4.80)$$

If we now redefine the constants K_8, K_9 and K_{10} as

$$K_8 = K_6^2 - 1$$
$$K_9 = 2K_6K_7 \qquad (7.4.80a)$$
$$K_{10} = K_7^2 - Y_C^2$$

Equation 7.4.80 may be expressed a

$$K_8X + K_9X + K_{10} = \qquad (7.4.81)$$

from which

$$X = \frac{-K_9 \pm \sqrt{K_9^2 - 4K_8K_{10}}}{2K_8} \qquad (7.4.82)$$

which is similar to Equation 7.4.12.

Equations 7.4.77 through 7.4.81 presented above are applicable to Case 1 and Case 3. For Case 2 and Case 4 the above methodology is also applied but with the difference that the right most term of the right side of Equation 7.4.78a has a minus and not a plus sign as indicated below.

$$X_1^2 - 2X_1X + Y_1^2 - 2Y_1Y_C = R_1^2 - 2R_1\sqrt{X^2 + Y_C^2} \quad (7.4.98a)$$

which changes the signs of the constants K_6 and K_7 to

$$K_7 = \frac{X_1}{R_1}$$
$$K_7 = \frac{2Y_1Y_C - X_1^2 - Y_1^2 + R_1^2}{2R_1} \qquad (7.4.99a)$$

All subsequent equations remain the same as above.

- It is noted that in Equation 7.4.12, as well as in Equation 7.4.82, one of their terms involves a square root. When the value under the square root becomes negative it is an indication that there is no real solution to the subject construction problem. Hence, when this is encountered by the procedure, the remaining steps are bypassed, and the condition that causes this effect constitutes a no solution.
- When the known point (P_3) through which the new tangent curve is to pass is supposed to be located on the circumference of one of the two known curves, the solution becomes very sensitive to the precision of the coordinates of the said point. For example, if the coordinates of the said point represent user input information, and their input specification is to two or three decimal points, it possible for this point to fall just within or just outside the circumference of a given curve if the coordinates of the center point of the said curve are computed to a greater precision. This may result in a condition in which there is no potential solution, or in a not too precise or rather desirable solution.

To overcome this problematic issue, the procedure determines whether point P_3 is close to the circumference of one of the two known curves within a tolerance of 0.005 feet (1.254 mm), and if so the procedure forces point P$_3$ to be positioned on the corresponding circumference.

```
Sub icCvt2CpX(CCN1, CCE1, R1, CCN2, CCN2, R2,
              PN, PE, _
              CCN3, CCE3, R3, NPT)
```

The input parameters to this procedure include the:

CCN1, CCE1	The north and east coordinates of the center point of the first given curve.
R1	The radius of the first given curve.
CCN2, CCE2	The north and east coordinates of the center point of the second given curve.
R2	The radius of the second given curve.
PN, PE	The north and east coordinates of the point through which the arc is to pass.

The parameters that are returned from this procedure include the:

CCN3, CCE3	The arrays of the north and east coordinates of the center points of the potential tangent curves dimensioned to eight (8).
R3	The array of the radii of the potential new tangent curves to be constructed dimensioned to eight (8).
NPT	An indicator, the value of which denotes the number of potential new curves that have been encountered.

Upon invocation, the procedure initializes the parameters to be returned to −9999999.0, makes certain that the given radii are positive, and then:

1. Compares the north and the east coordinates of the center points and radii of the two given curves to determine if the two given curves are the same within a tolerance of 0.005 feet (1.524 mm). If they
   - Are, NPT is set to be zero (0) and then the procedure terminates.
   - Are not, the procedure continues.
2. Preserves the coordinates of the given point through which the new arc is to pass, and then determines whether the said point is located on the circumference of one of the two given curves within a tolerance of 0.005 feet (1.524 mm) by calling the icforce procedure to determine the distance and azimuth from each center point towards the given point. This distance is then is compared to the respective radius. If the computed distance:
   - Is equal to a radius within the said tolerance, the procedure calls the icptl procedure to set the coordinates of the given point so as to be located a distance equal to the given radius along the said computed azimuth, and the procedure continues.

   - Is not equal to the radius within the said tolerance, the procedure accepts the coordinates of the given point as given and continues.
3. Translates the coordinates of the two given center points so that they may be located on a local coordinate system with its origin on the said given point.
4. Computes the values of the constants K_1 through K_{10}, and then solves by use of Equations 7.4.12 and 7.4.82 for the local coordinates of any valid constructions that may have been found for each of the four cases presented earlier in this section. In this process the procedure checks for a negative value under the radical of Equations 7.4.12 and 7.4.82, as well as for a K_1 value of zero. Any valid coordinates that have been found are saved in their respective arrays. It is possible that some of this solutions may be duplicates, or even a duplicate of one of the two given curves.
5. Translates the coordinates of any valid center points of new tangent curves back to the original coordinate system.
6. Computes the radii of any valid constructions.
7. Removes any duplicate solutions, and then terminates.

7.4.5　Arcs Tangent to an Arc and Passing Through Two Points

The icCrcTanCrc2Pts procedure constructs two circles that pass through two given points and are tangent to a given circle. The said points may be inside or outside the said circle, but if they straddle it then there is no solution. The left side of Figure 7.4k displays the case of the two given point being located outside the given circle and the right side of the figure displays the case of the two points being located inside the given circle. In the said figure point P$_1$ denotes the center point of the given circle of radius R$_1$, and points P$_2$ and P$_3$ denote the two given points, while points P$_4$ and P$_5$ denote the center points of the two sought after tangent circles of radii R$_4$ and R$_5$ respectively. The coordinates of the three known points are referred to by the indicated lower case letters, and those of the two new circle center points are referred to by the letters X$_4$, Y$_4$, and X$_5$, Y$_5$, respectively. In the construction methodology below because it is applicable to both new circles, the radii R$_4$ and R$_5$ are referred to as R$_2$, and the coordinates of the two new center points are referred to as x and y.

In reviewing the case of the left side of Figure 7.4k we see that the center points of the two new tangent circles must lie along the normal bisector line to the line connecting the two given points and be located a distance R$_2$ away from each of the two given points and a distance R$_1$+R$_2$ from the center point of the given circle. Similarly in the case of the right side of the said figure the center points of the two new tangent circles must lie along the normal bisector line to the line

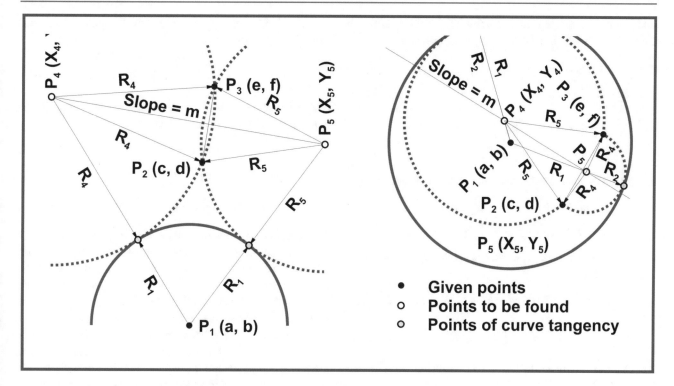

Figure 7.4k The Two Cases of an Arc Tangent to a Given Arc and Passing Through Two Given Points

connecting the two given points and be located a distance R_2 away from each of the two given points but it should be instead located a distance $R_1 - R_2$ from the center point of the given circle. Thus, we first connect the two given points with a line, we then construct its normal bisector line, and then we solve simultaneously three equations: (a) one representing the said bisector line, (b) one representing a circle of a radius of $R_1 + R_2$ (left figure side) or $R_1 - R_2$ (right figure side) having its center point at that of the given circle, and (c) another one representing a circle of radius of R_2 having its center point at either of the two unknown circles. So the coordinates of the said midpoint are

$$g = \frac{c+e}{2} \quad \text{and} \quad h = \frac{d+f}{2} \qquad (7.4.4a1)$$

The slope of the normal bisector line is

$$m = \frac{c-e}{f-d} \qquad (7.4.4a2)$$

and the Y intercept of the normal bisector line is then

$$Y_B = h - gm \qquad (7.4.4a3)$$

Therefore, the equations representing the normal bisector line, the circle of radius $R_1 + R_2$, and the two new tangent circles, respectively, may be expressed as

$$y = mx + Y_R \qquad (7.4.4b1)$$

$$(x-a)^2 + (y-b)^2 = (R_1 + R_2)^2 \qquad (7.4.4b2)$$

Note that at this time we will only address the case of $R_1 + R_2$ (left figure side). We will address the $R_1 - R_2$ (right figure side) later on.

$$(x-c)^2 + (y-d)^2 = R_2^2 \qquad (7.4.4b3)$$

$$(x-e)^2 + (y-f)^2 = R_2^2 \qquad (7.4.4b4)$$

Of the last two equations we will arbitrarily use Equation 7.4.4b3. If we now substitute Equation 7.4.4b1 for y in Equation 7.4.4b3 we get

$$(x-a)^2 + (mx + Y_B - b)^2 = (R_1 + R_2)^2 \qquad (7.4.4c1)$$

$$(x-c)^2 + (mx + Y_B - d)^2 = (R_2)^2 \qquad (7.4.4c2)$$

By expanding the terms of Equations 7.4.4c1 and 7.4.4c2 we get, respectively,

$$x^2 + a^2 - 2ax + m^2x^2 + (Y_B - b)^2 \\ + 2m(Y_B - b)x = R_1^2 + R_2^2 + 2R_1R_2 \qquad (7.4.4d1)$$

$$x^2 + c^2 - 2xc + m^2x^2 + (Y_B - d)^2 \\ + 2m(Y_B - d)x = R_2^2 \qquad (7.4.4d2)$$

and then rearranging and combining certain terms of Equations 7.4.4d1 and 7.4.4d2 we get, respectively,

$$\left(1+m^2\right)x^2 + 2\left[m\left(Y_B - b\right) - a\right]x + a^2 + \left(Y_B - b\right)^2$$
$$= R_2^2 + 2R_1R_2 + R_1^2 \qquad (7.4.4e1)$$

$$\left(1+m^2\right)x^2 + 2\left[m\left(Y_B - d\right) - c\right]x$$
$$+ c^2 + \left(Y_B - d\right)^2 = R_2^2 \qquad (7.4.4e2)$$

To simplify Equations 7.4.4e1 and 7.4.4e2 a bit, let us set

$$J = m\left(Y_B - b\right) - a \quad K = m\left(Y_B - d\right) - c$$
$$M = a^2 + \left(Y_B - b\right)^2 \quad N = c^2 + \left(Y_B - d\right)^2 \qquad (7.4.4f)$$

and then let us substitute J, K, M, and N into Equations 7.4.4e1 and 7.4.4e2 to get

$$\left(1+m^2\right)x^2 + 2Jx + M = R_2^2 + 2R_1R_2 + R_1^2 \qquad (7.4.4g1)$$

$$\left(1+m^2\right)x^2 + 2Kx + N = R_2^2 \qquad (7.4.4g2)$$

We may now subtract Equation 7.4.4g2 from Equation 7.4.4g1 to eliminate the x^2 terms and get

$$2\left(J - K\right)x + \left(M - N - R_1^2\right) = 2R_1R_2 \qquad (7.4.4h)$$

We may now solve for x, but before we do that let us set

$$Q = J - K \quad \text{and} \quad L = M - N - R_1^2 \qquad (7.4.4i1)$$

so that

$$x = \frac{2R_1R_2 - L}{2Q} = \frac{R_1}{Q}R_2 - \frac{L}{2Q} \qquad (7.4.4i2)$$

Again for the purpose of arithmetic simplification let us set

$$S = \frac{R_1}{Q} \quad \text{and} \quad T = \frac{L}{2Q} \qquad (7.4.4i3)$$

So that

$$x = SR_2 - T \qquad (7.4.4j)$$

We can now substitute Equation 7.4.4j in Equation 7.4.4c2 to solve for R_2 in terms of known terms as follows:

$$\left(SR_2 - T - c\right)^2 + \left(mSR_2 - mT + Y_B - d\right)^2 = R_2^2$$
$$\left[SR_2 - \left(T + c\right)\right]^2 + \left[mSR_2 - \left(mT - Y_B + d\right)\right]^2 = R_2^2$$
$$\left(SR_2\right)^2 + \left(T + c\right)^2 - 2\left(T + c\right)SR_2 + \left(mSR_2\right)^2$$
$$+ \left(mT - Y_B + d\right)^2 - 2\left(mT - Y_B + d\right)mSR_2 = R_2^2 \qquad (7.4.4k)$$
$$\left(S^2 + m^2S^2 - 1\right)R_2^2 - 2S\left[\left(T + c\right) + \left(mT - Y_B + d\right)m\right]R_2^2$$
$$+ \left[\left(T + c\right)^2 + \left(mT - Y_B + d\right)^2\right] = 0$$

Equation 7.4.4k may now be solved for R_2 by use of the standard quadratic equation as

$$R_2 = \frac{-B \pm \sqrt{B^2 - 4AC}}{2A} \qquad (7.4.4l1)$$

where A, B and C are not to be confused with the coordinates a, b and c, and they are defined as

$$A = S^2 + m^2S^2 - 1 \qquad (7.4.4l2)$$

$$B = 2S\left[\left(T + c\right) + \left(mT - Y_B + d\right)m\right] \qquad (7.4.4l3)$$

$$C = \left(T + c\right)^2 + \left(mT - Y_B + d\right)^2 \qquad (7.4.4l4)$$

The computation of the two values of R_2 from Equation 7.4.4l1 yields the radii of the two desired tangent circles R_4 and R_5 each of which can be substituted in Equation 7.4.4j to solve for the corresponding x coordinates X_4 and X_5, respectively, and which in conjunction with the radii R_4 and R_5 can be substituted in Equation 7.4.4b1 to solve for the y coordinates Y_4 and Y_5.

Presented above has been a procedure for solving the subject construction problem analytically for the case in which the two given points are located outside the given circle. We are now ready to address the case of the two given points being located within the given circle, but before doing that let us review the quadratic equation and certain others presented above. In so doing we notice that these certain equations contain divisions which for programming purposes create a problem if the divisor has a value of zero. Thus, prior to solving for the radii and for the center point coordinates of the two new tangent circles, it is wise to check for the existence of certain geometric configurations of the given data with respect to each other, and which are identified below.

7.4.5.1 Internal Position of the New Tangent Circles

If the two given points through which the new circles are to pass are positioned inside the given circle, then the two new circles will be located inside the given circle as indicated on the right side of Figure 7.4k. For this condition the right side of Equation 7.4.4b1 changes from $R_1 + R_2$ to $R_1 - R_2$. The net effect of this change is that the value of the term S in Equation 7.4.4i3 should be negative. That is, the said equation changes from $S = R_1/Q$ to $S = -R_1/Q$. The proof of this change is left up to the reader.

7.4.5.2 Orientation of the Two Points through which the New Circles Are to Pass

If the two given points through which the new circles are to pass are positioned in a **due north or south** direction then the slope of the normal bisector line is due east or west with the slope being zero ($m=0.0$). This does not create any issues with the above stated methodology.

However, if the two given points through which the new circles are to pass are located in a **due east or west direction** then the slope of the normal bisector line is due north or due south or having a value of infinity caused by a zero divide in Equation 7.4.4a2 (the north coordinates f and d of the two points through which the curve is to pass are the same within a tolerance of 0.00055 feet or 0.01524 mm), In this case, the two configurations of which are displayed in Figure 7.4l, the general method of solution is the same with one difference. The equation of the normal bisector line is not that of Equation 4.4.4b1 but that of the vertical line

$$x = \frac{c+e}{2} = g \qquad (7.4.4\text{m}1)$$

If we now substitute the above value g for x in Equations 7.4.4b2 and 7.4.4b3 we have

$$(g-a)^2 + (y-b)^2 = (R_1 + R_2)^2 \qquad (7.4.4\text{m}2)$$

$$(g-c)^2 + (y-d)^2 = R_2^2 \qquad (7.4.4\text{m}3)$$

If we next remove the squares in Equation 7.4.4m3, the unknown radius R_2 becomes

$$R_2 + \sqrt{(g-c)^2 + (y-d)^2} \qquad (7.4.4\text{n})$$

In Equation 7.4.4m2 we expand the terms in the parentheses with the unknowns y and R_2 to get

$$(g-a)^2 + y^2 + b^2 - 2by = R_1^2 + R_2^2 + 2R_1R_2 \qquad (7.4.4\text{o}1)$$

We can now substitute Equation 7.4.4n in Equation 7.4.4n and rearrange certain terms to get

$$\begin{aligned}
(g-a)^2 + b^2 - R_1^2 + y^2 - 2by = \\
(g-c)^2 + (y-d)^2 + 2R_1\sqrt{(g-c)^2 + (y-d)^2} \\
(g-a)^2 + (g-c)^2 + b^2 - R_1^2 + y^2 - (y-d)^2 - 2by = \\
2R_1\sqrt{(g-c)^2 + (y-d)^2}
\end{aligned} \qquad (7.4.4\text{o}2)$$

To simplify a bit the last equation let us set

$$W = (g-a)^2 - (g-c)^2 + b^2 - R_1^2 \qquad (7.4.4\text{p})$$

so that Equation 7.4.4o2 after certain expansions, and term combinations successively becomes

$$\begin{aligned}
W + y^2 - y^2 - d^2 + 2dy - 2by + 2R_1\sqrt{(g-c)^2 + (y-d)^2} \\
W - d^2 + 2(d-b)y = 2R_1\sqrt{(g-c)^2 + (y-d)^2}
\end{aligned} \qquad (7.4.4\text{q}1)$$

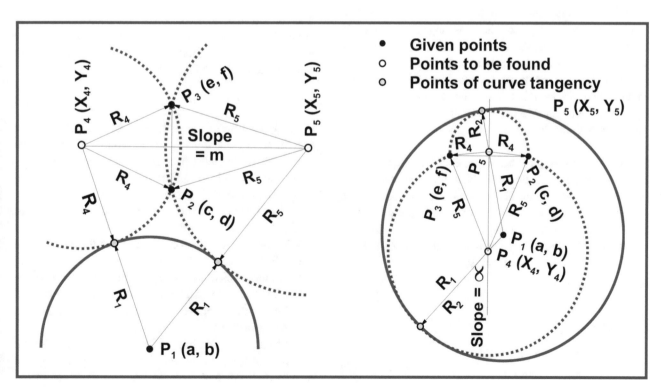

Figure 7.4l The Two Given Points Are Aligned Due East or Due West

$$\left(W-d^2\right)^2 + \left[4\left(d-b\right)^2 y^2\right] + 4\left[\left(W-d^2\right)\left(d-b\right)\right]y$$
$$= 4R_1^2\left[\left(g-c\right)^2 + y^2 + d^2 - 2dy\right]$$
$$= 4R_1^2\left(g-c\right)^2 + 4R_1^2 y^2 + 4R_1^2 d^2 - 8R_1^2 dy \qquad (7.4.4q2)$$

$$4\left(d-b\right)^2 y^2 - 4R_1^2 y^2 + 4\left[\left(W-d^2\right)\left(d-b\right)\right]y + 8R_1^2 dy$$
$$+ \left(W-d^2\right)^2 - 4R_1^2\left(g-c\right)^2 - 4R_1^2\left(g-c\right)^2 = 0 \qquad (7.4.4q3)$$

$$4\left[\left(d-b\right)^2 - R_1^2\right]y^2 + 4\left[\left(W-d^2\right)\left(d-b\right) + 2R_1^2 d\right]y$$
$$+ \left(W-d^2\right)^2 - 4R_1^2\left(g-c\right)^2 - 4R_1^2 d^2 = 0 \qquad (7.4.4q4)$$

Again for simplification if we let

$$A + 4\left[\left(d-b\right)^2 - R_1^2\right]$$
$$B = 4\left[\left(W-d^2\right)\left(d-b\right) + 2R_1^2 d\right] \qquad (7.4.4r)$$
$$C = \left[\left(W-d^2\right)^2 - 4R_1^2\left(g-c\right)^2 - 4R_1^2 d^2\right]$$

Equation 7.4.4q4 becomes

$$y = \frac{-B \pm \sqrt{B^2 - 4AC}}{2A} \qquad (7.4.4s)$$

Equation 7.4.4s determines the two values of the Y coordinates of the enter points of the new tangent circles while Equation 7.4.4m1 has set the corresponding X coordinates. These two pairs of coordinates may now be introduced in Equation 7.4.4n to compute the corresponding radii of the said two new circles. It is interesting to know that the above equations are applicable regardless whether the two given points are both inside or outside the given circle.

7.4.5.3 Colinearity of the Three Given Points
If the two given points through which the new circles are to pass are colinear with the center point of the given circle as shown in Figure 7.4m the value of Q in Equation 7.4.4i1 becomes zero, thus causing a division by zero in Equation 7.4.4i3. For Q to be zero the value of J would be K. For J to be equal to K, then from Equation 7.4.4f

$$m\left(Y_B - b\right) - a = m\left(Y_B - d\right) - c \qquad (7.4.4t)$$

from which

$$mY_B - mb - a = mY_B - md - c$$
$$mb + a = md + c \qquad (7.4.4u1)$$

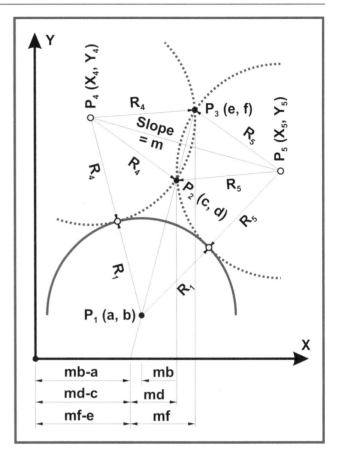

Figure 7.4m The Two Given Points Are Colinear with the Center Point of the Given Circle

Remembering that Equation 7.4.4f has been based on the arbitrary selection of Equation 7.4.4b3 and not of Equation 7.4.4b4, Equation 7.4.4u1 could be expanded to include Equation 7.4.4b4 and be written as

$$mb + a = md + c = mf + e \qquad (7.4.4u2)$$

To visually confirm this let as take a look at Figure 7.4m in which the two given points P_2 and P_3 through which the two new tangent circles must pass are colinear with the center point of the given circle P_1. The line passing through these three points is extended to meet the X axis. The component parts of Equation 7.4.4u2 are displayed below the X axis. In assessing these parts note that in the said figure the two given points are outside the given circle, and that the slope of the normal bisector line to the line connecting the said points is negative. Thus, each of the three parts of Equation 7.4.4u2 is equal to the other two parts. Similar results would be obtained if the slope is positive, and also if the two given point are inside the given circle.

Now back to the solution of the colinearity case. In the initial methodology presented above, having defined the constants J, K, M, and N, Equation 7.4.4g2 is subtracted from Equation 7.4.4g1 to eliminate the x^2 term and arrive at Equation 7.4.4h, in which there are two unknowns, x and $R2$. Since now J is equal to K, the left most term of this equation becomes zero, x drops out of the equation which may now be solved for $R2$ as

$$R_2 = \frac{M - N - R_1^2}{2R_1} \qquad (7.4.4v)$$

With the radius $R2$ now known and referring to Figure 7.4m we see that the coordinates of the new circle center points may be found by (a) intersecting a circle of radius R_1+R_2 with its center points at that of the given circle with the normal bisector line, or (b) intersecting two circles one having its center point at point P_2 and the other at point P_3 and both having the same radius.

Furthermore, it is worthwhile to note that the following configurations of colinearity may also be handled:

- The two given points, in addition to being colinear with the center point of the given circle and inside or outside the giver circle, they may also be located due east or west from each other, or due north or south of each other.
- The two given points may also be located due east or west of each other and be equidistant from the center point of the given circle, thus making the normal bisector line to the line connecting the two given points colinear with the center point of the given circle.

7.4.5.4 Positioning of the Two Points Through Which the New Circles Should Pass

It has been stated in the preamble to this construction problem that a solution is not possible if the two points straddle the given circle. That is, one point is located inside and the other point is located outside the said circle. But what happens if one or both of these points are located along the circumference of the said circle. It is obvious that if both points are located along the circumference then there is only one possible solution, and that is the given circle itself. Whenever then such a condition is encountered, the procedure so informs the user and terminates.

However, if one of the said two points is located on the circumference of the given circle and the other is either inside or outside the said circle, then there is but one solution as shown in Figure 7.4n. In this case, **(a)** the center point of the sole new circle must be located at the point of intersection of the bisector line of the line connecting the two given points and the line connecting the center point of the given circle with the given point located on the circumference of the given circle, and **(b)** the radius of the new circle is the distance from the new center point to either of the two given points. To satisfy programming considerations two identical circles are created.

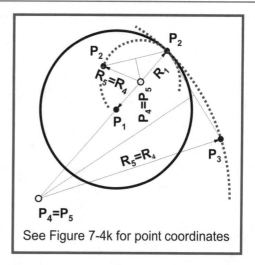

Figure 7.4n   One of the Two Given Point Lies on the Given Circle

7.4.5.5 The Two Points Through Which the New Circles Should Pass Are Coincident

If the two points through which the two new tangent circles are to pass are coincident within a tolerance of 0.005 feet (1.524 mm) they are considered to be the same point, referred as the common point, generating two new circles, one referred as the small circle and the other as the large circle, as indicated in Figure 7.4o. In this case, (a) the center point of either new circle must be located along the line from the center point of the given circle towards the given common point a distance of $B=D-R_2$ from the center point of the given circle, where D is the distance from the center point of the given circle to the common point, and (b) the radius of the new small circle is $R_2=(D-R_1)/2$, and that of the large one is $R_2=(D+R_1)/2$. Note that should the two given points be within the said tolerance but not identical, their average is used as the common point.

Shown in Figure 7.4o are the small and large circles for the case in which the common point is outside the given circle as well as inside the said circle. However, the radii and the distances B and D are shown only for the small circles. The reader may extend the above two equations and apply them to the construction of the large circles.

```
Sub icCrvTanCrv2pts(R1, CCE1, a, b, d, d,
                    e, f, _
                    R2, X, Y, IERR)
```

The input parameters to this procedure include the:

R1	The radius of the given circle.
a, b	The X (east) and Y (north) coordinates of the center point of the given circle.
c, d	The X (east) and Y (north) coordinates of one of the two given points through which the new circles are to pass.
e, f	The X (east) and Y (north) coordinates of the other of the two given points through which the new circles are to pass.

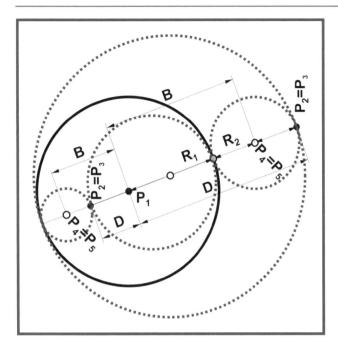

Figure 7.4o The Two Given Points Are Coincident

The parameters that are returned from this procedure include the:

R2   Two dimensional array of the radii of the new tangent circles to be created.

X, Y   Two dimensional arrays of the X (east) and Y (north) coordinates of the center points of the two tangent circles to be constructed.

IERR   An indicator, the value of which if it is:

      0   denotes that no errors have been detected.

      1   denotes that a negative radius of the given circle has been treated as being positive without altering the input value.

      2   denotes that the radius of the given circle has a zero value within a tolerance of 0.005 feet or 1.524 mm (fatal).

      3   denotes that both of the two given points through which the new tangent circles are to pass are located within a tolerance of 0.005 feet or 1.524 mm on the circumference of the given circle (fatal).

      4   denotes that both of the two given points through which the new tangent circles are to pass are not located within a tolerance of 0.005 feet or 1.524 mm are not located inside or outside the given circle (fatal).

      5   denotes that a division by zero or that the square root of a negative number has been detected when solving the quadratic equation most probably due to an input error (fatal).

Upon invocation, the procedure initializes the parameters to be returned to zero, and then:

1. Preserves the value of the radius of the given circle in a temporary parameter, and then checks it for being negative. If it is, IERR is set to 1 and continues.

2. Checks the value of the given radius for being zero within a tolerance of 0.005 feet (1.524 mm). If it is, IERR is set to 2 and the procedure aborts; else it continues.

3. Calls the icforce procedure to compute the distance and azimuth from the center point of the given circle towards the first given point, and then towards the second given point.

4. Compares the above two computed distances with the radius of the given circle, and if within a tolerance of 0.005 feet (1.524 mm):
   - Both of the two given points are outside the given circle, the temporary variable *InOut* is set to be "*OUTSIDE*" and continues.
   - Both of the two given points are inside the given circle, the temporary variable *InOut* is set to be "*INSIDE*" and continues.
   - The first of the two given points lies on the given circle and the second point:
     - Lies outside the said circle, *InOut* is set to be "*OUTSIDE 2*" and continues.
     - Lies inside the said circle, *InOut* is set to be "*INSIDE 2*" and continues.
   - The second of the two given points lies on the given circle and the first point:
     - Lies outside the said circle, *InOut* is set to be "*OUTSIDE 1*" and continues.
     - Lies inside the said circle, *InOut* is set to be "*INSIDE 1*" and continues.
   - Both of the two given points lie on the given circle, then IERR is set to 3 and the procedure aborts; else it continues.
   - Both of the two given points are not inside the given circle, and they are not both outside the given circle (they straddle it), then IERR is set to 4 and the procedure aborts; else it continues.

5. Determines whether the two given points are coincident or not. If so, the radius and the coordinates of the small circle and then those of the large circle are determined as described earlier, the value of the input radius of the given circle is restored to its original value, and the procedure terminates.

6. Computes the direction of the bisector line by adding 90° to the above computed azimuth,

7. Queries the *InOut* variable for not being "*INSIDE*" and not being "*OUTSIDE*".
   - If it is not, then one of the given two points lies on the circumference of the given circle and the other point is either inside or outside the said circle in which case:
     - Determines which one by querying the *InOut* variable for being "*INSIDE 1*", "*OUTSIDE 1*", "*INSIDE 2*", and/or "*OUTSIDE 2*," and calls the icforce procedure to compute the corresponding distance and azimuth from the center point of the given circle towards the appropriate point.

- Calls the `icintrs` procedure to intersect the selected line with the bisector line to find the coordinates of the center point of the new tangent circle [since the procedure is expected to return two center points X(2)=X(1) and Y(2)=Y(1)].
- Queries the *InOut* variable gain to determine which of the two given points does not lie on the given circle, and then calls the `icforce` procedure to determine the radius of the new tangent circle [since the procedure is expected to return two radii R2(2)=R2(1)].
- Restores the value of the input radius of the given circle to its original value, and the procedure terminates.
- If it is, proceeds to test whether the two points through which the new tangent curves are to pass are located due east or due west from each other, thus producing a slope of infinity for the bisector line.
8. Queries the Y coordinates of the two given points for being equal to each other within a tolerance of 0.0005 feet (0.1524 mm).
- If they are equal, the two given points are due east or west implying that the center points of the two new tangent circles are located along the bisector line and on opposite side of the line connecting the said two points, in which case the procedure:
  - Proceeds to compute the Y coordinates of the center points of the new tangent circles using Equations 7.4.4p, 7.4.4r, and 7.4.4s checking in the process whether a division by zero or the square root of a negative number has been encountered most probably due to an input error in which case IERR is set to be 5 and the procedure terminates.
  - Computes the X coordinates of the center points of the new tangent circles using Equation 7.4.4m1 keeping in mind that both of the new circles would have the same X coordinate.
  - Restores the value of the input radius of the given circle to its original value, and the procedure terminates.
- If they are not equal, it continues with the implicit solution of the construction problem as described earlier.
9. Computes the slope of the normal bisector line, the Y intercept of the bisector line, and the values of the constants J, K, M, N, and Q with Equations 7.4.4a2, 7.4.4a3, and 7.4.4f.
10. Queries the value of Q for being zero within a tolerance of 0.0005 feet (0.1524 mm) implying colinearity of the

center point of the given circle and the two given points through which the new tangent circles are to pass.
- If Q is zero, there is colinearity with the radii of the two tangent circles being the same and their center points being along the bisector line and on opposite side of the line connecting the two given points, in which case:
  - Computes both radii by Equation 7.4.4v.
  - Calls the `icxcvwcv` procedure to intersect two circles of the same radius R_2, one having its center point at point P_2 and the other at point P_3.
  - Restores the value of the input radius of the given circle to its original value, and the procedure terminates.
- If Q is not zero, proceeds to complete the explicit solution.
11. Computes the constant L using Equation 7.4.4i1, and queries the temporary parameter *InOut* to determine whether the given points are inside or outside the given circle.
- If *InOut*="OUTSIDE" computes the constant S using Equation 7.4.4i3, and
- If *InOut*="INSIDE" computes the constant S using Equation 7.4.4i3 multiplied by −1.
12. Computes the constant T using Equation 7.4.4i1, and then computes the parameter A using Equation 7.4.4l2.
13. Queries the value of A for being zero within a tolerance of 0.0005 (0.154 mm) and if so set IERR to be 5 and terminates; else continues.
14. Computes the constants B and C using Equation 7.4.4l2 and queries the value under the square root of Equation 7.4.4l1 for being negative and if so terminates; else continues.
15. Computes the value of the two radii R_2 using Equation 7.4.4l1, and queries the parameter *InOut* to compute the $X(1)$ and $X(2)$ coordinates of the new circles.
- If *InOut*="OUTSIDE" computes the $X(1)$ coordinate using Equation 7.4.4i1 with a positive radical, and
- If *InOut*="INSIDE" computes the $X(2)$ coordinate using Equation 7.4.4i1 with a negative radical.
16. Computes the value of the two radii R_2 using Equation 7.4.4l1, and queries the parameter *InOut* to compute the X(1) and X(2) coordinates of the new circles.
17. Computes the value of the two radii R_2 using Equation 7.4.4l1, and queries the parameter *InOut* to compute the X(1) and X(2) coordinates of the new circles.
18. Computes the absolute values of the two radii just in case either or both are negative, and terminates.

Compound and Reversed Curves

The preceding chapter presents procedures that construct tangent curves and lines in a variety of ways. In the design of roadways and other similar route related projects (levies, channels, and the like) the design civil engineer is faced with the need to create an alignment that is composed of a series of straight lines, circular arcs, and possibly spiral curves. When two or more circular arcs are connected to each other so as to be tangent at a point with the rotational direction remaining the same, clockwise or counterclockwise, these curves are referred to as compound curves. If the rotational direction from one curve into another curve changes direction, these curves are referred to as reversed curves.

In the preceding chapter we address the construction of straight lines being tangent to curves, as well as the reverse, and the construction of tangent curves to other curves. In this chapter we address the construction of compound and reversed curves as a single construction task under various conditions as they may arise as a result of certain impositions on the lines to which the compound and reversed curves must be tangent, and on the location of points through which the ending curves must start or end, or through which points one of the curves must pass. In these cases it is possible to encounter the introduction of spirals between a straight line and a curve, or between two curves. These conditions are addressed in Chap. 9.

8.1   Two Centered Compound Curves

In this section we consider the introduction of two curves tangent to each other, with each curve being tangent to a given line. It is assumed that each of these lines is defined by the north and east coordinates of a point (PN1, PE1) and (PN2, PE2), and its north azimuth AZ^1 and AZ^2. Unless otherwise noted, these are considered to be of known value.

In solving this construction problem, reference is made to Figures 8.1a and 8.1b in which the basic construction components are identified. In viewing these figures, the reader should note the following in both figures regarding the (a) local coordinate system used to which the various components are referenced, (b) deflection angles Δ, and (c) sign conventions:

- In Figure 8.1a, the origin of the local coordinates system is located at the PC point with the positive directions of Δ as indicated in the said figure.
- In Figure 8.1b, the origin of the local coordinates system is located at the PT point with the positive directions of Δ as indicated in the said figure.
- In both figures, the PC point is the point of tangency of the flatter curve (the one having the larger radius) with the first given line, and the PT point is the point of tangency of the sharper curve (the one having the smaller radius) with the second given line. That is to say that if an alignment has to advance from the sharper curve to the flatter curve, the user should reverse temporarily for solution purposes the direction of the alignment.
- If the overall deflection angle Δ between the two given tangent lines from AZ^1 towards AZ^2 is clockwise, the angle is considered as being positive; and if is counterclockwise, it is considered as being negative. The various equations to be presented below support both situations.
- The overall deflection angle Δ between the two given tangent lines from AZ^1 towards AZ^2 in both figures is shown to be less than 90° in absolute value. In the cases in which this angle is greater than 90°, the sign of the trigonometric functions that are used in the said equations accommodate the involved additions and/or subtractions.
- The overall deflection angle Δ between the two given tangent lines from AZ^1 towards AZ^2 in both figures is assumed to be less than 180° in absolute value, since the application of greater angles is not being considered to be too common, except in cloverleaf interchange type of ramps. In such cases a ramp's alignment could be broken into two or more parts.

© Springer International Publishing Switzerland 2016
E.C. Tonias, C.N. Tonias, *Geometric Procedures for Civil Engineers*, DOI 10.1007/978-3-319-24295-8_8

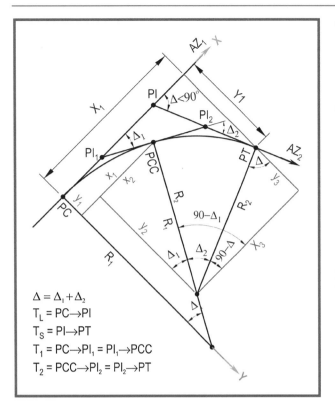

$$\Delta = \Delta_1 + \Delta_2$$
$$T_L = PC \rightarrow PI$$
$$T_S = PI \rightarrow PT$$
$$T_1 = PC \rightarrow PI_1 = PI_1 \rightarrow PCC$$
$$T_2 = PCC \rightarrow PI_2 = PI_2 \rightarrow PT$$

Figure 8.1a Two Centered Compound Curves $\left(\,|\Delta\,|<90^{\circ}\right)$ Clockwise Rotation—Local Axis at the PC

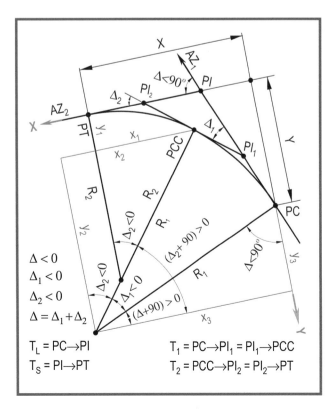

$$\Delta < 0$$
$$\Delta_1 < 0$$
$$\Delta_2 < 0$$
$$\Delta = \Delta_1 + \Delta_2$$

$T_L = PC \rightarrow PI$	$T_1 = PC \rightarrow PI_1 = PI_1 \rightarrow PCC$
$T_S = PI \rightarrow PT$	$T_2 = PCC \rightarrow PI_2 = PI_2 \rightarrow PT$

Figure 8.1b Two Centered Compound Curves $\left(\,|\Delta\,|<90^{\circ}\right)$ Clockwise Rotation—Local Axis at the PT

- The points that define each given tangent line may be located anywhere along its respective azimuth. However, the azimuth of the first tangent line should point towards the PI point, and the azimuth of the second given line should point away from the PI point (see the said figures).
- The sign conventions of the individual curve radii R1 and R2, and of their deflection angles Δ_1 and Δ_2 follow the sign of the overall deflection angle Δ. In the cases to be described below, in which R1, R2, Δ_1, and/or Δ_2 are input parameters, they may be entered as positive values. In such cases, the program changes these value to negative if the Δ angle is determined to be counterclockwise.

The solution to this construction problem presented below is quite similar to that presented by Thomas F. Hickerson in his *Route Location and Design* book, fifth edition, published by the McGraw-Hill Company in 1967.

The first task in solving this construction problem is to determine the north and east coordinates of the PI point of intersection (PIN, PIE) of the two given lines, and the angle of deflection Δ formed by these two lines may be found by use of the icintrs procedure. The returned value of the angle Δ is positive for a clockwise rotation and negative for a counterclockwise rotation.

In the aforesaid figures we note that

$$\Delta = \Delta_1 + \Delta_2 \tag{8.1a}$$

The procedure described later on in this section is generic and as such provides for several input parameters of which only selected ones should have a nonzero value in order to meet specific compound curve cases which are described in subsequent subsections. Thus, presented below are various groups of equations that solve for certain parameters in terms of other parameters that are considered to be known, or given. Although certain of the equations in these groups depend on a previously defined equation, the groups may be considered to be independent.

- In Figure 8.1a, the distance X_1 may be expressed as

$$x_1 = R_1 \sin\left(\Delta_1\right)$$
$$x_2 = R_2 \sin\left(\Delta_1\right)$$
$$x_3 = R_2 \sin\left(\Delta\right)$$

and $X_1 = x_1 - x_2 + x_3$

$$X_1 = R_1 \sin\left(\Delta_1\right) - R_2 \sin\left(\Delta_1\right) + R_2 \sin\left(\Delta\right)$$

or $X_1 = R_2 \sin\left(\Delta\right) + \left(R_1 - R_2\right)\sin\left(\Delta_1\right)$ (8.1b)

- In Figure 8.1a, the distance Y_1 may be expressed as

$$y_1 = R_1 - R_1 \cos\left(\Delta_1\right)$$
$$y_2 = R_2 \cos\left(\Delta_1\right)$$
$$y_3 = R_2 \cos\left(\Delta_1\right)$$

and $Y_1 = y_1 + y_2 - y_3$

$$Y_1 = R_1 - R_1 \sin\left(\Delta_1\right) + R_2 \cos\left(\Delta_1\right) - R_2 \cos\left(\Delta\right)$$

or $Y_1 = R_1 - R_2 \cos\left(\Delta\right) - \left(R_1 - R_2\right)\cos\left(\Delta_1\right)$ (8.1c)

- In Figure 8.1b, the distance X_2 may be expressed as

$$x_4 = R_2 \sin(\Delta_2)$$
$$x_5 = R_1 \sin(\Delta_2)$$
$$x_6 = R_1 \sin(\Delta)$$
and $\quad X_2 = x_4 - x_5 + x_6 \qquad (8.1d)$
$$X_2 = R_2 \sin(\Delta_2) - R_1 \sin(\Delta_2) + R_1 \sin(\Delta)$$
or $\quad X_2 = R_1 \sin(\Delta) - (R_1 - R_2)\sin(\Delta_2)$

- In Figure 8.1b, the distance Y_2 may be expressed as

$$y_4 = R_2 - R_2 \cos(\Delta_2)$$
$$y_5 = R_1 \cos(\Delta_2)$$
$$y_6 = R_1 \cos(\Delta)$$
and $\quad Y_2 = y_4 + y_5 - y_6 \qquad (8.1e)$
$$Y_2 = R_2 - R_2 \cos(\Delta_2) + R_1 \cos(\Delta_2) - R_1 \cos(\Delta)$$
or $\quad Y_2 = R_2 - R_1 \cos(\Delta) + (R_1 - R_2)\cos(\Delta_2)$

Although the given points that define the back and forward tangent lines of the overall compound curve arrangement may be located anywhere along their corresponding azimuths, the back azimuth should point towards the overall PI point, and the forward azimuth should point away from the said PI.

In the aforesaid figures the reader should note the distinction between the definitions of the long and short tangent lengths T_L and T_S, and the tangent lengths of the individual curves T_1 and T_2.

- In Figure 8.1a, the short tangent length T_S may be expressed as

$$T_s = Y_i / \sin(\Delta)$$
or $\qquad\qquad (8.1f)$
$$T_s = \frac{R_1 - R_2 \cos(\Delta) - (R_1 - R_2)\cos(\Delta_1)}{\sin(\Delta)}$$

- In Figure 8.1b, the long tangent length T_L may be expressed as

$$T_L = Y_2 / \sin(\Delta)$$

or

$$T_L = \frac{R_2 - R_1 \cos(\Delta) + (R_1 - R_2)\cos(\Delta_2)}{\sin(\Delta)} \qquad (8.1g)$$

- Solving Equation 8.1f for $cos(D_1)$, we get

$$\cos(\Delta_1) = \frac{R_1 - R_2 \cos(\Delta) - T_S \sin(\Delta)}{R_1 - R_2} \qquad (8.1h)$$

- Similarly, solving Equation 8.1g for $cos(\Delta 2)$, we get

$$\cos(\Delta_2) = \frac{R_1 \cos(\Delta) - R_2 + T_L \sin(\Delta)}{R_1 - R_2} \qquad (8.1i)$$

- Solving Equation 8.1b for $sin(\Delta 1)$, we get

$$\sin(\Delta_1) = \frac{X_1 \cos(\Delta) - R_2 \sin(\Delta)}{R_1 - R_2} \qquad (8.1j)$$

and noting from Figure 8.1a that $X1 = TL + TS\,cos(\Delta)$

$$\sin(\Delta_1) = \frac{T_L + T_S \cos(\Delta) - R_2 \sin(\Delta)}{R_1 - R_2} \qquad (8.1k)$$

- Solving Equation 8.1d for $sin(\Delta 2)$, we get

$$\sin(\Delta_2) = \frac{R_2 \sin(\Delta) - X_2}{R_1 - R_2} \qquad (8.1l)$$

and noting from Figure 8.1b that $X_2 = T_S + T_L cos(\Delta)$

$$\sin(\Delta_2) = \frac{R_1 \sin(\Delta) - T_L \cos(\Delta) - T_S}{R_1 - R_2} \qquad (8.1m)$$

- Now let us revisit Equation 8.1f. In so doing, let us multiply its both sides by $sin(\Delta)$ and expand its right most term to get

$$T_S \sin(\Delta) = R_1 - R_2 \cos(\Delta) - R_1 \cos(\Delta_1) + R_2 \cos(\Delta_1)$$

Since $vers(\Delta) = 1 - \cos(\Delta)$, or $\cos(\Delta) = 1 - vers(\Delta)$, the above equation may be written as

$$T_S \sin(\Delta) = R_1 - R_2[1 - vers(\Delta)] - R_1[1 - vers(\Delta_1)] + R_2[1 - vers(\Delta_1)]$$
$$T_S \sin(\Delta) = R_1 - R_2 + R_2 vers(\Delta) - R_1 + R_1 vers(\Delta_1) + R_2 - R_2 vers(\Delta_1)$$
$$T_S \sin(\Delta) = R_2 vers(\Delta) + (R_1 - R_2)vers(\Delta_1)$$
$$vers(\Delta_1) = \frac{T_S \sin(\Delta) - R_2 vers(\Delta)}{R_1 - R_2} \qquad (8.1n)$$

If we divide Equation 8.1n by Equation 8.1k, we get

$$\frac{vers(\Delta_1)}{sin(\Delta_1)} = \frac{\dfrac{T_S \sin(\Delta) - R_2 \, vers(\Delta)}{R_1 - R_2}}{\dfrac{T_L + T_S \cos(\Delta) - R_2 \sin(\Delta)}{R_1 - R_2}} \qquad (8.1o)$$

since $vers(\Delta_1)/\sin(\Delta_1) = \tan(\Delta_1/2)$, and if we replace $vers(\Delta)$ with $1 - \cos(\Delta)$, Equation 8.1o may be expressed as

$$\tan\left(\frac{1}{2}\Delta_1\right) = \frac{T_S \sin(\Delta) - R_2\left[1 - \cos(\Delta)\right]}{T_L + T_S \cos(\Delta) - R_2 \sin(\Delta)} \qquad (8.1p)$$

If we now apply the same methodology that we applied to Equation 8.1n to Equation 8.1g, we will get the following equations:

$$T_L \sin(\Delta) = R_2 - R_1 \cos(\Delta) + R_1 \cos(\Delta_2) - R_2 \cos(\Delta_2)$$
$$T_L \sin(\Delta) = R_2 - R_1\left[1 - vers(\Delta)\right] + R_1\left[1 - vers(\Delta_2)\right] - R_2\left[1 - vers(\Delta_2)\right]$$
$$T_L \sin(\Delta) = R_2 - R_1 + R_1 vers(\Delta) + R_1 - R_1 vers(\Delta_2) - R_2 + R_2 vers(\Delta_2)$$
$$T_L \sin(\Delta) = R_1 vers(\Delta) - (R_1 - R_2) vers(\Delta_2)$$
$$vers(\Delta_2) = \frac{R_1 vers(\Delta) - T_L \sin(\Delta)}{R_1 - R_2} \qquad (8.1q)$$

If we divide Equation 8.1p by Equation 8.1m, we get

$$\frac{vers(\Delta_2)}{sin(\Delta_2)} = \frac{\dfrac{R_1 vers(\Delta) - T_L \sin(\Delta)}{R_1 - R_2}}{\dfrac{R_1 \sin(\Delta) - T_S + T_L \cos(\Delta)}{R_1 - R_2}} \qquad (8.1r)$$

since $vers(\Delta_2)/\sin(\Delta_2) = \tan(\Delta_2/2)$, and if we replace $vers(\Delta)$ with $1 - cos(\Delta)$, Equation 8.1r may be expressed as

$$\tan\left(\frac{1}{2}\Delta_2\right) = \frac{R_1\left[1 - \cos(\Delta)\right] - T_L \sin(\Delta)}{R_1 \sin(\Delta) - T_S - T_L \cos(\Delta)} \qquad (8.1s)$$

• If we expand the right most term of Equation 8.1c, we get

$$Y_1 = R_1 - R_2 \cos(\Delta) - R_1 \cos(\Delta_1) + R_2 \cos(\Delta_1)$$
$$\text{or } Y_1 = R_1\left[1 - \cos(\Delta_1)\right] - R_2\left[\cos(\Delta) - \cos(\Delta_1)\right]$$

and if we solve for R_1, we get

$$R_1 = \frac{Y_1 + R_2\left[\cos(\Delta) - \cos(\Delta_1)\right]}{1 - \cos(\Delta_1)} \qquad (8.1t)$$

Noting in Figure 8.1a that $Y_1 = T_S \sin(\Delta)$, and substituting it in Equation 8.1t, R_1 becomes

$$R_1 = \frac{T_S \sin(\Delta) + R_2\left[\cos(\Delta) - \cos(\Delta_1)\right]}{1 - \cos(\Delta_1)} \qquad (8.1u)$$

• Similarly, if we expand the right most term of Equation 8.1e, we get

$$Y_2 = R_2 - R_1 \cos(\Delta) - R_1 \cos(\Delta_2) + R_2 \cos(\Delta_2)$$
$$\text{or } Y_2 = R_2\left[1 - \cos(\Delta_2)\right] - R_1\left[\cos(\Delta) - \cos(\Delta_2)\right]$$

• and if we solve for R_2, we get

$$R_2 = \frac{Y_2 + R_1\left[\cos(\Delta) - \cos(\Delta_2)\right]}{1 - \cos(\Delta_2)} \qquad (8.1v)$$

• Noting in Figure 8.1b that $Y_2 = T_L \, sin(\Delta)$, and substituting it in Equation 8.1r, R_2 becomes

$$R_2 = \frac{T_L \sin(\Delta) + R_1\left[\cos(\Delta) - \cos(\Delta_2)\right]}{1 - \cos(\Delta_2)} \qquad (8.1w)$$

In reviewing Figures 8.1a and 8.1b, we notice the following parameters that comprise the two centered compound curves:

• The two radii R_1 and R_2,
• The two tangent lengths T_L and T_S,
• The two intersection angles Δ_1 and Δ_2, and
• The overall intersection angle Δ.

Of these seven elements, if four are known, one of which must be one of the three angles Δ, Δ_1, or Δ_2, the remaining three parameters may be computed by use of certain of the above equations. It should be noted that due to the statement of the subject construction problem, the overall deflection angle Δ is a known, given, parameter since it can easily be determined by intersecting the two given lines.

The number of combinations of selecting four of the said seven parameters is

$$\binom{n}{r} = \frac{n!}{r!(n-r)!} = \frac{7!}{4!(n-r)!} = \frac{5040}{24.6} = 35 \quad (8.1x)$$

Of these 35 combinations, we are to consider in the subsections below only seven as being practical, or having a potential for use. In addition to these seven combinations, there is one other case that presents a special application by considering the long chord (the distance from the PC point to the PT point in Figures 8.1a and 8.1b) as a given parameter. The procedure to be called to handle these seven potential combinations is the one presented next.

```
Sub ic2cenCrv (PN1, PE1,AZ1,PN2, PE2,AZ2, Delt1,
               Delt2, R1, R2, TL, TS, _
               DELT, CCN1, CCE1, CCN2, CCE2,
               T1, T2, PIN, PIE, PIN1, PIE1, _
               PIN2, PIE2, PCN, PCE, PCCN, PCCE,
               PTN, PTE, IOK)
```

The input parameters to this procedure that are common to all cases include the:

PN1, PE1	The north and east coordinates of the point on the back tangent line.
AZ1	The north azimuth (in radians) of the back tangent line pointing towards the PI of the overall compound curves.
PN2, PE2	The north and east coordinates of the point on the forward tangent line.
AZ2	The north azimuth (in radians) of the forward tangent line pointing away from the PI of the overall compound curves.

The case dependent parameters include the ones listed below. Three or four of them (see Table 8.1) must have a nonzero value, and the others must have a value of zero (0.00).

Table 8.1 Combinations of Input Parameters

Case No.	Input Parameters With Nonzero Value	Chapter Subsection
1	R1, R2, Delt1 and/or Delt	8.1.1
2	R1, TL, TS	8.1.2
3	R2, TL, TS	8.1.3
4	R1, R2, TL	8.1.4
5	R1, R2, TS	8.1.5
6	R1, TL, Delt1	8.1.6
7	R2, TS, Delt2	8.1.7
8	Case with known chord	8.1.8

Delt1	The deflection angle (in radians) of the back (flatter) curve, or zero.
Delt2	The deflection angle (in radians) of the forward (sharper) curve, or zero.
R1, R2	The radii of the back and forward curves, respectively, or zero.
TL, TS	The back and forward tangent lengths, or zero.
NOTE	Regarding the Delt1, Delt2, R1, R2, TL, and TS input parameters, it is assumed that the calling program of this procedure has determined which parameters should have a nonzero value, and which should not.

The parameters that are returned from this procedure include the:

DELT	The deflection angle (in radians) of the overall compound curve alignment.
CCN1, CCE1	The north and east coordinates of the center point of the back (flatter) curve.
CCN2, CCE2	The north and east coordinates of the center point of the ahead (sharper) curve.
T1, T2	The tangent lengths of the back and forward curves, respectively.
PIN, PIE	The north and east coordinates of the PI point of the overall compound curve.
PIN1, PIE1	The north and east coordinates of the PI point of the back (flatter) curve.
PIN2, PIE2	The north and east coordinates of the PI point of the forward (sharper) curve.
PCN, PCE	The north and east coordinates of the PC point of the back (flatter) curve.
PCCN, PCCE	The north and east coordinates of the PCC point between the two curves.
PTN, PTE	The north and east coordinates of the PT point of the forward (sharper) curve.
IOK	An indicator, the value of which if it is: • 0 Denotes that no problems have been encountered. • 1 Denotes that AZ1 and AZ2 are colinear. • 2 Denotes that the sum of Delt1 and Delt2 is not equal to DELT. • 3 Denotes that an improper input group has been specified.

Upon invocation, the procedure initializes the parameters to be returned, and then:

1. Checks the difference of the AZ1 and AZ2 parameters for being equal to 0°, or to 180° within a tolerance of 0.0000048 radians (0.99″), and if within the said tolerance, sets IOK to be 1, and the procedure terminates; else it continues.

2. Calls the `icintrs` procedure to intersect the back and forward tangent lines and determine the coordinates of the PI point (`PIN`, `PIE`) and the deflection angle `DELT` of the overall compound curve. Having done so, the procedure checks to determine whether the two azimuths straddle north or not by querying the angle `GANG` that is returned from the `icintrs` procedure by adding 2π to, or subtracting 2π from `GANG` depending whether `GANG` is positive or negative and greater or less than π.

3. Checks the value of each of the `Delt1` and `Delt2` input parameters for a zero value within a tolerance of 0.0000048 radians (0.99″), and if within the said tolerance, sets the value to be zero.

4. Checks the value of each of the `Delt1` and `Delt2` input parameters for a nonzero value, and if either one, or both are found to have a nonzero value, the sign of the said value is set to be the same as the sign of the deflection angle `DELT`.

5. Checks the value of each of the `Delt1` and `Delt2` input parameters to determine whether both have a nonzero value, or not. If:
   - Both parameters have a nonzero value, the procedure checks their sum. If the sum is not equal to the value of `DELT`, within a tolerance of 0.0000048 radians (0.99″), the `IOK` is set to 2, and the procedure terminates; else it continues.
   - Only one has a nonzero value, the other is computed by subtracting the nonzero angle from `DELT`.

6. Checks the value of each of the `R1` and `R2` input parameters for a zero value within a tolerance of 0.005 feet (1.524 mm), and if within the said tolerance, sets the value to be zero.

7. Checks the value of each of the `R1` and `R2` input parameters for a nonzero value, and if either one, or both are found to have a nonzero value, the sign of the said value is set to be the same as the sign of the deflection angle `DELT`.

8. Forces the values of `TL` and `TS` to be positive, in case they are not.

9. Determines which of the six input parameters `Delt1`, `Delt2`, `R1`, `R2`, `TL`, and `TS` have a nonzero value, and which do not, and thus determines which of the first seven cases of Table 8.1 has been specified. Shown in this table by case number are the various combinations of the said input parameters that can be handled by the subject procedure, and the subsection of this chapter that describes the work to be done. Note that Case 8 is a special condition that if it is to be handled it should be addressed by the calling program of this procedure, and not by this procedure.

The query proceeds from left to right and from top to bottom in the sequence indicated in the said table. The first match to be made defines the applicable case for solution. If any other of the said six input parameters that are not shown in the selected case have been assigned a nonzero value, they are disregarded. For example, if each of the parameters `R1`, `R2`, `TL`, and `TS` have a nonzero value, the parameter combination case to be selected is Case 4 (see Table 8.1), and the parameter `TL` is disregarded. In this case the procedure branches to Sect. 8.1.4 of this chapter that describes the action to be taken.

10. Queries the `TheCase` parameter that identifies the encountered case and branches to carry out the case specific work that determines the values of the input parameters with a zero value as presented in the subsequent subsections. Thereafter the procedure continues as indicated below.

11. Calls the `icptl` procedure to compute the coordinates of the PC point (`PCN`, `PCE`) and the PT point (`PTN`, `PTE`) from the PI point along AZ1 and AZ2 using the `TL` and `TS` lengths.

12. Computes the individual curve tangent lengths `T1` and `T2` as being:

$$T1 = R1 * \tan\left(Delt1 / 2\right) \text{ and}$$
$$T2 = R2 * \tan\left(Delt2 / 2\right).$$

13. Computes the azimuth (`AZ3`) of the common tangent to the two curves at the PCC point by adding `Delt1` to `AZ1`.

14. Calls the `icptl` procedure to compute the coordinates of the:
   - PI points of the individual curves (`PIN1`, `PIE1`) and (`PIN2`, `PIE2`) by traversing a `T1` distance along `AZ1` from the PC point, and a `T2` distance along `AZ2` from the PT point.
   - Center points of the two curves (`CCN1`, `CCE1`) and (`CCN2`, `CCE2`) by traversing a zero plus distance and an offset distance of `R1` and `R2` along the normal to `AZ1` and `AZ2` from the PC and PT points, respectively.
   - PCC point (`PCCN`, `PCCE`) by traversing a plus distance `T1` along `AZ3` from the PI of the flatter curve.

   Having computed the above parameters, the procedure terminates. The computation of the zero input parameters are determined as indicated in the subsections below.

8.1.1   Case 1—Given R1, R2, Delt1, and Delt2

This portion of the `ic2cenCrv` procedure assumes that the direction of the common tangent line to the two curves is to be considered as being known and defined by the value of deflection angles of the two individual curves `Delt1` or `Delt2`. It is also assumed that the values of the two radii `R1` and `R2` are known as indicated in Figure 8.1c.

If `Delt1` and `Delt2` have been assigned a nonzero value, their sum must equal to `DELT`, the overall delta angle. If not, `IOK` is set to 2, and the procedure terminates. If only one of the said two input angles has a nonzero value, the value of the other is set to be the difference between the given value

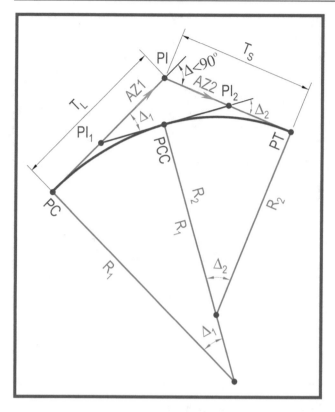

Figure 8.1c Given R1, R2, Delt1, and Delt2

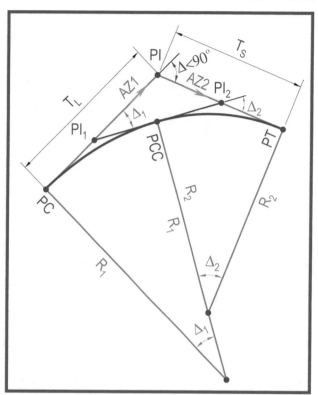

Figure 8.1d Given R1, TL, and TS

and that of DELT. Note that this work has been done prior to the procedure coming to this part of its path. With these parameters known, this part of the procedure:

10a. Computes the value of the long tangent length TL by use of Equation 8.1g.
10b. Computes the value of the short tangent length TS by use of Equation 8.1f.
10c. Returns to **Step 11** of the code of the main procedure ic2cenCrv.

8.1.2 Case 2—Given R1, TL and TS

This portion of the ic2cenCrv procedure assumes that the start and end points (PC and PT) are considered to be fixed at the specified distances TL and TS along AZ1 and AZ2, respectively, from the overall point of intersection (PIN, PIE). It is also assumed that the radius of the back (flatter) curve R1 is also known. Regarding this case of compound curves reference is made to Figure 8.1d. With the above parameters known, this part of the procedure:

10a. Computes the deflection angle of the sharper curve Delt2 by use of Equation 8.1s multiplied by two.
10b. Computes the radius of the sharper curve R2 by use of Equation 8.1w.

10c. Computes the deflection angle Delt1 of the flatter curve as being:

$$Delt1 = DELT - Delt2$$

10d. Returns to **Step 11** of the code of the main procedure ic2cenCrv.

A similar application is that in which the points that define the two given lines, point (PN1, PE1) and point (PN2, PE2), are actually the PC and PT points, respectively. In this case, the calling program of the ic2cenCrv procedure should:

- Determine the distances of the said two points from the point of intersection of the two given lines, the PI point, by use of the icintrs procedure;
- Find the larger of the two distances, thus determining the TL and the TS lengths; and
- Make certain that the first line corresponds to the TL length, and the second line corresponds to the TS length before calling the ic2cenCrv procedure.

8.1.3 Case 3—Given R2, TL and TS

This portion of the ic2cenCrv procedure assumes that the start point (PC) and the end point (PT) are considered to be fixed at the specified tangent distances TL and TS along the back azimuth AZ1 and along the forward azimuth AZ2, respectively,

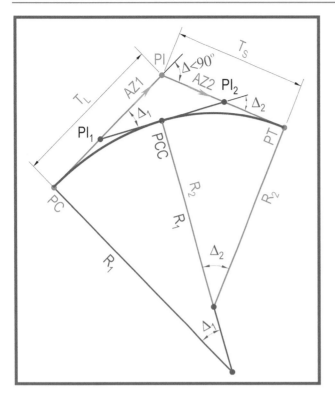

Figure 8.1e Given R2, TL, and TS

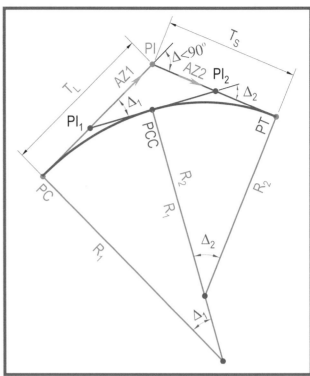

Figure 8.1f Given R1, R2, and TL

from the overall point of intersection (PIN, PIE) of the said tangent lines. It is also assumed that the radius of the sharper curve (R2) is also of known value. Regarding this case of compound curves reference is made to Figure 8.1e. With the above parameters known, this part of the procedure:

10a. Computes the deflection angle of the flatter curve Delt1 by use of Equation 8.1p.
10b. Computes the radius of the flatter curve R1 by use of Equation 8.1u.
10c. Computes the deflection angle Delt2 of the sharper curve as being:

$$Delt2 = DELT - Delt1$$

10d. Returns to **Step 11** of the code of the main procedure ic2cenCrv.

Refer to Sect. 8.1.2 for a known location of the start and end points PC and PT, respectively.

8.1.4 Case 4—Given R1, R2, and TL

This portion of the ic2cenCrv procedure assumes that the start point (PC) is to be considered as being fixed at the specified distance TL from the overall point of intersection (PIN, PIE) along AZ1. It is also assumed that the radii of both curves (R1 and R2) are also known. Regarding this case of compound curves reference is made to Figure 8.1f. With the above parameters known, this part of the procedure:

10a. Computes the deflection angle of the sharper curve (Delt2) by use of Equation 8.1i.
10b. Computes the deflection angle Delt1 of the flatter curve as being:

$$Delt1 = DELT - Delt2$$

10c. Computes the short tangent length TS by use of Equation 8.1f.
10d. Returns to Step 11 of the code of the main procedure.

A similar application is that in which the point that defines one of the two given lines, point (PN1, PE1) or point (PN2, PE2), is actually the PC point. In this case, the calling program of the ic2cenCrv procedure should:

- Determine the distances of the said two points from the point of intersection of the two given lines, the PI point, by use of the icintrs procedure; and
- Ascertain which of these two distances is the TL length before calling the ic2cenCrv procedure.

8.1.5 Case 5—Given R1, R2, and TS

This portion of the ic2cenCrv procedure assumes that the end point (PT) is considered to be fixed at the specified distance TS from the overall point of intersection (PIN, PIE) along AZ2. It is also assumed that the radii of both curves (R1 and R2) are also known. Regarding this case of compound curves reference is made to Figure 8.1g.

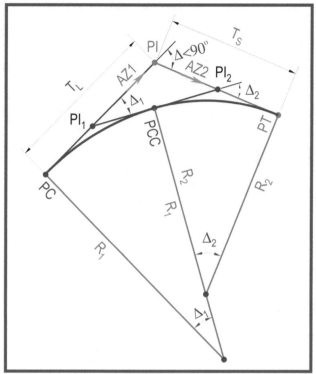

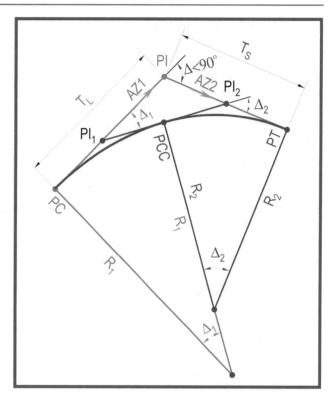

Figure 8.1g Given R1, R2, and TS

Figure 8.1h Given R1, TL, and Delt1

With the above parameters known, this part of the procedure:

10a. Computes the deflection angle of the flatter curve Delt1 by use of Equation 8.1h.
10b. Computes the deflection angle Delt2 of the sharper curve as being:
$$Delt2 = DELT - Delt1$$
10c. Computes the long tangent length TL by use of Equation 8.1g.
10d. Returns to **Step 11** of the code of the main procedure ic2cenCrv.

A similar application is that in which the point that defines one of the two given lines, point (PN1, PE1) or point (PN2, PE2), is actually the PT point. In this case, the calling program of the ic2cenCrv procedure should:

• Determine the distances of the said two points from the PI point, by use of the icintrs procedure; and
• Ascertain which of these two distances is the TS length before calling the ic2cenCrv procedure.

8.1.6   Case 6—Given R1, TL, and Delt1

This portion of the ic2cenCrv procedure assumes that the start point (PC) is considered to be fixed at the specified distance TL from the overall point of intersection (PIN, PIE) along the back azimuth AZ1. It is also assumed that the radius of the flatter curve (R1) and the deflection angle of the same curve (Delt1) are also known. Regarding this case of compound curves reference is made to Figure 8.1h. With the above parameters known, this part of the procedure:

10a. Computes the deflection angle Delt2 of the sharper curve as being:
$$Delt2 = DELT - Delt1$$
10b. Computes the radius R2 with Equation 8.1w.
10c. Computes the short tangent length TS with Equation 8.1f.
10d. Returns to **Step 11** of the code of the main procedure ic2cenCrv.

For a similar case refer to the end of Sect. 8.1.4.

8.1.7   Case 7—Given R2, TS, and Delt2

This portion of the `ic2cenCrv` procedure assumes that the end point (PT) is to be considered as being fixed at the specified distance `TS` from the overall point of intersection (`PIN`, `PIE`) along the forward azimuth `AZ2`. It is also assumed that the radius (`R2`) and the deflection angle of the sharper curve (`Delt2`) are also known. Regarding this case of compound curves reference is made to Figure 8.1i. With the above parameters known, this part of the procedure:

10a. Computes the deflection angle `Delt1` of the flatter curve as being:

$$Delt1 = DELT - Delt2$$

10b. Computes the radius of the flatter curve (`R1`) by use of Equation 8.1u.

10c. Computes the long tangent length `TL` by use of Equation 8.1g.

10d. Returns to Step 11 of the code of the main procedure `ic2cenCrv`.

For a similar case refer to the end of Sect. 8.1.5.

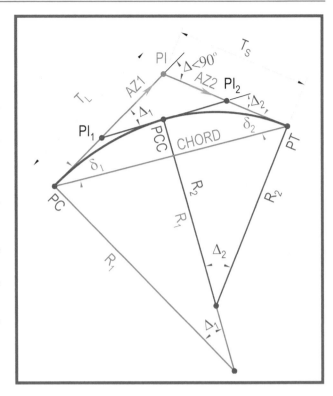

Figure 8.1j   Given R1, CHORD, Delt1, and Delt2

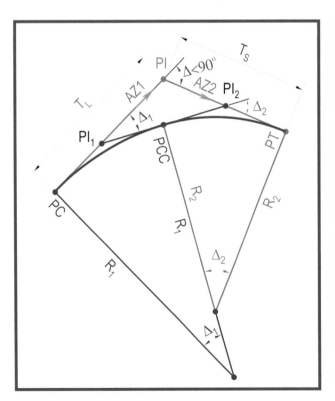

Figure 8.1i   Given R2, TS, and Delt2

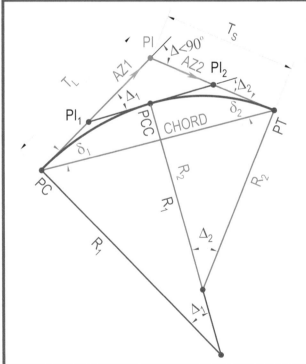

Figure 8.1k   Given R2, CHORD, Delt1, and Delt2

8.1.8   Case 8—Given the CHORD Length

A special condition may arise when the long chord length between the PC and the PT points is known in addition to certain other of the aforesaid input parameters. This condition would generally arise in the so called angle sight conditions at intersecting roadways, and at such other conditions at which a certain buffer area, or clearance distance from a specific control point has to be maintained from the center line of a roadway. Most always in such conditions the back and forward tangent lines have been established and thus their deflection angle DELT is known. In such cases enough of the previously said input parameters will most probably be also known so that one of the preceding seven cases could be used.

In angle sight conditions in which the back and forward tangent lines represent the center lines of two intersecting roadways, or offset lines therefrom, generally the TL and the TS distances are considered known, thus any of the preceding cases would also be called upon. However, if for some reason, or another it is the said long chord distance that is known, and if one of the two radii is known, as well as the angles δ_1 and δ_2 formed by the said long chord and the back and forward tangent lines as indicated in the last said two figures are known, than the law of signs could be used by the calling program of this procedure to determine the TL and TS lengths as follows:

$$TL = CHORD * \sin(\delta_2) / \sin d(DELT)$$
$$TS = CHORD * \sin(\delta_1) / \sin(DELT)$$

Thereafter the appropriate of the previously described case of this procedure could be specified.

8.2   Three Centered Compound Curves

In this section we will consider the introduction of three curves tangent to each other, and with the first and third curves being tangent to a given line referred to as the back and to the forward tangent lines, respectively, commonly referred as a three centered curve. It is assumed that each of these two lines is defined by the north and east coordinates of a point (PN1, PE1) and (PN2, PE2), and its north azimuth AZ1 and AZ2 referred to as the back and forward azimuths, respectively. It is also assumed that although each of the said points may be located anywhere along its corresponding tangent line, the back azimuth of the back tangent line should point towards the point of intersection of the said two lines (the PI point), and the azimuth of the forward

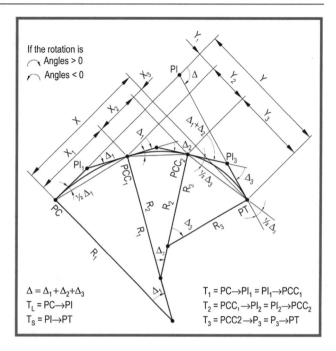

Figure 8.2a Three Centered Curvature—Decreasing Radii

tangent line should point away (ahead or forward) from the said PI point.

The three centered curve geometry is generally used to transition from a straight alignment into another straight alignment over a rather short total travel path and rather sharp, close to 90° deflection angle between the two straight alignments. Three centered curve alignments are usually introduced in low traffic level roadways, and particularly at major street intersections. For major highways, the introduction of spirals replaces the need for three centered curve geometries, although it is conceivable to introduce spirals between a series of compound curves.

To address the subject issue, let us take a look at four examples of a three centered compound curve geometry. Figure 8.2a displays three curves that transition successively from a flat curve towards a sharper curve, while the three curves of Figure 8.2b transition from a sharp curve towards a flatter curve. In Figure 8.2c the middle curve is sharper than its two neighbor curves, while in Figure 8.2d the middle curve is flatter than its neighbor curves. Of these four three centered curve arrangements, the one of Figure 8.2c is the one most commonly used at street intersections.

In each of these figures, a local coordinate system is assumed about the start point, the PC point, regardless whether this point represents the point of tangency of the first given line with the sharper curve or with the flatter curve. However, it is important to bear in mind that a clockwise

Figure 8.2b Three Centered
Curvature—Increasing Radii

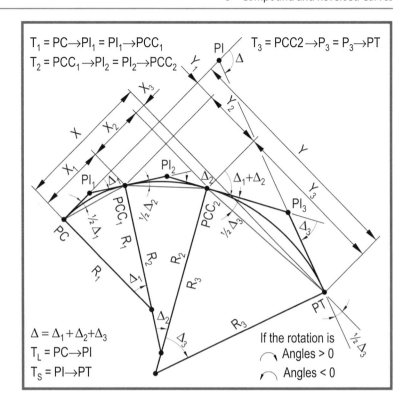

Figure 8.2c Three Centered
Compound Curves—Sharper
Center Curve

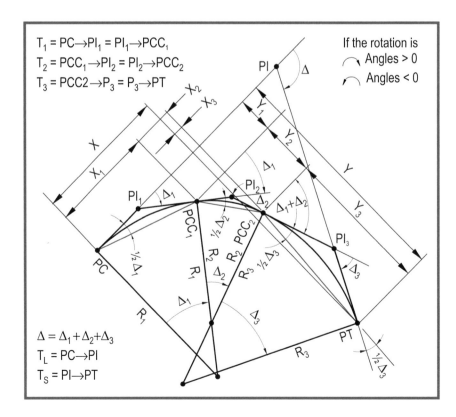

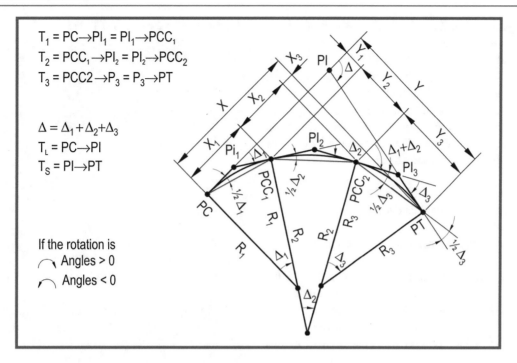

Figure 8.2d   Three Centered Compound Curves—Flatter Center Curve

deflection angle Δ is considered to be positive, and a counterclockwise deflection angle Δ is considered to be negative. The same applies to the deflection angles of the individual curves. Regardless of the rotation of these angles, all radii are considered to be positive. In reviewing these figures, it is seen that

$$\Delta = \Delta_1 + \Delta_2 + \Delta_3 \qquad (8.2a)$$

and that by using the chord length C of each individual curve as being $C = R\sin(0.5\Delta 1)$

$$X_1 = 2R_1 \sin\left(\frac{1}{2}\Delta_1\right)\cos\left(\frac{1}{2}\Delta_1\right) = R_1 \sin(\Delta_1) \qquad (8.2b)$$

$$X_2 = 2R_2 \sin\left(\frac{1}{2}\Delta_2\right)\cos\left(\Delta_1 + \frac{1}{2}\Delta_2\right) \qquad (8.2c)$$

$$X_3 = 2R_3 \sin\left(\frac{1}{2}\Delta_3\right)\cos\left(\Delta_1 + \Delta_2 + \frac{1}{2}\Delta_3\right) \qquad (8.2d)$$

and similarly

$$Y_1 = 2R_1 \sin\left(\frac{1}{2}\Delta_1\right)\sin\left(\frac{1}{2}\Delta_1\right) = 2R_1 \sin^2\left(\frac{1}{2}\Delta_1\right) \qquad (8.2e)$$
$$= R_1\left[1 - \cos(\Delta_1)\right]$$

$$Y_2 = 2R_2 \sin\left(\frac{1}{2}\Delta_2\right)\sin\left(\Delta_1 + \frac{1}{2}\Delta_2\right) \qquad (8.2f)$$

$$Y_3 = 2R_3 \sin\left(\frac{1}{2}\Delta_3\right)\sin\left(\Delta_1 + \Delta_2 + \frac{1}{2}\Delta_3\right) \qquad (8.2g)$$

Therefore

$$X = X_1 + X_2 + X_3 \qquad (8.2h)$$

and

$$Y = Y_1 + Y_2 + Y_3 \qquad (8.2i)$$

Two additional elements to be considered are the lengths of the tangent lines referred to as the back tangent (T_L) and the forward tangent (T_S) lines. The former represents the distance from the PC point towards the overall PI point and the latter represents the distance from the said PI point to the PT point. These two distances may be defined as

$$T_L = X + A \qquad (8.2j)$$

or

$$T_L = X - Y\left[\tan(\Delta)\right]$$

and

$$T_S = \frac{Y}{\sin(\Delta)} \qquad (8.2k)$$

where A is the distance from the projection point of the end PT point on the back tangent line T_L to the PI point, and it is computed to be positive when the deflection angle $\Delta < 90°$, and it is negative when $\Delta > 90°$. This is the reason for the minus sign in above equation.

It should be noted that if the sum $\Delta_1 + \frac{1}{2}\Delta_2$ or $\Delta_1 + \Delta_2 + \frac{1}{2}\Delta_3$. exceeds 90° in absolute value, the cosine of that angle sum is negative, thus holding the above equations true, as in the case of Figure 8.2c. In reviewing these four figures and the above equations, it is seen that there are nine components that contribute to the definition of the geometry of a three centered compound curve arrangement:

- The three curve radii (R1, R2, and R3);
- The overall and the individual deflection angles (Δ, $\Delta 1$, $\Delta 2$, and $\Delta 3$); and
- The back and forward tangent line lengths (TL and TS).

Since the angle Δ is controlled by the selection of the given back and forward tangent lines, this angle is fixed, and therefore it is considered to be a known quantity.

In constructing a three centered curve alignment, the procedure presented below considers the conditions of known information lists below and summarized in Table 8.2 by group and case within a group:

Group 1. The radii R1, R2, and R3 and two or three of the angles $\Delta 1$, $\Delta 2$, and $\Delta 3$ are known. If two angles are known, the third is computed by subtracting the sum of the two known angles from the Δ angle, thus constituting the **Cases A** through **D**.

Group 2. The radii R1, R2, and R3 are known, Δ_2 may, or may not be known, and Δ_1 and Δ_3 are not known. If Δ_2 is known (**Case A**) Δ_1 and Δ_3 are assumed to be equal to each other. If Δ_2 is not known (**Case B**) Δ_1, Δ_2, and Δ_3 are assumed to be equal to each other.

Group 3. The radii R1, R2, and R3 and X1 (**Case A**) or Y1 (**Case B**) are known, and none of the angles is known. The angle Δ_1 is calculated and Δ_3 is set to be equal to Δ_1 and Δ_2 is set to equal Δ minus the sum of the other two angles.

Group 4. The TL, R1, R2 and two or three of the angles $\Delta 1$, $\Delta 2$, and $\Delta 3$ are known. If two angles are known, the third is computed as stated for **Group 1** and corresponding to **Classes A** through **D**.

Group 5. The TL, R1, R2, and X1 (**Case A**) or Y1 (**Case B**) are known, and none of the angles $\Delta 1$, $\Delta 2$, and $\Delta 3$ is known. The angle $\Delta 1$ is calculated and Δ_3 is set to be equal to Δ_1 and Δ_2 is set to equal Δ minus the sum of the other two angles.

Similar conditions to the above five groups could also be generated if the direction of the two known tangent lines is reversed. Shown in Table 8.2 for each group and case are the input parameters that should have a nonzero value; all other input parameters should have a zero value. It is assumed that the calling program of this procedure has made the appropriate input value assignments.

Table 8.2 Combinations of Input Parameters

Group	Case	Input Parameters With Nonzero Value	Chapter Subsection
1	A	R1, R2, R3, Delt1, Delt2, Delt3	8.2.1
1	B	R1, R2, R3, Delt1, Delt2	8.2.1
1	C	R1, R2, R3, Delt1, Delt3	8.2.1
1	D	R1, R2, R3, Delt2, Delt3	8.2.1
2	A	R1, R2, R3, Delt2	8.2.2
2	B	R1, R2, R3	8.2.2
3	A	R1, R2, R3, X1	8.2.3
3	B	R1, R2, R3, Y1	8.2.3
4	A	TL, R1, R2, Delt1, Delt2, Delt3	8.2.4
4	B	TL, R1, R2, Delt1, Delt2	8.2.4
4	C	TL, R1, R2, Delt1, Delt3	8.2.4
4	D	TL, R1, R2, Delt2, Delt3	8.2.4
5	A	TL, R1, R2, X1	8.2.5
5	B	TL, R1, R2, Y1	8.2.5

```
Sub ic3cenCrv  (PN1, PE1, AZ1, PN2, PE2, AZ2,
                Delt1, Delt2, Delt3, R1, R2, R3, _
                TL, TS, X1, X3, Y1, Y3, _
                DELT, Delt1x, Delt2x, Delt3x, _
                R1x, R2x, R3x, T1, T2, T3, _
                PIN, PIE, PIN1, PIE1, PIN2, PIE2,
                PIN3, PIE3, _
                CCN1, CCE1, CCN2, CCE2, CCN3,
                CCE3, _
                PCN, PCE, PC1N, PC1E, PC2N, PC2E,
                PTN, PTE, IOK)
```

The input parameters to this procedure include the:

PN1, PE1	The north and east coordinates of the point on the back tangent line.
AZ1	The north azimuth (in radians) of the back tangent line pointing towards the PI of the overall compound curve.
PN2, PE2	The north and east coordinates of the point on the forward tangent line.
AZ2	The north azimuth (in radians) of the forward tangent line pointing away from the PI of the overall compound curve.

Regarding the following nine input parameters refer to the note below.

Delt1	The deflection angle (in radians) of the back curve, or zero.
Delt2	The deflection angle (in radians) of the center curve, or zero.
Delt3	The deflection angle (in radians) of the forward curve, or zero.
R1, R2, R3	The radii of the three curves, or zero.
TL	The back tangent length, or zero.
X1	The projection distance of the chord length of the back curve on the back tangent line, or zero (see Figures 8.2a through 8.2d).
Y1	The offset distance of the end point of the back curve from the back tangent line, or zero (see Figures 8.2a through 8.2d).

Although the given points that define the back and forward tangent lines of the overall compound curve arrangement may be located anywhere along their corresponding azimuths, the back azimuth should point towards the overall PI point, and the forward azimuth should point away from.

NOTES

1. All nonzero radii and delta angle parameter values are assumed to be positive. If not, they are first forced by the procedure to be positive, and thereafter they are assigned the sign of DELT. Also, if they are within a certain tolerance (see later on) of being equal to zero, they are set to be zero for computation purposes. However, their original input value is not altered, and their computed value is returned under a different parameter name as indicated below.

2. Any nonzero TL, X1, and Y1 parameter values are assumed to be positive. If not, they are forced by the procedure to be positive for computation purposes. Also, if they are within a certain tolerance (see later on) of being zero, they are forced to be zero for computation purposes. However, their original input value is not altered.

The parameters that are returned from this procedure include the:

DELT	The deflection angle (in radians) of the overall compound curve alignment.
Delt1x	The deflection angle (in radians) of the back curve as computed.
Delt2x	The deflection angle (in radians) of the center curve as computed.
Delt3x	The deflection angle (in radians) of the forward curve as computed.
R1x, R2x, R3x	The radii of the three curves as computed.
T1, T2, T3	The tangent lengths of the back, center and forward curves, respectively.
PIN, PIE	The north and east coordinates of the PI point of the overall alignment.
PIN1, PIE1	The north and east coordinates of the PI point of the back curve.
PIN2, PIE2	The north and east coordinates of the PI point of the center curve.
PIN3, PIE3	The north and east coordinates of the PI point of the forward curve.
CCN1, CCE1	The north and east coordinates of the center point of the back curve.
CCN2, CCE2	The north and east coordinates of the center point of the center curve.
CCN3, CCE3	The north and east coordinates of the center point of the forward curve.
PCN, PCE	The north and east coordinates of the PC point of the back curve.
PC1N, PC1E	The north and east coordinates of the PCC point between the back and center curves.
PC2N, PC2E	The north and east coordinates of the PCC point between the center and forward curves.
PTN, PTE	The north and east coordinates of the PT point of the forward curve.
IOK	An indicator, the value of which if it is:

- 0 Denotes that no problems have been encountered.
- 1 Denotes that AZ1 and AZ2 are colinear.
- 2 Denotes that each of the Delt1, Delt2, and Delt3 has a nonzero value and their sum is not equal to Δ.
- 3 Denotes that an improper input group has been specified.

Upon invocation, the procedure initializes the parameters to be returned, and then:

1. Checks the difference of the AZ1 and AZ2 parameters for being equal to 0°, or to 180° within a tolerance of 0.0000048 radians (0.99″), and if within the said tolerance, sets IOK to be 1, and the procedure terminates; else it continues.

2. Calls the icintrs procedure to intersect the back and forward tangent lines and determine the coordinates of the PI point (PIN, PIE) and the deflection angle DELT of the overall compound curve.

3. Checks the value of each of the Delt1, Delt2, and Delt3 input parameters for a zero value within a tolerance of 0.0000048 radians (0.99″), and if within the said tolerance, sets the value to be zero. Whether modified or not, these values are assigned to the Delt1x, Delt2x, and Delt3x parameters so that the original input angle values are preserved.

4. Checks the value of the Delt1x, Delt2x, and Delt3x parameters for a nonzero value, and if anyone has a nonzero value, the sign of the said value is set to be the same as the sign of the deflection angle DELT.

5. Checks the value of the Delt1, Delt2, and Delt3 parameters to determine whether all three have a nonzero value, or not. If:
   - All three parameters have a nonzero value, the procedure checks their sum. If the sum is not equal to the value of DELT, within a tolerance of 0.0000048 radians (0.99″), the indicator IOK is set to 2, and the procedure terminates.
   - Only one has a zero value, its value is computed by subtracting the other two angles from DELT.
   - All three parameters as modified, or not, by the procedure are assigned to new parameter names to be returned so that the input values are to be preserved.

6. Checks the value of the R1, R2, and R3 input parameters for being equal to zero within a tolerance of 0.005 feet (1.524 mm), and if so sets the value to be zero. If the input value is not zero, its sign is set to be that of the deflection angle DELT. All three parameters

as modified by the procedure, or not, are assigned to the `R1x, R2x, and R3x` parameters to be returned so that the input values are to be preserved.

7. Preserves the input value of the `TL, X1, and Y1` parameters as `TLx, X1x, and Y1x`, and then checks their value for being equal to zero value within a tolerance of 0.005 feet (1.524 mm), and if within the said tolerance, sets the value to be zero.

8. Queries the values of the `Delt1x, Delt2x, Delt3x, R1x, R2x, R3x, TL, X1,` and `Y1` parameters to determine which of the combinations of nonzero values are in accord with the combinations shown in Table 8.2, and accordingly assigns the appropriate group and class identification. The query proceeds from left to right and from top to bottom in the sequence indicated in the said table. The first match to be made defines the applicable case for solution. If a match:
   - Is not made, `IOK` is set to be 3 and the procedure terminates.
   - Is made, the procedure:
     - For the classes of which one or more of the `Delt1x, Delt2x,` and `Delt3x` parameters has a zero value computes the said value.
     - Branches to subsection of this chapter as indicated in Table 8.2 and described below. Thereafter the work returns to the main code and continues as indicated below.

9. Computes the individual curve tangent lengths `T1, T2,` and `T3` as being:

$$T1 = R1 * \tan\left(\frac{1}{2} Delt1 \right).$$
$$T2 = R2 * \tan\left(\frac{1}{2} Delt2 \right).$$
$$\text{and } T3 = R3 * \tan\left(\frac{1}{2} Delt3 \right).$$

10. Calls the `icpt1` procedure to compute the coordinates of the PC point (`PCN, PCE`) from the PI point by traversing from the overall PI point (`PIN, PIE`) along AZ1 a negative `TL` plus distance and a zero offset distance.

11. Calls the `icpt1` procedure to compute the coordinates of the PI1 point (`PIN1, PIE1`) from the PC (`PCN, PCE`) point by traversing along AZ1 a `T1` plus distance and a zero offset distance.

12. Computes the azimuth AZ3 of the common tangent line to the back and center curves at the PCC1 point by adding `Delt1` to AZ1, and then calls the `icpt1` procedure to compute the coordinates of the PC1 point (`PC1N, PC1E`) by traversing from the PI1 point (`PIN1, PIE1`) along AZ3 a `T1` plus distance and zero offset distance.

13. Calls the `icpt1` procedure to compute the coordinates of the PI2 point (`PIN2, PIE2`) by traversing from the

PC1 point (`PC1N, PC1E`) along AZ3 a `T2` plus distance and zero offset distance.

14. Computes the azimuth AZ4 of the common tangent line to the center and ahead curves at the PCC2 point by adding `Delt2` to AZ3, and then calls the `icpt1` procedure to compute the coordinates of the PC2 point (`PC2N, PC2E`) by traversing from the PI1 point (`PIN2, PIE2`) along AZ4 a `T2` and zero offset distance.

15. Calls the `icpt1` procedure to compute the coordinates of the PI3 point (`PIN3, PIE3`) by traversing from the PC2 point (`PC2N, PC2E`) along AZ4 a `T3` plus distance and zero offset distance.

16. Calls the `icpt1` procedure to compute the coordinates of the PT point (`PTN, PTE`) by traversing from the PI point (`PIN, PIE`) along AZ2 a `TS` plus distance and zero offset distance.

17. Calls the `icpt1` procedure to compute the coordinates of the center points of the three curves (`CCN1, CCE1`), (`CCN2, CCE2`), and (`CCN3, CCE3`) by traversing from the respective curve start points (`PCN, PCE`), (`PC1N, PC1E`), and (`PC2N, PC2E`) along AZ1. AZ3 and AZ4 a zero plus distance and an `R1x, R2x,` and `R3x` offset distance, respectively.

Having computed the above parameters, the procedure terminates. The computation of the zero input parameters are determined as indicated in the subsections below.

8.2.1 Group 1—Given All Radii and Two or Three Central Angles

This portion of the `ic3cenCrv` procedure assumes that in addition to the back and forward tangent lines, the value of each of the three radii (`R1, R2, R3`) is also given. It is also assumed that the value of at least two of the central angles (`Delt1, Delt2,` and `Delt3`) of the three curves is known as indicated in Figure 8.2e. If only two of the central angles are known, the third is computed by subtracting the sum of the two known central angles from the overall deflection angle Δ.

When this group is encountered, all angles have been computed prior to arriving to this point of the procedure. Thus, at this time the procedure:

8a. Computes the value of `X1, X2, X3, Y1, Y2, Y3, X,` and `Y` with Equations 8.2b through 8.2i, respectively.

8b. Returns to **Step 9** of the code of the main procedure `ic3cenCrv`.

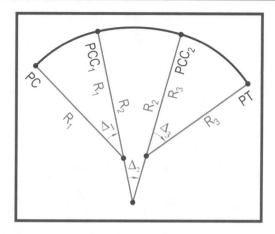

Figure 8.2e Given Three Radii and Three Angles

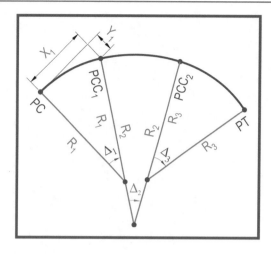

Figure 8.2g Given Three Radii and Either X1 or Y1

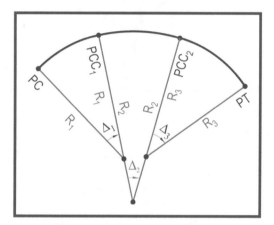

Figure 8.2f Given Three Radii and Center Curve Angle

8.2.2 Group 2—Given All Radii and the Angle of the Center Curve

This portion of the `ic3cenCrv` procedure assumes that in addition to the back and forward tangent lines, the value of each of the three radii (`R1`, `R2`, `R3`) is also given. It is also assumed that the value of the central angle of the center curve (`Delt2`) is known as indicated in Figure 8.2f, and that the central angles of the back and forward curves are equal to each other, with each one being equal to one half the difference of the overall deflection angle Δ minus the value of the known central angle of the center curve.

When this group is encountered, all angles have been computed prior to arriving to this point of the procedure. Thus, this part of the proceed becomes the same as that of Group 1, and it is handles by the procedure together with Group 1.

8.2.3 Group 3—Given All Radii and X1 or Y1

This portion of the `ic3cenCrv` procedure assumes that in addition to the back and forward tangent lines, the value of each of the three radii (`R1`, `R2`, `R3`) is also given. It is also assumed that the value of either the plus distance X1 or the offset distance Y1 (Class B) of the back curve along the back tangent line is also known as indicated in Figure 8.2g. It should be noted that this assumes that the central angle of the first curve Δ_1 is less than 90° in absolute value. Reference is made to the next section of this chapter regarding the magnitude of these two parameters.

With these parameters known, this part of the procedure:

8a. Determines whether `X1` or `Y1` has been specified. If:
 • `X1` has been specified (Class A), the procedure computes the value of:
  – The central angle of the back curve as

$$Delt_1 = \sin^{-1}\left(X_1 / R_1\right) \qquad (8.2l1)$$

  and then
  – The value of `Y1` by use of Equation 8.2e.
 • `Y1` has been specified (Class B), the procedure assigns to `Y1` the sign of `DELT` and then computes the value of
  – The central angle of the back curve as

$$Delt1 = \cos^{-1}\left(\frac{|R_1| - Y_1}{|R_1|}\left(Test1\right)\right) \qquad (8.2l2)$$

  – where *Test1* is −1 if `DELT` is negative and +1 if `DELT` is positive, and then
  – The value of `X1` by use of Equation 8.2b.

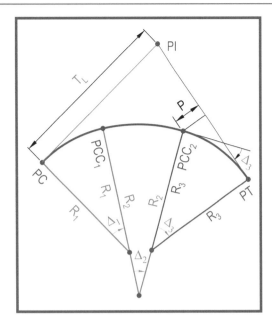

Figure 8.2h Given TL, R1, R2, and Two or Three Angles

8b. Sets the value of `Delt3` to be equal to that of `Delt1`, and computes the value of `Delt2` by subtracting the sum of `Delt1` and `Delt3` from `DELT`, computes the value of `X2`, `X3`, `Y2`, `Y3`, `X`, and `Y` with Equations 8.2c, 8.2d, 8.2f, 8.2g, 8.2h, and 8.2i, and then returns to **Step 9** of the code of the main procedure `ic3cenCrv`.

8.2.4   Group 4—Given TL, R1, R2, and Two or Three Central Angles

This portion of the `ic3cenCrv` procedure assumes that the PC point of the compound curves is to be fixed at a distance `TL` from the point of intersection (PI) of the two tangent lines. It is also assumed that at least two central angles are also known as shown in Figure 8.2h. When this group is encountered, all angles have been computed prior to arriving to this point of the procedure. Thus, at this time the procedure:

8a. Computes the value of `X1`, `X2`, `Y1`, and `Y2` with Equations 8.2b, 8.2c, 8.2e, and 8.2f, respectively.

8b. Calls the `icptl` procedure to compute the coordinates of the PC (`PCN`, `PCE`) by traversing from the PI (`PIN`, `PIE`) along `AZ1` a negative distance of `TL` and an offset distance of zero.

8c. Calls the `icptl` procedure to compute the coordinates of the PCC₁ point (`PC1N`, `PC1E`) by traversing from the said PC (`PCN`, `PCE`) along `AZ1` a plus distance of `X1` and an offset distance of `Y1` for a clockwise deflection angle

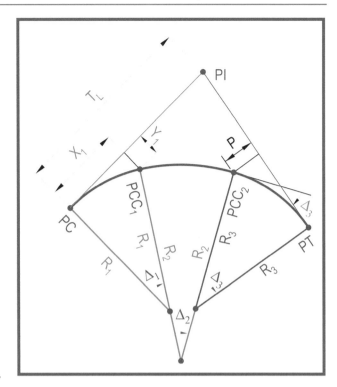

Figure 8.2i Given TL, R1, R2, and X1 or Y1

`DELT`, or of minus `Y1` for a counterclockwise deflection angle `DELT`.

8d. Calls the `icptl` procedure to compute the coordinates of the PCC₂ point (`PC2N`, `PC2E`) by traversing from the said PCC₁ along `AZ1` a plus distance of `X2`, and an offset distance of:
  • `Y2`, for a clockwise deflection angle `DELT`, or
  • Minus `Y2`, for a counterclockwise deflection angle `DELT`.

8e. Calls the `icprjct` procedure to project the PCC₂ point (`PC2N`, `PC2E`) on the second given tangent line (`AZ2`) to find the offset distance `P` (see Figure 8.2h).

8f. Computes the tangent length (`T3`) and radius (`R3`) of the exiting curve as follows:

$$T3 = P \, / \sin\left(Delt3\right)$$
$$R3 = T3 \, / \tan\left(\tfrac{1}{2} \, Delt3 \right)$$

8g. Computes the values of the `X3`, `Y3`, `X`, and `Y` with Equations 8.2d, 8.2g, 8.2h, and 8.2i, respectively.

8h. Calls the `icptl` procedure to compute the coordinates of the PT (`PTN`, `PTE`) by traversing from the said PCC₂ along `AZ1` a distance of `X3`, and an offset distance of:
  • `Y3`, for a clockwise deflection angle `DELT`, or
  • Minus `Y3`, for a counterclockwise deflection angle `DELT`.

8i. Returns to **Step 9** of the code of the main procedure `ic3cenCrv`.

8.2.5   Group 5—Given TL, R1, R2, and X1 or Y1

This portion of the `ic3cenCrv` procedure assumes that the PC point of the compound curves is to be fixed at a distance TL from the point of intersection (PI) of the two given tangent lines. It is also assumed that either X1 or Y1 is also known as indicated in Figure 8.2i. With these parameters known, this part of the procedure:

8a. Finds which of X1 or Y1 has been specified, and computes the values of Delt1x and Y1 or X1 in the same manner as that of **Step 8a** of Group 3 in Sect. 8.2.3 above.
8b. Sets the value of Delt3 to be equal to that of Delt1, and computes the value of Delt2 by subtracting the sum of Delt1 and Delt3 from DELT.
8c. Computes the value of X2 and Y2, with Equations 8.2c and 8.2f, respectively.
8d. Carries out the **Steps 8b** through 8i of Sect. 8.2.4 described above.

8.3   Multi Centric Compound Curves

If we are to review Equations 8.2a through 8.2i, and the approach of the solution of the three centered compound curve of the preceding section, it should be evident that the individual curve X and Y increments are dependent on the radii and central angles of the individual curves. Hence, Equations 8.2d and 8.2g could be generalized as

$$X_i = 2R_1 \sin\left(\frac{1}{2}\Delta_i\right)\cos\left(\sum_{i=2}^{i=n-1}\Delta_i + \frac{1}{2}\Delta_i\right) \quad (8.3a)$$

and Equations 8.2f and 8.2g could be generalized as

$$Y_i = 2R_1 \sin\left(\frac{1}{2}\Delta_i\right)\sin\left(\sum_{i=2}^{i=n-1}\Delta_i + \frac{1}{2}\Delta_i\right) \quad (8.3b)$$

Correspondingly, Equations 8.2h and 8.2i could be generalized as

$$X = \sum_{i=1}^{i=n-1} X_i \quad (8.3c)$$

and

$$Y = \sum_{i=1}^{i=n-1} Y_i \quad (8.3d)$$

where "i" denotes the sequential number of a curve, and "n" denotes the total number of curves. Similarly

$$\Delta = \sum_{i=1}^{i=n-1} \Delta_i \quad (8.3e)$$

The reader is alerted to the fact that:

- The summation of the X and Y increment terms in Equations 8.3a and 8.3b is from i = 2 to i = n − 1.
- The Equations 8.2a and 8.2d that determine the X and Y increments of the first curve hold true for this application.

Extending the above generalization one step further, if we are to apply the convention that clockwise curve rotations are

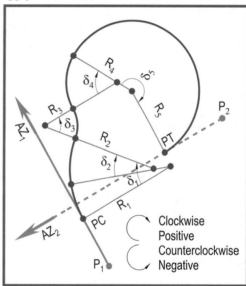

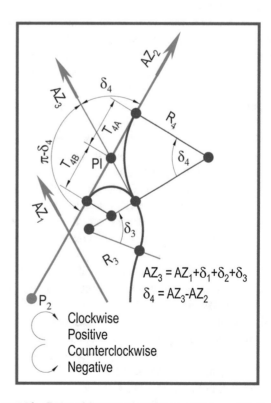

Figure 8.3a   Series of Compound and Reversed Curves

$$AZ_3 = AZ_1 + \delta_1 + \delta_2 + \delta_3$$
$$\delta_4 = AZ_3 - AZ_2$$

Figure 8.3b   Series of Compound and Reversed Curves—Floating PC at AZ2 Overall Tangent Deflection Angle Less than 180°

positive, and counterclockwise curve rotations are negative, we can apply the Equations 8.3a through 8.3d to a series of compound and reversed curves.

As the number of curves to be introduced between the first and the last curve increases, the number of the necessary parameters for solving the geometry of the curved alignment, as well as the number of design controls (points through which the alignment must pass or avoid) that the design alignment must meet increase many fold. Thus, one may justifiably question the need for a procedure to handle a multitude of curves since a variety of procedures has been presented in this publication to create a series of compound and reversed curves that meet various control conditions. However, for a relatively small number of curves as may be encountered in the design of a ramp in a cloverleaf, or similar type of highway interchange, the `icMcenCrv` procedure is presented below subject to the following conditions:

1. Any number of curves (N) can be introduced to generate an alignment of compound and reversed curves in series. Thus, in an alignment there could be a mixture of clockwise and counterclockwise curves.
2. Each curve is to be defined, except as qualified below, by its radius (R) and central angle (δ). A positive central angle denotes a clockwise rotation, and a negative central angle denotes a counterclockwise rotation.
3. The alignment may be such that the overall deflection angle (Δ) formed between the back and the forward tangent lines may exceed 180°. Since, in addition, there can be both clockwise and counterclockwise curves in an alignment, the procedure for intersecting the said two tangent lines cannot be called upon to determine the overall Δ angle. If this flexibility were to be removed, that is if an alignment is to contain only clockwise or only counterclockwise curves, then that alters the case.
4. An alignment composed of a series of compound and reversed curves may be generated in one of the ways presented below. The alignment may:
   (a) Start at a fixed point (PC) along the back tangent line and have its end point (PT) float in the two dimensional space. In this arrangement the values of all δ angles and radii must be known except as otherwise noted below. In Figure 8.3a the point P_2 and the forward tangent azimuth AZ_2 are of no concern. They are unknown. In this arrangement the value of any δ angle may exceed 180°.
   (b) Start at the said PC point and have its PT point float along the forward tangent line of which the direction (north azimuth), and not its position, is known. In this arrangement the values of the:
   • Radii of all curves must be known.
   • Central angles of all curves, except that of the last curve, must be known. The value of the angle δ of the last curve is computed by the procedure by adding all given δ angles to the azimuth of the back

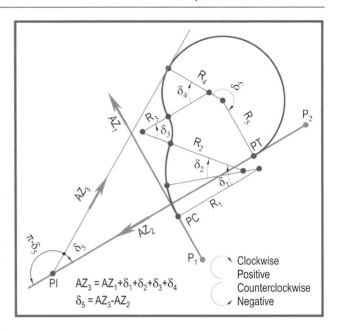

Figure 8.3c Series of Compound and Reversed Curves—Floating PC at AZ_2 Overall Tangent Deflection Angle Greater than 180°

tangent line (AZ_1) and subtracting the sum from the azimuth of the forward tangent line (AZ_2).

In Figure 8.3a the point P_2 is of no concern. It is unknown. However the azimuth AZ_2 of the forward tangent line is known in value only. Its position is dependent upon the values of the radii and δ angles of the curves. That is in the said figure its position and orientation could change from what is portrayed to be. In this arrangement the value of any given δ angle may exceed 180°.

Figure 8.3b portrays a blow up of the end portion for an alignment that terminates at the end of the fourth curve of Figure 8.3a, and in which the two tangent lines form a deflection angle of less than 180°, while Figure 8.3c portrays the overall alignment of Figure 8.3a when the said deflection angle exceeds 180°.

Regarding Figure 8.3b and Figure 8.3c, it should be noted that there are two potential solutions of the last curve, a clockwise and a counterclockwise curve. Of these two solutions, the procedure assumes one. If the other one is desired, the user will need to introduce it as an afterwards operation.

   (c) Start at the said PC point and have its PT point float along the forward tangent of which the direction (north azimuth) and position is known. That is in Figures 8.3a through 8.3c the coordinates of the PT (P_2) point and the azimuth of the forward tangent line (AZ_2) are known. In this arrangement the values of the radii and of the δ angles of all curves, except those of the last curve, must be known.

The angle δ of the last curve is computed as indicated in paragraph 4b above by adding all given δ

angles to the azimuth of the back tangent line (AZ_1) and subtracting the sum from the azimuth of the forward tangent line (AZ_2).

Having found the central angle of the last curve (δ_4 in Figure 8.3b), the procedure:

- Computes the coordinates of the end point of the second from end curve (the start point of the last curve).
- Intersects the common tangent line (AZ_3 in Figure 8.3b) to the last two curves (its azimuth was determined in the solution of the central angle of the last curve) with the forward tangent line to determine the length of the tangent of the last curve (T_{4A} in Figure 8.3b).
- Computes the value of the radius of the last curve (R_4) as being

$$R_4 = T_{4A} / \tan\left(\% \delta_4\right) \qquad (8.3f)$$

As it was stated above regarding Figure 8.3b there are two potential solutions for the last curve, one determined by the angle δ_4, and the other as determined by the angle $\pi - \delta_4$. The procedure disregards the latter solution.

It is also noted that whereas in the previous two arrangements the value of any δ angle may exceed 180°, in this arrangement, any δ angle but not that of the last curve may exceed 180°.

(d) Be fitted between the back (entering) and forward (exiting) tangent lines, both of which are fixed (given) with the start (PC) and end (PT) points being free to float along their respective tangent lines. In this arrangement the two tangent lines cannot be parallel, for if they are, the start point must be fixed in which case we have one of the preceding arrangements. Thus, if an alignment is to be fitted between two given tangent lines, then:

- If all central angles are specified, their sum when added to the azimuth of the back tangent must equal to the azimuth of the forward tangent.
- If the central angle of the last curve is unknown, it is computed as stated above.
- The radii of all curves must be known.
  To generate the alignment, the procedure:
- Assumes a start point (PC) located 100 feet (meters) back of the PI point of the overall alignment and along the back tangent line (see Figures 8.3b and 8.3c);
- Computes the coordinates of all alignment points to be returned;
- Intersects an imaginary line passing through the end point of the alignment as computed above and having the azimuth AZ_1 of the back tangent line with the forward tangent line to determine the

distance of the said end point from the forward tangent line; and then

- Translates the above computed coordinates along the back tangent line with the last said distance.

5. If the start tangent point (PC) of the series of curves is to be fixed at the back tangent line, and the other end is to float, then the coordinates of the point that defines the back tangent line (PT_1) must be those of the PC point. If the other tangent point (PT) is to be fixed, then the progression of the curves should be reversed. Both tangent points cannot be fixed.

6. At times it may be desirable to have the first (back) curve pass through a specific point the coordinates of which are known. Such condition has been presented in the preceding section of this book in Group 3 and Group 4 of the three centered curve alignment. For those cases the central angle δ of the first curve could be determined by the X distance (the projection of the first curve on the back tangent) or the Y distance (the offset distance of the end of the first curve from the back tangent) using Equation 8.211 and Equation 8.212, respectively. In these cases, the said δ angle is restricted to being in the first Cartesian quadrant, that is in the range $0 < \delta \leq 90°$.

In addition to specifying either the X or the Y distances, the subject procedure of this section provides for the specification of both the X and Y distances to define the end point of the first curve. Figure 8.3d displays a clockwise and a counterclockwise curve that start at the PC point

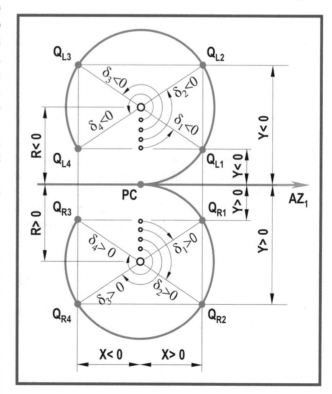

Figure 8.3d   First Curve Through a Fixed Point

and progress along the indicated back azimuth AZ_1. In this figure note the following:

- On each curve there are four points Q_{R1} through Q_{R4} on the clockwise curve and four points Q_{L1} through Q_{L4} on the counterclockwise curve denoting the end point of the first curve.
- The X and Y distances and central of each said point are identified with their positive or negative values denoting a clockwise or counterclockwise rotation.
- The sign and magnitude of each angle is dependent on the magnitude of the specified X and Y distances, and whether they are positive or negative as indicated in the said figure.
- Of the said eight points, points Q_{R3} and Q_{R4} and Q_{L3} and Q_{L4} are of rather improbable application in the design of an alignment and are presented solely from an academic point of view.

If only the X or the Y distance is to be specified:

- The radius R of the first curve is assumed to be known.
- The central angle δ is restricted to the range $0 < |\delta| \leq 90°$, and it is computed by Equation 8.2l1 or Equation 8.2l2, respectively.

If both, the X and the Y distances are specified, then:

- The radius R of the first curve is computed as being

$$R^2 = X^2 + (R - Y)^2$$
$$R^2 = X^2 + Y^2 - 2RY \qquad (8.3g)$$

or

$$R = \frac{X^2 + T^2}{2Y} \qquad (8.3h)$$

Note that the sign of R would be the same as that of Y. Also note that Equation 8.3h is the same as Equation 7.2.4a of the `iccvta2q` procedure in Chap. 7.

- The central angle δ is restricted to the range $0 < |\delta| \leq 360°$, and it is computed by first computing the angle δ' of the circle of radius R as being

$$\delta' = \sin^{-1}\left(\frac{X}{R}\right) \qquad (8.3i)$$

The value of the central angle δ is dependent upon the values of R, Y and X, and the value of K which is defined as

$$K = \frac{R}{|R|} \qquad (8.3j)$$

The value of δ is then determined as indicated below and summarized in Table 8.3 for the eight points of Figure 8.3d.

Table 8.3 Determination of angle δ for the conditions of Figure 8.3d

Point	X	Y	R		K	δ'	δ
Q_{R1}	$X>0$	$Y>0$	$R>0$	$|Y| \leq |R|$	1	$\delta'>0$	$\delta=\delta'$
Q_{R2}	$X>0$	$Y>0$	$R>0$	$|Y| > |R|$	1	$\delta'>0$	$\delta=K\pi-\delta'$
Q_{R3}	$X<0$	$Y>0$	$R>0$	$|Y| > |R|$	1	$\delta'<0$	$\delta=K\pi-\delta'$
Q_{R4}	$X<0$	$Y>0$	$R>0$	$|Y| \leq |R|$	1	$\delta'<0$	$\delta=2K\pi+\delta'$
Q_{L1}	$X>0$	$Y<0$	$R<0$	$|Y| \leq |R|$	-1	$\delta'<0$	$\delta=\delta'$
Q_{L2}	$X>0$	$Y<0$	$R<0$	$|Y| > |R|$	-1	$\delta'<0$	$\delta=K\pi-\delta'$
Q_{L3}	$X<0$	$Y<0$	$R<0$	$|Y| > |R|$	-1	$\delta'>0$	$\delta=K\pi-\delta'$
Q_{L4}	$X<0$	$Y<0$	$R<0$	$|Y| \leq |R|$	-1	$\delta'>0$	$\delta=2K\pi+\delta'$

- If $|Y| \leq |R|$ and $X > 0$　then $\delta = \delta'$
- If $|Y| > |R|$ and $X > 0$　　　$\delta = K\pi - \delta'$
- If $|Y| > |R|$ and $X < 0$　　　$\delta = K\pi - \delta'$
- If $|Y| \leq |R|$ and $X < 0$　　　$\delta = K\pi + \delta'$

The specification of the X and/or Y distances implies that the first curve is to terminate at the point defined by the said distances. If the first curve is to extend beyond such point, then the curve should be broken into two curves of the same radius, and with the first terminating at the said fixed point and the second curve beginning therefrom.

For any of the cases in Table 8.4, if any of the parameters that are supposed to be unknown (have a zero value) they have instead a nonzero value, that value is set by the program to be zero. However, the original input data is not altered.

The `icMcenCrv` procedure has been designed to handle the cases that are presented below and summarized in Table 8.4. Additional cases can be introduced in this procedure as the need may arise from time to time. In this table the symbol X denotes a known parameter with a nonzero value, and its absence denotes an unknown parameter with a zero value.

Case 1　The series of curves is to start at a fixed point (PC) on the back tangent line, and the end point (PT) is to float in the two dimensional space. There are four sub-cases to this case, 1A, 1B, 1C and 1D depending on the use of parameters X and/or Y.

Case 2　The series of curves is to start at a fixed point (PC) as in Case 1. However, the end point (PT) of the last curve is to float along the specified azimuth for the forward tangent line of the alignment. Thus, the central angle of the last curve must be unknown and it is determined by the procedure by adding all given central angles to the azimuth of the back tangent line (AZ_1) and then subtracting the sum from the azimuth of the forward tangent line (AZ_2). There are four sub-cases to this case, 2A, 2B, 2C and 2D depending on the use of parameters X and/or Y.

Table 8.4 Multicentered Compound Curve Cases

		Conditions												
Given	Known	1A	1B	1C	1D	2A	2B	2C	2D	3A	3B	3C	3D	4
	PT_1	X	X	X	X	X	X	X	X	X	X	X	X	X
	$PC=PT_1$	X	X	X	X	X	X	X	X	X	X	X	X	
	AZ_1	X	X	X	X	X	X	X	X	X	X	X	X	X
	PT_2									X	X	X	X	X
	AZ_2					X	X	X	X	X	X	X	X	X
	R_{ALL}	X	X	X		X	X	X						X
	R_1-R_{N-1}									X	X	X		
	R_2-R_N					X			X					
	R_2-R_{N-1}												X	
	δ_{ALL}	X												
	$\delta_1-\delta_{N-1}$					X				X				X
	$\delta_2-\delta_N$		X	X	X									
	$\delta_2-\delta_{N-1}$						X	X	X		X	X	X	
	X		X		X		X		X		X		X	
	Y			X	X			X	X			X	X	

Case 3 This case is similar to Case 2 above with the exception that (a) the coordinates of the point that positions the forward tangent line must be known, and (b) the radius and the central angle of the last curve are unknown and computed by the procedure. The radius is computed by Equation 8.3f as discussed previously under condition 4(c), and the central angle is computed as in Case 2 above. There are four sub-cases to this case, 3A, 3B, 3C, and 3D depending on the use of parameters X and/or Y.

Case 4 The series of curves is to be fitted between the back and forward tangent lines with the PC and PT points of the alignment floating along their respective tangent lines. In this case it is assumed that the central angle of the last curve is considered to be unknown, and computed by the procedure.

In each of the sub-cases of Cases 1, 2, and 3 the two tangent lines may be parallel and they may have the same or opposite directions. However, in Case 4 the two tangent lines may not be parallel since in such conditions there could be an infinite number of solutions.

The input parameters of the radii and of the central angles are introduced in dynamic arrays of equal size. The size of each dynamic array is defined by the number of specified radii, including known and unknown radii. A nonzero value denotes a known parameter, while a zero value denotes an unknown parameter. For any of the said cases, if any of the parameters that are supposed to be unknown have instead a nonzero value, that value is set by the program to be zero. However, the original input data are not altered.

The parameters to be returned include the coordinates of the point of intersection of the two given tangent lines (the overall PI point of the alignment), the deflection angle of the said two lines (the overall Δ angle of the alignment), the north azimuth of the forward tangent line as it may have been computed by the procedure (the azimuth of the back tangent line is either an input value, or computed from the X and/or Y components of the first curve), and a dynamic array called `TheData` in which the radius, central angle, and the north and east coordinates of the PI point, the center point, and the start and end points of each curve are saved sequentially as indicated in Table 4.5. For example, if there are five curves, the above ten parameters are saved in the following positions:

- `Rx` in positions 1, 11, 21, 31, and 41,   Radii of the curves
- `Deltx` in positions 2, 12, 22, 32, and 42,   Central angles of the curves
- `PINx` in positions 3, 13, 23, 33, and 43,   North coordinates of the curve PIs
- `PIEx` in positions 4, 14, 24, 33, and 44,   East coordinates of the curve PIs
- `CCN` in positions 5, 15, 25, 35, and 45,   North coordinates of the curve center points
- `CCE` in positions 6, 16, 26, 36, and 46,   East coordinates of the curve center points
- `PCN` in positions 7, 17, 27, 37, and 46,   North coordinates of the curve start points
- `PCE` in positions 8, 18, 28, 38, and 48,   East coordinates of the curve start points
- `PTN` in positions 9, 19, 29, 39, and 49,   North coordinates of the curve end points
- `PTE` in positions 10, 20, 30, 40, and 50,   East coordinates of the curve end points

The reader should note that whereas in the preceding procedures that involved a number of different operational conditions a procedure identified the operational condition based on the values of various input parameters, in the `icMcenCrv` procedure the calling program should specify the desired

operational condition. Based on this specification the procedure performs certain input data verifications to ascertain the
validity of the input information.

```
Sub icMcenCrv (PTN1,  PTE1, AZ1, PTN2,  PTE2, AZ2,
              Ncrv, RN, DeltN, X, Y, TheCase, _
              PIN, PIE, DELTA, AZF, TheData, IOK)
```

The input parameters to this procedure include the:

PTN1, PTE1   The north and east coordinates of the point that
             defines the back tangent line. For Cases 1A, 2A,
             and 3A this point is assumed to be the PC of the
             first curve.
AZ1   The north azimuth in radians of the back tangent
             line.
PTN2, PTE2   The north and east coordinates of the point that
             defines the forward tangent line. These parameters
             are disregarded for Cases 1A, 1B, 1C, 1D, 2A,
             2B, 2C and 2D these parameters are disregarded.
AZ2   The north azimuth in radians of the forward
             tangent line.
Ncrv   The number of curves to be generated. Since the
             calling program of this procedure is to prepare the
             input data, this number should be known to it.
RN   The dynamic array that contains the radii of the
             curves to be generated. For Cases 3A, 3B, 3C, and
             3D this parameter is disregarded. For all other
             cases a radius may be positive or negative. The
             procedure assigns to the absolute value of each
             radius the sign of its corresponding central angle.
DeltN   The dynamic array that contains in radians the
             central angles of the curves to be generated. Refer
             to Table 8.3 for the required angles for each case.
             Note that for Case 1B, 2B or 3B in which the X
             parameter is specified the central angle of the first
             curve must have a nonzero positive or negative
             value to denote whether the first curve is to be
             clockwise or counterclockwise. Thus, a 1.0 or
             −1.0 will suffice.
X   The value of the X parameter of the first curve
             (the projection of the first curve on the back
             tangent as shown in Figure 8.3d) if the first curve
             is to end at a fixed point. Refer to Table 8.3 for the
             cases that require this parameter. Note that for:
             • Case 1B, 2B or 3B this parameter is always
               assumed to be positive.
             • Case 1D, 2D or 3D this parameter could be
               positive or negative.
Y   The value of the Y parameter of the first curve
             (the offset distance of the end point of the first
             curve from the back tangent as shown in
             Figures 8.2a through 8.2d) if the first curve is to
             end at a fixed point. Refer to Table 8.3 for the
             cases that require this parameter. Note that a
             positive value produces a clockwise curve, while
             a negative value produces a counterclockwise
             curve.
TheCase   The identifier 1A, 1B, 1C, 1D, 2A, 2B,
             2C, 2D, 3A, 3B, 3C, 3D, or 4 denoting
             the desired operational condition as per Table 8.3.

The parameters that are returned from this procedure
include the:

PIN, PIE   The north and east coordinates of point of
             intersection of the two given lines. Note that it is
             possible for these two lines to be parallel. Thus, in
             such a case the said parameters are each set to the
             value of 99999.0.
DELTA   The deflection angle (in radians) formed by the two
             given tangent lines. Note that for Cases 2A, 2B, 2C,
             2D, 3A, 3B, 3C, and 3D it is possible for these two
             lines to be specified as being parallel. Thus, in such
             a case this parameter is set to be zero, if the said
             lines have the same direction, or to 180° if the said
             lines are opposite in direction. For Cases 1A, 1B and
             1C it possible for these lines to become (be
             computed) parallel. The value of DELTA is
             computed as the sum of all central angles as given or
             as being computed and not as the difference between
             AZ1 and AZ2.
AZF   The north azimuth of the forward tangent line as it
             may have been computed by the procedure.
TheData   The dynamic array that contains the radius, central
             angle, and the north and east coordinates of the PI,
             center, start and end points of each curve as
             described previously and in Table 8.5.
IOK   An indicator, the value of which denotes the following:
             0   No problems have been encountered.
             1   The number of radii in the RN array does not
                 equal Ncrv.
             2   The number of angles in the DeltN array does
                 not equal Ncrv.
             3   The number of radii and central angle in the RN
                 and DeltN arrays, respectively, does not equal
                 Ncrv.
             4   A curve has been assigned a zero radius.
             5   A curve has been assigned a zero central angle.
             6   The X and/or Y have been assigned a zero value.
             7   Case 3 or 4 has been encountered and the two
                 tangent lines are parallel.
             8   Case 3 has been encountered and the sum of all
                 input angles does not equal the overall deflection
                 angle DELTA.
             9   Improper input data has been encountered other
                 than the above identified six error conditions.

Upon invocation, the procedure initializes the parameters
to be returned, and then:

1. Checks if the specified TheCase is valid. If not, IOK is set
   to be 9 and the procedure aborts; else makes certain it is
   in capital letters and continues.
2. Checks if the number of contents of the RN and DeltN
   arrays equal to Ncrv, and if
   • They are each equal, the contents of the RN and DeltN
     arrays are preserved in the RNx and DeltNx arrays,
     respectively, and the program continues.
   • Either or both are unequal, IOK is set to be 1, 2, or 3,
     respectively, and the procedure aborts.

3. Determines whether the back and forward tangent lines are parallel, or not, and if they are sets the PI coordinates (PIN, PIE) to be 99999.0 each, the overall deflection angle (DELTA) to be 0 or π, and the exiting azimuth to be AZ1 or AZ1 + π.

4. Proceeds to check the input parameters for conformance with the requirements of the conditions case specified by the TheCase input parameter.
   - Check of the radii (RN array): For
     - Case 1A, 1B, 1C, 2A, 2B, 2C and 4 all radii must be nonzero.
     - Cases 3A, 3B, and 3C all radii but that of the last curve must be nonzero.
     - Cases 1D and 2D all radii but that of the first curve must be nonzero.
     - Case 3D all radii but that of the first and last curve must be nonzero.

Failure to meet the above requirements sets IOK to 4, and aborts the procedure.
   - Check for central angles (DeltN array): For
     - Case 1A all curves must have a nonzero angle value.
     - Cases 2A, 3A and 4 all angles but that of the last curve must be nonzero.
     - Cases 1B, 1C and 1D all angles but that of the first curve must be nonzero.
     - Cases 2B, 2C, 2D, 3B, 3C and 3D all angles but that of the first and last curves must be nonzero.

Failure to meet the above requirements sets IOK to 5, and aborts the procedure.
   - Check for the X and Y distances: For
     - Cases 1B, 2B and 3B X must be nonzero. If so, the central angle of the first curve is computed by Equation 8.2l1, and then if X is negative the value of π multiplied by the signum of X is added to the computed angle.
     - Cases 1C, 2C, and 3C Y must have a nonzero value. If so, the central angle of the first curve is computed by Equation 8.2l2, and then if the absolute value of Y is greater than the absolute value of the radius the value of 2π multiplied by the signum of Y is added to the computed angle.
     - Cases 1D, 2D and 3D X and Y must both have a nonzero value. If so, the central angle of the first curve is computed as described earlier in this section under condition 6.

Failure to meet the above requirements sets IOK to 6, and aborts the procedure.
   - Check for parallel back and forward tangent lines as may have been determined in Step 3 above: For
     - Case 4 if the said two lines are parallel (PIN = 99999 and PIE = 99999), IOK is set to 7, and the procedure aborts.
     - All cases if the said two lines are not parallel (PIN <> 99999 and PIE <> 99999):

- Calls the icintrs procedure to intersect the said two lines and compute the coordinates of the overall PI point and the overall deflection angle.
- Sets to zero any radius within a tolerance of 0.005 feet (1.524 mm), and any central angle within a tolerance of 0.0000048 radians (0.99″).
- Computes the sum of all central angles.
- For Cases 2A, 2B, 2C, 2D, 3A, 3B, 3C, 3D, and 4 computes the central angle of the last curve by subtracting from the forward azimuth the back azimuth and the sum of the angles (the last angle is zero), and accounting for the two azimuths straddling north.
- Assigns to each radius the sign of the corresponding central angle.

5. Sets the coordinates of the PC point of the alignment as follows:
   - For Case 4 assumes that the coordinates of the PC point is to be located 100 feet or meters back of the above computed overall PI point. Later on the alignment is to be translated to its proper position.
   - For all other cases the coordinates of the PC point are set to be those of the point (PTN1, PTE1) that defines the back tangent line.

6. Computes the values of the remaining unknown parameters as follows:
   - Initializes a loop setting its counter j = 1, the temporary back azimuth azB = AZ1, and the temporary PC point coordinates PCNx = PCN and PCEx = PCE.
   - Within this DO loop the procedure:
     - Introduces the value of the radius of each curve in the TheData array as indicated in Table 8.5 and below:
     - For all but the last curve uses the radius of the input array RN.
     - For the last curve and for Cases 1A, 1B, 1C, 1D and 4 uses the radius of the input array RN.
     - For the last curve and for Cases 2A, 2B, 2C, 2D, 3A, 3B, 3C, and 3D computes the radius by calling the icintrs procedure to intersect the last computed back azimuth with the input value of the forward tangent line azimuth AZF to determine the tangent T length of the last curve. Having found T, the radius of the last curve is then computed with Equation 8.3f using the central angle of the last curve as it has been previously computed. The reader is reminded that for these cases there could be two solutions, one for the delta angle as computed, and the other for its 180° supplement which is not being considered by this procedure.
     - Introduces the value of the central angle of each curve, as it has been specified in the input array DeltN or as it may have been computed as indicated previously, in the TheData array as indicated in Table 8.5.

Table 8.5 Structure of the `TheData` Dynamic Array

Rx	Deltx	PINx	PIEx	CCN	CCE	PCN	PCE	PTN	PTE	Rx	Deltx	PINx	PIEx	CCN	CCE	PCN	PCE	PTN	PTE
CURVE #1										CURVE #2									
1	2	3	4	5	6	7	8	9	10	11	12	13	14	15	16	17	18	19	20
CURVE #3										CURVE #4									
21	22	23	24	25	26	27	28	29	30	31	32	33	34	35	36	37	38	39	40
CURVE #5										CURVE #6									
41	42	43	44	45	46	47	48	49	50	51	52	53	54	55	56	57	58	59	60
CURVE #n−1										CURVE #n									
i-19	i-18	i-17	i-16	i-15	i-14	i-13	i-12	i-11	i-10	i-9	i-8	i-7	i-6	i-5	i-4	i-3	i-2	i-1	i

where n = Number of curves; $i = 10n$
The numbers in the rows indicate the position of an attribute within the dynamic array

- Computes the tangent length T and the forward azimuth azF of a curve using its radius and central angle.
- Computes the coordinates of the PI, PC, center, and PT points of each a by calling the `icptl` procedure and traversing along azB or azF, a plus distance of T or zero, and an offset distance of zero or the radius.
- Introduces the said computed coordinates in the `TheData` array as indicated in Table 8.5.
- Sets the back azimuth of the next curve azB to be the forward azimuth of the current curve azF, and increments the loop counter j by 10.

7. For Case 4 translates the coordinates to their proper position by:
   - Calling the `icintrs` procedure to intersect an imaginary line passing through the last computed PT point with the forward tangent line to compute the length of translation.
   - Calling the `icptl` procedure repetitively for each curve to compute the coordinates of the PI, PC, center, and PT points from their current position to their final position by traversing along the input azimuth of the back tangent line with a plus distance equal to the translation length computed above and an offset distance of zero.
   - Preserving the coordinates of the last translated PT for use in the last step below.

8. Resets the values of the input parameters in case some of them have been computed.

9. Calls the `icintrs` procedure to intersect (a) the input azimuth of the back tangent line as specified through the PC point as specified or as it may have been computed, with (b) the azimuth of the forward tangent line as specified or as it may have been computed through the last computed PT point.

Having computed the above parameters, the procedure terminates.

8.4 Reversed Curves

In the preceding section we introduced the `icMcenCrv` procedure with the ability to introduce a curve that reverses the rotation of two tangent curves. In this section we address the introduction of two reversed curves under certain control conditions. If the need arises to introduce three or more curves successively reversing direction, it is recommended that use be made of the `icMcenCrv` procedure. Although there are certain similarities between compound and reversed

curves, one basic difference is the fact that there are more instances of reversed curves in which the two given lines to which the curves are tangent can be parallel as indicated in Figure 8.4. In each of the procedures presented below:

- The direction AZ1 of the back tangent line is of known value, and it should point towards the progression of the two curves.
- The direction AZ2 of the forward tangent line is of known value, and should point away from the progression of the two curves. If that is not the case, the procedures below may generate two compound curves or undesirable results.

The procedures that are presented in this section of the book are composed of two main cases, (a) one in which the start point (PC) and, or the end point (PT) is, or are known, and (b) the other in which the two curves float along the two given two tangent lines. It is evident that in the latter case the two given tangent lines cannot be parallel because there is an infinite number of solutions. The corollary to this is that if the two given tangent lines are parallel, one or both the PC and, or the PT must be known. To simplify matters, if the PT and not the PC is to be known, then the sequence of entry of the two tangent lines and of any curve data should be reversed so that the PT becomes the PC and the PC becomes the PT. Additionally, note that:

- PRC denotes the point of reversed tangency of the two curves to be introduced between the two given tangent lines.
- PP denotes a point through which the first curve (back or entering curve) must pass. If a point through which the second curve (forward or exiting curve) must pass, the sequence of the two given tangent lines and any associated curve data should be reversed.
- A positive central angle of a curve denotes a clockwise rotation, and a negative central angle denotes a counter-clockwise rotation.
- A positive deflection angle between the two tangent lines denotes an overall clockwise rotating alignment, and a negative deflection angle denotes a counterclockwise rotating alignment.
- If the two given tangent lines are parallel within a tolerance of 0.000000048 radians (0.99 s of a degree), the said deflection angle is set to be zero or 180°, and the coordinates of the overall PI are set to be 99999.0 feet (meters) each.
- The algebraic sum of the central angles of the two reversed curves is equal to the said deflection angle.

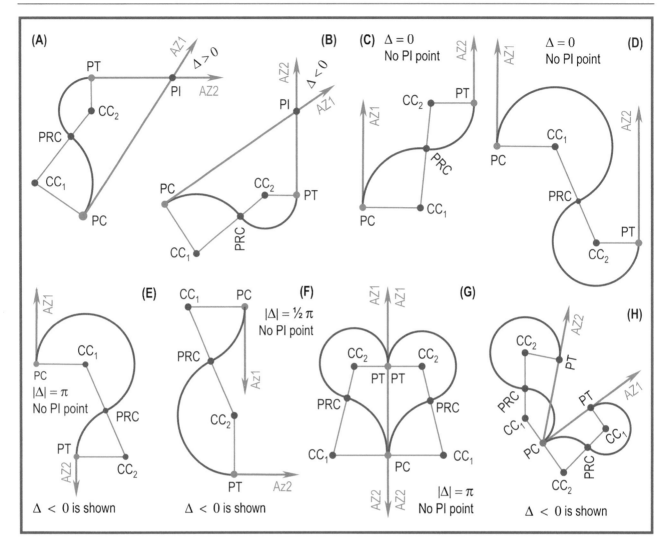

Figure 8.4 Potential Reversed Curves

The conditions table of Table 8.6 summarizes the various conditions addressed by the procedures presented below, and the elements that should be considered as being known, taking into account that each of the two reversed curves is defined by two variables, the curve radius, and its central angle. Theoretically, one of these two variables may be implied in terms of the length of a curve's individual tangent length, or excess distance. However, such cases are not considered as being practical, or rather potential conditions, and therefore are not considered herein. A known parameter is identified by a nonzero value, while an unknown parameter id identified by a zero value.

For each of the potential conditions addressed in this section, Table 8.6 identifies the chapter subsection in which the procedure that addresses a potential condition is described in detail. In the said table there are certain conditions in which

one, or both central angles are supposed to be known parameters. Also shown in this table are points PT_1, PT_2, PC and PT. Regarding these angles and points note that:

- If one central angle is known, the other is computed by subtracting the given angle from the deflection angle formed by the two given tangent lines.
- If both are central angles are known, their sum must equal the said deflection angle.
- PT_1 and PT_2 denote a known point through which the back and forward tangent lines must pass, respectively.
- If $PC = PT_1$ they denote that the point that defines the back tangent line must be the start point of the first curve.
- If $PT = PT_2$ they denote that the point that defines the forward tangent line must be the end point of the second curve.

Table 8.6 Chapter Subsection of Reversed Curve Cases

	Conditions									
	KNOWN	1	2	3	4	5	6	7	8	9
Elements	AZ_1	X	X	X	X	X	X	X	X	X
	AZ_2	X	X	X	X	X	X	X	X	X
	PT_1	X	X	X	X	X	X	X	X	X
	PT_2	X	X	X	X	X	X	X	X	X
	$PC = PT_1$	X	X	X	X	X	X	X		
	$PT = PT_2$					X	X	X		
	PP			X	X			X	X	X
	R_1	X	X				X		X	X
	R_2		X		X				X	X
	δ_1	X		X		X				X
Subsections	8.4.1	X								
	8.4.2		X							
	8.4.3			X						
	8.4.4				X					
	8.4.5					X				
	8.4.6						X			
	8.4.7							X		
	8.4.8								X	
	8.4.9									X

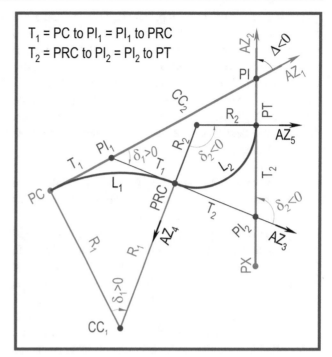

T_1 = PC to PI_1 = PI_1 to PRC
T_2 = PRC to PI_2 = PI_2 to PT

Figure 8.4a Reversed Curves Given PC, R_1, and $\delta_1 > 0$

8.4.1 Reversed Curves Given the PC, R_1, and δ_1

The `icRcrv1` procedure enables the programmer to generate two reversed curves when the given parameters, in addition to the two tangent lines (back and forward tangent lines) include the starting point (PC), the radius (R_1) and the central angle (δ_1) of the back curve as indicated in Figures 8.4a and 8.4b. The difference between the reversed curves in these two figures is that the central angle of the back curve is clockwise in Figure 8.4a, and counterclockwise in Figure 8.4b. To construct the pair of reversed curves, the procedure:

- Constructs the back curve since its radius and central angle are known quantities. In so doing the coordinates of CC_1, PC, PI_1, and PRC, as well as the azimuth AZ_3 of the common tangent line between the two curves are computed.
- Constructs the forward curve by:
  - Computing the length of the tangent T_2 and the coordinates of PI_2 and of the PT of the forward curve,
  - Computing the coordinates of CC_2 and the radius R_2 of the forward curve by intersecting the normal AZ_4 to the said common tangent line AZ_3 with the normal AZ_5 to the forward tangent line AZ_2, and then by
  - Computing the central angle δ_2 by calling the `iccenang` procedure (see Chap. 3) and using the coordinates of PRC, CC_2 and PT.

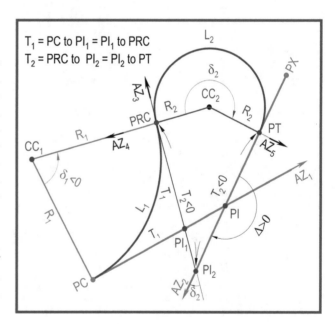

T_1 = PC to PI_1 = PI_1 to PRC
T_2 = PRC to PI_2 = PI_2 to PT

Figure 8.4b Reversed Curves Given PC, R_1, and $\delta_1 < 0$

The forward curve could have been constructed by computing the central angle of the forward curve and then computing the radius of the said curve as $R_2 = T_2 / \tan(0.5d_2)$, but this would have complicated matters to some extent when straddling north by the two given tangent lines and when trying to handle certain special conditions that could be encountered.

There is a special condition that needs to be addressed with reference to the back curve in Figure 8.4a, and to Figure 8.4b if the back curve progressed in a clockwise and not in a counterclockwise rotation as it is shown. If the radius and/or the central angle of the back curve were to increase in the clockwise rotation it is possible that this curve could meet and even cross the forward tangent line. Thus, if the radius R_1 and central angle δ_2 were to be such that the PRC would be located:

- On the forward tangent line, it is not possible for a reversed curve to exist. In this case the procedure returns an error condition message and aborts.
- Beyond the forward tangent line, the forward curve might oppose the input direction of the forward tangent line. In this case a forward curve would be created that would be tangent to the back curve and to the forward tangent line, but would create an improper reversed curve. Thus, in such cases the direction of the forward tangent line will need to be reversed by 180°. This is a rather rare if not improbable condition to be encountered, and hence an effort to account for it has not been expended.

```
Sub icRcrv1 (PCN, PCE, AZ1, PXN, PXE, AZ2,
             R1, DEL1, _
             RR1, R2, DEL2, DELTA, CCN1, CCE1,
             CCN2, CCE2, PRCN, PRCE, _
             PTN, PTE, PIN, PIE, PIN1, PIE1,
             PIN2, PIE2, T1, T2, L1, L2, IOK)
```

The input parameters to this procedure include the:

PCN, PCE	The north and east coordinates of the PC point on the back tangent line.
AZ1	The north azimuth (in radians) of the back tangent line pointing towards the reversed curves.
PXN, PXE	The north and east coordinates of a point on the forward tangent line. It could be located anywhere along the said line.
AZ2	The north azimuth (in radians) of the forward tangent line pointing away from the reversed curves.
R1	The radius of the first (back) curve. Could be positive or negative.
DEL1	The central angle (δ_1 in Figures 8.4a and 8.4b) of the back curve.

The parameters that are returned from this procedure include the:

RR1	The radius of the first curve having the sign of the central angle DEL1.
R2	The radius of the second curve always positive.
DEL2	The central angle in radians of the second curve.
DELTA	The deflection angle in radians of the two given tangent lines.
CCN1, CCE1	The north and east coordinate of the first curve's center point.

CCN2, CCE2	The north and east coordinate of the second curve's center point.
PRCN, PRCE	The north and east coordinate of the PRC point.
PTN, PTE	The north and east coordinate of the PT point.
PIN, PIE	The north and east coordinate of the overall PI point.
PIN1, PIE1	The north and east coordinate of the PI point of the first curve.
PIN2, PIE2	The north and east coordinate of the PI point of the second curve.
T1, T2	The tangent length of the first and second curves.
L1, L2	The arc length of the first and second curves.
IOK	An indicator, the value of which denotes that:

- 0 No problems have been encountered.
- 1 The radius and/or the central angle of the back curve have a zero value.
- 2 The PRC lies on the 2nd line and the forward curve cannot be created.
- 3 The forward curve cannot be created due to an erroneous input value.

Upon invocation, the procedure initializes the parameters to be returned, and then:

1. Queries the values of the radius and central angle of the back curve for being equal to zero within a tolerance of 0.005 feet (1.524 mm) and 0.0000048 radians (0.99 s of a degree), respectively. If either or both:
   - Are equal to zero, sets IOK to be 1 and aborts.
   - Are not equal to zero, forces the radius of the back curve to have the sign of the central angle of the said curve.
2. Queries the directions of the two given tangent lines, and if they are within a tolerance of 0.0000048 radians (0.99 s of a degree):
   - Equal to each other, sets the coordinates of the overall PI to be 99999.0 each and the overall deflection angle Δ to be 0°.
   - Of opposite direction, sets the coordinates of the overall PI to be 99999.0 each and the overall deflection angle Δ to be 180°.
   - Nonparallel, calls the icintrs procedure to intersect the two given lines to compute the coordinates of the overall PI and the overall deflection angle Δ.
3. Computes the tangent distance (T1) of the back curve to be $T_1 = R_1 \tan(\%\Delta_1)$
4. Calls the icpt1 procedure to compute the coordinates of the:
   - Center point of the back curve (CCN1, CCE1) by traversing a zero distance and an offset distance of R1 along AZ1 from the PC point (PCN1, PCE1).
   - PI point of the first curve (PIN1, PIE1) by traversing a T1 distance and a zero offset distance along AZ1 from the PC point (PCN1, PCE1).
5. Computes the north azimuth of the common tangent line as being $AZ3 = AZ1 + DEL1$.

6. Calls the `icptl` procedure to compute the coordinates of the PRC point (`PRCN`, `PRCE`) by traversing a `T1` plus distance and a zero offset distance from the said point along the azimuth of `AZ3`.
7. Calls the `icprjct` procedure to project the PRC point on the forward tangent line and determine the length of the normal offset distance of the said point from the said line, and then compares the said offset distance for being equal to zero within a tolerance of 0.005 feet (1.524 mm), and if so sets `IOK` to be 2 and aborts; otherwise the procedure continues.
8. Calls the `icintrs` procedure to intersect the common tangent line (`AZ3`) through the PRC point with the forward tangent line (`AZ2`) through the PX point, and return the coordinates of the PI of the second curve (`PIN2`, `PIE2`) and the tangent length (`T2`) of the second curve.
9. Calls the `icptl` procedure to compute the coordinates of the PT point (`PTN`, `PTE`) by traversing a `T2` distance and a zero offset distance along the azimuth of `AZ2`.
10. Determines the normal azimuth `AZ4` to the azimuth of the common tangent line `AZ3` by adding 90° if δ_l is positive and subtracting 90° if δ_l is negative.
11. Determines the normal azimuth `AZ5` to the forward tangent line `AZ2` by adding to it 90° if δ_l is positive and subtracting 90° from it if δ_l is negative.
12. Calls the `icintrs` procedure to intersect the normal azimuth `AZ3` through the PRC with the normal azimuth `AZ5` through the PT, and return the coordinates of the center point of the forward curve (`CCN2`, `CCE2`) and the radius (`R2`) of the forward curve. Regarding this intersection it should be pointed out that the `icintrs` procedure returns two distances (`D2` and `D2`) which should be theoretically equal to each other and represent the radius of the forward curve. If these two distances:
    - Are equal to each other within a tolerance of 0.0000048 radians (0.99 s), the radius `R2` is set to be the overage of these two distances, and if not
    - `IOK` to be 3 and the procedure aborts.
13. Calls the `iccenang` procedure to compute the central angle δ2 of the forward curve and the arc length (`L2`) of the said curve.
14. Computes the arc length of back curve (`L1`) as $L_1 = R_1 D_1 2$.

8.4.2 Reversed Curves Given the PC, R₁, and R₂

The `icRcrv2` procedure enables the programmer to construct two reversed curves when the given parameters, in addition to the two tangent lines (back and forward tangents) include the starting point (PC) and the radius of each of the

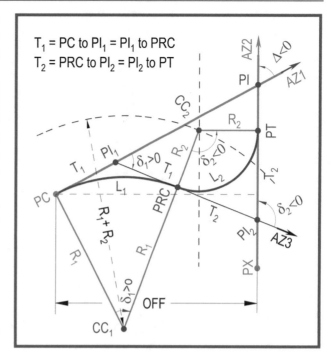

Figure 8.4c Reversed Curves Given PC, R₁, and R₂

two curves as indicated in Figures 8.4c and 8.4d. The difference between the reversed curves in these two figures is that the absolute value of the central angles of each curve of Figure 8.4c is less than 180°, while in Figure 8.4d the central angle of the second curve is greater than 180°. Both of these two conditions are included in Figure 8.4d. In essence, Figure 8.4c is a partial enlargement of the upper portion of Figure 8.4d. The possibility, even though the probability is rather low, exists that the absolute value of both central angles may exceed 180°.

In both of the said two figures, the back (entering) curve is a clockwise curve while the forward (exiting) curve is a counterclockwise curve. The possibility exists for the condition in which the back curve is a counterclockwise curve and the forward curve is a clockwise curve. To denote the appropriate rotational direction the procedure utilizes the sign of the given radii. Thus, a positive radius denotes a clockwise curve, while a negative radius denotes a counterclockwise curve. The two given radii cannot both be positive, or negative.

For the methodology to construct the two reversed curves reference is made to either of the said two figures, but particularly to Figure 8.4d. With the PC and the radius of the back curve being given, the center point of the back curve may be constructed. If we can construct the center point of the forward curve, then the two curves can easily be constructed. A review of the said two figures indicates that this center point should be located at the intersection of (a) a circle having its center point at the center point of the back

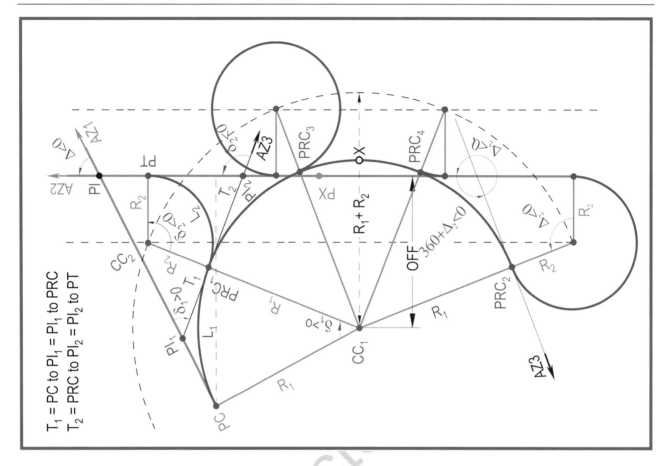

Figure 8.4d   Potential Reversed Curves Given PC, R_1, and R_2

curve and a radius equal to the sum of the absolute values of the two given radii, with (b) an imaginary line located at an offset distance from the forward tangent line equal to the absolute value of the radius of the forward tangent line.

Thus, since there are two potential intersections between a circle and a line, and since the imaginary parallel line could be to the left and to the right of the given forward tangent line, there could be four potential center points or four sets of potential reversed curves as indicate in Figure 8.4d identified by their respective PRC_1, PRC_2, PRC_3, and PRC_4 points. At each of these PRC points it is possible to construct two arcs, the one that is displayed in the said figure, and its 360° supplement. Since the said supplement would not yield a proper alignment configuration (one cannot possibly travel at some reasonable speed along a path and suddenly reverse travel 180°) it is not considered and it is not displayed in the said figure.

The two parallel lines have been addressed above as being left and right of the forward tangent line. For the said two figures this is appropriate, but what about a mirror image of

the said two figures about the vertical axis. Thus, a more appropriate reference for these parallel lines would be the outer and inner rather than the right and left parallel lines.

Of the four potential sets of reversed curves displayed in Figure 8.4d, the:

- PRC_1 and PRC_2 represent viable solutions since R_1 and R_2 are opposite in sign and the ahead curve in each case progresses in the direction of the given forward azimuth AZ_2.
- PRC_3 and PRC_4 do not represent viable solutions since although R_1 and R_2 are opposite in sign, the forward curve in each case progresses in the opposite direction of AZ_2.

Therefore a choice needs to be made the desired set of the potential sets of reversed curves. Hence, the procedure returns the viable solutions, and the calling program can display their center points from which the user may select the appropriate set of reversed curves. In addition, to simplify the procedure, since a choice has to be made, all four potential solutions are being returned by the subject procedure.

It should be noted that if the radius of the back curve is such that it would position the center point of the said curve a normal offset distance OFF from the forward tangent line equal to the said radius (see point X in Figure 8.4d), then there can be no viable potential solution to the subject construction problem. In addition, it should also be noted that, if:

• The forward tangent line may pass through the given PC point provided its azimuth is not the same or opposite to that of the back tangent line. In such case there are only two potential solutions, as the other two would be similar to the case of the said point X.
• The forward tangent lines may pass through the given PC point, but in such case the azimuth of the forward tangent line may not be the same as, or 180° opposite to that of the back tangent line.

Let us now turn our attention to Figure 8.4d1 in which the forward tangent line passes through the PC point of the back curve. If we draw the offset parallel lines and apply the same construction methodology as before we see that we obtain

two potential solutions. In the said figure the radius of the back curve is positive, and the direction of the forward azimuth is pointing upwards (north). Therefore the two solutions are located to the right of the forward tangent line. Now had the direction of the:

• Forward tangent line been pointing downwards (south) we would get the same exact solutions with the only difference being that they would belong to the left parallel line, and not to the right one.
• Radius of he back curve been negative similar results would have been obtained with the difference that the reversed curves would have been located to the opposite side of the forward tangent line.

Now let us visualize what happens as the back azimuth is rotated about the PC point until it coincides with the forward tangent line. The two solutions that would be obtained would be similar to those shown in Figure 8.4d1. Actually, the center point of the back curve would be located to the right of the two colinear tangent lines at 90° and the two curves would move towards the top of the said figure. The comments pertaining to the direction of the forward tangent line and to the sign of the back curve would also hold true.

```
Sub icRcrv2 (PCN, PCE, AZ1, PXN, PXE, AZ2,
             R1, R2, _
             CCN1, CCE1, CCN2, CCE2, PRCN, PRCE,
             PTN, PTE, IOK)
```

The input parameters to this procedure include the:

PCN, PCE	The north and east coordinates of the PC point on the back tangent line.
AZ1	The north azimuth (in radians) of the back tangent line pointing towards the PI of the reversed curves.
PXN, PXE	The north and east coordinates of the point on the forward tangent line. It could be located anywhere along the said line.
AZ2	The north azimuth (in radians) of the forward tangent line pointing away from the PI of the reversed curves.
R1, R2	The radius of the first (back) curve, and of the second (forward) curve, respectively.

The parameters that are returned from this procedure include the:

CCN1, CCE1	The array that contains the north and east coordinate of the back curve's center point (see note below).
CCN2, CCE2	The array that contains the north and east coordinate of the forward curve's center point.
PRCN, PRCE	The array that contains the north and east coordinate of the PRC point.
PTN, PTE	The array that contains the north and east coordinate of the PT point of the forward curve.

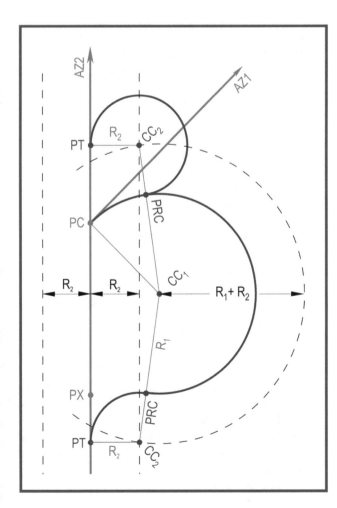

Figure 8.4d1   Potential Reversed Curves Given PC, R_1, and R_2

IOK   An indicator, the value of which if it is:

• 0   Denotes that no problems have been encountered.

• 1   Denotes that R1 and/or R2 have a zero value.

• 2   Denotes that R1 and R2 have the same sign.

• 3   Denotes that there is no potential solution to the left of the forward tangent line. No intersection of the left parallel line with the arc of radius R1 plus R2.

• 4   Denotes that there is no potential solution to the left of the forward tangent line. The left parallel line is tangent to the arc of radius R1 plus R2.

• 5   Denotes that there is no potential solution to the right of the forward tangent line. No intersection of the right parallel line with the arc of radius R1 plus R2.

• 6   Denotes that there is no potential solution to the right of the forward tangent line. The right parallel line is tangent to the arc of radius R1 plus R2.

• 7   Denotes that the outer parallel line to the right forward tangent line is tangent to the arc of radius R1 plus R2.

NOTES

1. Due to the definition of this construction problem, the center point of the back curve is the same for all potential sets of reversed curves.
2. The IOK messages 3, 4, 5 and 6 are not fatal errors. The warn the user that there are only two and not four potential solutions.

Upon invocation, the procedure initializes the parameters to be returned, and then:

1. Queries the contents of R1 and R2 for (a) either or both being equal to zero within a tolerance of 0.005 feet (1.524 mm) and (b) for having the same sign, in which case IOK is set to be 1 or 2, respectively, and the procedure aborts.
2. Calls the icptl procedure to establish the center points of the back curves for each of the four potential sets of reversed curves by traversing from the PC point (PCN, PCE) a plus distance of zero and an offset distance equal to the radius of the back curve (R1). Again note that this point would be the same for all potential sets of solutions.
3. Calls the icptl procedure to establish a temporary point by traversing from the PX point (PXN, PXE) a plus distance of zero and an offset distance of −|R2| in order to

create a temporary parallel line to the left of the forward tangent line, and then repeats the same operation with an offset distance of |R2| in order to create a temporary parallel line to the right of the forward tangent line. Note that the procedure does not distinguish between inner and outer parallel lines. It always creates one line to the left and one to the right of the forward tangent line.

4. Calls the iclincir procedure twice to intersect an imaginary circle of radius |R1|+|R2| having its center at the center point of the back curve with each of the said two temporary parallel lines to determine the coordinates of the center points of the potential forward curves (CCN2, CCE2).

This procedure returns two center points for each call if an intersection exists, one center point if the parallel line is tangent to the arc of radius |R1|+|R2|, and no center point if there is not intersection. These three conditions are denoted by the parameters INTRSC1 for the left parallel line and the INTRSC2 for the right parallel line. For each of these two parameters a value of 2 denotes a pair of intersections, a value of 1 denotes a tangency point, and a value of zero denotes no intersection.

5. Queries the INTRSC1 and INTRSC2 parameters for having a value of 2. If either of the two has a value of 2, the procedure continues as indicated below. If not, that is their values are either zero or 1, the procedure sets IOK to be 7 and aborts.

6. Queries the INTRSC1 parameter for having a value of zero and 1, and:
• If so, skips to the next step.
• If not, for each of the two intersection of the left parallel line with the arc of radius $|R1|+|R2|$:
   – Calls the icforce procedure to determine the direction from the center point of the forward curve towards the center point of the back curve, and then
   – Calls the icprjct procedure to project the center point of the forward curve on the forward tangent line to determine the coordinates of the PT (PTN, PTE) point.

7. Queries the INTRSC2 parameter for having a value of zero and 1, and:
• If so, terminates the procedures.
• If not, for each of the two intersection of the right parallel line with the arc of radius $|R1|+|R2|$:
   – Calls the icforce procedure to determine the direction from the center point of the forward curve towards the center point of the back curve, and then
   – Calls the icprjct procedure to project the center point of the forward curve on the forward tangent line to determine the coordinates of the PT (PTN, PTE) point and terminates.

8.4.3 Reversed Curves Given the PC, a Point, and δ_1

The `icRcrv3` procedure enables the programmer to construct two reversed curves when the given parameters, in addition to the two tangent lines (back and forward tangents) include the starting point (PC), a point through which the first (back) curve is to pass, and the central angle of the back curve as indicated in Figure 8.4e. For all intents and purposes this procedure is the same as the `icRcrv1` procedure presented in Sect. 8.4.1 of this chapter with the sole difference being the definition of the radius of the back curve. Whereas in the `icRcrv1` procedure the said radius is considered to be of a known input value, in the `icRcrv3` procedure the radius is computed by use of Equations 8.3g through 8.3j and described in Sect. 8.3 of this chapter. Presented below is the list of the procedure's calling parameters, a description of the input parameters and an abbreviated description of the procedure's returned parameters and operational steps. A full description may be found in Sect. 8.4.1.

```
Sub icRcrv3 (PCN, PCE, AZ1, PXN, PXE, AZ2,
             PPN, PPE, DEL1, _
             R1, R2, DEL2, DELTA, CCN1, CCE1,
             CCN2, CCE2, PRCN, PRCE, _
             PTN, PTE, PIN, PIE, PIN1, PIE1,
             PIN2, PIE2, T1, T2, L1, L2, IOK)
```

$T_1 = PC - PI_1 = PI_1 - PRC$
$T_2 = PRC - PI_2 = PI_2 - PT$

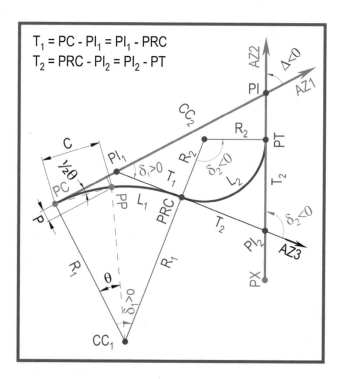

Figure 8.4e Reversed Curves Given PC, PP, and δ_1 or δ_2 $\left(|\delta_1| < 180 \quad \text{and} \quad |\delta_2| < 180 \right)$

The input parameters to this procedure include the:

PCN, PCE	The north and east coordinates of the PC point on the back tangent line.
AZ1	The north azimuth (in radians) of the back tangent line pointing towards the PI of the reversed curves.
PXN, PXE	The north and east coordinates of the point on the forward tangent line. It could be located anywhere along the said line.
AZ2	The north azimuth (in radians) of the forward tangent line pointing away from the PI of the reversed curves.
PPN, PPE	The north and east coordinates of the point through which the back curve is to pass.
DEL1	The central angle (δ_1 in Figure 8.4e) of the back curve. A positive value denotes a clockwise rotation, and a negative value denotes a counterclockwise rotation.

NOTES

1. If the given point (PPN, PPE) through which the back curve is to pass is to the right of the back tangent line with respect to the said line's direction, thus implying a clockwise rotation, then the central angle of the back curve (δ_1) should be positive. Similarly, if the said point is to the left of the said back tangent line, thus implying a counterclockwise rotation, then the said central angle (δ_2) should be negative. The reverse in either case aborts the procedure.

2. The given point (PPN, PPE) through which the back curve is to pass forms a chord from the PC point which subtends a central angle θ. This angle should be equal to or less than the specified central angle of the back curve δ_1. This is an arbitrary limitation of the program, and it could be removed if so desired.

Some of the parameters that are returned from this procedure (see Sect. 8.4.1 for the remaining parameters) include the:

R1, R2	The radii of the back and forward curves, respectively.
DEL2 to L2	The same as those identified in the `icRcrv1` procedure.
IOK	An indicator, the value of which if it is: • 0 Denotes that no problems have been encountered. • 1 Denotes that the central angle of the back curve has a zero value. • 2 Denotes that the point through which the back curve is to pass (PPN, PPE) is located on the back tangent line. • 3 Denotes that the central angle of the back curve (DEL1) and the point through which the back curve is to pass (PPN, PPE) are on opposite side of the back tangent line.

- 4 Denotes that the point through which the back curve is to pass (PPN, PPE) is outside the limit of the central angle (DEL1).
- 5 Denotes that the PRC is located on the second line and the forward curve cannot be created.
- 6 Denotes that the forward curve cannot be created due to some erroneous input value.

Upon invocation, the procedure initializes the parameters to be returned, and then:

1. Queries the value of the central angle of the back curve for being equal to zero within a tolerance of 0.0000048 radians (0.99 s of a degree), and if so, sets IOK to be 1 and aborts; else it continues.
2. Calls the icprjct procedure to project the PP point (PPN, PPE) on the back tangent line, to obtain the values of the chord distance (C) from PC to PP, the offset distance P from the said tangent line and the angle formed by the said chord and the back tangent line. Note that this angle is one half the angle formed by PC → CC1 → PP (θ).
3. Queries the value of the said offset distance for being zero within a tolerance of 0.005 feet (1.524 mm), and if so, sets IOK to be 2 and aborts; else it continues.
4. Compares the central angle of the back curve (DEL1) with the angle (0.5θ) that has been returned by the icprjct procedure to determine whether they are of the same sign, or not. If not, sets IOK to be 3 and aborts; else it continues.
5. Compares the central angle of the back curve (DEL1) with the said computed angle (θ) to determine whether DEL1 is equal to or less then θ, or not. If not, sets IOK to be 4 and aborts; else it continues.
6. Computes the value of the radius of the back curve by use of Equation 8.3h.
7. From this point on the icRcrv3 procedure operate the same as the icRcrv1 procedure to which reference is made.

8.4.4 Reversed Curves Given the PC, a Point, and R₂

The icRcrv4 procedure enables the programmer to construct two reversed curves when the given parameters, in addition to the two tangent lines (back and forward tangents) include the starting point (PC), a point (PP) through which the first (back) curve is to pass, and the radius of the second of the two curves.

To construct this set of reversed curves, this procedure operates in a manner similar to that of the icRcrv3 procedure to determine the radius of the back curve by using the chord from the PC point towards the PP point (see Figure 8.4e). With the two radii now known this procedure becomes the same as the icRcrv2 procedure as indicated in Figures 8.4c,

8.4d, and 8.4d1, and to which procedure reference is made. This implies that the icRcrv4 procedure returns up to four potential sets of pairs of reversed curves from which the call-

```
Sub icRcrv4 (PCN, PCE, AZ1, PXN, PXE, AZ2, _
             R2, PPN, PPE, _
             R1, RR2, CCN1, CCE1, CCN2, CCE2, _
             PRCN, PRCE, PTN, PTE, IOK)
```

ing program may select the desired pair.
The input parameters to this procedure include the:

PCN, PCE	The north and east coordinates of the PC point on the back tangent line.
AZ1	The north azimuth (in radians) of the back tangent line pointing towards the PI of the reversed curves.
PXN, PXE	The north and east coordinates of the point on the forward tangent line. It could be located anywhere along the said line.
AZ2	The north azimuth (in radians) of the forward tangent line pointing away from the PI of the reversed curves.
R2	The radius of the forward curve.
PPN, PPE	The north and east coordinates of the point through which the back curve is to pass.

NOTES

1. The position of the point through which the back curve is to pass (PPN, PPE) with respect to the back tangent defines the orientation of the two reversed curves.
2. The input value for the radius of the forward curve (R2) could be either positive or negative. However, if point (PPN, PPE) is located to the right of the back tangent line, then the radius of the back curve (R1) is considered to be positive, and the radius of the forward curve (R2) is considered to be negative regardless of its input sign. The signs of the two radii are reversed if (PPN, PPE) is located to the left of the back tangent line.

The parameters that are returned from this procedure include the:

R1	The radius of the first (back) curve.
RR2	The radius of the second (forward) curve having the value of R2 but with the appropriate sign to indicate the rotation of the said curve.
CCN1 to PTE	The same as those identified for the icRcrv2 procedure.
IOK	An indicator, the value of which if it is:

- 0 Denotes that no problems have been encountered.
- 1 Denotes that R2 have a zero value.
- 2 Denotes that the point through which the back curve is to pass (PPN, PPE) is located on the back tangent line.
- 3–7 Refer to the error messages of the icRcrv2 procedure.

Upon invocation, the procedure initializes the parameters to be returned, and then:

1. Queries the absolute value of the radius of the forward curve for being equal to zero within a tolerance of 0.05 feet (1.524 mm), and if so, sets IOK to be 1 and aborts; else it continues.
2. Calls the icprjct procedure to project the point through which the back curve is to pass (PPN, PPE) on the back tangent line, to obtain the values of the chord distance (C) from PC to PP, the offset distance P from the said tangent line and the angle formed by the said chord and the back tangent line. Note that this angle is one half the angle formed by PC → CC1 → PP (θ).
3. Queries the value of the said offset distance for being:
   • Zero within a tolerance of 0.005 feet (1.524 mm), and if so, sets IOK to be 2 and aborts; else it continues.
   • Negative or positive and assigns the corresponding sign to the absolute value of the R2 parameter to the RR2 parameter.
4. Computes the value of the radius of the back curve by use of Equation 8.3h.
5. Performs the same step as Steps 2 through 7 of the icRcrv2 procedure to determine the remaining parameters to be returned.

8.4.5   Reversed Curves Given the PC, PT, and δ_1

The icRcrv5 procedure enables the programmer to construct two reversed curves when the given parameters, in addition to the two tangent lines (back and forward tangents) include the starting point (PC), the ending point (PT), and the central angle of the back curve (δ_1) as indicated in Figures 8.4f and 8.4g. The points that define the position of the back and forward tangent lines are the known PC and PT points, and the two curves are forced to start and end at these two points, respectively.

To construct the two reversed curves, we intersect the two tangent lines and determine the overall deflection angle Δ, from which we can subtract the value of the given central angle of the back curve δ_1 to determine the value of the central angle of the forward curve δ_2. Referring to the said two figures, and remembering Equations 8.2a through 8.2i, if we project the PT point on the back tangent line we can obtain the values of the X and Y distances, and we can express the X_1, Y_1, X_2, and Y_2 distance as

$$X_1 = R_1 \sin(\delta_1) \qquad (8.4a1)$$

$$X_2 = 2R_2 \sin\left(\frac{1}{2}\delta_2\right)\cos\left(\delta_1 + \frac{1}{2}\delta_2\right) \qquad (8.4a2)$$

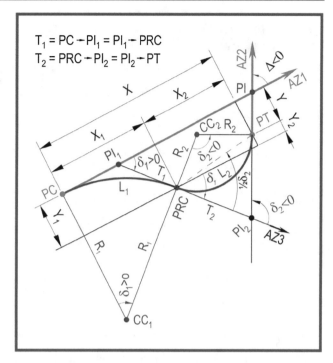

$$T_1 = PC \div PI_1 = PI_1 \div PRC$$
$$T_2 = PRC \div PI_2 = PI_2 \div PT$$

Figure 8.4f Reversed Curves Given PC, PT, and δ_1 or δ_2 $\left(\delta_1 + \frac{1}{2}\delta_2 < 0\right)$ and $\left(Y = Y1 - Y2\right)$

$$Y_1 = R_1 - R_1 \cos(\delta_1) = R_1 \left[1 - \cos(\delta_1)\right] \qquad (8.4a3)$$

$$Y_1 = R_1 - R_1 \cos(\delta_1) = R_1 \left[1 - \cos(\delta_1)\right] \qquad (8.4a4)$$

$$Y_2 = 2R_2 \sin\left(\frac{1}{2}\delta_2\right)\sin\left(\delta_1 + \frac{1}{2}\delta_2\right)$$

from which

$$X = X_1 + X_2 \qquad (8.4a5)$$

and

$$Y = Y_1 + Y_2 \qquad (8.4a6)$$

Regarding the Y_2 term in Equations 8.4a4 and 8.4a6, in Figures 8.4f and 8.4g the angle δ_1 is positive and the angle δ_2 is negative. Thus, the sum $\delta_1 + \delta_2$ yields in essence the difference rather than the sum of these two angles, and the value of Y_2 in Equation 8.4a4 is negative because the sine of a negative angle ($\frac{1}{2}\delta_2$) is negative. Hence, Equation 8.4a6 would hold true for the case of either of the said figures. If said curves had been located on the opposite side of the back tangent line, the signs of angles δ_1 and δ_2 would be reversed, thus yielding the proper results.

Another issue regarding the distance Y_2 in the said two figures is the case in which the PC and the PT are interchanged, and the azimuths AZ1 and AZ2 are interchanged and reversed in direction (if they are not reversed there would be no solution). In this case the overall deflection angle Δ

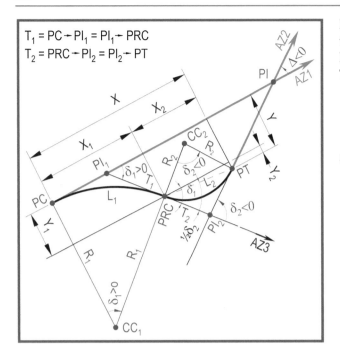

$T_1 = PC \rightarrow PI_1 = PI_1 \rightarrow PRC$
$T_2 = PRC \rightarrow PI_2 = PI_2 \rightarrow PT$

Figure 8.4g Reversed Curves Given PC, PT, and δ_1 or δ_2 $\left(\delta_1 + \tfrac{1}{2}\delta_2 < 0 \right)$ and $\left(Y = Y1 + Y2 \right)$

would exceed 180°, but the signs of the angles in the trigonometric functions would accommodate the situation.

We may now expand Equations 8.4a5 and 8.4a6 by substituting for X_1, X_2, Y_1, and Y_2 their respective Equations 8.4a1 through 8.4a4 to obtain

$$ X = R_1 \sin(\delta_1) + 2R_2 \sin\left(\frac{1}{2}\delta_2\right)\cos\left(\delta_1 + \frac{1}{2}\delta_2\right) \quad (8.4a7)$$

$$ Y = R_1\left[1 - \cos(\delta_1)\right] + 2R_2 \sin\left(\frac{1}{2}\delta_2\right)\sin\left(\delta_1 + \frac{1}{2}\delta_2\right) \quad (8.4a8)$$

If we now let

$$ A = \sin(\delta_1) \quad (8.4b1)$$

$$ A_2 = 2\sin\left(\frac{1}{2}\delta_2\right)\cos\left(\delta_1 + \frac{1}{2}\delta_2\right) \quad (8.4b2)$$

$$ A_3 = 1 - \cos(\delta_1) \quad (8.4b3)$$

$$ A_4 = 2\sin\left(\frac{1}{2}\delta_2\right)\sin\left(\delta_1 + \frac{1}{2}\delta_2\right) \quad (8.4b4)$$

the Equations 8.4a7 and 8.4a8 may be expressed respectively as

$$ X = A_1 R_1 + A_2 R_2 \quad (8.4b5)$$

$$ Y = A_3 R_1 + A_4 R_2 \quad (8.4b6)$$

In Equations 8.4b1 through 8.4b6, the angles δ_1 and δ_2 and the distances R_1 and R_2 are by now known. Thus, in Equations 8.4b5 and 8.4b6 R_1 and R_2 are the only unknowns. Thus, solving Equation 8.4b5 for R_1 we have

$$ R_1 = \frac{X - A_2 R_2}{A_1} \quad (8.4c1)$$

and substituting it in Equation 8.4b6 we get

$$ Y = A_3\left(\frac{X - A_2 R_2}{A_1}\right) + A_4 R_2 $$

$$ Y = \frac{A_3 X - A_2 A_3 R_2}{A_1} + A_4 R_2 $$

$$ Y = \frac{A_3 X}{A_1} - \frac{A_2 A_3 R_2}{A_1} + A_4 R_2 $$

$$ Y = \frac{A_3 X}{A_1} - R_2\left(\frac{A_2 A_3}{A_1} - A_4\right) $$

$(8.4c2)$

If we now solve Equation 8.4c3 for R_2 we get

$$ R_2 = \frac{\dfrac{A_3 X}{A_1} - Y}{\dfrac{A_2 A_3}{A_1} - A_4} = \frac{A_3 X - A_1 Y}{A_1}\frac{A_1}{A_2 A_3 - A_1 A_4} = \frac{A_3 X - A_1 Y}{A_2 A_3 - A_1 A_4} $$

$(8.4c3)$

We may now substitute R_2 of Equation 8.4c3 in Equation 8.4b5 to find R_1 as

$$ R_1 = \frac{X - A_2 R_2}{A_1} \quad (8.4c4)$$

which is the same as Equation 8.4c1. With both radii and both central angles known, we may now proceed to determine the remaining elements of the reversed curves as in Sect. 8.4.1.

Now let us assume that the PC is to remain fixed at its shown position in the said figures, and that the PT is to be moved forward along its fixed azimuth a certain small distance. A pair of reversed curves can be constructed but in the process the back curve is reduced in radius and the forward curve is increased in radius. As this distance is increased, the size of the two curves changes until at some specific distance the back curve attains a radius of zero. Continuing to increase the said distance the above equations do not hold true and the shape of the two curves changes to an unconventional shape of two compound curves or to an undesirable shape. That is, *any fixed position of the PC and PT points, and any fixed direction of the associated back and forward tangent lines can not necessarily create a pair of reversed curves.*

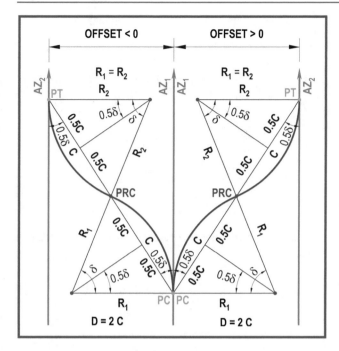

Figure 8.4g1 Reversed Curves between Two Parallel and Non-colinear Tangent Lines

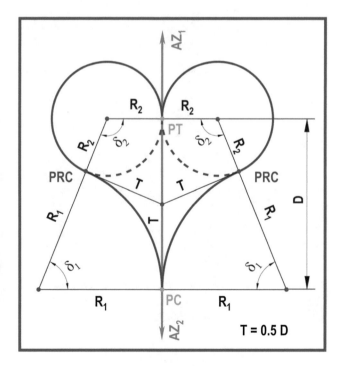

Figure 8.4g2 Reversed Curves Along Two Colinear Tangent Lines

Now let us consider the two special cases of Figures 8.4g1 and 8.4g2 which most probably will never be encountered in most roadways except perhaps in certain special interchanges and possibly in multistory garages.

Any fixed position of the PC and PT points, and any fixed direction of the associated back and forward tangent lines can not necessarily create a pair of reversed curves.

Figure 8.4g1 depicts three parallel tangent lines with the one in the middle being the back tangent line and each of the others being a case of a forward tangent line, one to the right and the other to the left of the back tangent line. This construction problem assumes the following:

- The back azimuth is the controlling azimuth, and if the forward azimuth opposes the back azimuth, it is reversed.
- The reversed curves to be create would be equal to each other but of reversed rotation.
- The PC and PT points are to remain fixed.
- The back curve's central angle is not used.

Based on these assumptions the chord length C of each curve would equal one half the distance D between the PC and PT points, and the absolute value of the radius would be:

$$R = \frac{C}{2\sin\left(\frac{1}{2}\delta\right)} \quad (8.4c5)$$

The distance D (and from it the value of C) may be determined by projecting the PT point on the back azimuth by calling the `icprjct` procedure which in addition returns the half of the central angle δ. With d and R having being determined the radii R_1 and R_2, and the central angles δ_1 and δ_2 of the two reversed curves are set to be

$$R_1 = R \quad \text{and} \quad R_2 = -R$$
$$\delta_1 = \delta \quad \text{and} \quad \delta_2 = -\delta$$

In addition the forward azimuth is forced to be the same as the back azimuth.

Now let turn our attention to Figure 8.4g2 which depicts two collinear tangent lines with opposing azimuths. This construction problem assumes the following:

- The back azimuth is the controlling azimuth, and if the ahead azimuth is the same as the back azimuth, it is reversed.
- The pair of reversed curves to be introduced would be equal to each other but of reversed rotation.
- The PC and PT points are to remain fixed.

- The back curve's central angle is used and controls the positioning of the reversed curves to the right or to the left of the back tangent line.

Based on these assumptions and with reference to the last said figure it is seen that:

- The center points of each of the curves are to be located at 90° to the right or left of the PC and PT points depending on the sign of the central angle of the back curve δ_1.
- The line between the two center points is equal to the sum of R_1 and R_2 in length.
- The length of the tangent lines that are common to each of the two curves are equal in length and direction, and actually they are one and the same line. Thus, the PI of each curve is the same and located midway between the PC and the PT points.
- There are two curves shown in the said figure pertaining to the forward curve, one represented by a solid line, which is the one desired, and one represented by a dash line, which is the 360° supplement of the former. The PI point and the tangent length of each of these two curves would be the same.
- The central angle of the ahead curve represented by the dash line δ_2 is equal to the 180° supplement of δ_1 but opposite in sign.
- The central angle of the ahead curve represented by the solid line δ_2 would then be dependent to the sign of δ_1 and equal to:

$$-\left(\pi + \delta_1\right) \quad if \quad \delta_1 > 0$$
$$\pi - \delta_1 \quad if \quad \delta_1 < 0$$

Thus, by being able to determine the distance D between the PC and the PT points we can compute the common tangent length T as being $0.5D$, and he radius R as

$$R = \frac{C}{\tan\left(\frac{1}{2}\delta\right)} \qquad (8.4c6)$$

The distance D between the PC and PT points, and hence the value of the common tangent to the two curves, is determined by inversion with the `icforce` procedure between the said two points.

```
Sub icRcrv5 (PCN, PCE, AZ1, PTN, PTE, AZ2,
             DEL1, _
             R1, R2, DEL2, DELTA, CCN1, CCE1,
             CCN2, CCE2, PRCN, PRCE, _
             PIN, PIE, PIN1, PIE1, PIN2, PIE2,
             T1, T2, L1, L2, IOK)
```

The input parameters to this procedure include the:

PCN, PCE	The north and east coordinates of the PC point on the back tangent line.
AZ1	The north azimuth (in radians) of the back tangent line pointing towards the PI of the reversed curves.
PTN, PTE	The north and east coordinates of the PT point on the forward tangent line.
AZ2	The north azimuth in radians of the forward tangent line pointing away from the PI of the reversed curves.
DEL1	The central angle in radians of the back curve. A positive value denotes a clockwise curve rotation, while a negative value denotes a counterclockwise curve rotation.

The parameters that are returned from this procedure include the:

R1, R2	The radius of the first (back), and of the second (forward) curves, respectively.
DEL2	The central angle in radians of the second curve.
DELTA	The deflection angle in radians formed by the two given tangent lines.
CCN1, CCE1	The north and east coordinate of the first curve's center point.
CCN2, CCE2	The north and east coordinate of the second curve's center point.
PRCN, PRCE	The north and east coordinate of the PRC point.
PIN, PIE	The north and east coordinate of the overall PI point.
PIN1, PIE1	The north and east coordinate of the PI point of the first curve.
PIN2, PIE2	The north and east coordinate of the PI point of the second curve.
T1, T2	The tangent length of the first and second curves.
L1, L2	The arc length of the first and second curves.
IOK	An indicator, the value of which denotes:
	0 No problems have been encountered.
	1 The central angle of the back curve has a zero value.
	2 The PRC is lies on the ahead tangent line and the ahead curve cannot be created, or for any other reason the reversed curves cannot be created.
	3 The ahead curve radius cannot be computed. A condition has arisen that the denominator of Equation 8.4c3 is zero (division by zero).

The user of this procedure is alerted to the fact that the input forward azimuth AZ2 and the central angle of the back curve DEL1 are preserved when the two given tangent lines are parallel and not colinear.

Upon invocation, the procedure initializes the parameters to be returned, and then:

1. Queries the value of the central angle of the back curve (DEL1) for being equal to zero within a tolerance of 0.0000048 radians (0.99 s of a degree), and if so sets IOK to be 1 and aborts; else it continues.
2. Queries the directions of the two given tangent lines, and if they are within a tolerance of 0.0000048 radians (0.99 s of a degree):
   • Equal to each other, sets the coordinates of the overall PI to be 99999.0 each and the overall deflection angle DELTA to be 0°.
   • Of opposite direction, sets the coordinates of the overall PI to be 99999.0 each and the overall deflection angle DELTA to be 180°.
   • Nonparallel, calls the icintrs procedure to intersect the two given lines to compute the coordinates of the overall PI and deflection angle DELTA taking into account the potential of the two azimuths straddling north, and specifying a rotation that would oppose the sign of the central angle of the back curve.
3. Queries DELTA whether it is equal to zero or 180° in order to determine whether the two tangent lines are parallel and opposite in direction, or not.
   • If they are, continues with the next step.
   • If they are not, branches to **Step 8**.
4. Projects the PT point on the back tangent to compute the plus and offset distances of the PT from the PC along AZ1, the azimuth AZ from the PC towards the PT and the angle of its inclination from AZ1, the said angle being $0.5\ \delta_1$, if it is not zero.
5. Queries the value of the said offset distance for being equal to zero within a tolerance of 0.005 feet (0.154 mm).
   • If so, continues with the next step.
   • If not so, branches to **Step 7a**.
6a. Begins to handle the case of the two tangent lines being colinear by dividing the distance between the PC and PT points in two to determine the length of the common tangent (T) of the two curves.
6b. Calls the icptl procedure to establish the coordinates of the common PI point along AZ1, and then calls the same procedure to establish the coordinates of the PRC along AZ.
6c. Computes the central angle of the forward curve taking into account the sign of its rotation, computes the radii of the two curves by Equation 8.4c6, then forces AZ2 to oppose AZ1 in case it does not and then branches to **Step 13**.
7a. Begins to handle the case of the two tangent lines being parallel and offset from each other by dividing the distance between the PC and the PT in two to determine the length of the chord of the two equal curves to be created.

7b. The angle of Step 4 as the central angle of each curve, computes the radii of the curves by Equation 8.4c5, forces AZ2 to be the same as AZ1, and then branches to **Step 13**.
8. Begins to handle the case of the two tangent lines being nonparallel and non-colinear by computing the value of the central angle (DEL2) of the forward curve by subtracting from the just computed overall deflection angle (DELTA) the central angle of the back curve (DEL1).
9. Computes the constants A_1, A_2, A_1, and A_4 of Equations 8.4b1 through 8.4b4.
10. Computes the value of the denominator of Equation 8.4c3, and checks whether it is zero, or not. If this denominator is equal to zero, sets IOK to be 3 and aborts; else it continues.
11. Computes the radius of the forward curve by Equation 8.4c3.
12. Computes the radius of the back curve by Equation 8.4c4.
13. It performs the same Steps 5 through 12 of the icRcrv1 procedure to determine the remaining parameters to be returned. During this process checks whether the PRC is located on the ahead tangent line within a tolerance of 0.005 feet (1.524 mm), or not, and if so sets IOK to be 2. In either case it continues with the next step.

8.4.6   Reversed Curves Given the PC, PT, and R_1

The icRcrv6 procedure enables the programmer to construct two reversed curves when the given parameters, in addition to the two tangent lines (back and forward tangents) include the starting point (PC), the ending point (PT), and the radius of the back curve (R1) as indicated in Figure 8.4h. The points that define the position of the back and forward tangent lines are the known PC and PT points, and the two curves are forced to start and end at these two points, respectively. This construction problem is very similar to that of the preceding subsection (the icRcrv5 procedure), and reservations and qualifications stated for that construction problem apply to this one as well. The basic difference is that instead of the central angle of the back curve, that curve's radius is known. This necessitates a bit different computational approach.

To construct the two reversed curves, we assume a local coordinate system about the center of the back curve CC_1 (see Figure 8.4h), the coordinates of which may be computed since we know the position of the PC and the value of R_1. Thus, the local north and east coordinates of the PT point are:

$$A = PTE - CCE1 \qquad (8.4d1)$$

$$B = PTN - CCN1 \qquad (8.4d2)$$

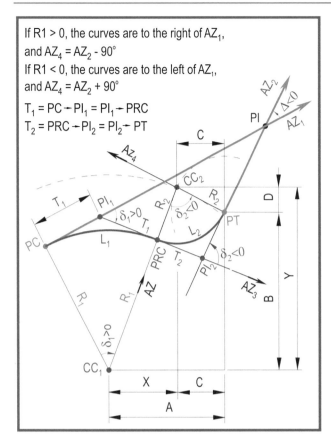

If R1 > 0, the curves are to the right of AZ_1, and $AZ_4 = AZ_2 - 90°$
If R1 < 0, the curves are to the left of AZ_1, and $AZ_4 = AZ_2 + 90°$
$T_1 = PC \dotdiv PI_1 = PI_1 \dotdiv PRC$
$T_2 = PRC \dotdiv PI_2 = PI_2 \dotdiv PT$

Figure 8.4h   Reversed Curves Given the PC, PT, and R_1

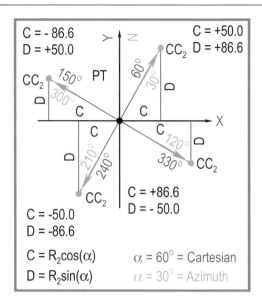

Figure 8.4h1   Relationship of CC_2 to PT

The normal azimuth (AZ_4) to that of AZ_2 is dependent on the sign of R_1. So, if $R_1 > 0$ then

$$AZ_4 = AZ_2 - 0.5\pi \qquad (8.4d3)$$

and if $R_1 < 0$ then

$$AZ_4 = AZ_2 + 0.5\pi \qquad (8.4d4)$$

In either case AZ_4 is reduced to be positive within the range of 0 and 2π.

In Figure 8.4h we see that the center of the forward curve CC_2 is located a distance D north of the PT point, and a distance C to the west. If we now wish to express the location of CC_2 with respect to point PT, and account for CC_2 to be located in any of the four quadrants about the PT point, we can translate the north azimuth AZ_4 into a Cartesian angle α, and then let:

$$C = R_2 \cos(\alpha) \qquad (8.4e1)$$

$$D = R_2 \sin(\alpha) \qquad (8.4e2)$$

As an example consider Figure 8.4h1 in which R_2 is assumed to be 100, and the north azimuths and corresponding Cartesian angles as shown in the said figure, as well as the corresponding C and D distances.

In Figure 8.4h we also see that the coordinates of the center point of the forward curve CC2 would be located at the intersection of a:

- Circle of radius $R_1 + R_2$ having its center at CC1 and a circle of radius R_2 having its center at the PT, or
- Circle of radius $R_1 + R_2$ having its center at CC1 and a line parallel to AZ_2 and a distance of R_2 away from it.

Of the above two solution methods the former is chosen because the latter would require the knowledge of side (left or right) of the forward tangent line the offset parallel should be positioned. This is doable, but would necessitate a few extra steps.

The equations of the said two circles in the local coordinate system with origin at the center point of the back curve may be expressed by Equations 8.4f1 and 8.4f2 in which x and y represent the coordinates of the intersection point of the said circles. Thus

- The equation of the circle of radius $R_1 + R_2$ with its center point at the center point of the back curve at local coordinates of (0.00, 0.00) is

$$x^2 + y^2 = (R_1 + R_2)^2 \qquad (8.4f1)$$

- The equation of the circle of radius R2 with its center point at the PT point at local coordinates of (A, B) is

$$(x - A)^2 + (y - B)^2 = R_2^2 \qquad (8.4f2)$$

Also, the local coordinates of the center point of the second curve may be expressed as

$$X = A + C = A + R_2 \cos(a) \tag{8.4f3}$$

$$Y = B + D = B + R_2 \sin(a) \tag{8.4f4}$$

Note that in Equations 8.4f3 and 8.4f4 the magnitude of α causes the cosine and sine to yield the correct plus or minus sign. In Figure 8.4h the azimuth AZ4 is in the fourth northeast quadrant or in the second Cartesian quadrant about the PT point, thus yielding a negative C distance and a positive D distance. Expanding Equations 8.4f1 and 8.4f2, we get:

$$x^2 + y^2 = R_1^2 + R_2^2 + 2 R_1 R_2 \tag{8.4g1}$$

$$x^2 + A^2 - 2Ax + y^2 + B^2 - 2By = R_2^2 \tag{8.4g2}$$

Solving Equation 8.4g1 for R22, substituting it for R22 in Equation 8.4g2 and then moving it to the left side of Equation 8.4g2, we get:

$$\begin{aligned} R_2^2 &= x^2 + y^2 - R_1^2 - 2R_1R_2 \\ &= x^2 + A^2 - 2Ax + y^2 + B^2 - 2By \end{aligned} \tag{8.4h1}$$

or

$$-R_1^2 - 2R_1R_2 = A^2 + B^2 - 2Ax - 2By \tag{8.4h2}$$

Substituting Equations 8.4f3 and 8.4f4 in Equation 8.4h2 for x and y, we get:

$$\begin{aligned} -R_1^2 - 2R_1R_2 &= A^2 + B^2 \\ &\quad - 2A\left[A + R_2\cos(\alpha)\right] - 2B\left[B + R_2\sin(\alpha)\right] \end{aligned} \tag{8.4h3}$$

or

$$\begin{aligned} -R_1^2 - 2R_1R_2 &= A^2 + B^2 - 2A^2 \\ &\quad - 2AR_2\cos(\alpha) - 2B^2 - 2BR_2\sin(\alpha) \end{aligned} \tag{8.4h4}$$

or

$$\begin{aligned} -2R_1R_2 + 2AR_2\cos(\alpha) + 2BR_2\sin(\alpha) \\ = R_1^2 + A^2 + B^2 - 2A^2 - 2B^2 \end{aligned} \tag{8.4h5}$$

or

$$-2R_1R_2 + 2AR_2\cos(\alpha) + 2BR_2\sin(\alpha) = R_1^2 - A^2 - B^2 \tag{8.4h6}$$

Solving Equation 8.4h6 for R2, we get:

$$R_2 = \frac{R_1^2 - A^2 - b^2}{2\left[-R_1 + A\cos(\alpha) + B\sin(\alpha)\right]} \tag{8.4h7}$$

With radius R_2 now known we can locate the center point CC_2 of the forward curve by traversing along AZ_2 from the PT

point a zero distance and an offset distance of $-R_2$ if R_1 is positive, or a distance of $+R_2$ if R_1 is negative. Thereafter we can:

• Inverse from the CC_1 point towards the CC2 point to find its azimuth AZ;
• Locate the coordinates of the PRC point a distance R_1 from the CC_1 point;
• Compute the normal to AZ1 and subtract it from AZ to determine the central angle of the back curve δ_1, taking into account the rotational direction of the back curve, and the possibility of these two azimuths straddling north;
• Compute the central angle of the forward curve δ_2 by subtracting it from the overall deflection angle, Δ, which Δ can be determined by intersecting the two given azimuths;
• Compute the lengths of the tangents of each curve as

$$T_1 = R_1 \tan\left(\%\delta_1\right) \tag{8.4h8}$$

$$T_2 = R_2 \tan\left(\%\delta_1\right) \tag{8.4h9}$$

• Compute the lengths of the tangents of each curve as

$$L_1 = R_1\left(\delta_2\right) \tag{8.4h10}$$

$$L_2 = R_1\left(\delta_2\right) \tag{8.4h11}$$

Before we proceed there are two issues that need to be addressed, division by zero and a zero radius of the back curve.

• **Division by zero** Let us take a look at Equation 8.4h7 and particularly at its denominator $2[-R1 + A\cos(\alpha) + B\sin(\alpha)]$. Under certain conditions that cause the Cartesian angle α to be zero or 180°, and the distance A to be equal to R_1, $\sin(\alpha)$ becomes zero and $\cos(\alpha)$ becomes 1 or -1, depending on the value of α. This forces the said denominator to become zero, thus causing computing problems. To overcome this problem, the procedure rotates counterclockwise the coordinates of the PC and PT points about the origin of their north/east coordinate system an arbitrary amount of 30°, and recomputes the radius of the forward curve by Equation 8.4h7. If this recomputation produces a denominator of:

  – Zero again, then that is considered to be an issue of some erroneous data specification and the procedure aborts.
  – A nonzero value, then that is treated as the radius of the back curve, the input value of the forward tangent line is forced to be the same as that of the back tangent line, and the procedure continues.

The rotation effect on the parameters used in the said equation is illustrated in Figure 8.4h2.

Figure 8.4h2 Reversed
Curves with Collinear
Tangents and Fixed PC, PT,
and R_1

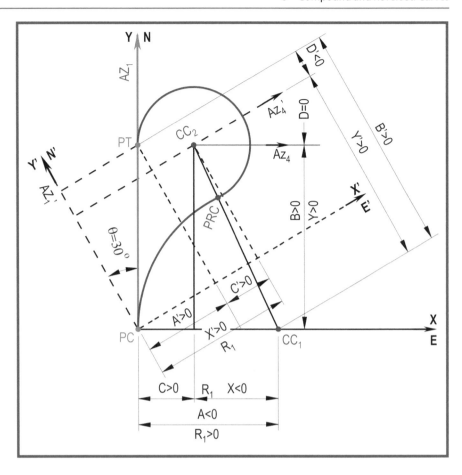

- **Zero Radius**   The specification of a zero radius for the back curve (R_1) should be considered an improper entry. *This holds true when the two tangent lines are collinear, and in which case the said radius must not be zero.* However, an exception is allowed when the back tangent line and the forward tangent line are not collinear, in which a case the specification of a zero radius causes the procedure to generate two circular arcs of equal radius but opposite in rotational direction (the central angles and arc lengths will not necessarily be equal to each other). When the procedure senses that a zero radius has been specified for the back curve, the action taken depends on whether the two tangent lines are parallel and offset from each other, or not.
  - If the said lines are parallel, the procedure creates the reversed curves in a manner like that of the icRcrv5 procedure, and forces the input value of the forward azimuth (AZ_2) to be the same as that of the back azimuth (AZ_1).
  - If the said two lines are not parallel, the procedure creates the reversed curves as indicated in Figure 8.4h3 and based on the following assumptions:
    - The overall deflection angle Δ is positive and less than 180°.

- The directions of the back and forward tangent lines adhere to the convention that the back one points towards the PI and ahead one away from the PI.
- The reversed curves are to be located in the quadrant formed by the tangent lines in which the triangle with vertices at the PC, PI, and PT points is located.
- If the absolute value of each of the two equal radii is R, than the distance between the center points of the two curves is $2R$.
- The center point CC_1 of the back curve is located to the same side of the back tangent line that the PT point is located with respect to the said line, and a normal distance of R from the PC point.
- The center point CC_2 of the forward curve is located to the same side of the forward tangent line that the PC point is located with respect to the said line, and a normal distance of R from the PT point.
- If the PT point is to the right of the back tangent, then $R1 > 0$ and $R2 < 0$.
- If the PT point is to the left of the back tangent, then $R1 < 0$ and $R2 > 0$.

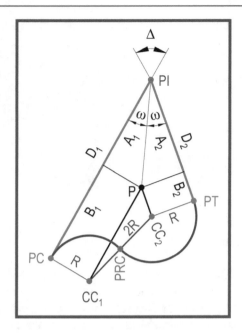

Figure 8.4h3 Reversed Curves with Zero Input Radius Value

Based on the above assumptions as displayed in Figure 8.4h3 a line from CC_1 may be drawn parallel to the back tangent line, and a line may be drawn from CC_2 parallel to the forward tangent line intersecting the first line at point P. Point P may then be projected on each of the two tangent lines dividing the back tangent line into segments B_1 and A_1, and the forwards tangent line into segments B_2 and A_2. With the coordinates of the PC and PT points and the azimuths AZ_1 and AZ_2 of the two tangent lines being given, the PI point and the distances D_1 and D_2 as well as the angle Δ can be computed. Because of the symmetry of the two parallel lines with respect to their corresponding tangent lines, it is seen that:

$$\omega = \frac{1}{2}\Delta \quad and \quad A_1 = A_2 = A = \frac{R}{\tan(\omega)}$$

$$B_1 = D_2 - A = D_1 - \frac{R}{\tan(\omega)} \quad (8.4h12)$$

$$B_2 = D_2 - A = D_2 - \frac{R}{\tan(\omega)}$$

In the triangle CC1, CC2 and P we now know the three sides, in terms of R, and the actual value of the angles Δ and ω. Therefore we can use the law of cosines to solve for the value of R as follows:

$$(2R)^2 = B_1^2 + B_2^2 - 2B_1B_2\cos(\Delta) \quad (8.4h13)$$

By substituting the values of B_1 and B_2 from Equation 8.4h12 in Equation 8.4h13 and expanding the squared terms Equation 8.4h13 becomes

$$4R^2 = \left[D_1 - \frac{R}{\tan(\omega)}\right]^2 + \left[D_2 - \frac{R}{\tan(\omega)}\right]^2$$
$$-2\left[D_1 - \frac{R}{\tan(\omega)}\right]\left[D_2 - \frac{R}{\tan(\omega)}\right]\cos(\Delta) \quad (8.4h14)$$

or

$$4R^2 = D_1^2 + \frac{R^2}{\tan^2(\omega)} - \frac{2D_1R}{\tan(\omega)} + D_2^2 + \frac{R^2}{\tan^2(\omega)} - \frac{2D_2R}{\tan(\omega)}$$
$$-2\left\{D_1D_2 - \frac{D_1R}{\tan(\omega)} - \frac{D_2R}{\tan(\omega)} + \left[\frac{R^2}{\tan^2(\omega)}\right]\right\}\cos(\Delta)$$

(8.4h15)

By rearranging certain terms the above equation becomes

$$4R^2 = D_1^2 + D_2^2 + 2\left[\frac{R^2}{\tan^2(\omega)}\right] - \frac{2D_1R}{\tan(\omega)} - \frac{2D_2R}{\tan(\omega)}$$

$$-2D_1D_2\cos(\Delta) + 2\left[\frac{D_1R}{\tan(\omega)}\right]\cos(\Delta) + 2\left[\frac{D_2R}{\tan(\omega)}\right]\cos(\Delta) - 2\left[\frac{R^2}{\tan^2(\omega)}\right]\cos(\Delta)$$

$$4R^2 = D_1^2 + D_2^2 + 2\left[\frac{R^2}{\tan^2(\omega)}\right]\left[1 - \cos(\Delta)\right] - 2(D_1 + D_2)\left[\frac{R}{\tan(\omega)}\right]$$

$$-2D_1D_2\cos(\Delta) + 2(D_1 + D_2)\left[\frac{R}{\tan(\omega)}\right]\cos(\Delta) \quad (8.4h16)$$

We can now transfer all unknown terms to the left of the equation to get

$$R^2\left\{4-\left[\frac{2\left[1-\cos(\Delta)\right]}{\tan^2(\omega)}\right]\right\}+2\left[\frac{(D_1+D_2)\left[1-\cos(\Delta)\right]}{\tan(\omega)}\right]R=$$
$$D_1^2+D_2^2-2D_1D_2\cos(\Delta) \qquad (8.4h17)$$

For simplification if we define the constants F_1, F_2 and F_3 as

$$F_1=4-\left[\frac{2\left[1-\cos(\Delta)\right]}{\tan^2(\omega)}\right]$$
$$F_2=\frac{(D_1+D_2)\left[1-\cos(\Delta)\right]}{\tan(\omega)} \qquad (8.4h18)$$
$$F_3=D_1^2+D_2^2-2D_1D_2\cos(\Delta)$$

Equation 8.4h16 becomes

$$F_1R^2+2F_2R=F_3 \qquad (8.4h19)$$

Divide the equation by F_1, and completing the square the equation becomes

$$R^2+\frac{2F_2}{F_1}R=\frac{F_3}{F_1}$$
$$R^2+\frac{2F_2}{F_1}R+\frac{F_2^2}{F_2^2}=\frac{F_3}{F_1}+\frac{F_2^2}{F_2^2} \qquad (8.4h20)$$
$$\left(R+\frac{F_2}{F_1}\right)^2=\frac{F_3}{F_1}+\frac{F_2^2}{F_2^2}$$

From which the radius maybe expressed as

$$R=\pm\sqrt{\frac{F_3}{F_1}+\frac{F_2^2}{F_2^2}}-\frac{F_2}{F_1} \qquad (8.4h21)$$

which yields an answer for each of the two signs of the square root. The procedure selects first the one with the positive root, and locates the two center points. If the distance between the said two points is equal to *2R*, the solution is accepted. If it is not, the other root is selected, and the process is repeated. If this one also does not yield a distance or *2R* between the said points, there has to be an input error, and the procedure aborts.

```
Sub icRcrv6 (PCN, PCE, AZ1, PTN, PTE, AZ2, R1, _
             R2, DEL1, DEL2, DELTA, CCN1, CCE1,
             CCN2, CCE2, _
             PRCN, PRCE, PIN, PIE, PIN1, PIE1,
             PIN2, PIE2, T1, T2, L1, L2, IOK)
```

In using the zero radius option remember that the azimuth of the back tangent line must point towards the overall PI point and that of the forward tangent line must point away from it.

The input parameters to this procedure include the:

PCN, PCE	The north and east coordinates of the PC point on the back tangent line.
AZ1	The north azimuth (in radians) of the back tangent line pointing towards the PI of the reversed curves.
PTN, PTE	The north and east coordinates of the PT point on the forward tangent line.
AZ2	The north azimuth (in radians) of the forward tangent line pointing away from the PI of the reversed curves.
R1	The radius of the back curve, or zero.

NOTE that:

- A positive radius denotes a clockwise rotation and a negative radius denotes a counterclockwise rotation.
- The positive or negative value of R1 is used only for the location of the reversed curves and not for the solution of the construction problem addressed above, and in which the absolute value of R1 is used.
- The value of R1 could be zero if the back and forward tangent lines are not colinear. If they are colinear, an error message is generated and the procedure aborts.
- When a valid zero R2 is specified it implies that the reversed curves should be of the same radius. The arcs will be equal only if the lines are parallel and offset from each other.

The parameters that are returned from this procedure include the:

R2	The radius of the second (forward) curve.
DEL1, DEL2	The central angles of the back and forward curves in radians.
DELTA to L2	The same as those identified for the icRcrv1 procedure.
IOK	An indicator, the value of which if it is:

0 Denotes that no problems have been encountered.

1 Denotes that either:
- The radius of the back curve has a zero value when the tangent lines are colinear; or that
- The tangent lines are parallel and not colinear with R1 = 0 and there can be no solution. Check the input data.

2 Denotes that the forward and back tangent line are either:
- Colinear and the two azimuths are the same, in which case AZ2 has been changed to oppose AZ1; or
- Parallel but not colinear and the azimuths are not the same, in which case AZ2 has been changed to be the same as AZ1.

Note that this is only for information purposes and not a fatal error.

3 Division by zero in solving for R2.

Upon invocation, the procedure initializes the parameters to be returned, and then:

1. Queries the directions of the two given tangent lines, and if they are within a tolerance of 0.0000048 radians (0.99 s of a degree):
   - Equal to each other, sets the coordinates of the overall PI to be 99999.0 each and the overall deflection angle DELTA to be 0°.
   - Of opposite direction, sets the coordinates of the overall PI to be 99999.0 each and the overall deflection angle DELTA to be 180°.
   - Nonparallel, calls the icintrs procedure to intersect the tangent lines to compute the coordinates of the overall PI and the overall deflection angle DELTA taking into account the potential of the two azimuths straddling north, and specifying a rotation that would oppose the sign of the central angle of the back curve.
2. Queries the value of DELTA whether it is equal to 0° or 180° in order to determine whether the two tangent lines are colinear, or not. To do so, it projects the PT point on the back tangent line and checks the offset distance for being equal to zero within a tolerance of 0.005 feet (1.524 mm), or not. If they are, a flag is set up to indicate colinearity, AZ2 is forced to oppose AZ1, and continues; else continues. Note the input value of AZ2 is altered.
3. Queries the value of R1, when the two tangent lines are colinear, for being equal to zero within a tolerance of 0.005 feet (1.524 mm), or not.
4. Checks the colinearity flag, DELTA and R1 for parallel non-colinear lines and zero radius.
   - If so, computes the value of R1 by use of Equation 8.4c5 and shown in Figure 8.4g1, and branches to Step 11.
   - If not so, continues.
5. Queries the value of R1 for being equal to zero within a tolerance of 0.005 feet (1.524 mm), or not.
   - If so:
     - Makes certain that DELTA is positive and less or equal to 180°.
     - Sets the value of the angle ω to be one half of DELTA, and makes certain that the distances from the PC and PT to the overall PI are positive.
     - Computes the values of two radii by use of Equations 8.4h18 through 8.4h21. One of these radii (Rplus) is using the positive result of the square root of Equation 8.4h21, and the other (Rminus) the negative result.
     - Calls the icprjct procedure to project the PT point on the back tangent line and determine whether the PT is located to the right or to the left of the said line. The returned offset distance B2 is the indicator of the side on which the PT is located.

     - Assumes that the Rplus solution might be the proper radius and calls the icpt1 procedure to compute the coordinates of the center points of the two curves to be constructed, one from the PC and the other from the PT point using (a) their respective azimuths, (b) a plus distance of zero and (c) the Rplus radius as offset taking into account the sign of B2, and then calls the icforce procedure to determine the distance between the said two center points.
     - Checks the above distance for being equal to two times Rplus within a tolerance of 0.005 feet (1.524 mm), or not.
       - If so, accepts the value of Rplus for R1, and its negative for R2, and branches to branches to Step 11.
       - If not so, repeats the preceding and this current step using the Rminus.
         If a proper two times Rminus distance is found, it is accepted and branches to Step 11. If not, IOK is set to be 1 and aborts.
   - If not so, continues with the next step.
6. Calls the icpt1 procedure to compute the coordinates of the center point of the back curve by traversing from the PC along the azimuth of the back tangent a plus distance of zero and R1 offset distance.
7. Computes the distance A and B using Equations 8.4d1 and 8.4d2, and the azimuth AZ4 using Equations 8.4d3 and 8.4d4.
8. Calls the icaz3car procedure to convert the azimuth AZ4 to the Cartesian angle Alpha.
9. Computes the numerator D1 and denominator D2 of Equation 8.4h7 using the absolute value of the back radius RA1.
10. Queries the value of D2 for being equal to zero within a tolerance of 0.005 feet (1.524 mm), or not.
    - If so, rotates the coordinates of the PC and PT points and arbitrary angle THETA of 30° by calling the icdrot2s procedure, and the azimuth AZ4 by the same angle, and then repeating Steps 6 through 10.
      - If D2 is still equal to zero within the said tolerance, sets IOK to 2 and aborts.
      - If not, computes the radius of the forward curve by completing Equation 8.4h7, computes the coordinates of the center point of the forward curve and continues with the next step.
11. Calls the icforce procedure to determine the azimuth from the center point of the back curve towards that of the forward curve, and then calls the icpt1 procedure to compute the coordinates of the PRC.
12. Computes the normal azimuth to the azimuth of the forward tangent line so as to point from the center point of the forward curve towards the PT point taking into account the sign of the radius and straddling of the north.

13. Computes the central angles of the two curves by subtracting the relevant radial azimuths and accounting for straddling north.
14. Computes the values of the tangent lengths to each of the two individual curves as

$$T_1 = R_1 \tan\left(\%\delta_1\right)$$
$$T_2 = R_2 \tan\left(\%\delta_1\right)$$

15. Computes the coordinates of the PI points of the two curves, calls the `iccenang` procedure to compute the arc lengths of the two curves, and terminates.

8.4.7 Reversed Curves Given the PC, PT, and Point

The `icRcrv7` procedure enables the programmer to perform two functions, to construct two reversed curves in a manner similar to that presented in the preceding procedures, or to construct three reversed curves so as to break a long stretch of a straight alignment.

First, two reversed curves can be constructed when the given parameters, in addition to the two tangent lines (back and forward tangents) include the starting point (PC), the ending point (PT), and a point through which the back curve is to pass as indicated in Figure 8.4i1. The points that define the position of the back and forward tangent lines are the

known PC and PT points, and the two curves are forced to start and end at these two points, respectively.

In looking at the said figure it is seen that it is basically the same as Figure 8.4h with the only difference being the presence of the point PP through which the back curve is to pass. With the coordinates of the PC and PP points known, the chord distance E, the plus and offset distances F and G of point PP from the back tangent line, and the chord deflection angle 0.5Θ can be determined. With these parameters now known, the radius of the back curve R_1 can be computed by Equation 8.3h in *Condition 6 of Sect. 8.3*. Since we now know the radius of the back curve the problem becomes the same as that of the preceding `icRcrv6` procedure which is called to complete the solution. Regarding this last said procedure it is noted that an option exists for specifying a zero radius. This can be accomplished by letting the PP point be located on the back tangent line, provided of course that all other requirements of the said procedure are met.

Second, in certain cases, particularly in low speed residential or rural roadways, it becomes necessary to brake a straight alignment by implanting pairs of reversed curves as displayed in Figure 8.4i2. This could become necessary in order to bypass an obstacle such as a big old tree, a historic monument or some environmental feature, or to break the monotony of a long straight run. There are several ways that this can be accomplished, each of which can be solved by calling two of the procedures presented in this chapter, one to commence the alignment deviation, and the other to return back to the original alignment. The subject procedure provides for the simplest of the constructions, that of two pairs of reversed curves of equal radius with the forward curve of the first pair being concentric and compound with the back curve of the second pair, thus in essence constituting one curve. With reference to Figure 8.4i2, the subject procedure provides for the handling of two cases:

1. Three distinct points, the PC, PT, and the PP are known. In this case:
   - The coordinates of point PP are used to determine the offset distance A from the back tangent (AZ1) at which a line parallel to AZ1 is constructed and along which the PCC point is to be located.
   - The PCC point is positioned along the said parallel line and midway between the PC and PT points.
   - The coordinates of the PT point are held as such.
2. Two distinct points, the PC and the PP are known. In this case:
   - The coordinates of point PT are the same as those of the PP point.
   - Point PP is projected on the back tangent (AZ1) and the distance B is determined.
   - Point PT is positioned along AZ_1 a distance of 2B from the PC point.

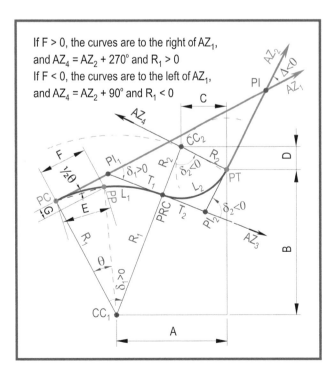

Figure 8.4i1 Reversed Curves Given PC, PT, and POC

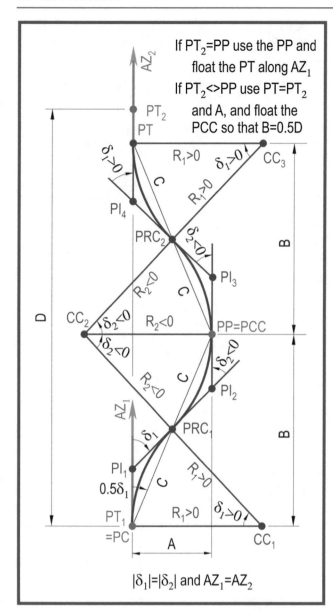

In the figure:

If PT$_2$=PP use the PP and float the PT along AZ$_1$
If PT$_2$<>PP use PT=PT$_2$ and A, and float the PCC so that B=0.5D

$R_1>0$ $\delta_1>0$ CC$_3$

$\delta_1>0$

PT$_2$
PT

$R_1>0$

PI$_4$

C

PRC$_2$ $\delta_2<0$

$R_2<0$ C PI$_3$

B

CC$_2$ $\delta_2<0$ $R_2<0$ PP=PCC

$\delta_2<0$

$R_2<0$ C $\delta_2<0$

PI$_2$

AZ$_1$

δ_1 PRC$_1$

PI$_1$

0.5δ_1 C $R_1>0$

B

PT$_1$
=PC $R_1>0$ $\delta_1>0$ CC$_1$

A

D

$|\delta_1|=|\delta_2|$ and AZ$_1$=AZ$_2$

Figure 8.4i2 Equal Reversed Curves for Bypassing an Obstacle

In either of the above two cases it is assumed that point PP is located between the PC and the PT and either to the right or left of the line connecting these two points (the back tangent line). Thus, the distance A should be positive if point PP is to the right, and negative if to the left. The distance B should always be positive.

Once the above case has been identified and the distances A and B have been determined, it follows that $\delta = |\delta_1| = |\delta_2|$ and that chord length of each curve is

$$C = \frac{1}{2}\sqrt{A^2 + B^2} \qquad (8.4i1)$$

and the central angle of each curve is

$$\delta = 2\sin^{-1}\left(\frac{C}{2C}\right) \qquad (8.4i2)$$

where C is always positive and the sign of δ is dependent on the sign of A. If the PP point is located to the right of the back tangent line AZ_1, than A would be positive, and if it is to the left than A would be negative. It should be noted that the sign of A also controls the sign of R and δ as computed below. From the above then the value of the radius R_1 would be

$$R = \frac{C}{2\sin\left(\frac{1}{2}\delta\right)} \qquad (8.4i3)$$

where the sign of R is dependent on the sign of δ. The next four unknowns, $R1$, $R2$, δ_1, and δ_2, then become

$$R_1 = R \quad \text{and} \quad R_2 = -R_1 \qquad (8.4i4)$$

and

$$\delta_1 = \delta \quad \text{and} \quad \delta_2 = -\delta_1 \qquad (8.4i5)$$

Having determined the radii and the central angles, the coordinates of the various points to be returned can be computed as has been shown previously.

Regarding the coordinates of the points to be returned it is necessary to return to the preamble of this procedure in which it was stated that there are types of reversed curves that can be constructed, those that are portrayed by Figure 8.4i1 and those that are portrayed by Figure 8.4i2. A comparison of these figures indicates that the latter requires the return of a greater number of points to be computed and

```
Sub icRcrv7 (PCN, PCE, AZ1, PT2N, PT2E, AZ2,
            PPN, PPE, TheCASE, _
            R1, R2, DEL1, DEL2, DELTA, PRC1N,
            PRC1E, PRC2N, PRC2E, PCCN, _
            PCCE, CCN1, CCE1, CCN2, CCE2,
            CCN3, CCE3, PTN, PTE, PIN, PIE, _
            PIN1, PIE1, PIN2, PIE2, PIN3, PIE3, _
            PIN4, PIE4, T1, T2, L1, L2, IOK)
```

returned. To handle both cases, the procedure returns certain unnecessary parameters for the case of Figure 8.4i1 as is to be seen below.

The input parameters to this procedure include the:

PCN, PCE	The north and east coordinates of the PC point on the back tangent line.
AZ1	The north azimuth (in radians) of the back tangent line pointing towards the PI of the reversed curves.
PT2N, PT2E	The north and east coordinates of the PT point on the forward tangent line for the case of Figure 8.4i1. For the case of Figure 8.4i2 this is a point on the back tangent line ahead of the PC point and could be the PT point, or it could be the same point as the (PPN, PPE) point if the PT point is to float along the back tangent line.

AZ2   The north azimuth (in radians) of the forward tangent line pointing away from the PI of the reversed curves for the case of Figure 8.4i1. For the case of Figure 8.4i2 this is azimuth is not used, and it is returned having the same value as AZ1.

PPN, PPE   The north and east coordinates of the point through which the back curve is to pass for the case of Figure 8.4i1. For the case of Figure 8.4i2 if this point:

- Is the same as (PT2N, PT2E) denotes that it is to be used as the PCC point, and that the PT point is to float along the back tangent line.
- Is not the same as (PT2N, PT2E) denotes that it is to be used only for the computation of the A and B distances in Figure 8.4i2, and that the (PT2N, PT2E) point is to be used as the PT point.

TheCASE   An indicator the value of which if it is:

- 1   Denotes that the case of Figure 8.4i1 is to be used.
- 2   Denotes that the case of Figure 8.4i2 is to be used.

The parameters that are returned from this procedure include the:

R1, R2   The radius of the back and forward curves for the case of Figure 8.4i1, and the radius of each curve for the case of Figure 8.4i2, in which $|R1| = |R2|$ but opposite in sign.

DEL1, DEL2   The central angles of:

(a)   The back and forward curves for the case of Figure 8.4i1, or

(b)   The first and last curve or of the two center curves for the case of Figure 8.4i2 with $|DEL1| = |DEL2|$ but opposite in sign.

DELTA   The overall deflection angle of the two given tangent lines for the case of Figure 8.4i1, and zero for the case of Figure 8.4i2.

PRC1N, PRC1E   The north and east coordinates of the PRC between the back and forward curves for the case of Figure 8.4i1, and the north and east coordinates of the PRC point between the first two curves for the case of Figure 8.4i2.

PRC2N, PRC2E   Zeros for the case of Figure 8.4i1, and the north and east coordinates of the PRC point between the last two curves for the case of Figure 8.4i2.

PCCN, PCCE   Zeros for the case of Figure 8.4i1, and the north and east coordinates of the PCC point between the second and the third curve for the case of Figure 8.4i2.

CCN1, CCE1   The north and east coordinates of the center point of the back curve for case of Figure 8.4i1 and of Figure 8.4i2.

CCN2, CCE2   The north and east coordinates of the center point of the forward curve for the case of Figure 8.4i1, and the north and east coordinates of the center point of the second and third curve for case of Figure 8.4i2.

CCN3, CCE3   Zeros for the case of Figure 8.4i1, and the north and east coordinates of the center point of the fourth curve for case of Figure 8.4i2.

PTN, PTE   The north and east coordinates of the PT point for either case. It could be a replica of the (PTN, PTE) or (PT2N, PT2E) or not depending on the case being used.

PIN, PIE   The overall PI point of the two given tangent lines for the case of Figure 8.4i1, and zeros for the case of Figure 8.4i2.

PIN1, PIE1   Zeros for the case of Figure 8.4i1, and the north and east coordinates of the PI point of the back or first curve for case of Figure 8.4i2.

PIN2, PIE2   Zeros for the case of Figure 8.4i1, and the north and east coordinates of the PI point of the second curve for case of Figure 8.4i2.

PIN3, PIE3   Zeros for the case of Figure 8.4i1, and the north and east coordinates of the PI point of the third curve for case of Figure 8.4i2.

PIN4, PIE4   Zeros for the case of Figure 8.4i1, and the north and east coordinates of the PI point of the fourth or forward curve for case of Figure 8.4i2.

T1, T2   The lengths of the tangents of the back and forward curves for the case of Figure 8.4i1. For case of Figure 8.4i2 they are equal to each other and denote the tangent length of each of the four curves.

L1, L2   The curve lengths of the back and forward curves for the case of Figure 8.4i1. For case of Figure 8.4i2 they are equal to each.

IOK   An indicator the value of which if it is:

- 0   Denotes that no problems have been encountered.
- 1, 2, and 3   Denotes the error conditions of the icRcrv6 procedure of the same number.

For the case of Figure 8.4i2 the middle curve is considered as two separate but equal curves. Thus, the three reversed curves may be considered as two sets of two reversed curves.

- 4   Denotes that the (PPN, PPE) point is located on the back tangent line. This is not fatal for when TheCASE = 1, but it is fatal for when TheCASE = 2.
- 5   Denotes that the (PT2N, PT2E) point is not located on the back tangent line for when TheCASE = 2. Abort.

Upon invocation, the procedure initializes the parameters to be returned, and then:

1. Calls the `icprjct` procedure to project the (`PPN`, `PPE`) on the back tangent line to determine the offset distance of the said point from the said line. If this offset distance is equal to zero within a tolerance of 0.005 feet (1.524 mm) `IOK` is set to be 4.

2. Queries the indicator `TheCASE` to determine which of the two cases is applicable. If it is 1, continues with the next step, else it branches to Step 6.

3. Queries the indicator `IOK` for being equal to 4 or not. If it is equal to 4 sets the radius of the back curve to be zero and continues with the next step. For the case of Figure 8.4i1 under certain conditions the said radius could have a value of zero (refer to the `icRcrv6` procedure).

4. Computes the radius `R1` of the back curve by use of Equation 8.3h, and then calls the `icRcrv6` procedure to construct the two curves.

5. Sets the coordinates of the PT (`PTN`, `PTE`) point to be those of the PT2 (`PT2N`, `PT2E`) point, and terminates.

6. Queries the indicator `IOK` for being equal to 4 or not. If it is equal to 4 it aborts, else continues with the next step. For the case of Figure 8.4i2 the PP (`PPN`, `PPE`) point cannot be located on the back tangent line as that would not modify the alignment of the back tangent line.

7. Calls the `icforce` procedure to inverse from the PC point (`PCN`, `PCE`) towards the PT point (`PTN`, `PTE`) and determine the distance *B* of Figure 8.4i2 and direction of the line between the said two points.

8. Sets the distance *A* of Figure 8.4i2 to be the offset distance of the PP (`PPN`, `PPE`) point from the back tangent as computed in Step 1, and then compares the coordinates of the PT2 (`PT2N`, `PT2E`) with those of the PP (`PPN`, `PPE`) to determine whether they are the same within a tolerance of 0.005 feet (1.524 mm). If the said two points:
   - Are the same, the PCC is assumed to be fixed and the distance *B* of Figure 8.4i2 is set to be the plus distance computed in Step 1 and branches to Step 10.
   - Are not the same, the PT is assumed to be fixed and continues with Step 9.

9. Compares the azimuth between the PC and the PT points computed in Step 7 with that of the input `AZ1` to determine whether these two azimuths are not equal within a tolerance 0.0000048 radians (0.99 s of a degree), or not.
   - If they are not the same, `IOK` is set to be 5 and aborts.
   - If they are the same, the distance *B* of Figure 8.4i2 is set to be the one half the distance computed in Step 7, and `AZ2` is forced to be the same as `AZ1`.

10. Computes the values of the common radius *R* and central angle δ of the four curves by use of Equations 8.4i1 through 8.4i5. Again note that the sign of the above said

distance *A* dictates the sign of the said radius and of the central angle of the first curve, and that `R1` and `DEL1` are equal to their values, while `R2` and `DEL2` are equal to them respectively in absolute value but of opposite sign.

11. Computes the values of the common tangent lengths of the curves by multiplying the radius by the tangent of one half of the central angle and using the absolute value for each of T1 and T2.

12. Computes the azimuths of the radial lines that connect the tangency points with their corresponding curve center points starting with AZ1, subtracting 90° and then adding `DEL1` and subtracting `DEL2` successively. In computing the azimuths, the `icaz0_2pi` procedure is called to reduce each azimuth to the range of zero to 360° in absolute value.

13. Calls the `icptl` procedure to compute the coordinates of the tangency points and of the arc center points by starting from the PC points and traversing along the respective radial and tangential azimuths the corresponding radius or tangent lengths.

14. Calls the `iccenang` procedure to compute the common arc length L1 of each of the four curves using the two radial azimuths of the first curve and its radius and central angle.

8.4.8 Reversed Curves Given Curve Radii and Point

The `icRcrv8` procedure enables the programmer to construct two reversed curves when the given parameters, in addition to the two tangent lines (back and forward tangents) include the two radii and the coordinates of a point (PP) through which the back curve is to pass as indicated in Figure 8.4j (the top part of this figure displays the various parameters used in the solution, while the bottom part is used to identify the default and additional solutions referenced below). Noted that with this procedure the PC and PT points are to float along their respective tangent lines depending on the position of the said point and size of the two radii.

To construct the reversed curves, with reference to Figure 8.4j, the procedure projects the point PP on the back tangent to determine the said point's offset distance *F* from the said tangent which is positive when PP is to the right of the back tangent and negative when to the left. In the said figure it is seen that the center point of the back curve is to be located at the intersection of a line parallel to the back tangent and a distance *F* away therefrom on the side of point PP, and a curve of radius R_1 (the radius of the back curve) having its center point at point PP.

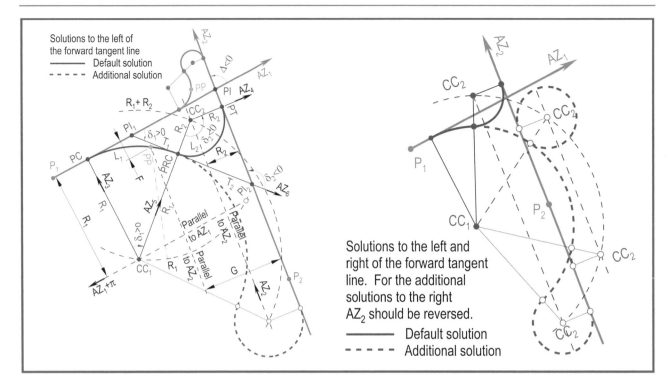

Figure 8.4j Reversed Curves Given R₁, R₂, and PP

With the center point of the back curve known, the center point of the ahead curve is located at the intersection of a circle of radius equal to the sum of the two radii and having its center point at the center point of the back curve with a line parallel to and offset from the ahead tangent line a distance of R_2 (radius of the ahead curve). This defines the general methodology of solving this construction problem. However, we must take into consideration the following:

- The fact that the intersection of a circular curve with a line can produce two, one or no intersection points. This issue is addressed in the construction steps below.
- The side, left or right with respect to AZ_2, of the forward tangent line to which the parallel line is to be constructed.
- The general premise of the reversed curves that the each of the two curves is to advance towards the direction of its respective tangent line.

As stated above, the position of point PP with respect to the back tangent dictates the side of the said tangent that center point of the back curve is to be located and therefore the side on which the parallel line is to be constructed. To determine the coordinates of the center point of the back curve the procedure calls the `iclincrc` procedure which returns as the intersection point the second intersection point in the direction of the line being intersected. In Figure 8.4j one of the two intersection points is shown with a solid red

dot annotated as CC_1, and the other with a hollow dot. The latter is not chosen because the curve that would be generated tangent to the back tangent line would pass through point PP would progress in a direction that would oppose AZ_1. Thus, the direction of the offset and parallel line should oppose AZ_1 since the second intersection is desired.

Now we are faced with the issue of deciding on which side of the forward tangent line we are to construct the parallel and offset line, and which of the two potential intersections we are to select. Thus, we could have two solutions to the left and two to the right of the forward tangent line. To help with this dilemma, the procedure assumes what is referred to as the default solution for which the two curves are constructed and their pertinent parameters are returned to the calling program. In addition to this default solution the procedure returns the coordinates of the other three intersections (the additional center points). With this information the calling program can accept the default solution, or select one of the other three center points and complete the two curves.

For the default solution, the point PP is projected on the forward tangent line to obtain the offset distance G. A line parallel to the said tangent line and offset therefrom a distance G on the side of PP is then constructed. The sign of G is dependent upon which side of the forward tangent line PP is located. The center point of the forward curve is then to be located at the intersection of the said parallel line with a curve of radius $R_1 + R_2$ having its center point at the center point of the back curve. The direction of the said parallel line is that of AZ_2, and the second intersection is selected by use of the `iclincrc` procedure.

With the center points of the back and forward curves now known, the points of tangency may now be determined, and then the central angles, the PI points, the tangent lengths and the arc lengths may then be computed as has been done previously with the other reversed curve procedures.

The input parameters to this procedure include the:

```
Sub icRcrv8 (P1N, P1E, AZ1, P2N, P2E, AZ2, _
             R1, R2, PPN, PPE, _
             DEL1, DEL2, DELTA, PCN, PCE, PRCN,
             PRCE, PTN, PTE, CCN1, CCE1, _
             CCN2, CCE2, PIN, PIE, PIN1, PIE1,
             PIN2, PIE2, T1, T2, L1, L2, IOK, _
             CCN2a, CCE2a, CCN2b, CCE2b, _
             CCN2c, CCE2c)
```

P1N, P1E	The north and east coordinates of a point anywhere on the back tangent line.
AZ1	The north azimuth (in radians) of the back tangent line pointing towards the PI of the reversed curves.
P2N, P2E	The north and east coordinates of a point anywhere on the ahead tangent line.
AZ2	The north azimuth (in radians) of the forward tangent line pointing away from the PI of the reversed curves.
R1, R2	The radii of the back and forward curves. They should be positive for a clockwise rotation and negative for a counterclockwise rotation, and opposite in sign. If they are not so, their input values are altered by the procedure to correspond with the position of the (PPN, PPE) point with respect to the back tangent line.
PPN, PPE	The north and east coordinates of the point through which the back curve is to pass.

The parameters that are returned from this procedure include the:

DEL1 to DELTA	The same as those identified for the icRcrv1 procedure.
PCN, PCE	The north and east coordinates of the PC point on the back tangent line.
PRCN to L2	The same as those identified for the icRcrv1 procedure.
IOK	An indicator the value of which denotes: • 0 No problems have been encountered. • 1 Either of the radii has a zero value. • 2 The back curve cannot be constructed. Check the input data. • 3 The forward curve cannot be created. Probably one or both radii is too small. Check the input data.

CCN2a, CCE2a	The north and east coordinates of the additional center point located on the same side of the forward tangent line as the default solution.
CCN2b, CCE2b	The north and east coordinates of the additional center point located on the opposite side of the forward tangent line as the default solution.
CCN2c, CCE2c	The north and east coordinates of the other additional center point located on the opposite side of the forward tangent line as the default solution.

Upon invocation, the procedure initializes the parameters to be returned, and then:

1. Sets the absolute value of each of the two radii in RR1 and RR2, and queries each one of them for being equal to zero within a tolerance value of 0.005 feet (1.524 mm). If so, sets IOK to 1 and aborts; else it continues.
2. Calls the `icprjct` procedure to project the PP point (PPN, PPE) on the back and forward tangent lines to find the value of the offset distances F and G, respectively, with positive denoting the right side.
3. Compares the back and forward tangent lines for being parallel with the same or with opposite azimuth within a tolerance of 0.0000048 radians (0.99 s of a degree).
   • If so, sets DELTA equal to zero or 1π and the coordinates of the PI to 99999.0, and continues with the next step.
   • If not so, calls the `icintrs` procedure to intersect the back and forward tangent lines and compute the PI coordinates and the deflection angle DELTA.
4. Calls the `icptl` procedure to construct a line parallel to:
   • The back tangent line and opposite in direction to AZ1, located a distance RR1 therefrom on the side of point PP (PPN, PPE), and referred herein as Line 1.
   • The forward tangent line in the direction of AZ2, located a distance RR2 therefrom on the side of the PP point (PPN, PPE), and referred herein as Line 2.
5. Calls the `iclincrc` procedure to intersect Line 1 with an imaginary circle of radius R1 having its center point at the PP point (PPN, PPE) to locate the center point of the back curve (CCN1, CCE1). The reader is reminded that the `iclincrc` procedure returns the second of the two potential intersections. Thus, Line 1 should oppose AZ1 to return the proper intersection. The last said procedure also returns the indicator INTRS denoting the number of solutions that have been encountered.
6. Queries INTRS for being equal to 2 which denotes that an intersection has not been found, and if so sets IOK to be 2, beeps and aborts.

7. Calls the `iclincrc` procedure to intersect Line 2 with an imaginary circle of radius `R1 + R2` having its center point at the center point of the back curve (`CCN1`, `CCE1`) to locate the center point of the forward curve (`CCN2`, `CCE2`). For the reasons stated previously Line 2 should have the direction of `AZ2` to return the proper intersection. This intersection returns the indicator `INTRSA` denoting the number of solutions that have been encounter.

8. Queries `INTRSA` for being equal to 1, 2, or 3. If it is:
   - 2, there is no intersection, and thus, there is no default solution and most probably there are no additional center points. The indicator `IOK` is set to be 2, beeps and aborts.
   - 3, there is one intersection point (Line 2 is tangent to the imaginary circle), the coordinates (`CCN2a`, `CCE2a`) to be returned are set to be zero, and continues.
   - 1, there are two solutions, the procedure:
     – Reverses the azimuth of Line 2, and calls the `iclincrc` procedure to intersect Line 2 with the same imaginary circle to determine the coordinates of the additional center point on the same side of the forward tangent line. The indicator `INTRSA` is not been queried because it already has been established that there are two intersections.
     – Calls the `icptl` procedure to construct a new Line 2 on the opposite side of the forward tangent line in the direction of `AZ2`.
     – Calls the `iclincrc` procedure to intersect the new Line 2 with the same imaginary circle to determine the coordinates of the additional center point on the opposite side of the forward tangent line. At this time the indicator `INTRSB` is returned.
     – Queries `INTRSB` for being equal to 1, 2, or 3. If it is:
       - 2, there are no additional center points of the opposite side of the forward tangent line, sets to zero the coordinates of (`CCN2b`, `CCE2b`) and (`CCN2c`, `CCE2c`), and continues.
       - 3, there in only one additional center points of the opposite side of the forward tangent line, sets to zero the coordinates of (`CCN2c`, `CCE2c`), and continues.
       - Reverses the azimuth of the new Line 2, and calls the `iclincrc` procedure to intersect the new Line 2 with the same imaginary circle to determine the coordinates of the second additional center point on the opposite side of the forward tangent line. The indicator `INTRSB` is not been queried because it already has been established that there are two intersections.

9. Calls the `icprjct` procedure to project the center point of the back curve (`CCN1`, `CCE1`) computed above on the back tangent line to obtain the coordinates of the PC point (`PCN`, `PCE`), and computes the radial azimuth from the said center point towards the said PC point.

10. Calls the `icprjct` procedure to project the center point of the forward curve (`CCN2`, `CCE2`) computed above on the forward tangent line to obtain the coordinates of the PT point (`PTN`, `PTE`), and computes the radial azimuth from the said center point towards the said PT point.

11. Calls the `icforce` procedure to inverse from the center point of the back curve (`CCN1`, `CCE1`) towards the center point of the forward curve (`CCN2`, `CCE2`) to determine the common radial azimuth, and then calls the `icptl` procedure to determine the coordinates of the PRC point (`PRCN`, `PRCE`) by traversing from the center point of the back curve toward that of the forward curve a distance of `RR1`.

12. Modifies the input values of the two radii to have the proper sign denoting the rotational direction of the curves.

13. Calls the `icstrdln` procedure to determine the central angle of the back curve using the pertinent radial azimuths, and assigns the same sign as that of the radius.

14. Reverses the direction of the common radial azimuth and then calls the `icstrdln` procedure to determine the central angle of the forward curve using the pertinent radial azimuths, and assigns the same sign as that of the radius.

15. Computes the lengths of the tangents of the two curves as $T = RTan(\delta/2)$.

16. Computes the arc lengths of the two curves as $L = R(\delta/2)$.

17. Calls the `icptl` procedure to determine the coordinates of the PI points (`PIN1`, `PIE1`) and (`PIN2`, `PIE2`) by traversing from the PC point along `AZ1` and then along the normal direction to the common radial line.

This concludes the work of the subject procedure.

8.4.9   Reversed Curves Given the Curve Radii and δ_1

The `icRcrv9` procedure enables the programmer to construct two reversed curves when the given parameters, in addition to the two tangent lines (back and forward tangents) include the radii of the two curves and the central angle of the back curve as indicated in Figure 8.4k1. Under these conditions the start (PC) and end (PT) points of the back and forward curves, respectively, will float along their respective tangent lines. In a way this procedure is similar to the preceding one with the basic difference being that under this procedure the end point of the back curve is fixed by the specification of the PRC point (since a central angle is specified as being given) rather than by the specification of a point

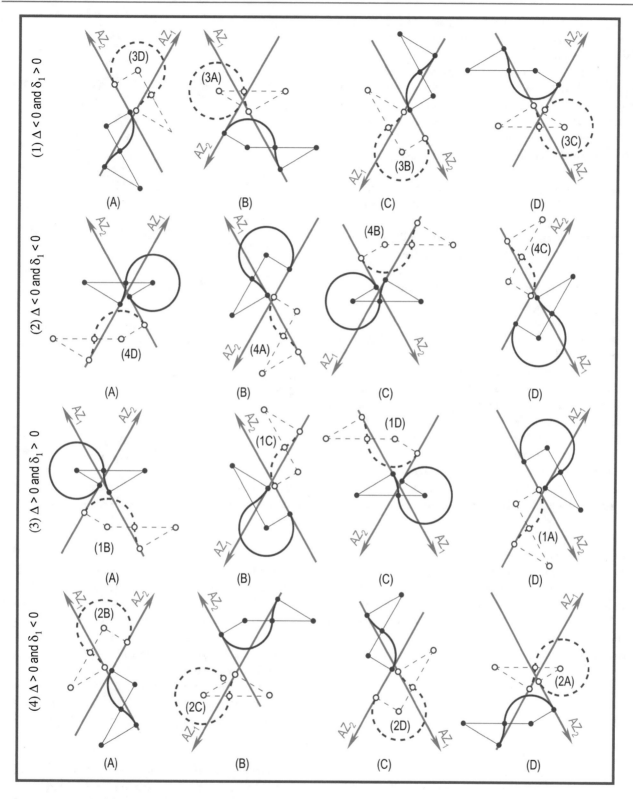

Figure 8.4k1 Reversed Curves Given R_1, R_2, and δ_1 or δ_2 (*red dash lines* oppose the direction of AZ_2)

through which the back curve is to pass. It should be noted that although the PT is fixed by the PRC, the position of the PRC is not known, and has to be computed.

In Figure 8.4k1 there are four rows (numbered 1 through 4) of potential solutions with two pairs of reversed curves that can be generated for each of the four quadrants (labeled A through D) formed by the given tangent lines. Referring to these reversed curve conditions from top to bottom in the said figure and from left to right it is seen that:

- In top two rows 1 and 2 the two tangent lines develop a counterclockwise (negative) overall deflection angle (Δ) when rotating from the back tangent line (AZ_1) towards the forward tangent line (AZ_2), while in the bottom two rows 3 and 4 the said angle Δ is clockwise (positive).
- For rows 1 and 3 the central angle (δ_1) of the back curve is clockwise (positive), and for the rows 2 and 4 it is counterclockwise (negative).
- In each of the four quadrants of each row there are two potential solutions, one of which is represented by a solid red line, and the other represented with a dashed red line. In both sets of curves the back curve advances from the PC along the direction of the back tangent line AZ_1. However, the forward curve represented with the solid line advances in the same direction of AZ_2, while that with the dashed line advances in an opposite direction of forward tangent line AZ_2.
- In each set of curves represented by dashed lines there is an identifier denoting the row and quadrant in which the curves with the dashed lines become curves with solid lines.

Since the position of the tangent lines is known, the overall PI point and deflection angle Δ can easily be computed as done in the preceding procedures, and with the central angle of the back curve δ_1 being given that of the forward curve δ_2 can be computed by subtracting δ_1 from Δ. With this information now known, the lengths of the tangents of each of the back curve T_1 and of the forward curve T_2 may now be determined as being

$$T_1 = \left| R_1 \tan\left(\frac{1}{2}\delta_1 \right) \right| \qquad (8.4\text{j}1)$$

and

$$T_2 = \left| R_2 \tan\left(\frac{1}{2}\delta_2 \right) \right| \qquad (8.4\text{j}2)$$

In the triangle formed by the points PI_1, PI, and PI_2 in Figure 8.4k2 we have the relationship

$$\frac{T_1 + T_2}{\sin(\Delta)} = \frac{A}{\sin(\delta_2)} \qquad (8.4\text{j}3)$$

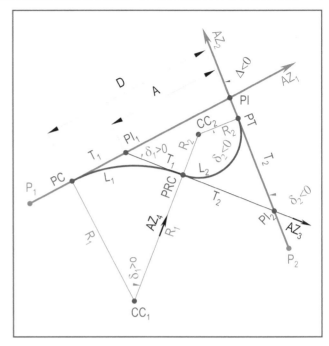

Figure 8.4k2 Reversed Curves Given R_1, R_2, and δ_1 or δ_2 $\left(\delta_1 > 0 \right)$ Tangents Method

or

$$A = \left(T_1 + T_2\right)\frac{\sin(\Delta)}{\sin(\delta_2)} \qquad (8.4\text{j}4)$$

from which we can located the coordinates of the PC point by traversing from the PI point along AZ_1 the instance of $-D$ which is

$$D = A + T_1 \qquad (8.5\text{j}5)$$

We now have all the necessary information to construct the two reversed curves. This methodology is sound as long as the geometry of the reversed curves stays simple as that of the last said figure. However what about certain of the cases of Figure 8.4k1 in which δ_1 and/or δ_2 exceed 90° or even 180°, and what about the case in which δ_1 or δ_2 equals 180° in which T_1 or T_2 becomes infinity?

To generalize the procedure let us take a different approach by referring to Figure 8.4k3 which is a slight modification of the previous figure. In perusing this figure we see that:

- The center point of the back curve C_1 is located on a line parallel to AZ_1 and offset therefrom an R_1 distance.
- The center point of the ahead curve C_2 is located on a line parallel to AZ_2 and offset therefrom an R_2 distance.
- The line from C_1 to C_2 is $R = R_1 + R_2$ long and is inclined at an angle $\alpha = 90° - \delta_1$ from the parallel line to the back tangent.

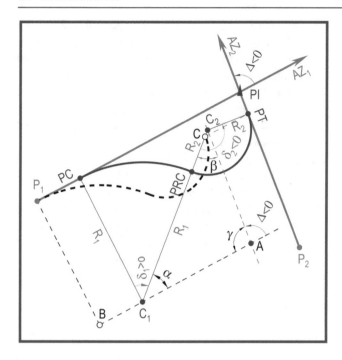

Figure 8.4k3 Reversed Curves Given R_1, R_2, and δ_1 or δ_1 $(\delta_1 > 0)$ Method of the Center Points

- The line from C_1 to C_2 is also inclined at an angle $\beta = \delta_1 - 90°$ from the parallel line to the ahead tangent.
- The angle at the point of intersection A of the said two parallel lines is $\gamma = 180° - \Delta$.

So far we have not referred to any sign conventions of the angles α, β, and γ. With these three angles and the distance R now known, the law of sines may be invoked to compute the other two sides of the triangle C, C_2, A as

$$C_1A = R\frac{\sin(\beta)}{\sin(\Delta)} \qquad (8.5j6)$$

$$C_2A = R\frac{\sin(\alpha)}{\sin(\Delta)} \qquad (8.5j7)$$

Point A is found by intersecting the parallel lines, to AZ_1 and AZ_2 and the sides of the C, C_2, A triangle can be computed and the curves generated. But there is one issue that has to be settled, that of the signs of the angle α, β and γ. It is not a difficult issue but it can be a bit messy if the procedure is to accommodate the conditions of Figure 8.4k1. Therefore to address this issue the procedure utilizes the rather simplistic approach of:

- Assuming that the center point of the back curve B in Figure 8.4k3 is to be offset from back tangent at the P_1 a distance R_1 with the sign of R_1 denoting the side of the tangent.

- Developing the curves and their control points using the given values of R_1, R_2, δ_1 and the computed value of δ_2, with point C in Figure 8.4k3 denoting the temporary PT.
- Intersecting a line of azimuth AZ_1 through C with the ahead tangent AZ_2, and using the distance from C to the point of intersection to translate all curve points along AZ_1.

```
Sub icRcrv9 (P1N,  P1E,  AZ1,  P2N,  P2E,  AZ2,
             R1,     R2, A1, A2, ROT, _
             DEL1, DEL2, DELTA, PCN, PCE, PRCN,
             PRCE, PTN, PTE, CCN1, CCE1, _
             CCN2, CCE2, PIN, PIE, PIN1, PIE1,
             PIN2, PIE2, T1, T2, L1, L2, IOK)
```

The input parameters to this procedure include the:

P1N, P1E	The north and east coordinate of a point anywhere on the back tangent line.
AZ1	The north azimuth (in radians) of the back tangent line pointing towards the PI of the reversed curves.
P2N, P2E	The north and east coordinates of a point anywhere on the ahead tangent line.
AZ2	The north azimuth (in radians) of the forward tangent line pointing away from the PI of the reversed curves.
R1, R2	The radius of the back and forward curves. The radius of the back curve R1 should be positive for a clockwise curve, and negative for a counterclockwise curve. The radius of the forward curve R2 could be either positive or negative. The procedure forces it to have the opposite sign of R1.
DEL1	The central angle of the back curve. Should have the same sign as R1.
ROT	An indicator denoting the rotational direction of the overall deflection angle DELTA from AZ1 towards AZ2. It is used in testing if AZ1 and AZ2 straddle north. Any integer value:
	>0: clockwise rotation.
	<0: counterclockwise rotation.

The parameters that are returned from this procedure include the:

DEL2 to L2	The same as those identified for the `icRcrv8` procedure.
IOK	An indicator the value of which if it is:
	• 0 Denotes that no problems have been encountered.
	• 1 Either of the radii has a zero value.
	• 2 DEL1 has a zero value.
	• 3 R1 and DEL1 are of opposite sign.
	• 4 The tangent lines are parallel. Infinite solutions.

Upon invocation, the procedure initializes the parameters to be returned, and then:

1. Forces R2 to have the opposite sign to R1, and then queries R1 and R2 for being equal to zero within a tolerance of 0.005 feet (1.524 mm). If either is zero, sets IOK to 1 and aborts, else it continues.
2. Queries DEL1 for being equal to zero within a tolerance of 0.0000048 radians (0.99 s of a degree). If so sets IOK to 2 and aborts, else it continues.
3. Queries R1 and DEL1 for having the same sign. If their signs are not the same, sets IOK to 3 and aborts, else it continues.
4. Queries AZ1 and AZ2 for being parallel with the same or opposite direction. If they are, sets IOK to 4 and aborts, else it continues. In the process of checking calls the icintrs procedure to intersect AZ1 and AZ2 and computes the overall DELTA angle (adjusted for straddling north) and the PI coordinates.
5. Computes the value of DEL2 by subtracting DEL1 from DELTA.
6. Calls the icptl procedure to compute the coordinates of the temporary center point of the back curve (CC1) to the right or to the left side of AZ1 at the P1 point (P1N, P1E).
7. Determines the azimuth of the common radial line from CC1 towards CC2 by adding DEL1 to the normal azimuth to AZ1 and then adding or subtracting 90° depending on the sign of R1.

8. Calls the icptl procedure to compute the coordinates of the temporary PRC and of the temporary center point of the ahead curve (CC2) along the said common radial azimuth by traversing a distance R1 and R = R1 + R2.
9. Determines the normal azimuth to AZ2 from CC2 and calls the icptl procedure to compute the coordinates of the temporary PT by traversing along the said normal azimuth a distance R2.
10. Calls the icintrs procedure to intersect a line parallel to AZ1 and through the temporary PT point with the AZ2 line to determine the distance of translation.
11. Calls the icptl procedure five times to compute the final coordinates of the PT, CC1, PRC, CC2, and PT points by traversing from their temporary positions along AZ1 the said distance of translation.
12. Computes the azimuth of the common tangent line by adding 90° to that of the common radial line.
13. Calls the icintrs procedure to intersect the said common tangent line with the back tangent line AZ1 and forward tangent line AZ2 to determine the coordinates of the PI of the back and forward curves, respectively, as well as the corresponding tangent length of each of the said curves.
14. Computes the arc length of each of the said curves by multiplying the corresponding central angle with the respective radius.

The procedure now terminates.

As an automobile travels along a straight stretch of a highway, or as a train speeds along a long straight pair of rails, or as a bicyclist or motorcyclist for that matter does the same, they all occasionally have to turn left or right. We all know that as that happens, the body that travels along such a straight course at high speeds builds a centrifugal force that pushes the body away from its path when the path has to veer to the left or right. The higher the speed and the higher the body's mass, the higher the centrifugal force. To compensate for this force, a cyclist leans left or right in the direction of the turn. Since an automobile or train cannot lean in such a manner, the engineers and contractors elevate one end of the travel path to induce such a tilt to the traveling vehicle.

When a travel path has to change direction, the change is provided by a circular curve. For safety and passenger comfort the radius of this curve has to be as large as possible. But a circular curve does not provide a smooth transition from a straight line having a radius of infinite value to a curve of a specific radius. Thus, the need arises in providing a travel path the radius of which changes directly with respect to the distance from the start of the turn. This is provided by a special curve, the transition spiral.

Many an automobile driver has noticed that when turning even at low or average speeds, one makes his or her own spiral within the pavement ribbon of the travel path. However, when traveling at high speeds, the driver may not be able to keep the automobile within the ribbon of the travel path, which could be hazardous. Thus, spiral curves are generally encountered on expressways and arterial routes and not on city streets and rural roadways.

Before the advent of the computer the use of spirals involved a tedious operation. A variety of tables and approximate formulas had been developed to assist an engineer in easily solving for the basic elements of a spiral. However, the intersection of a spiral and lines equidistant from it with other lines (straight, circular, or spiral) was a long laborious process of trial and error, while graphical solution within

proper scales was not accurate enough for bridge geometry. The computer has alleviated this process.

In the preceding chapters we address the construction of straight lines and circular curves, as well as the interactions between them such as intersections and tangencies. In this chapter we address the introduction of a spiral curve between a straight line and a circular curve, or between two circular curves. In addition, we address (a) the intersection between a spiral and a straight line, circular curve, and another spiral curve, (b) the construction of a tangent line to a spiral curve at a point or from a point, (c) the projection of a point on a spiral curve, and (d) the construction of lines offset , or equidistant from a spiral curve.

Regarding offset lines to a spiral, it is important to note that such offset lines may not be represented by a pure mathematical equation. Thus, they are constructed as polylines that connect points that are created along a spiral at a uniform spacing and at a constant offset distance thereof.

In the chapter sections below we address the development of the general or basic spiral equation, followed by the various operations that may be performed with the said equation.

The contents of this chapter are a modification of the paper by Elias C. Tonias, P.E. as published in the Journal of the Highway Division Proceedings of the American Society of Civil Engineers, Volume 91, Issue HW2, December, 1965, pages 101–127 and titled *Spiral Problems For An Electronic Computer*, with discussions by Thomas F. Hickerson M. ASCE (HW2, October 1966) and by M. Gonzales, A.M. ASCE (HW2, October 1966).

Reference is also made to the paper by Mr. Jack C. Cluff as published in the Journal of Surveying Engineering of the American Society of Civil Engineers, Volume 121, No. 3, August, 1995, pp. 118–127. Mr. Cluff's paper introduces certain valuable additions to the original work, one being the application of Horner's rule to the solution of Equations 15a and 15b, and the expansion of the number of terms in Equation 15c. The said three equations refer to Equations 9.1.15a, 9.1.15b, and 9.1.15c of this publication.

9.1   The Basic Spiral

When a vehicle travels along a given path, the angle of its steering mechanism has to be changed each time a change in the path's degree of curvature occurs. For passenger safety and comfort such a change has to be small and smooth so as to practically be unnoticeable. Thus, at the proximity of the point of tangency between two curvatures of considerable variance in degree, a transition has to be offered in the path to produce such an effect. Such a transition is to have throughout its entire length a constantly changing degree of curvature which should be directly dependent on the distance from its beginning.

The said transition may be necessary when entering a sharper curvature from a flatter curvature (entering or entry spiral), or when entering a flatter curvature from a sharper curvature (exiting or exit spiral) in the direction of travel. With regard to computation, these two spirals are treated identically, by reversing the direction of the exiting spiral. In this discussion, the said transition is always between two curvatures, with a straight line being considered as a curve with an infinite radius. It is noted that for programming purposes, the radius, or the degree of curvature as addressed below, of a straight line is expressed as zero (0.00). It is also noted that throughout this publication all angles are expressed in radians for all angles, and all linear mensurations are expressed in feet. However, the procedures included with this publication can accommodate the SI linear units of measure.

Let us consider a transition curve, such as that shown in Figure 9.1a (herein called spiral or main spiral), which begins at point A and ends at point B. The path up to point A

(progressing in this figure clockwise from left to right) has a constant radius of R_1 and a corresponding degree of curvature of D_1, and from point B on forward a constant radius of R_2 and a corresponding degree of curvature of D_2. It is noted that in highway terminology the degree of curvature in radians equals the reciprocal of the corresponding radius multiplied by one hundred ($D = 100/R$). In the same figure, let $R_1 > R_2$ or $D_1 < D_2$. Had the path up to point A been a straight line, then the radius $R_1 = \infty$ (for programming purposes infinity is expressed herein as zero) and $D_1 = 0$.

The degree of curvature in radians equals the reciprocal of the corresponding radius multiplied by one hundred ($D = 100/R$).

For a one degree curve the radius is equal to $R = 100/D$ or $R = 100 * 180/\pi$ $R = 5729.577951$ feet, or $R = 1746.375359$ m, and for a $10°$ curve $R = 572.9577951$ feet, or $R = 174.6375359$.

Now let P represent a point on the spiral, L be the distance from point A to point P along the spiral with an instantaneous radius of R, and the corresponding degree of curvature of D at point P. Since the spiral to be selected is such that its degree of curvature is to vary directly with the distance from point A, it follows that

$$D = D_1 + K L \tag{9.1.1a}$$

where K is the constant of variance in the degree of curvature.

At the start and end extremities of the spiral

$$D = D_1 \tag{9.1.1b}$$

and

$$D = D_1 + K L_S \tag{9.1.1c}$$

respectively, where L_s represents the curve's distance along the spiral from point A to point B. Solving Equation 9.1.1c for K, and substituting its value in Equation 9.1.1a, the instantaneous degree of curvature becomes

$$D = D_1 + \frac{(D_2 - D_1)L}{L_S} \tag{9.1.2a}$$

or

$$\frac{1}{R} = \frac{1}{R_1} + \left(\frac{1}{R_2} - \frac{1}{R_1}\right)\frac{L}{L_S} \tag{9.1.2b}$$

In the differential segment of Figure 9.1a,

$$dL = Rd\theta \quad or \quad d\theta = dL / R.$$

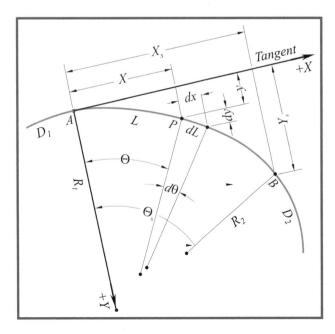

Figure 9.1a The Basic Spiral

By substituting the value of the reciprocal of R in Equation 9.1.2b

$$d\theta = \left[\frac{L}{R_1} + \left(\frac{1}{R_2} - \frac{1}{R_1}\right)\frac{L}{L_S}\right]dL \qquad (9.1.3)$$

which if integrated it becomes

$$\theta = \frac{L}{R_1} + \frac{L^2}{2L_S}\left(\frac{1}{R_2} - \frac{1}{R_1}\right) \qquad (9.1.4a)$$

or

$$\theta = \frac{L}{100}\left[D_1 + \frac{L}{2L_S}(D_2 - D_1)\right] \qquad (9.1.4b)$$

in which the angle θ is referred to herein as the instantaneous spiral angle, and which at the end of the spiral loses the identification of being instantaneous, and is

$$\theta = \frac{L_S}{200}(D_1 + D_2) \qquad (9.1.5)$$

In this publication the spiral angle, may it be instantaneous or not, is defined as the angle formed by the radius at the beginning of the spiral (at point A) and the radius at a point on the spiral (P, B, or any other point between point A and point B) and is always positive.

Expanding Equation 9.1.4b to remove the brackets, multiplying both of its sides by $(200L_s)/(D_2 - D_1)$ and interchanging the resultant two equation sides, Equation 9.1.4b becomes

$$L^2 + \frac{2D_1 L_S}{D_2 - D_1}L = \frac{200 L_S \theta}{D_2 - D_1} \qquad (9.1.6a)$$

By completing the square, obtaining the square roots and transposing the terms, Equation 9.1.6a becomes

$$L = \sqrt{\frac{(200 L_S \theta)}{(D_2 - D_1)} + \frac{(D_1 L_S)^2}{(D_2 - D_1)^2}} - \frac{D_1 L_S}{(D_2 - D_1)} \qquad (9.1.6b)$$

Multiplying and dividing the left most term under the square root sign with $L_s(D_2 - D_1)(D_2 + D_1)$, recognizing θ_s as expressed by Equation 9.1.5, and then by factoring

$$L = \frac{L_S}{D_2 - D_1}\left[\sqrt{\frac{\theta}{\theta_S}\left((D_2^2 - D_1^2)\right) + D_1^2} - D_1\right] \qquad (9.1.6c)$$

Thus far, all work presupposed that $D_1 < D_2$, a transition from a flat to a sharp curvature. Had this condition been reversed, then Equation 9.1.1a should have been written as

$$D = D_1 + K L \qquad (9.1.7a)$$

because the curvature will tend to decrease in a forward direction. At the forward end of the spiral (point B), the degree of curvature will be

$$D = D_1 - K L_S - D_2 \qquad (9.1.7b)$$

and

$$K = \frac{(D_1 - D_2)}{L_S} \qquad (9.1.7c)$$

In a manner similar to that used previously,

$$D = D_1 - \frac{(D_1 - D_2)L}{L_S} \qquad (9.1.8a)$$

or

$$D = D_1 + \frac{(D_2 - D_1)L}{L_S} \qquad (9.1.8b)$$

which is identical to Equation 9.1.2a. Hence, Equations 9.1.1a through 9.1.6c will hold true for all spirals provided the subscript 1 refers to the curvature of the low station and the subscript 2 refers to the curvature of the high station (beginning and end of the spiral respectively) in a station ascending order.

Equation 9.1.8b is identical to Equation 9.1.2a because in the arc definition of a circular curve $D_i = 1/R_i$.

The variables L and θ identify a point on a spiral, but they do not offer all information needed to locate it easily. In Figure 9.1a, let the tangent to the spiral at the point of beginning (A) represent the x-axis of a Cartesian coordinate system (positive direction forward in station ascending order) and the radius at point A, the y-axis (positive direction to the right or towards the center of the theoretical circle). The tangent to the spiral at point A henceforth is called, for the sake of brevity, as the *Tangent*. From the differential segment again

$$dy = dL \ \sin\theta \qquad (9.1.9a)$$

and

$$dx = dL \ \cos\theta \qquad (9.1.9b)$$

The trigonometric functions of Equations 9.1.9a and 9.1.9b may be expanded into their series forms. By maintaining powers of θ not greater than the seventh

$$dy = \left(\theta - \frac{\theta^3}{3!} + \frac{\theta^5}{5!} - \frac{\theta^7}{7!} + \ldots\right)dL \qquad (9.1.10a)$$

$$dx = \left(1 - \frac{\theta^2}{2!} + \frac{\theta^4}{4!} - \frac{\theta^6}{6!} + \ldots\right) dL \qquad (9.1.10b)$$

$$\theta = \frac{D_1 L}{100} + \frac{L^2 (D_2 - D_1)}{200 L_s} \qquad (9.1.11)$$

Equation 9.1.4b gave an expression for q which may be substituted in Equations 9.1.10a and 9.1.10b to express dy and dx in terms of L, as

By expanding each term binomially, and integrating with respect to L, values for y and x are obtained as follows:

$$
\begin{aligned}
y = & \left[\left(\frac{D_1}{100}\right)\frac{L^2}{2} + \left(\frac{D_2 - D_1}{200 L_s}\right)\frac{L^3}{3}\right] - \frac{1}{3!}\left[\left(\frac{D_1}{100}\right)^3\frac{L^4}{4} + 3\left(\frac{D_1}{100}\right)^2\left(\frac{D_2 - D1}{200 L_s}\right)\frac{L^5}{5}\right. \\
& + 3\left(\frac{D_1}{100}\right)\left(\frac{D_2 - D_1}{200 L_s}\right)^2\frac{L^6}{6} + \left.\left(\frac{D_2 - D_1}{200 L_s}\right)^3\frac{L^7}{7}\right] + \frac{1}{5!}\left[\left(\frac{D_1}{100}\right)^5\frac{L^6}{6}\right. \\
& + 5\left(\frac{D_1}{100}\right)^4\left(\frac{D_2 - D_1}{200 L_s}\right)\frac{L^7}{7} + 10\left(\frac{D_1}{100}\right)^3\left(\frac{D_2 - D_1}{200 L_s}\right)^2\frac{L^8}{8} \\
& + 10\left(\frac{D_1}{100}\right)^2\left(\frac{D_2 - D_1}{200 L_s}\right)^3\frac{L^9}{9} + 5\left(\frac{D_1}{100}\right)\left(\frac{D_2 - D_1}{200 L_s}\right)^4\frac{L^{10}}{10} + \left.\left(\frac{D_2 - D_1}{200 L_s}\right)^5\frac{L^{11}}{11}\right] \\
& - \frac{1}{7!}\left[\left(\frac{D_1}{100}\right)^7\frac{L^8}{8} + 7\left(\frac{D1}{100}\right)^6\left(\frac{D_2 \, D_1}{200 L_s}\right)\frac{L^9}{9} + 21\left(\frac{D_1}{100}\right)^5\left(\frac{D_2 - D_1}{200 L_s}\right)^2\frac{L^{10}}{10}\right. \\
& + 35\left(\frac{D_1}{100}\right)^4\left(\frac{D_2 - D_1}{200 L_s}\right)^3\frac{L^{11}}{11} + 35\left(\frac{D_1}{100}\right)^3\left(\frac{D_2 - D_1}{200 L_s}\right)^4\frac{L^{12}}{12} \\
& + 21\left(\frac{D_1}{100}\right)^2\left(\frac{D_2 - D_1}{200 L_s}\right)^5\frac{L^{13}}{13} + 7\left(\frac{D_1}{100}\right)\left(\frac{D_2 - D_1}{200 L_s}\right)^6\frac{L^{14}}{14} \\
& + \left.\left(\frac{D_2 - D_1}{200 L_s}\right)^7\frac{L^{15}}{15}\right]
\end{aligned}
$$

$$(9.1.12a)$$

and

$$
\begin{aligned}
X = & L - \frac{1}{2!}\left[\left(\frac{D_1}{100}\right)^2\frac{L^3}{3} + 2\left(\frac{D_1}{100}\right)\left(\frac{D_2 - D_1}{200 L_s}\right)\frac{L^4}{4} + \left(\frac{D_2 - D1}{200 L_s}\right)^2\frac{L^5}{5}\right] \\
& + \frac{1}{4!}\left[\left(\frac{D_1}{100}\right)^4\frac{L^5}{5} + 4\left(\frac{D_1}{100}\right)^3\left(\frac{D_2 - D_1}{200 L_s}\right)\frac{L^4}{4} + 6\left(\frac{D_1}{100}\right)^2\left(\frac{D_2 - D_1}{200 L_s}\right)^2\frac{L^7}{7}\right. \\
& + 4\left(\frac{D_1}{100}\right)\left(\frac{D_2 - D_1}{200 L_s}\right)^3\frac{L^8}{8} + \left.\left(\frac{D_2 - D_1}{200 L_s}\right)^4\frac{L^9}{9}\right] \\
& - \frac{1}{6!}\left[\left(\frac{D_1}{100}\right)^6\frac{L^7}{7} + 5\left(\frac{D_1}{100}\right)^5\left(\frac{D_2 \, D_1}{200 L_s}\right)\frac{L^8}{8} + 15\left(\frac{D_1}{100}\right)^4\left(\frac{D_2 - D_1}{200 L_s}\right)^2\frac{L^9}{9}\right. \\
& + 20\left(\frac{D_1}{100}\right)^3\left(\frac{D_2 - D_1}{200 L_s}\right)^3\frac{L^{10}}{10} + 15\left(\frac{D_1}{100}\right)^2\left(\frac{D_2 - D_1}{200 L_s}\right)^4\frac{L^{11}}{11} \\
& + 6\left(\frac{D_1}{100}\right)\left(\frac{D_2 - D_1}{200 L_s}\right)^5\frac{L^{12}}{12} + \left.\left(\frac{D_2 - D_1}{200 L_s}\right)^6\frac{L^{13}}{13}\right]
\end{aligned}
$$

$$(9.1.12b)$$

Equations 9.1.12a and 9.1.12b may be simplified by letting

$$\frac{D_1}{100} = \left(\frac{D_2 - D_1}{200 L_S}\right) L \cdot M \tag{9.1.13a}$$

and

$$U = \left(\frac{D_2 - D_1}{200 L_S}\right) L^2 \tag{9.1.13b}$$

Hence, by substituting Equations 9.1.13a and 9.1.13b in Equations 9.1.12a and 9.1.12b,

$$y = L\left[U\left(\frac{1}{3} + \frac{1}{2}M\right) - \frac{1}{3!}U^3\left(\frac{1}{7} + \frac{3}{6}M + \frac{3}{5}M^2 + \frac{1}{4}M^3\right)\right.$$
$$+ \frac{1}{5!}U^5\left(\frac{1}{11} + \frac{5}{10}M + \frac{10}{9}M^2 + \frac{10}{8}M^3 + \frac{5}{7}M^4 + \frac{1}{6}M^5\right)$$
$$\left. - \frac{1}{7!}U^7\left(\frac{1}{15} + \frac{7}{14}M + \frac{21}{13}M^2 + \frac{35}{12}M^3 + \frac{35}{11}M^4 + \frac{21}{10}M^5 + \frac{7}{9}M^6 + \frac{1}{8}M^7\right)\right] \tag{9.1.14a}$$

and

$$X = L[1 - \frac{1}{2!}U^2\left(\frac{1}{5} + \frac{2}{4}M + \frac{1}{3}M^2\right) + \frac{1}{4!}U^4\left(\frac{1}{9} + \frac{4}{8}M + \frac{6}{7}M^2 + \frac{4}{6}M^3 + \frac{1}{5}M^4\right)$$
$$- \frac{1}{6!}U^6\left(\frac{1}{13} + \frac{6}{12}M + \frac{15}{11}M^2 + \frac{20}{10}M^3 + \frac{15}{9}M^4 + \frac{6}{8}M^5 + \frac{1}{7}M^6\right)] \tag{9.1.14b}$$

in which

$$M = \frac{2D_1 L_S}{(D_2 - D_1)L} \tag{9.1.14c}$$

Since certain terms within the parentheses resemble the coefficients of a binomial expansion

$$y = L\left(UF_1 - \frac{1}{6}U^3 F_3 + \frac{1}{120}U^5 F_5 - \frac{1}{5040}U^7 F_7\right) \tag{9.1.15a}$$

and

$$x = L\left(1 - \frac{1}{2}U^2 F_2 + \frac{1}{24}U^4 F_4 - \frac{1}{720}U^6 F_6\right) \tag{9.1.15b}$$

in which

$$F_n = \frac{1}{(n+1)}M^n + \frac{n}{(n+2)}M^{n-1} + \frac{n(n-1)}{(n+3)2!}M^{n-2} + \frac{n(n-1)(n-2)}{(n+4)3!}M^{n-3}$$
$$+ \frac{n(n-1)(n-2)(n-3)}{(n+5)4!}M^{n-4} + \frac{n(n-1)(n-2)(n-3)(n-4)}{(n+6)5!}M^{n-5}$$
$$+ \frac{n(n-1)(n-2)(n-3)(n-4)(n-5)}{(n+7)6!}M^{n-6}$$
$$+ \frac{n(n-1)(n-2)(n-3)(n-4)(n-5)(n-6)}{(n+8)7!}M^{n-7} \tag{9.1.15c}$$

in which n = 1, 2, 3, 4, 5, 6, and 7.

To introduce a spiral between two curves or between a curve and a straight line most often than not the degrees of curvature of the two curves and the length of the spiral are known (again a straight line has as a degree of curvature the value of zero). However, by reviewing Equation 9.1.5 we see that there are four distinct parameters, the two degrees of curvature D_1 and D_2, the length of the spiral L_S, and the spiral angle θ_S. Thus, knowing any three of them the other one may be determined. To find the spiral coordinates X_S and Y_S, we must first find M and U employing Equations 9.1.14c and 9.1.13a respectively, and then we may use Equations 9.1.15a and 9.1.15b, letting $L = L_S$ to determine the said coordinates.

To generate the graphic or the physical representation of a spiral it is similar in concept to developing any curve. That is to break it down into a concatenated series of small chords by defining the coordinates at the endpoints of these chords. In addition there are various geometric constructions, presented later on in this chapter, that express the location of a computed point in terms of its distance from the start point of the spiral. Thus, it becomes necessary to compute the coordinates of these points along the spiral (POS). To do that we need to define certain additional spiral elements or components. These are addressed in the next section, but before doing so we need to say that a highway spiral:

(a) Commences at a point which is commonly referred to as the
  • TS (when a spiral transitions from a line into a curve), or as the
  • CS (when a spiral transitions from a curve into another curve), and
(b) Terminates at a point commonly referred to as the
  • ST (when a spiral transitions from a spiral into a line), or as the
  • SC (when a spiral transitions from a spiral into a curve).

In Figures 9.2a and 9.2b:
For any point on the spiral P_x

  • *C is the chord from A to P_x,*
  • *ϕ is the angle formed by the tangent at the start of the spiral and the said chord, and*
  • *θ is the instantaneous spiral angle at the POS*

9.2   Additional Spiral Elements

Figure 9.2a may be considered as an extension of Figure 9.1a without the differential segment while Figure 9.2b shows the same spiral for $D_1 = 0$. The reader should note the similarities and differences between the two figures as the subsequent work applies to both.

In Figure 9.2a the curvatures D_1 and D_2 are extended towards each other until they are both intercepted by a common radius. The distance between these two extensions along this common radius is to be called p (the shift or throw of the spiral). Since a straight line may be thought of as having an infinite number of radii of infinite length, the distance p is to be measured along the radius of D_2 which is normal to the straight line D_1. The intersection point of the extension of the D_2 curve and the said common radius when projected on the Tangent produces a plus (distance along the Tangent from the start of the spiral) and an offset distance termed k and k', respectively. Note that when $D_1 = 0$, $k' = p$. These distances have been defined in the aforesaid manner in order to make the equations below applicable to either Figure 9.2a or Figure 9.2b. For this reason and to make the angles θ and θ_S compatible with each other and defined as the angle formed by the intersection of an instantaneous spiral radius and a line parallel to the spiral radius at the start point of the spiral.

A line may be drawn tangent to the spiral to any point on the spiral P_X between A and B whose X, Y, and θ values have been found. To simplify the pictorial presentation, point B, the end of the spiral, is selected. When this tangent is extended to intersect the Tangent, the two tangents form an angle equal to θ. The chord connecting this point on the spiral with the point of the spiral beginning A is represented as C. In Figures 9.2a and 9.2b the subscript S refers to the elements of the entire spiral (for a point located at the end of the spiral); this subscript is omitted from all items in the formulas below. From the above

$$T_2 = \frac{Y}{\sin\theta} \tag{9.2.1}$$

and

$$T_1 = X - \frac{Y}{\sin\theta} \tag{9.2.2}$$

while

$$C = \frac{X}{\cos\varphi} \tag{9.2.3}$$

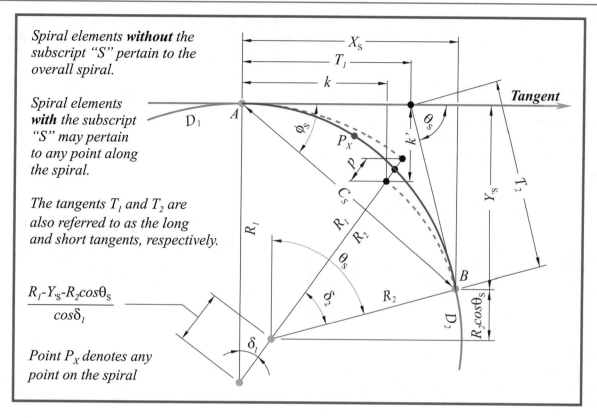

Figure 9.2a   Elements of a Spiral Between Two Curves (Spiraled Compound Curves)

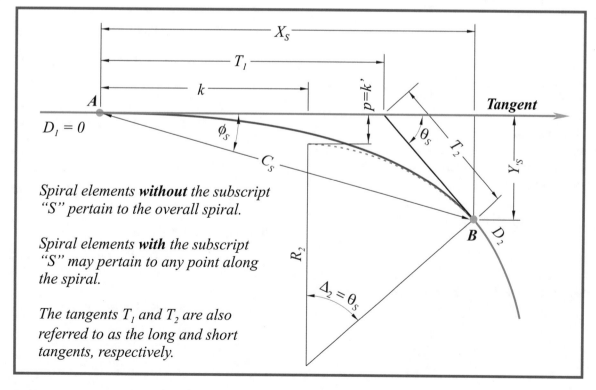

Figure 9.2b   Elements of a Spiral Between a Line and a Curve

where

$$\varphi = \arctan \frac{Y}{X} \qquad (9.2.4)$$

The elements expressed by Equations 9.2.1 through 9.2.4 inclusive may refer to any point on the spiral; the elements solved below refer only to the entire spiral.

From Figure 9.2a we see that the spiral angle qS is equal to the sum of d1 and d2 and it is also defined by Equation 9.1.5. Thus

$$\theta_S = \frac{L_S}{200}(D_1 + D_2) = \delta_1 + \delta_2 \qquad (9.2.5)$$

Equations 9.2.6a to 9.2.9c pertain to Figure 9.2a while Equations 9.2.10 to 9.2.12b pertain to Figure 9.2b.

In the solution for the parameters p and k and k' presented below, Equations 9.2.6a through 9.2.9c pertain to Figure 9.2a, while Equations 9.2.10 through 9.2.12b pertain to Figure 9.2b.

In Figure 9.2a, the circular arcs from the point of beginning of the spiral to the common radius, and from the common radius to the ending point of the spiral are respectively

$$R_1\delta_1 = \frac{100\delta_1}{D_1} = L_D k_1 \quad or \quad \delta_1 = \frac{D_1}{100}L_S k_1 \qquad (9.2.6a)$$

and

$$R_2\delta_2 = \frac{100\delta_2}{D_2} = L_S k_2 \quad or \quad \delta_2 = \frac{D_2}{100}L_S k_2 \qquad (9.2.6b)$$

in which k_1 and k_2 are constants of proportionality of the circular arcs to the spiral length. If we now substitute the values of δ_1 and δ_2 of Equations 9.2.6a and 9.2.6b in the right side of Equation 9.2.5 we get

$$\frac{L_S}{200}(D_1 + D_2) = \frac{D_1}{100}L_S k_1 + \frac{D_2}{100}L_S k_2 = \frac{L_S}{100}(D_1 k_1 + D_2 k_2) \qquad (9.2.7a)$$

or

$$(D_1 + D_2) = 2(D_1 k_1 + D_2 k_2) = 2D_1 k_1 + 2D_2 k_2 \qquad (9.2.7b)$$

Hence, for Equation 9.2.7b to hold true k_1 must equal k_2 and they should be equal to 1/2 which implies that the aforesaid two circular arcs are equal to each other and that

$$\delta_1 = \frac{L_S}{2R_1} \qquad (9.2.8a)$$

and

$$\delta_2 = \frac{L_S}{2R_2} \qquad (9.2.8b)$$

With δ_1 and δ_2 known, p, the difference between R_1 and R_2 minus the distance between the centers of the two curves along the common directions of the said radii, may be expressed as (reference is made to Figure 9.2a):

$$p = R_1 - R_2 - \frac{R_1 - Y_S - R_2 \cos\theta_S}{\cos\delta_1} \qquad (9.2.9a)$$

or

$$p = \frac{100}{D_1}\left(1 - \frac{1}{\cos\delta_1}\right) - \frac{100}{D_2}\left(1 - \frac{\cos\theta_S}{\cos\delta_1}\right) + \frac{Y_S}{\cos\delta_1} \qquad (9.2.9b)$$

or

$$p = \frac{R_1(\cos\delta_1 - 1) + R_2(\cos\theta_S - \cos\delta_1) + Y_S}{\cos\delta_1} \qquad (9.2.9c)$$

For Figure 9.2b $\delta_1 = 0$ and thus Equation 9.2.9c becomes

$$p = Y_S - R_2(1 - \cos\theta_S) \qquad (9.2.10)$$

Also

$$k = X_S - R_2(\sin\theta_S - \sin\delta_1) \qquad (9.2.11a)$$

and

$$k' = Y_S - R_2(\cos\delta_1 - \cos\theta_S) \qquad (9.2.11b)$$

which for Figure 9.2b becomes

$$k = X_S - R_2 \sin\theta_S \qquad (9.2.12a)$$

and

$$k' = Y_S - R_2(1 - \cos\theta_S) = p \qquad (9.2.12b)$$

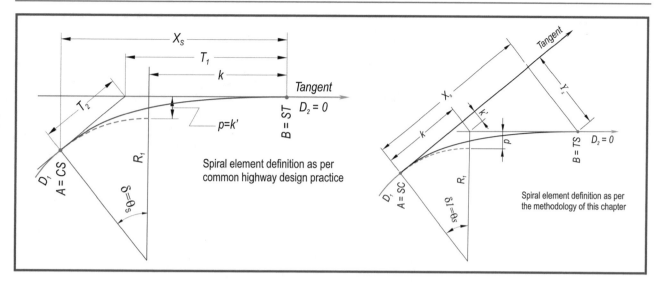

Figure 9.2c Element Definition of an Existing Spiral

All equations that have been presented apply to spirals that transition from one curve into another, or from a straight line into a curve (entering spiral), or from a curve into a straight line (exiting spiral). However, exiting spirals could present a special issue in an actual design environment as illustrated in Figure 9.2c in which the top part indicates the various elements of the spiral as used in common highway design practice, while the bottom part indicates the corresponding elements as of the methodology of this publication.

In the most common case, that of introducing a curve with a back and an ahead spiral tangent to two fixed lines, there is no issue to programmatically design this part of the alignment in one step. On the other hand, if a spiral has to transition from a curve to a tangent line the methodology as used herein would require the knowledge of the instantaneous tangent to the circular curve at the CS point. Although not impossible this would not be the most efficient processes for the designer. For this reason in this case an exiting spiral should be treated as an entering spiral.

9.3 Points Along a Spiral and Offset Spirals

Many times it is necessary to establish points along a spiral and/or at specified offsets therefrom. Also at lines it is necessary to establish lines that are "parallel to" or "offset from" a given main spiral (see Figure 9.3a). Such lines could be the curbs or fascia lines of a bridge the center line of which is located on a spiral, the retaining wall of a ramp at a highway interchange, or the right-of-way acquisition lines of a highway with spiral curves. Such "parallel" or "offset" spirals as they may be called represent the locus of points equidistant from a spiral at a specified offset constant distance W. In addition to such "parallel" spirals the need arises occasionally to vary the offset distance over the entire length of a main spiral, or over a portion thereof (see Figure 9.3b).

It is noted that Equations 9.1.1a through 9.1.15c that define a main spiral do not apply to offset spirals. To create an offset spiral, it is necessary to create its main spiral and then locate point on the offset spiral by creating points along the main spiral and at such points use the instantaneous θ angle to define the instantaneous radial line and along that line create a point a W distance away. This point is addressed later on in this chapter.

Figure 9.3a represents a main spiral and two offset spirals each at a constant offset distance W. The main spiral progresses from the left towards the right in a clockwise rotation, and one of the offset spirals is to the left of the main spiral and the other to the right. In this case the spiral on the left side (with respect to the progression of the main spiral) is referred to as the outside spiral while the one to the right is referred to as the inside spiral. Had the rotation of the main spiral been

Figure 9.3a Offset Spirals

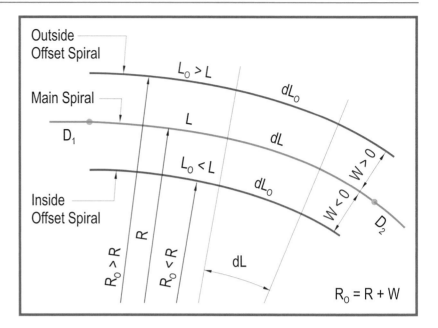

Figure 9.3b Offset Spirals with Variable Offsets

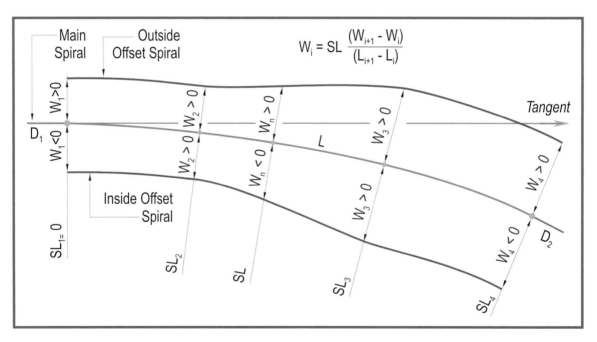

counterclockwise, than the left offset spiral would have been the inside offset spiral, and the one on the right side of the main spiral would have been the outside spiral. If R_O is to represent the instantaneous radius of an offset spiral and W the offset distance from the main spiral then

and at any point

$$W = R_o - R \qquad (9.3.1a)$$

$$R_o = R + W \qquad (9.3.1b)$$

Thus, Equation 9.3.1b establishes the following sign convention to be followed from now on:

The spiral offset is positive when the offset spiral is located "outside" the main spiral, and negative when located "inside" the main spiral.

In the differential segment of Figure 9.3a

$$dL_O = (R + W)d\theta \tag{9.3.2a}$$

Since $Rd\theta = dL$, then

$$dL_O = dL + Wd\theta \tag{9.3.2b}$$

If the degree of curvature is substituted for the radius in Equation 9.1.3, and then this equation is substituted for $d\theta$ in Equation 9.3.2b,

$$dL_O = dL + \left(\frac{W}{100}\right)\left[D_1 + (D_2 - D_1)\left(\frac{L}{L_S}\right)\right]dL \tag{9.3.3a}$$

which if integrated becomes

$$L_O = L + \left(\frac{W}{100}\right)\left[D_1 L + (D_2 - D_1)\left(\frac{L^2}{2L_S}\right)\right] \tag{9.3.3b}$$

or

$$L_O = L + W\theta \tag{9.3.3c}$$

Thus, Equations 9.3.3a, 9.3.3b, and 9.3.3c hold true for either a positive or negative W offset, in finding the length of an offset spiral from a main spiral.

Figure 9.3a is similar to Figure 9.3b in concept but with the difference that the offset lines in the former figure are "parallel" to, or rather they are at uniform offset distance from the main spiral while in the latter they are at a variable offset. To define such offset lines it is necessary to define the location and offset at specific points along the main spiral where the offset distance changes are desired. Thus, an offset spiral may be comprised of a series of "parallel" spiral segments and straight line transition segments between them.

The two offset spirals in Figure 9.4a, one on each side of the main spiral, are each comprised of two "parallel" segments and one in-between transition segment. If each station SL of a change in offset distance W is denoted with the subscript "i" with "$i = 1$" denoting the station of the start of the spiral and "$i = n$" denoting the station of the end of the spiral, where "n" is the total number of transition points including the start and end points of the main spiral, then the offset distance W at station SL between stations SL_{i-1} and SL_i may be expressed as

$$W = (SL - SL_1)\left(\frac{W_i - W_{i-1}}{SL_i - SL_{i-1}}\right) + W_{i-1} \tag{9.3.3d}$$

9.4 Support Procedures

The `icComspr`, `icDeriv`, `icCeg86`, `icCeg8Y` and `icPntsOnSprl` support procedures have been created to define the spiral elements identified in the preceding sections for use by the geometric construction procedures of this chapter. Of these procedures the `icCeg86` and `icCeg8Y` are quite similar. Figure 9.4a presents a general flow chart identifying the relationship and operation of the said construction procedures. The first four of the above support procedures assume that the prime calling program performs the necessary error detection of the input data. The `icPntsOnSprl` procedure is not called by any of the procedures in this publication but it is called by certain test procedures included in the accompanying CD of all procedures. This procedure performs several input data error detection operations as described later on in this section.

9.4.1 Computation of the Spiral Components

The `icComspr` procedure returns the parameters of a spiral given the degrees of curvature of the back and forward curves, and the distance to a point on the spiral (POS) from the spiral's start point. In perusing the description of this procedure reference is made to Figure 9.2b (transition from a line to a curve), and to Figure 9.2a (compound curves that transition from one curve to another curve).

```
Sub icComspr (DC1, DC2, SPRL, SL, _
              THETA, XC, YC, A, SP, SK, SKP,
              SCL, STS, STL, PHI)
```

The input parameters (see Figures 9.1d and 9.1e) to this procedure are the:

DC1	The degree of curvature in radians of the curve back of the spiral (back curve), which should be zero if it is a straight line.
DC2	The degree of curvature in radians of the curve ahead of the spiral (forward curve), which should be zero if it is a straight line.
SPRL	The length of the spiral in feet (meters).
SL	The distance in feet (meters) of a point on the spiral (POS) from its start point.

The returned parameters (see Figures 9.2a and 9.2b) from this procedure are the:

THETA	The instantaneous spiral angle θ at the POS.
XC	The X (plus) distance of the POS along the instantaneous tangent at the start point.
YC	The Y (offset) distance of the POS from the instantaneous tangent at the start point.
A	The overall spiral angle if DC1 = 0.

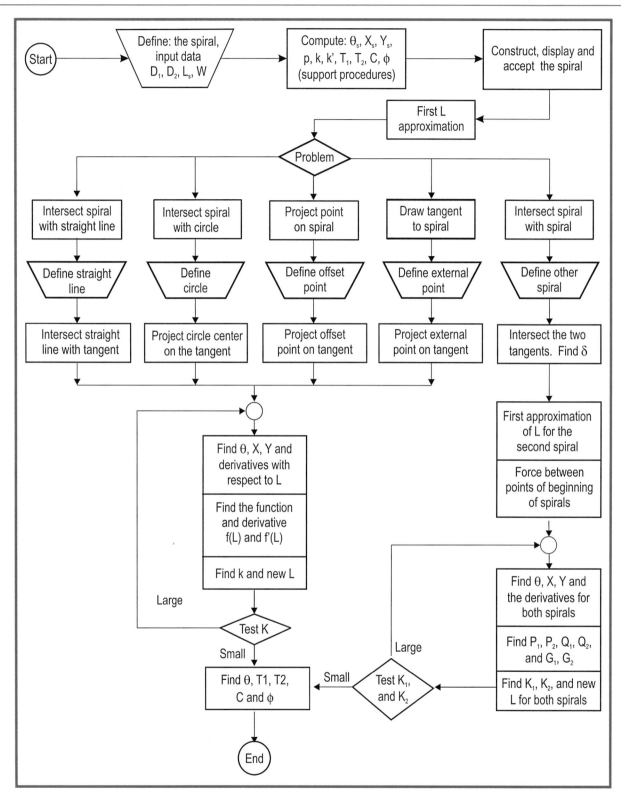

Figure 9.4a Offset Spirals with Variable Offsets

SP	The throw distance *p* of the overall spiral in feet (meters) along the line:

SP — The throw distance *p* of the overall spiral in feet (meters) along the line:
- Where the radius of the curve when extended backward is normal to the back tangent (Figure 9.2b) measured between the curve and the tangent.
- Where the radii of the back and forward curves when extended forward and backward, respectively, are colinear (Figure 9.2a) measured between the curves.

SK — The distance *k* in feet (meters) along the back tangent from the start point "A" of the spiral to the point of projection of the endpoint of the:
- Curve extended backward until its radius is normal to the back tangent (Figure 9.2b).
- Ahead curve until its radius is colinear with the back curve extended forward (Figure 9.2a).

SKP — The distance *k'* shown in the two last said figures.

SCL — The chord length *C* in feet (meters) from the spiral's start point to the POS.

STS — The short tangent T_2 from the point of intersection of the instantaneous tangents at the start point of the spiral and at the POS to the end point of the spiral.

STL — The long tangent T_1 from the start point of the spiral to the point of intersection of the instantaneous tangents at the start point of the spiral and at the POS.

PHI — The angle in radians formed by the tangent line to the spiral at the start point of the spiral and the abovesaid chord (SCL).

Note that the parameters A, SP, SK, SKP, SCL, STS, STL, PHI are valid only when SL=SPRL at which time THETA=A and the parameters XC and YC are those of the endpoint of the spiral.

Upon invocation the procedure initializes to zero the parameters to be returned and the temporary collection F which is used to evaluate Equations 9.1.15a and 9.1.15b, and sets the constant RD to be $36{,}000/(2\pi)$ to convert the degrees of curvature into radians while accounting for the 100 denominator constant of the spiral angle equation which is used in the assessment of Equation 9.1.15c and in the computation of the two radii. Thereafter the procedure:

1. Queries DC1 and DC2 for both being zero within a tolerance of 0.0000001 radians (0.02 s), and if so sets the value of IERR to be 1 and aborts; else continues.
2. Queries the length of the spiral SPRL and the distance SL to the POS from the start point for beings to zero within a tolerance of 0.005 feet (1.524 mm), and if so sets the value of IERR to be 2 or 3 respectively and aborts; else continues.
3. Solves Equations 9.1.13a and 9.1.13b, and the factorial components of Equations 9.1.15a and 9.1.15b utilizing the above mentioned F collection (the *1/7!* factorial is not being computed because its contribution is considered to be minuscule, if not unimportant).

4. Computes the local coordinates (the plus and offset para meters XC and YC) and the θ (THETA) angle at the POS.
5. Queries the values of the degrees of curvature DC1 for being equal to zero within a tolerance of 0.0000001 radians (0.02 s), and if so
   - Sets the radius R1 and the spiral angle A each to be equal to zero; else
   - Sets the radius R1 equal to the above referenced constant RD divided by the degree of curvature DC1, and the value of the overall spiral angle A is set to be SPRL/2/R1.
6. Queries the values of the degrees of curvature DC2 for being equal to zero within a tolerance of 0.0000001 radians (0.02 s), and if so
   - Sets the radius R2 to be equal to zero; else
   - Sets the radius R2 equal to the above referenced constant RD divided by the degree of curvature DC2.
7. Computes the values of the spiral throw SP and its two offsets from the back tangent line SK and SKP
8. Computes the values of the angle PHI, its corresponding chord length SCL and the lengths of the long STL and short STS tangent lines, and then terminates.

9.4.2 Computation of the First Derivative of the Spiral Function

The icDeriv procedure computes the first derivatives $d\theta/dL$, dy/dL, and dx/dL as defined by Equations 9.1.3, 9.1.10a, and 9.1.10b, respectively. In solving for the derivative $d\theta/dL$ the procedure expresses Equation 9.1.3 in terms of the degrees of curvature D1 and D2 instead of the corresponding radii R1 and R2. These derivatives are used by the various procedures to be presented in the subsequent sections of this chapter, and which:

(a) intersect a spiral with a line, circular curve, or another spiral,
(b) project a point on a spiral, and
(c) construct a tangent line to a spiral from a point which is not located on the spiral.

This support procedure, in addition to the required two degrees of curvature and the length of the length spiral that define the main spiral, requires as input the distance, or the first approximation of the distance, from the start point of the spiral to the point in question on the main spiral (POS). The returned spiral parameters listed below are those at the POS.

```
Sub icDeriv (DC1, DC2, SPRL, SL, _
             THETA, XC, YC, DTHE, DX, DY)
```

The input parameters (see Figures 9.2a and 9.2b) to this procedure are the:

DC1   The degree of curvature in radians of the curve back of the spiral (back curve), which should be zero if it is a straight line.
DC2   The degree of curvature in radians of the curve ahead of the spiral (forward curve), which should be zero if it is a straight line.
SPRL   The length of the spiral in feet (meters).
SL   The distance to any specific point or the first approximation of the distance to the point of intersection, projection or tangency with, on or to the spiral in feet (meters) from the spiral's start point.

The returned parameters from this procedure are the:

THETA   The instantaneous spiral angle θ at the approximation point on the spiral.
XC   The local X coordinate of the approximation point on the spiral.
YC   The local Y coordinate of the approximation point on the spiral.
DTHE   The derivative of the spiral angle equation.
DX   The DX/DL derivative.
DY   The DY/DL derivative.

Having been called, this procedure defines the constant RD to convert the degrees of curvature into radians while accounting for the 100 denominator constant of the spiral angle equation, and then creates a temporary collection named T to compute and store the factorial components of the DX/DL and DY/DL equations. Thereafter the procedure:

1. Calls the icComspr procedure to determine and return for use the values of THETA, and IERR.
2. Computes the factorial components of the expansion of THETA into its trigonometric form, and then computes the differentials dx and dy of Equations 9.1.10a and 9.1.10b.

9.4.3 Points Along a Spiral without Stations

The icCeg86 procedure returns (**a**) the coordinates of a point on a spiral (POS) located a given distance from the start point of the spiral, (**b**) the azimuth of the instantaneous radial line towards the spiral and the instantaneous tangent line to the spiral at the said POS, and (**c**) the chord distance from the start point of the spiral to the said POS and the angle that the said chord forms with the tangent line at the start of the spiral. Reference is also made to the icCeg8Y procedure which is slightly different than the icCeg86 procedure.

```
Sub icCeg86 (DC1, DC2, SPRL, SL, STARTN, STARTE,
             AZ1, DELTA, _
             NORTH1, EAST1, CHORD, PHI, AZ, AZONE)
```

The input parameters (see Figures 9.1d and 9.1e) to this procedure are the:

DC1   The degree of curvature in radians of the curve back of the spiral (back curve), which should be zero if it is a straight line.
DC2   The degree of curvature in radians of the curve ahead of the spiral (forward curve), which should be zero if it is a straight line.
SPRL   The length of the spiral in feet (meters).
SL   The distance in feet (meters) from the spiral's start point of the first approximation of the point of intersection, projection or tangency with, on or to the spiral.
STARTN The north coordinate of the start point of the spiral.
STARTE The east coordinate of the start point of the spiral.
AZ1   The north azimuth in radians of the tangent line at the spiral's start point.
DELTA   An indicator denoting the spiral's rotation. A positive value denotes a clockwise rotation and a negative value denotes a counterclockwise rotation.

The returned parameters from this procedure are the:

NORTH1 The north coordinate of the point to be determined along the spiral.
EAST1   The east coordinate of the point to be determined along the spiral.
CHORD   The chord length in feet (meters) from the spiral's start point to the point to be determined along the spiral (see Figures 9.1d and 9.1e).
PHI   The angle in radians formed by the tangent line to the spiral at the start point of the spiral and the chord from the said point to the point to be determined along the spiral (see Figures 9.1d and 9.1e).
AZ   The instantaneous tangent azimuth in radians at the new point pointing to the spiral's endpoint.
AZONE   The instantaneous radial azimuth in radians at the new point pointing towards the spiral.

Having been called, this procedure:

1. Sets the constant ONE to be ±1 depending on the sign of the indicator DELTA. DELTA is the default entry parameter denoting the deflection angle of the overall alignment of line and curve between which the spiral is inserted. Since this angle carries a plus of minus sign it may be used as the rotation indicator.
2. Calls on the icComspr procedure to determine the values of THETA, XC, YC, SCL, and PHI at the new point located and SL distance from the spiral's start point.
3. Assigns to the YC and THETA values the sign of ONE. The value of THETA is not changed. Its signed value is assigned to the temporary parameter ANGLE.
4. Calls on the icptl procedure to compute the north and east coordinates (NORTH2, EAST2) of the new point on the spiral by traversing along AZ1 a plus distance of XC and an offset distance of YC.
5. Computes the value of the instantaneous radial line at the POS as AZ1+ANGLE, and the value of the instantaneous tangent azimuth by adding half a π to the said radial azimuth.

9.4.4 Points Along a Spiral with Stations

The `icCeg8Y` procedure returns (**a**) the coordinates of a point on a spiral (POS) located a given distance from the start point of the spiral, (**b**) the azimuth of the instantaneous radial line towards the spiral at the said POS, (**c**) the chord distance from the said start point to the said POS, and (**d**) the angle that the said chord forms with the tangent line to the spiral at the said start point. Reference is also made to the `icCeg86` procedure which is slightly different than the `icCeg8Y` procedure.

```
Sub icCeg8Y (DC1, DC2, SPRL, SL, PTNN, PTNE,
             AZ1, DELTA, STAT, _
             NORTH2, EAST2, STAT2, AZTWO, PHI, SCL)
```

The input parameters (see Figures 9.1d and 9.1e) to this procedure are the:

DC1	The degree of curvature in radians of the curve back of the spiral (back curve), which should be zero if it is a straight line.
DC2	The degree of curvature in radians of the curve ahead of the spiral (forward curve), which should be zero if it is a straight line.
SPRL	The length of the spiral in feet (meters).
SL	The distance in feet (meters) from the spiral's start point of the first approximation of the point of intersection, projection or tangency with, on or to the spiral.
PTNN	The north coordinate of the start point of the spiral.
PTNE	The east coordinate of the start point of the spiral.
AZ1	The north azimuth in radians of the tangent line at the spiral's start point.
DELTA	An indicator denoting the spiral's rotation. A positive value denotes a clockwise rotation and a negative value denotes a counterclockwise rotation.
STAT	The station value of the spiral's start point without the plus sign.

The returned parameters from this procedure are the:

NORTH2	The north coordinate of the point to be determined along the spiral.
EAST2	The east coordinate of the point to be determined along the spiral.
STAT2	The station value without the plus sign of the new point to be determined.
AZTWO	The forward pointing north azimuth of the tangent line to the spiral at the point to be determined along the spiral.
PHI	The angle in radians formed by the tangent line to the spiral at the start point of the spiral and the chord from the said point to the point to be determined along the spiral (see Figures 9.1d and 9.1e).
SCL	The chord length in feet (meters) from the spiral's start point to its endpoint.

Having been called, this procedure:

1. Carries out Steps 1 through 4 of the `icCeg86` procedure.
2. Computes the value of the instantaneous radial line as being equal to AZ1 plus ANGLE, and the station value of the new point by adding SL to the station value STAT of the spiral's start point.

9.4.5 Point Collection Along a Main or Offset Spiral

The preceding two subsections presented procedures for determining the coordinates of a point along a main or offset spiral, and the subsequent section will address the positioning of a main spiral along the confines of an alignment. To do so we will have to (**a**) first position the control points of the main spiral or spirals and associated circular curve or curves and tangent lines and (**b**) then generate the actual line work that represents the said lines, curves and spirals. Lines and circular curves have been addressed in other sections of this publication. Thus, this section is to address the preparation of the line work of a spiral. This is accomplished by creating a collection (array) of point coordinates along the spiral at a specified interval. The `icPntsOnSprl` procedure helps create such a collection of points along a main or offset spiral.

To define the main spiral `icPntsOnSprl` requires knowledge of the coordinates of the start point of the spiral, the length of the spiral, the two associate degrees of curvature, and the interval at which coordinates are to be computed. To define an offset spiral, since such a spiral may involve the entire or only a portion of the main spiral, and it may be at a constant or at a variable offset distance from the main spiral, the said procedure also requires a series distances (stations) from the start point of the main spiral and offsets therefrom to define the points along the desired offset spiral (see Figures 9.3b and 9.4a).

It is noted that none of the procedures in this publication call the `icPntsOnSprl` procedure. The said procedure is called by certain spiral related test procedures in the accompanying CD. It could also be called by any user produced procedures to generate a spiral.

```
Sub icPntsOnSprl (TSN, TSE, AZ1, TSSTA, DC1,
                  DC2, SPRL, OFFSPR, _
                  NOFF, SCL, ROTN, InOut_
                  NUMB, CURVE, IERR)
```

The input parameters (see Figures 9.3a, 9.3b, and 9.4a) to this procedure are the:

TSN, TSE	The north and east coordinate of the start point of the main spiral.
AZ1	The north azimuth in radians of the tangent line at the spiral's start point.
TSSTA	The station value (without the plus sign) of the start point of the main spiral. It could be either positive or negative or zero.
DC1	The degree of curvature in degrees of the curve back of the spiral (back curve), which should be zero if it is a straight line.
DC2	The degree of curvature in degrees of the curve ahead of the spiral (forward curve), which should be zero if it is a straight line.
SPRL	The length of the spiral in feet (meters).
OFFSPR(i, j)	Two dimensional dynamic array of pairs of stations along the main spiral and offsets therefrom and in which:

i = 1 Station (without the plus sign) along the main spiral at which an offset value is specified. Stations are relative to TSSTA and they should be in an increasing or decreasing sequence.

i = 2 Specified offset at the said station. Positive offsets are considered to be outside the main spiral and negative offsets are considered to be inside the main spiral.

j = 1 to NOFF See below.

There could be two and only two identical stations with different offsets.

NOFF	The number of station and offset pairs in the OFFSPR(i, j) array, and in which if

NOFF = 0 Create a main spiral only and disregard any data that may be in the OFFSPR(i, j) array.

NOFF = 1 Create an offset spiral along the entire spiral using as a constant offset the value of the OFFSPR(2, 1) array.

It is assumed that the calling program has checked the validity of NOFF with respect to the contents of the OFFSPR(i, j) array.

SCL	The interval along the main spiral at which points are to be computed to define the main or offset spiral. It should be noted that if the offset spiral is at a relatively great offset from the main spiral consideration should be given to reducing SCL for the offset spiral if a smooth curve is desired.
ROTN	Any positive integer value denotes a clockwise alignment rotation and any negative value denotes a counterclockwise alignment rotation (use ±1).
InOut	A string which if it contains the word ENTERING indicates a back spiral, and if it contains the word EXITING indicates an ahead or forward spiral. It is used to handle the stationing of an offset spiral if a negative station is used for the start point of the spiral. Note that the simplest way to create the line work of a main or offset spiral is to:

• Let TSSTA = 0 and the stations of OFFSPR be distances from TSSTA.
• Treat the exiting spiral as entering spiral, and change the sign of ROTN and reverse the direction of the ahead tangent line.

The OFFSPR array stations are relative to TSSTA and advance towards the end of the spiral. Thus:

• *For the back (entering) spiral the stations of the said array should proceed forward along the alignment, and*
• *For the head (exiting) spiral the stations of the said array should proceed backward along the alignment.*

It might be prudent to have the calling program set TSSTA *to be zero and measure the stations of the said array as distances from the start point towards the end point of the spiral.*

Also the OFFSPR array stations and offsets create a continuous offset spiral without any discontinuances. If discontinuous offsets are needed, separate procedure calls need be made.

The returned parameters from this procedure are the:

NUMB	The number of points comprising the main or offset spiral. The actual number of points used is determined as indicated in the operational steps below.
CURVE (i, j)	The two dimensional dynamic array of pairs of north and east coordinates of points comprising the main or offset spiral and in which:

i = 1 The north coordinate of a point.
i = 2 The east coordinate of a point.
j = 1 to NUMB.

IERR	An indicator the value of which denotes the following:

0 No input errors encountered.
1 Negative or zero SPRL value (abort).
2 Negative DC1 value (abort).
3 Negative DC2 value (abort).
4 Both DC1 and DC2 have zero value (abort).
5 Zero ROTN value (abort).
6 Negative or zero SCL value (abort).
7 More than two identical offset stations (abort).
8 The stations in the OFFSPR(i, j) array are not in an ascending or descending order (abort).
9 The first station in the OFFSPR(i, j) array is outside the spiral's limits (abort).
10 The last station in the OFFSPR(i, j) array is outside the spiral's limits (abort).
11 The last station in the OFFSPR(i, j) array is outside the spiral's limits (abort).

Upon invocation the procedure initializes the number of points in the array to be returned and the type of input error encountered, if any, and then:

Note that in the icPntsOnSprl *procedure it is the* **ROTN** *parameter that denotes a clockwise or counterclockwise rotation and not the radius or degrees of curvature or the **DELTA** angle.*

1. Queries the SPRL parameter for being negative or equal to zero within a tolerance of 0.005 feet (1.524 mm) and if so sets IERR to be 1 and then aborts; else continues.
2. Queries the DC1 and the DC2 parameter for being negative and if so sets IERR to be 2 or 3, respectively, and then aborts; else continues.
3. Queries the DC1 and DC2 parameters for both being equal to zero within a tolerance of 0.005 feet (1.524 mm) and if so sets IERR to be 4 and then aborts; else continues.
4. Queries the ROTN and the SCL parameters for being negative or equal to zero within a tolerance of 0.005 feet (1.524 mm) and if so sets IERR to be 5 or 6, respectively, and then aborts; else continues.
5. Queries the NOFF parameter for being negative and if so sets IERR to be 11 and then aborts; else queries the same parameter for being zero or greater than zero. If it is
   - Zero sets the SPRLtype indicator to be "MAIN" denoting a main spiral;
   - Greater then zero sets the SPRLtype indicator to be "OFFSET" denoting an offset spiral;
     and then continues.
6. Queries the SPRLtype indicator for being "MAIN" or "OFFSET". If it is:
   - "OFFSET" continues with **Step 7** for additional input data verification, and
   - "MAIN" branches to **Step 12**.
7. Queries the NOFF parameter for being greater than 1 and if so continues with the next step; else branches to **Step 12**.
8. Queries the first two stations of the OFFSPR(i, j) array for being the same within a tolerance of 0.005 feet (1.524 mm), increasing or decreasing. If they are:
   - The same sets the AhdBck indicator to be "SAME".
   - Increasing sets the AhdBck indicator to be "AHEAD".
   - Decreasing sets the AhdBck indicator to be "BACK".
9. Queries successive pairs of the remaining stations of the OFFSPR array successively, and if at any time during this query two stations are:
   - The same, within a tolerance of 0.005 feet (1.524 mm), and the AhdBck indicator is "SAME", the indicator IERR is set to be 7 and aborts; else continues.
   - Increasing and the AhdBck indicator is "BACK", the indicator IERR is set to be 8 and aborts; else continues.
   - Decreasing and the AhdBck indicator is "AHEAD", the indicator IERR is set to be 8 and aborts; else continues.
10. Queries the NOFF and the InOut parameters. If:
    - NOFF is greater than 1 and InOut is "ENTERING" and the first station in the OFFSPR array is less than the TSSTA parameters sets the indicator IERR to be 9 and aborts; else continues.

- NOFF is greater than 1 and InOut is "EXITING" and the first station in the OFFSPR(i, j) array is greater than the TSSTA parameters sets the indicator IERR to be 9 and aborts; else continues.
11. Queries the NOFF and the InOut parameters. If:
    - NOFF is greater than 1 and InOut is "ENTERING" and the last station in the OFFSPR array is greater than TSSTA + SPRL parameters sets the indicator IERR to be 10 and aborts; else continues.
    - NOFF is greater than 1 and InOut is "EXITING" and the last station in the OFFSPR(i, j) array is less than the TSSTA − SPRL parameters sets the indicator IERR to be 11 and aborts; else continues.
12. Queries the SPRLtype indicator and the NOFF parameter. If:
    - SPRLtype = "MAIN" and NOFF = 0 or if SPRLtype = "OFFSET" and NOFF = 1, that is the spiral is a main spiral or a uniform offset spiral,
      (a) Computes the number of points NUMB to populate the CURVE array by first dividing the spiral length SPRL by the specified uniform spacing of points SCL and extracting the integer portion (rounding down), and then dividing SPRL by NUMB to find an adjusted uniform spacing of points (the input value is not altered).
      (b) Adds 1 to NUMB to account for the starting point of the spiral and dimensions the CURVE array to (2, NUMB).
      (c) Sets a temporary station back of TSSTA by a value equal to the adjusted uniform spacing, and then in a loop calls the icCeg86 procedure to compute the coordinates of a point on the main spiral.
      (d) Queries the value of NOFF for being 0 or 1. If it is:
         - 0 (main spiral), populates the CURVE array with the said coordinates.
         - 1 (uniform offset spiral), calls the icptl procedure to compute the coordinates of the point on the offset spiral by traversing from the coordinates of the main spiral along the instantaneous radial azimuth a distance equal to the specified offset, and then populates the CURVE array with the said offset point coordinates.
      (e) Terminates.
    - SPRLtype = "OFFSET", that is the spiral is an offset spiral with variable offsets,
      (f) Computes the number of points NUMB to populate the CURVE array in a manner similar to **Steps 12(a)** and **12(b)** but by parsing through the various segments of the OFFSPR array and computing the number of points within each segment of the said array and accumulating them to deter-

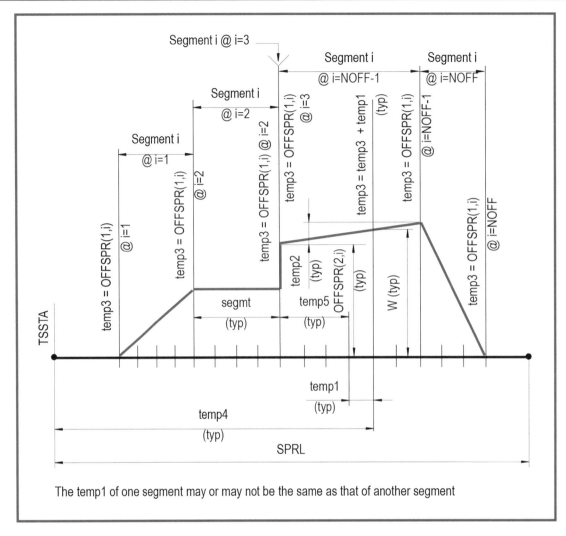

The temp1 of one segment may or may not be the same as that of another segment

Figure 9.4b Parameters Used in the Determination of an Offset for a Variable Offset Spiral

mine the overall number of points. In this process the specified uniform spacing of points is not adjusted.

(g) Computes the coordinates of the first point of the offset spiral and populates the `CURVE` array with the said coordinates as per **Steps 12(c)** and **(d)**.

(h) Parses through the `OFFSPR` array and for each segment

Computes the number of points and the adjusted uniform spacing of points for the segment.

For each point to be created within the segment

Computes the station value of a point within the segment by adding the uniform point spacing of the segment.

Interpolates the corresponding offset using Equation 9.3.3d.

Computes the coordinates of the offset point and then populates the `CURVE` array with the said coordinates as per **Steps 12(c)** and **12 (d)**.

The various parameters used in the process of this step are depicted in Figure 9.4b in which the main spiral is represented for pictorial purposes as a straight line, and then terminates.

Although the various offsets of a variable offset spiral are interpolated using straight line interpolation, the offset spiral will not be a straight line.

9.5   Establishing the Main Spiral

The first operation in establishing a transition spiral is to identify its relationship and position within the overall alignment of a project. That is to identify the "from which" and the "to which" alignment component the transition is to occur. As discussed previously there are three types of transition: (a) from a straight line or tangent to a circular curve or just curve, (b) from a curve to a tangent and (c) from a curve to another curve which may be divided further into two types from a flatter curve to a sharper curve and vice versa. Samples of these relationships are illustrated in Figure 9.5 and they are addressed individually by a unique procedure later on in this section. These unique procedures assume certain conditions as given and may not necessarily address all possible

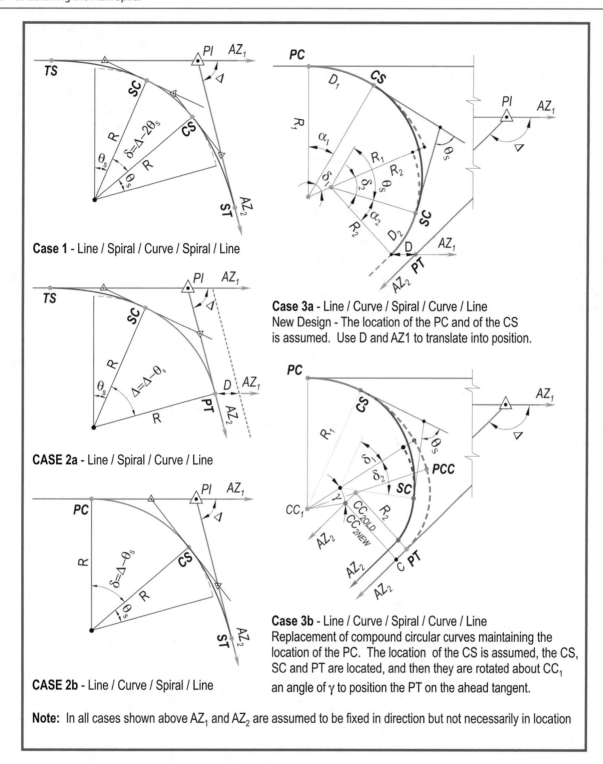

Case 1 - Line / Spiral / Curve / Spiral / Line

CASE 2a - Line / Spiral / Curve / Line

CASE 2b - Line / Curve / Spiral / Line

Case 3a - Line / Curve / Spiral / Curve / Line
New Design - The location of the PC and of the CS
is assumed. Use D and AZ1 to translate into position.

Case 3b - Line / Curve / Spiral / Curve / Line
Replacement of compound circular curves maintaining the
location of the PC. The location of the CS is assumed, the CS,
SC and PT are located, and then they are rotated about CC_1
an angle of γ to position the PT on the ahead tangent.

Note: In all cases shown above AZ_1 and AZ_2 are assumed to be fixed in direction but not necessarily in location

Figure 9.5 Spiral–Circular Curve–Spiral Between Two Tangent Lines

conditions as the may be encountered in an actual design project. This is considered outside the scope of this publication and is left on the design of the overall computer aided design system or calling program.

The second operation in establishing a transition spiral is the creation of the polyline that graphically can represent the main or offset spiral. The `icPntsOnSprl` procedure presented previously does this.

9.5.1   Case 1—Curve with Entering and Exiting Spiral

Case 1 of Figure 9.5 is expanded in Figure 9.5a. This is the most common occurrence of spiral transition from a tangent line to a circular arc (curve) and then to another tangent line. Generally the two tangent lines are fixed (intersecting at the PI) and the two spirals and the curve are inserted between them and tangent thereto at the spiral endpoints. The deflection angle Δ formed by the said tangent lines, various vehicle travel and safety related criteria and topographic controls (all beyond the scope of this publication) dictate the radius of the curve and the length of the two spirals to be used. Usually the lengths of the two spirals are the same as indicated in Figure 9.5a but they can be different. In Figure 9.5a all indicated parameters of the alignment are symmetrical about the line connecting the center of the circular curve CC with the PI. In this case the distance T_1 and T_2 are equal to each other and equal to

$$T = k + \left[(R+P)\tan\left(\frac{"}{2}\right) \right] \qquad (9.5)$$

If the two spirals are not of the same length, then the above referenced symmetry would not exist and T_1 and T_2 would be unequal. To determine their lengths let us consider the triangles with the vertices PK_1-PK_2-CC and PK_1-PK_2-PI. In the

PK_1-PK_2-CC triangle the angle at CC is Δ and its two adjacent sides are $R_1 + p_1$ and $R_2 + p_2$ which can be computed when the parameters of the two spirals are defined as indicated previously (note that the subscripts 1 and 2 pertain to the back and ahead spirals, respectively). If we let the angle of the said triangle at PK_1 be Δ_1 and the angle at PK_2 be Δ_2, then by the law of cosines the unknown length W between PK_1 and PK_2 may be expressed as

$$W = \sqrt{\left(R+p_1\right)^2 + \left(R+p_2\right)^2 - 2\left(R+p_1\right)\left(R+p_2\right)\cos\left(\text{"}\right)} \qquad (9.5a)$$

and by the law of sines the two unknown angles at the endpoints of the said W long side may be expressed as

$$"_1 = \sin^{-1}\left[\frac{\left(R+p_2\right)\sin\left(\text{"}\right)}{W} \right] \qquad (9.5b)$$

and

$$"_2 = \sin^{-1}\left[\frac{\left(R+p_1\right)\sin\left(\text{"}\right)}{W} \right] \qquad (9.5c)$$

Since R_1 and R_2 are each normal to the back and ahead tangent lines, respectively, the angles of the triangle PK_1-PK_2-PI are $90°$-Δ_1 at the PK_1, $90°$-Δ_2 at the PK_2 and $180°$-Δ at the

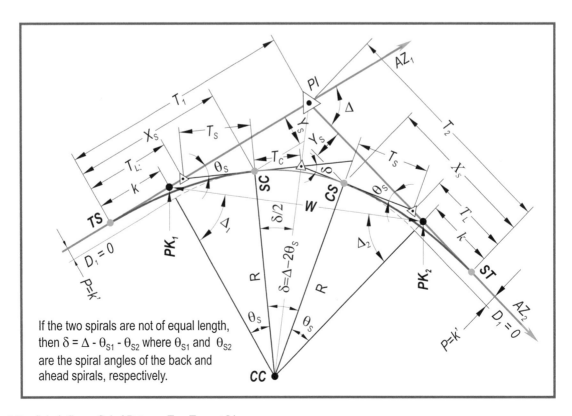

If the two spirals are not of equal length, then $\delta = \Delta - \theta_{S1} - \theta_{S2}$ where θ_{S1} and θ_{S2} are the spiral angles of the back and ahead spirals, respectively.

Figure 9.5a   Spiral–Curve–Spiral Between Two Tangent Lines

PI, and the distances T_1 and T_2 along their respective tangent lines would then, by the use of the law of sines, be

$$T_1 = k_1 + W \left[\frac{\sin\left(0.5\pi - "_2\right)}{\sin\left(\pi - "\right)} \right] \qquad (9.5d)$$

and

$$T_2 = k_2 + W \left[\frac{\sin\left(0.5\pi - "_1\right)}{\sin\left(\pi - "\right)} \right] \qquad (9.5e)$$

The `icCase1spiral` procedure establishes this type of alignment assuming that the spiral lengths, the radius of the circular curve and the two tangent line directions are known. This implies three alignment arrangements:

1. The tangent line directions are known and so is the positioning of these lines. That is each line is defined by the coordinates of a known point and a known north azimuth. In this case the tangent lines are intersected to locate their point of intersection (PI), and the computed spirals and circular curve are aligned about the said PI with the TS and ST being located somewhere along their respective tangent lines (see commentary below).
2. The direction and location of the back tangent line are known, but only the direction of the ahead tangent line is known. In this case the point that defines the location of the back tangent needs to be the location of the TS, and it is about this point that the alignment is positioned.
3. The direction and location of the ahead tangent line are known, but only the direction of the back tangent line is known. In this case the point that defines the location of the ahead tangent needs to be the location of the ST and it is about this point that the alignment is positioned.

The operational steps below outline the procedure used.

The preceding discussion on the tangent lines is based on the assumption that they are not parallel and that they form a deflection *Δ* angle of less than 180°. In most main line alignments this is the case. However, it is possible at interchange ramps to have a *Δ* angle that equals or exceeds 180°, and thus to require a bit more scrutiny. Therefore, if the *Δ* angle:

- Equals 180°, then the two tangent lines cannot intersect, there cannot be a PI, and the two tangent lengths T_1 and T_2 are each of infinite length (Equations 9.5d and 9.5e will encounter a division by zero, $\pi - \Delta$). Thus, either the TS or the ST must be known as is to be shown in the operational steps below.
- Exceeds 180°, then Equations 9.5a through 9.5e should hold true as the signs of the sine and cosine functions of the involved angle will act appropriately.

Thus, it is possible for the *Δ* angle to exceed 180° provided the back tangent line points towards the PI and the ahead tangent line points away from the PI.

```
Sub icCase1spiral (PTN1, PTE1, AZ1, PTN2, PTE2,
                   AZ2, SPRL1, SPRL2, RAD, ROTN,
                   _
                   PIN, PIE, TSN, TSE, SCN, SCE,
                   CSN, CSE, STN, STE, CCN, CCE,
                   _
                   DELTA, THETA1, THETA2, DELTAC,
                   DC1, DC2, IERR)
```

The input parameters (see Figure 9.5a) to this procedure are the:

`PTN1, PTE1`	The north and east coordinates of a point on the back tangent line.
`AZ1`	The north azimuth in radians of the back tangent line pointing from the TS towards the PI.
`PTN2, PTE2`	The north and east coordinates of a point on the ahead tangent line.
`AZ2`	The north azimuth in radians of the ahead tangent line pointing from the PI towards the ST.
`SPRL1`	The length of the entering spiral in feet (meters).
`SPRL2`	The length of the exiting spiral in feet (meters).
`RAD`	The radius of the circular curve in feet (meters) should be positive.
`ROTN`	Any positive value denotes a clockwise alignment rotation and any negative value denotes a counterclockwise alignment rotation.

Note the following regarding the procedures to follow:

- *Any reference to an alignment refers to the lines and curves of a procedure advancing within the context of the overall alignment.*
- *Unless they define a control point such as a TS or PC, or a ST or PT, the points that define the back tangent line (PTN1, PTE1) and the ahead tangent line (PTN2, PTE2) could be located anywhere along their respective tangent line.*

NOTE that if:

- `PTN1`, `PTE1`, `PTN2`, and `PTE2` each has a value other than 99999.9 then the two tangent lines are intersected to locate the PI about which the alignment is positioned.
- `PTN2` and `PTE2` each has a value of 99999.9 then `PTN1` and `PTE1` should each have a value other than 99999.9 denoting the TS and the alignment is positioned about the TS.
- `PTN1` and `PTE1` each has a value of 99999.9 then `PTN2` and `PTE2` should each have a value other than 99999.9 denoting the ST and the alignment is positioned about the ST.

The returned parameters (see Figure 9.5a) from this procedure are the:

PIN, PIE	The north and east coordinates of the PI.
TSN, TSE	The north and east coordinates of the TS.
SCN, SCE	The north and east coordinates of the SC.
CSN, CSE	The north and east coordinates of the CS.
STN, STE	The north and east coordinates of the ST.
CCN, CCE	The north and east coordinates of the CC.
DELTA	The deflection angle formed by the two tangent lines.
THETA1	The spiral angle of the entering spiral.
THETA2	The spiral angle of the exiting spiral.
DELTAC	The central angle of the circular curve.
DC1	The degree of curvature (decimal degrees) of the back segment (zero if a line).
DC2	The degree of curvature (decimal degrees) of the ahead segment (zero if a line).
IERR	An indicator the value of which denotes the following:

> 0 No input errors encountered.
> 1 Negative or zero SPRL1 value (abort).
> 2 Negative or zero SPRL2 value (abort).
> 3 Negative or zero RAD value (abort).
> 4 The two given points are the same (abort).
> 5 Zero ROTN value (abort).
> 6 Parallel tangents with the same azimuth (abort).
> 7 Parallel tangents with opposite azimuths and the TS or ST has not been specified (abort).

Upon invocation the procedure initializes the parameters to be returned, and then:

1. Queries SPRL1 and SPRL2 for being negative or equal to zero within a tolerance of 0.005 feet (1.524 mm) and if so sets IERR to be 1 or 2, respectively, and aborts; else continues.
2. Queries RAD for being negative or equal to zero within a tolerance of 0.005 feet (1.524 mm) and if so sets IERR to be 3 and aborts; else continues.
3. Queries the parameters of the pair PTN1 and PTN2 and of the pair PTE1 and PTE2 for both pairs being equal to each other within a tolerance of 0.005 feet (1.524 mm) and if so sets IERR to be 4 aborts; else continues.
4. Queries the parameter ROTN for being equal to zero within a tolerance of 0.005 feet (1.524 mm) and if so sets IERR to be 5 and aborts; else continues.
5. Calls the icstrdln procedure to compute the DELTA angle by rotating from AZ1 towards AZ2, and then sets DELTA to be positive since ROTN is to denote a clockwise or counterclockwise rotation.
6. Sets the degrees of the curvature (in decimal degree form) of:
   - DC1 to be zero regardless whether TheSP = 1 or 2 for the line to which the spiral is to be tangent.
   - DC2 to be 18,000/(RAD π) for the circular curve.

7. Calls the icComspr procedure twice, once to compute the various parameters of the back spiral, and then those of the ahead spiral.
8. Computes the central angle of the circular curve DELTAC by subtracting from the DELTA angle the spiral angle THETA1 and THETA2, the spiral angles of the back and ahead spirals.
9. Queries DELTA for being (a) equal to zero, (b) equal to 180° (π) within a tolerance of 0.00009 radians (18.56 s of a degree) or (c) greater than zero and not 180° within the said tolerance.
   (a) If DELTA is equal to zero, sets IERR to be 6 and aborts.
   (b) If DELTA is equal to 180° (π):
      (b1) Queries whether either each of the pair PTN1 and PTE1 or of the pair PTN2 and PTE2 does not have a value of 99999.9 and if so sets IERR to be 7 and aborts; else continues.
      (b2) Queries the parameters of each of the pair PTN1 and PTE1 or of the pair PTN2 and PTE2 to determine which pair or parameters has values of 99999.9 and which do not. If one pair has values of 99999.9 and the other does not signifies that either the TS or the ST has been specified, and the other should float, and T_1 and T_2 are set to 99999.9 because the two tangent lines are parallel.
   (c) If DELTA is greater than zero and not 180°, assumes that the two spirals are unequal and computes the distance W, the angles Δ_1 and Δ_2, and the tangent lengths T_1 and T_2 (see Figure 9.5a) using Equations 9.5a, 9.5b, 9.5c, 9.5d, and 9.5e, respectively.
10. Queries the parameters of each of the pair PTN1 and PTE1 and/or of the pair PTN2 and PTE2 to determine about which point the alignment is to be aligned.
   (a) If each of PTN1, PTE1, PTN2, and PTE2 has a value of other than 99999.9 (the alignment is to be fixed about the PI),
      (a1) Calls the icintrs procedure to intersect the two tangent lines and compute the coordinates of the PI.
      (a2) Calls the icptl procedure to locate the TS by traversing from the PI along AZ1 backwards a distance of $-T_1$.
      (a3) Calls the icptl procedure to locate the ST by traversing from the PI along AZ2 forward a distance of T_2.
   (b) If each of PTN2 and PTE2 has a value of 99999.9 (the alignment is to be fixed about the TS),
      (b1) Sets the coordinates of the TS to be those of PTN1, PTE1.

(b2) Calls the `icptl` procedure to locate the PI by traversing from the TS along `AZ1` forward a distance of T_1.

(b3) Calls the `icptl` procedure to locate the ST by traversing from the PI along `AZ2` forward a distance of T_2.

(c) If each of `PTN1` and `PTE1` has a value of 99999.9 (the alignment is to be fixed about the ST),

(c1) Sets the coordinates of the ST to be those of `PTN2`, `PTE2`.

(c2) Calls the `icptl` procedure to locate the PI by traversing from the ST along `AZ2` backwards a distance of $-T_2$.

(c3) Calls the `icptl` procedure to locate the TS by traversing from the PI along `AZ1` backwards a distance of $-T_1$.

11. Queries `ROTN` for being positive (clockwise) or negative (counterclockwise) and then calls the `icptl` procedure to locate the

(a) SC by traversing from the TS along `AZ1` forward a distance of XS_1 and an offset of YS_1 (see Figure 5a), and the CS by traversing from the ST along `AZ2` backwards a distance of $-XS_2$ and an offset of YS_2.

(b) Center point of the circular curve CC_1 and CC_2 by traversing first from the TS along `AZ1` forward a distance of k_1 and an offset of $R+p_1$ (see Figure 5a), and then by traversing from the ST along `AZ2` backwards a distance of $-k_2$ and an offset of $R+p_2$. The two sets of coordinates should be the same, but due to inherent precision some minor differences may exist. Thus, the two sets of coordinates are averaged.

Note that in the above two steps if `ROTN` is negative the offset signs in the `icptl` procedure are to be reversed.

12. Terminates and returns to the calling program.

9.5.2   Case 2—Curve with only One Spiral

Cases 2a and 2b in Figure 9.5 are basically the same except for the end of the circular curve at which the spiral is to be applied. Case 2 is also in essence similar to Case 1 of the same figure with one spiral missing. A single procedure could be have been made to accommodate both cases, but the `icCase2spiral` procedure was created so as not to overburden the `icCase1spiral` procedure.

The left part of Figure 9.5b is an expansion of Case 2a and the right part is a mirror image of Case 2b of Figure 9.5. Both of these two cases could be encountered in a new design when the transition is from an expressway onto a local roadway or vice versa. The possibility also exists in which an existing circular curve as indicated by the dash blue line and

the center point labeled CC$_{OLD}$ in Figure 9.5b, under similar conditions, is to be replaced by a slightly smaller radius curve and maintain the location of the PT or PC.

The methodology of the subject procedure is similar to that of Case 1 with two differences: (a) one being the computation the distances W, T_1, and T_2, with the subscript 1 referring to the side of the spiral and 2 to the other end, and (b) the other being the number of control points to be established. Regarding the said three distances, whereas Case 1 has two θ angles, Case 2 has only one such angle. Thus, using the same approach as in Equations 9.5a through 9.5e we have

$$W = \sqrt{(R+p)^2 + (R)^2 - 2(R+p)(R)\cos(\Delta)} \quad (9.5f)$$

and by the law of sines the two unknown angles at the endpoints of the said W long side may be expressed as

$$\Delta_1 = \sin^{-1}\left[\frac{R\sin(\Delta)}{W}\right] \quad (9.5g)$$

and

$$\Delta_2 = \sin^{-1}\left[\frac{(R+p)\sin(\Delta)}{W}\right] \quad (9.5h)$$

Since R is normal to the back and ahead tangent lines, the angles of the triangle *PK-PT-PI* are *90°-Δ_1* at the *PK*, *90°-Δ_2* at the *PT* and *180°-Δ* at the *PI*, and the distances T_1 and T_2 along their respective tangent lines would then, by the use of the law of sines, be

$$T_1 = k + W\left[\frac{\sin(0.5\pi - \Delta_2)}{\sin(\pi - \Delta)}\right] \quad (9.5i)$$

and

$$T_2 = W\left[\frac{\sin(0.5\pi - \Delta_1)}{\sin(\pi - \Delta)}\right] \quad (9.5j)$$

The assumptions and commentary in the `icCase1spiral` procedure regarding the tangents, directions and given points are all also applicable in the `icCase2spiral` procedure. Any differences are referred to in the operational steps below.

```
Sub icCase2spiral (PTN1, PTE1, AZ1, PTN2, PTE2,
           AZ2, SPRL, RAD, ROTN, TheSPR,
           ‾
           PIN,  PIE,  TSN,  TSE,  SCN,
           SCE, PTN, PTE, CCN, CCE,  _
           DELTA,  THETA,  DELTAC, DC1,
           DC2, IERR)
```

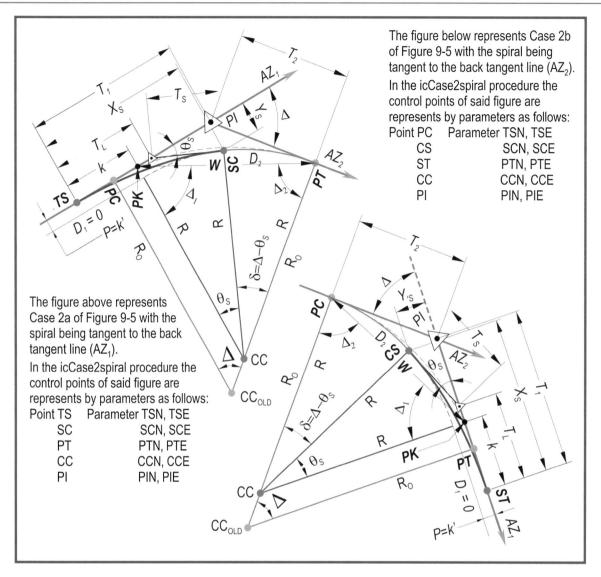

The figure below represents Case 2b of Figure 9-5 with the spiral being tangent to the back tangent line (AZ$_2$).
In the icCase2spiral procedure the control points of said figure are represents by parameters as follows:

Point PC	Parameter TSN, TSE
CS	SCN, SCE
ST	PTN, PTE
CC	CCN, CCE
PI	PIN, PIE

The figure above represents Case 2a of Figure 9-5 with the spiral being tangent to the back tangent line (AZ$_1$).
In the icCase2spiral procedure the control points of said figure are represents by parameters as follows:

Point TS	Parameter TSN, TSE
SC	SCN, SCE
PT	PTN, PTE
CC	CCN, CCE
PI	PIN, PIE

Figure 9.5b   Spiral and Circular Curve with Fixed PC or PT (Line/Spiral/Curve/Line)

The following should be noted for the icCase2spiral procedure when the parameter TheSPR is 2:

- *The entire Figure 9.5b is reversed except for the indicator ROTN and the two tangent lines. That is the local alignment of the spiral; and circular curve is reversed.*
- *When the spiral or any offset spiral therefrom is to be generated with the icPntsOnSprl procedure the azimuth AZ2 should be reversed and so should be the sign of ROTN.*

The input parameters (see Figures 9.5b) to this procedure are the:

PTN1, PTE1	The north and east coordinates of a point on the back tangent line.
AZ1	The north azimuth in radians of the back tangent line pointing from the start of the alignment (TS or PC) towards the PI.
PTN2, PTE2	The north and east coordinates of a point on the ahead tangent line.
AZ2	The north azimuth in radians of the ahead tangent line pointing from the PI towards the end of the alignment (PT or ST).
SPRL	The length in feet (meters) of an entering or exiting spiral.

RAD	The radius of the circular curve in feet (meters) should be positive.
ROTN	Any positive value denotes a clockwise alignment rotation and any negative value denotes a counterclockwise alignment rotation.
TheSPR	An indicator denoting the following: 1 The spiral is to be tangent to the back tangent (entering spiral). 2 The spiral is to be tangent to the ahead tangent (exiting spiral).

NOTE that if:

- PTN1, PTE1, PTN2, and PTE2 each has a value other than 99999.9 then the two tangent lines are intersected to locate the PI about which the alignment is positioned.
- PTN2 and PTE2 each has a value of 99999.9 then each of PTN1 and PTE1 should have a value other than 99999.9 denoting the TS or the PC about which the alignment is to be positioned.
- PTN1 and PTE1 each has a value of 99999.9 then each of PTN2 and PTE2 should have a value other than 99999.9 denoting the ST or the PT about which the alignment is to be positioned.

The returned parameters (see Figure 9.5b) from this procedure are the:

PIN, PIE	The north and east coordinates of the PI.
TSN, TSE	The north and east coordinates of the TS if TheSP=1 or if the ST if TheSP=2.
SCN, SCE	The north and east coordinates of the SC if TheSP=1 or if the CS if TheSP=2.
PTN, PTE	The north and east coordinates of the PT if TheSP=1 or if the PC if TheSP=2.
CCN, CCE	The north and east coordinates of the CC.
DELTA	The deflection angle formed by the two tangent lines.
THETA	The spiral angle of the entering or exiting spiral.
DELTAC	The central angle of the circular curve.
DC1	The degree of curvature (decimal degrees) of the back segment (zero if a line).
DC2	The degree of curvature (decimal degrees) of the ahead segment (zero if a line).
IERR	An indicator the value of which denotes the following: 0 No input errors encountered. 1 Negative or zero SPRL value (abort). 2 Negative or zero RAD value (abort). 3 The two given points are the same (abort). 4 Zero ROTN value (abort). 5 The TheSP value is neither 1 nor 2 (abort). 6 Parallel tangent lines with the same azimuth (abort). 7 Parallel tangent lines with opposite azimuths and the TS or the ST has not been specified (abort).

Upon invocation the procedure initializes the parameters to be returned, and then:

Remember that in the icCase2spiral procedure the parameters T_1 and D_1 pertain to the end of the alignment at which the spiral is located, and the parameters T_2 and D_2 pertain to the other end of the alignment.

Refer to the notes in Figure 9.5b regarding the parameters in which the control points TS and PC and the control points ST and PT are saved.

1. Queries the parameter SPRL for being negative or equal to zero within a tolerance of 0.005 feet (1.524 mm) and if so sets IERR to be 1 and aborts; else continues.
2. Queries the parameter RAD for being negative or equal to zero within a tolerance of 0.005 feet (1.524 mm) and if so sets IERR to be 2 and aborts; else continues.
3. Queries the parameters of the pair PTN1 and PTN2 and of the pair PTE1 and PTE2 for both pairs being equal to each other within a tolerance of 0.005 feet (1.524 mm) and if so sets IERR to be 3 aborts; else continues.
4. Queries the parameter ROTN for being equal to zero within a tolerance of 0.005 feet (1.524 mm) and if so sets IERR to be 4 and aborts; else continues.
5. Queries the parameter TheSPR for being either 1 or 2 and if neither sets IERR to be 5 and aborts; else continues.
6. Calls the icstrdln procedure to compute the DELTA angle by rotating from AZ1 towards AZ2, and then sets DELTA to be positive since ROTN is to denote a clockwise or counterclockwise rotation.
7. Sets the degrees of the degree of curvature (in decimal degree form) of:
   - DC1 to be zero regardless whether TheSP=1 or 2.
   - DC2 to be 18,000/(RAD π) for the circular curve.
8. Calls the icComspr procedure to compute the various parameters of the sole spiral which could be either an entering or and exiting spiral, and then computes the central angle of the circular curve DELTAC by subtracting from DELTA the spiral angle THETA.
9. Queries DELTA for being (**a**) equal to zero, (**b**) equal to 180° (π) within a tolerance of 0.00009 radians (18.56 s of a degree) or (**c**) greater than zero and not 180° within the said tolerance.

(a) If `DELTA` is equal to zero, sets `IERR` to be 6 and aborts.

(b) If `DELTA` is equal to 180° (π):

(b1) Queries whether either each parameter of the pair `PTN1` and `PTE1` or of the pair `PTN2` and `PTE2` does not have a value of 99999.9 and if so sets `IERR` to be 7 and aborts; else continues.

(b2) Queries the parameters of each of the pair `PTN1` and `PTE1` or of the pair `PTN2` and `PTE2` to determine which pair of parameters has values of 99999.9 and which do not. If one pair has values of 99999.9 and the other does not signifies that either the TS or PC, or the ST or PT has been specified, and the other should float, and T_1 and T_2 are set to 99999.9 because the two tangent lines are parallel.

(c) If `DELTA` is greater than zero and not 180°, computes the distance W, the angles Δ_1 and Δ_2, and the tangent lengths T_1 and T_2 (see Figure 9.5b) using Equations 9.5f, 9.5g, 9.5h, 9.5i, and 9.5j, respectively.

10. Queries the parameters of each of the pair `PTN1` and `PTE1` and/or of the pair `PTN2` and `PTE2` to determine about which point the alignment is to be aligned.

(a) If each of `PTN1`, `PTE1`, `PTN2`, and `PTE2` has a value of other than 99999.9 (the alignment is to be fixed about the PI),

(a1) Calls the `icintrs` procedure to intersect the two tangent lines and compute the coordinates of the PI.

(a2) Calls the `icptl` procedure to locate the TS by traversing from the PI along `AZ1` backwards a distance of:
  • $-T_1$ if `TheSPR` is 1.
  • $-T_2$ if `TheSPR` is 2.

(a3) Calls the `icptl` procedure to locate the PT by traversing from the PI along `AZ2` forward a distance of:
  • T_2 if `TheSPR` is 1.
  • T_1 if `TheSPR` is 2.

(b) If each of `PTN2` and `PTE2` has a value of 99999.9 (the alignment is to be fixed about the TS),

(b1) Sets the coordinates of the TS to be those of `PTN1`, `PTE1`.

(b2) Calls the `icptl` procedure to locate the PI by traversing from the TS along `AZ1` forward a distance of:
  • T_1 if `TheSPR` is 1.
  • T_2 if `TheSPR` is 2.

(b3) Calls the `icptl` procedure to locate the PT by traversing from the PI along `AZ2` forward a distance of:
  • T_2 if `TheSPR` is 1.
  • T_1 if `TheSPR` is 2.

(c) If each of `PTN1` and `PTE1` has a value of 99999.9 (the alignment is to be fixed about the ST),

(c1) Sets the coordinates of the ST to be those of `PTN2`, `PTE2`.

(c2) Calls the `icptl` procedure to locate the PI by traversing from the ST along `AZ2` backwards a distance of:
  • $-T_2$ if `TheSPR` is 1.
  • $-T_1$ if `TheSPR` is 2.

(c3) Calls the `icptl` procedure to locate the TS by traversing from the PI along `AZ1` backwards a distance of:
  • $-T_1$ if `TheSPR` is 1.
  • $-T_2$ if `TheSPR` is 2.

11. Queries `ROTN` for being positive (clockwise) or negative (counterclockwise) and then calls the `icptl` procedure to locate the

(a) SC by traversing from the:
  • TS, if `TheSPR` is 1, along `AZ1` forward a distance of XS and an offset of YS (see Figure 9.5b).
  • PT if `TheSPR` is 2, along `AZ2` backwards a distance of $-XS$ and an offset of YS.

(b) Center point of the circular curve CC_1 by traversing from the TS along `AZ1` forward a distance of:
  • k and an offset of $R + p$ (see Figure 9.5b), if `TheSPR` is 1.
  • $Zero$ and an offset of `RAD` (see Figure 9.5b), if `TheSPR` is 1.

(c) Center point of the circular curve CC_2 by traversing first from the PT along `AZ2` backwards a distance of:
  • Zero and an offset of `RAD` (see Figure 9.5a), if `TheSPR` is 1.
  • $-k$ and an offset of $R + p$ (see Figure 9.5b), if `TheSPR` is 2.

    Note that in the above three steps if `ROTN` is negative the offset signs in the `icptl` procedure are to be reversed.

12. Computes the coordinates of the center point of the circular curve as being the average of CC1 and CC2, and then terminates and returns to the calling program.

9.5.3   Case 3—One Spiral Between Two Curves

Cases 3a and 3b in Figure 9.5 are essentially the same except for the conditions under which they are encountered. Whereas in Case 3a the start of the spiral (CS) is assumed be fixed at a given distance from the PC, in Case 3b (to be discussed later on) the CS floats so that the PC and the PT are located on the two given and fixed tangents.

The said Case 3a has been expanded in Figure 9.5c into the three sub-cases under which the `icCase3spiral` procedure introduces a spiral between two circular curves. Each of these sub-cases is composed of two groups, are characterized

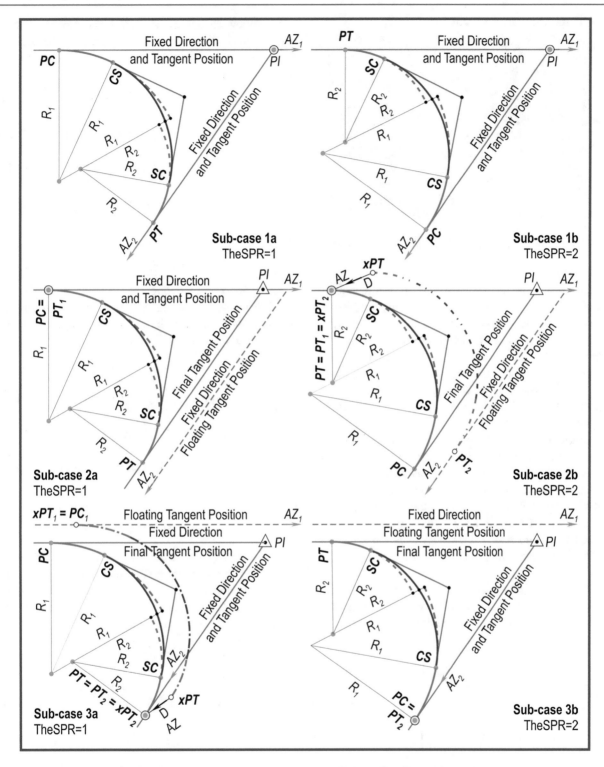

Figure 9.5c Expanded Case 3a of Figure 9.5: Line–Curve–Spiral–Curve–Line Between Two Tangent Lines

by the conditions listed below and they are summarized in Table 9.1. The said two groups are:

(a) The PC is located on the back tangent and the PT is located on the ahead tangent.
(b) The PC is located on the ahead tangent and the PT is located on the back tangent.

In each group the spiral transitions from the PC towards the PT. That is the curve at the PC is flatter than that of the PT. In all sub-cases the direction of both tangent lines is fixed (considered to be known)

Sub-case 1 The two tangents are fixed in direction and position, and the PC and PT float along the two tangents with the PI being fixed at the intersection of the two tangents.

Table 9.1 Spiral Between Two Curves

Elements	Conditions						
Sub-case		1		2		3	
Group		a	b	a	b	a	b
Back Tangent	Fixed	×	×	×	×		
Back Tangent	Floats					×	×
Ahead Tangent	Fixed	×	×			×	×
Ahead Tangent	Floats			×	×		
Fix PC on the	Bck.Tng	×		×		×	
Fix PC on the	Ahd.Tng		×		×		×
Fix at PT1 the	PC			×			
Fix at PT1 the	PT				×		
Fix at PT2 the	PC						×
Fix at PT2 the	PT					×	
TheSPR =		1	2	1	2	1	2

Sub-case 2   The back tangent is fixed in direction and position, the PC or PT (depending on the subgroup) is fixed on the back tangent at the point defining the said tangent, and the PT or PC floats along the ahead tangent direction.

Sub-case 3   The ahead tangent is fixed in direction and position, the PT or PC (depending on the subgroup) is fixed on the ahead tangent at the point defining the said tangent, and the PC or PT floats along the back tangent direction.

The introduction of the spirals for Case 1 and Case 2 of Figure 9.5 was accomplished by computing the lengths of the overall tangents T_1 and T_2. The equations for these two parameters were kind of similar. The introduction of the spiral of Case 3a of Figure 9.5 follows the same approach of computing T_1 and T_2 but in a somewhat different manner as indicated in Figure 9.5d, which represents the three sub-cases and their two groups of Figure 9.5c, and described below.

In the said last figure let us consider a local coordinate system at point A, the said point being the PC of the three curves. From the said figure we see that the tangents of the two circular curves are

$$T_A = R_1 \tan\left(\frac{1}{2}\alpha_1\right) \qquad (9.5\text{k}1)$$

and

$$T_B = R_2 \tan\left(\frac{1}{2}\alpha_2\right) \qquad (9.5\text{k}2)$$

Before we continue it should be said that in Figure 9.5d the back tangent is displayed as being located along a due east direction, and the three curves are developed in a clockwise rotational direction. This is done strictly for display purposes. The graphics of figure could have been inclined along any direction and the curves could be developed in a counterclockwise rotation. The signs of the trigonometric functions of the directions in the equations presented below will accommodate any such orientation.

If we now commence at point A and traverse along the tangents to the three curves and accumulate the horizontal projections of the said tangents their sums will produce the value of X_E. Thus:

$$X_A = T_A \cos\left(AZ_1\right) \qquad (9.5\text{m}1)$$

$$X_B = \left(T_A + T_L\right)\cos\left(AZ_1 + \alpha_1\right) \qquad (9.5\text{m}2)$$

$$X_C = \left(T_S + T_B\right)\cos\left(AZ_1 + \alpha_1 + \theta_S\right) \qquad (9.5\text{m}3)$$

$$X_D = T_B \cos\left(AZ_2\right) \qquad (9.5\text{m}4)$$

and

$$X_E = X_A + X_B + X_C + X_D \qquad (9.5\text{m}5)$$

where T_L and T_S are the long and short tangent lengths of the spiral. If we next traverse similarly from point A and accumulate the vertical projections of the said tangents their sums will produce the value of Y_E. Thus,

$$Y_A = T_A \sin\left(AZ_1\right) \qquad (9.5\text{n}1)$$

$$V_B = \left(T_A + T_L\right)\sin\left(AZ_1 + \alpha_1\right) \qquad (9.5\text{n}2)$$

$$Y_C = \left(T_S + T_B\right)\sin\left(AZ_1 + \alpha_1 + \theta_S\right) \qquad (9.5\text{n}3)$$

$$Y_D = T_B \sin\left(AZ_2\right) \qquad (9.5\text{n}4)$$

and

$$Y_E = Y_A + Y_B + Y_C + Y_D \qquad (9.5\text{n}5)$$

In Figure 9.5d we notice that X_E and Y_E represent the *dx* and *dy* components of the line AB which can help compute the length and azimuth of the said line, *AB* and AZ_X, respectively. Thus, with AZx known, angles b1 and b2, and distance AB, T1, and T2 may now be computed as follows:

$$AB = \sqrt{X_E^{\,2} + Y_E^{\,2}} \qquad (9.5\text{o}1)$$

$$AZ_X = \frac{1}{2}\pi - \left(\frac{Y_E}{X_E}\right) \qquad (9.5\text{o}2)$$

$$\beta_1 = AZ_X - AZ_1 \qquad (9.5\text{o}3)$$

$$\beta_2 = -\beta_1 \qquad (9.5\text{o}4)$$

$$T_1 = AB\left[\frac{\sin\left(\beta_2\right)}{\sin\left(\Delta\right)}\right] \qquad (9.5\text{o}5)$$

and

$$T_2 = AB\left[\frac{\sin\left(\beta_1\right)}{\sin\left(\Delta\right)}\right] \qquad (9.5\text{o}6)$$

Figure 9.5d Generalization
of the Sub-cases of Case 3a of
Figure 9.5: Line–Curve–
Spiral–Curve–Line Between
Two Tangent Lines

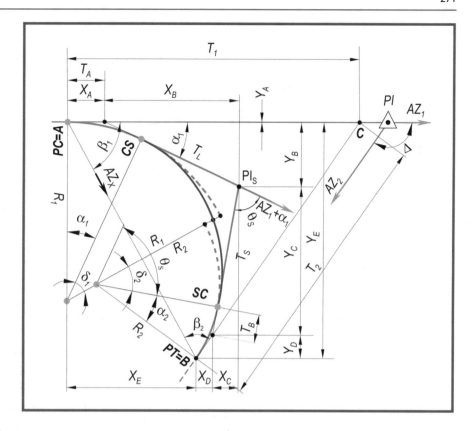

With the above information the coordinates of the control points of the three curves may now be determined. The procedure of how the above equations are applied to attain the desired results is addressed in the operational steps below.

The above work assumes that the position of the CS with respect to the PC is known in terms of an arc distance or subtending an angle between the said two points. Now let us consider the case in which the location of the CS is to float and that the PC is to be fixed on its fixed tangent and the PT is to float along the other fixed tangent. The primary intent of this condition is to provide for the replacement of two compound curves with two curves of the same radius and a transition spiral between them maintaining the position of the original PC of the compound curves. If the maintenance of the original PC is not an issue, then the construction reverts to the above described sub-case conditions. The methodology addressing this condition in the operational steps below is associated only with sub-case 2a. Similar conceptual methodology may also be used to accommodate the other sub-cases.

It is noted that the position of both the PC and PT of the original compound curves cannot be maintained with the original radii. As shown in Figure 9.5e there are two fixed tangent lines and two compound circular curves with their PC and PT. If the PC is to be maintained, the introduction of the spiral causes the sharper curve to shift towards the inside of the curves by the spiral throw distance. Thus, it is impossible to maintain both the PC and the PT while maintaining

the radii of the original two compound curves. If the PT is to be maintained then the flatter curve would shift outwards of the two curves a distance equal to the said throw distance. Regardless of which point is maintained the effect on the other point is minimal.

To provide for this condition the procedure:

- Assumes that the CS is first positioned using an assumed arc length from the PC,
- Computes the angle γ shown in Figure 9.5e as noted below, and then
- Rotates the CS, SC, PT, and CC_2 a γ angle about CC_1 to their final position.

To compute angle γ, the procedure:

- Intersects an arc of radius R_3 being equal to R_1 minus R_2 minus the said throw distance having its center at CC_1 with a line parallel to the tangent line on which the PT is to lie and offset therefrom towards the inside of the curves a distance equal to the R_2 radius to locate point B in Figure 9.5e, point B being the new location of CC_2.
- Inverts from CC_1 to point B to determine that azimuth AZ_3.
- Computes the angle ε as being the angle between the normal line to the tangent on which the PC is located and the said azimuth AZ_3.
- Computes the angle γ by subtracting angles $\alpha1$ and $\delta1$ from angle ε.

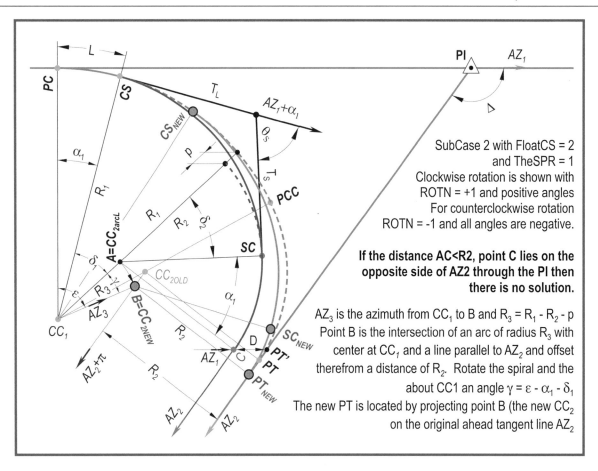

Figure 9.5e Spiral between Two Compound Curves Maintaining the PC or PT

Now let take us take a look at the positional relationship of point C to the ahead tangent line. Point C is the PT of the ahead curve prior to applying the above rotation to force the PT on the ahead tangent line. In Figure 9.5e the said point is located to the left of the ahead tangent line or inside the towards the three curves. Looking at this in a different manner, the distance from point A to point C is R_2 which is less than the normal offset of point A from the given ahead tangent line AZ_2 passing through the PI, or conversely the normal offset is greater than R_2. However, what if the said offset were less than R_2? Then a solution could not exist. In reviewing Figure 9.5e such a condition could be encountered if:

- The deflection angle of the two tangent lines D is too large,
- One or both radii of the circular curves is too large, and/or
- The spiral length is too large, and/or
- The assumed arc length to position the CS is too large.

When this condition arises, a remedial step would be to reduce the assumed arc length to zero (refer to the commentary in the input data). If so, then a solution is possible the CS would be properly positioned, and if not so, then one or more of the other parameters should be reduced.

There is yet another option which is not handled by the subject procedure, and that is to introduce the two curves and a spiral between them starting at any point along the back tangent line. Once the curves have been constructed, the center point of the sharper curve (point A in Figure 9.5e) is projected on the new location of the ahead tangent line to get point C. From here on a line passing through C and having a direction of AZ_1 may be intersected with the original ahead tangent line to locate point PT' the new location of the PT. The remaining points of the three curves may then be translated a distance D from the PC along AZ_1 to their new location.

```
Sub icCase3spiral (PTN1, PTE1, AZ1, PTN2, PTE2,
                   AZ2, SPRL, ArcL, RAD1, RAD2,
                   _
                   FloatCS,    ROTN,    SubCASE,
                   TheSPR, _
                   PIN, PIE, PCN, PCE, CSN,
                   CSE, SCN, SCE, PTN, PTE, _
                   CCN1,   CCE1,   CCN2,   CCE2,
                   DELTA, THETA, ALPHA1, ALPHA2,
                   _
                   DC1, DC2, IERR)
```

The input parameters (see Figure 9.5e) to this procedure are the:

PTN1, PTE1	The north and east coordinates of a point on the back tangent line.
AZ1	The north azimuth (radians) of the back tangent from the PC to the PI.
PTN2, PTE2	The north and east coordinates of a point on the ahead tangent line.
AZ2	The north azimuth (radians) of the ahead tangent from the PI to the PT.
SPRL	The length of the spiral in feet (meters) that transitions from the flatter towards the sharper circular curve.
ArcL	The arc length along the flatter curve in feet (meters) from the PC to the CS. This value would most probably change if the value of FloatCS is 2 so that the CS is to float along the tangent line on which it is to lie. If ArcL is set to be zero, the spiral would start at the PC in which case if the PC is: • A temporary assumed point to locate the spiral, then there is no further concern. • The actual PC point of the back curve, that is the preceding course is a line, then the corresponding radius should be zero.
RAD1	The radius of the flatter circular curve in feet (meters) should be positive.
RAD2	The radius of the sharper circular curve in feet (meters) should be positive.
FloatCS	An indicator denoting the following: 1 The CS is to be fixed an ArcL arc length from the PC. 2 The CS is to float along the sharper curve as described above. The use of this parameter requires an assumed value of ArcL.
ROTN	1 Denotes a clockwise alignment rotation of the curves. −1 Denotes a counterclockwise alignment rotation of the curves.
SubCASE	An indicator the value of which should be 1, 2, or 3 denoting the applicable sub-case condition as per Table 9.1.
TheSPR	An indicator denoting the following: 1 The PC is to be on the back tangent line. 2 The PC is to be on the ahead tangent line.

If FloatCS = 2 it is possible that there may not be a solution. For this refer to the commentary preceding the input data and to the operational steps.

The returned parameters (see Figure 9.5e) from this procedure are the:

PIN, PIE	The north and east coordinates of the PI.
PCN, PCE	The north and east coordinates of the PC.
CSN, CSE	The north and east coordinates of the CS.
SCN, SCE	The north and east coordinates of the SC.
PTN, PTE	The north and east coordinates of the PT.

CCN1, CCE1	The north and east coordinates of the center point of the flatter curve.
CCN2, CCE2	The north and east coordinates of the center point of the sharper curve.
DELTA	The deflection angle formed by the two tangent lines to the two circular curves.
THETA	The spiral angle of the spiral.
ALPHA1	The central angle of the flatter circular curve.
ALPHA1	The central angle of the sharper circular curve.
DC1	The degree of curvature (decimal degrees) of the flatter circular curve.
DC2	The degree of curvature (decimal degrees) of the sharper circular curve.
IERR	An indicator the value of which denotes the following: 0 No input errors encountered. 1 Negative or zero SPRL value (abort). 2 Negative or zero RAD1 or RAD2 value (abort). 3 The RAD2 value is greater than or equal to the RAD1 value (abort). 4 The two given points are the same (abort). 5 Negative ArcL value (abort). 6 Zero ROTN value (abort). 7 The SubCASE value is not 1, 2, or 3 (abort). 8 The TheSPR value is neither 1 nor 2 (abort). 9 Parallel tangent lines with the same azimuth (abort). 10 Parallel tangent lines and SubCASE is 1 (abort). 11 No intersection to locate the center point of the sharper curve when the FloatCS is 2 (abort). 12 No solution when the FloatCS is 2 (abort).

NOTE that:

• If TheSPR is 1, the spiral transition is from the back tangent (AZ1) towards the ahead tangent (AZ2). That is, the PC, CS, SC and PT advance from the back tangent towards the ahead tangent.

• If TheSPR is 2, the spiral transition is from the ahead tangent (AZ2) towards the back tangent (AZ1). That is, the PC, CS, SC and PT advance from the ahead tangent towards the back tangent.

• If FloatCS is 2, the value of ArcL would most probably change to replace the assumed (input) arc length from the PC to the CS to its floating value.

• If FloatCS is 2, there are two potential causes for no solution as addressed previously and further discussed in the operational steps below.

Upon invocation the procedure initializes the parameters to be returned, and then:

1. Queries SPRL, RAD1, and RAD2 for being negative or equal to zero within a tolerance of 0.005 feet (1.524 mm), and whether RAD2 is greater or equal to RAD1, and if so sets IERR to be 1, 2, or 3, respectively, and aborts; else continues.

2. Queries the parameters of the pair PTN1 and PTN2 and of the pair PTE1 and PTE2 for both pairs being equal to each other within a tolerance of 0.005 feet (1.524 mm) and if so sets IERR to be 4 and aborts; else continues.
3. Queries ArcL for being negative and if so sets IERR to be 5 and aborts; else continues.
4. Queries ROTN for being equal to zero within a tolerance of 0.005 feet (1.524 mm) and if so sets IERR to be 6 and aborts; else continues.
5. Queries SubCASE for being 1, 2, or 3 and if not sets IERR to be 7 and aborts; else continues.
6. Queries TheSPR for being either 1 or 2, and if neither sets IERR to be 8 and aborts; else continues.
7. Sets the degrees of the degree of curvature (in decimal degree form) of:
   - DC1 to be 18,000/(RAD1 π) for the flatter circular curve.
   - DC2 to be 18,000/(RAD2 π) for the sharper circular curve.
8. Calls the icComspr procedure to compute the various parameters of the spiral.
9. Queries the parameter TheSPR for being either 1 or 2 and if:
   - TheSPR = 1, sets temporary parameters to represent the input coordinates and azimuth of the point and azimuth that define the back tangent line (PTN1, PTE1, AZ1), as well as the parameter ROTN.
   - TheSPR = 2, sets the said temporary parameters to represent the input coordinates of the point that defines the ahead tangent line (PTN2, PTE2), reverses the azimuth of the direction of the ahead tangent line (AZ2 + π) and reverses the sign of the parameter ROTN.

     The temporary parameters are labeled with the same name as the permanent parameters preceded by the lower case letter x (xROTN for ROTN).
10. Calls the icstrdln procedure to compute the DELTA angle by rotating from AZ1 towards AZ2, and then sets DELTA to be positive since ROTN is to denote a clockwise or counterclockwise rotation.
11. Computes the central angle:
    - ALPHA1 of the flatter circular curve by dividing the given ArcL by the radius of the flatter curve RAD1.
    - ALPHA2 of the sharper curve by subtracting from DELTA the values of the spiral angle θ_S and the central angle of the flatter curve ALPHA1.
12. Queries DELTA for being (a) equal to zero, (b) equal to 180° (π) within a tolerance of 0.00009 radians (18.56 s of a degree), or (c) greater than zero and not 180° within the said tolerance.
    (a) If DELTA is equal to zero, sets IERR to be 9 and aborts.
    (b) If DELTA is equal to 180° (π) queries SubCASE for being 1 and if so sets IERR to be 10 and aborts; else sets the length of the overall tangent lines T_1 and T_2

from the PC to the PI and from the PI to the PT, respectively, to be 99999.9 and continues.
    (c) If DELTA is greater than zero and not 180°, computes the lengths of the said tangent line T_1 and T_2 using Equations 9.5m1 through 9.5o6.
13. Queries SubCASE for being for being 1, 2, or 3.
    (a) If SubCASE is equal to 1, both tangent lines are fixed in direction and position and the three curves are oriented about the PI.
       (a1) Calls the icintrs procedure to intersect the two tangent lines and compute the coordinates of the PI.
       (a2) Calls the icptl procedure to determine the coordinates of the PC by traversing along AZ1 a negative T_1 distance.
       (a3) Calls the icCase3spr1 procedure (see later on after the description of this procedure) to determine the coordinates of the SC, CS, PT and those of the center points of both circular curves. It is noted that the icCase3spr1 procedure is called using the temporary coordinates defined at the above **Step 9**. This way the said procedure need not worry whether the flatter curve is tangent to the back (TheSPR = 1) or to the ahead (TheSPR = 2) tangent line.
    (b) If SubCASE is 2 or 3, one tangent is to be fixed in direction and position, the other is to float, and the three curves are oriented about the PC.
       (b1) Sets the coordinates of the PC to be those of the temporary point corresponding to (PTN1, PTE1) or (PTN2, PTE2) defined in the above **Step 9**.
       (b2) Calls the icCase3spr1 procedure to determine the coordinates of the SC, CS, PT and those of the center points of both circular curves (see commentary in **Step 12 (a) (a3)** above).
       (b3) Calls the icintrs procedure to intersect the two tangent lines and compute the coordinates of the PI.
       (b4) Queries the values of SubCASE and TheSPR as follows:
          - If SubCASE = 2 and TheSPR = 1 or if SubCASE = 3 and TheSPR = 2, then the coordinates as returned by the icCase3spr1 procedure are accepted as returned.
          - If SubCASE = 2 and TheSPR = 2 or if SubCASE = 3 and TheSPR = 1, then the procedure:
             – Calls the icforce procedure to determine the azimuth AZ and distance D from the PT as computed in Step 12 (b) **(b2)** above and the temporary point to be fixed on the ahead tangent as defined in the above **Step 9**.

- Calls the `icptl` procedure to relocated the PC, CS, SC, PT, PI and center points of the two circular curves as computed above by a distance *D* along the azimuth *AZ* as compute in the preceding step.

14. Queries the parameter `FloatCS` for being for being 1 or 2. If it is 1, the procedure terminates; else if it is 2 continues to rotate the spiral and sharper curve to their final position.

15. Calls the `icprjct` procedure to project the center point of the sharper curve on the ahead curve as represented by the temporary parameters to determine the said point's normal offset distance OFF, and compares OFF with `RAD2`. If OFF is less than `RAD2` sets `IERR` to be 12 and aborts, else continues.

16. Calls the `icptl` procedure to establish a point on the inside of the curves and offset from the ahead tangent a distance of `RAD2` to define a line parallel and offset from the ahead tangent line using the temporary parameters. Note that the azimuth of this parallel line is set to oppose the input direction of the tangent line in order to get the second intersection in **Step 18** below.

17. Sets the radius R3 equal to `RAD1` minus `RAD2` minus the spiral throw distance `SP`.

18. Calls the `iclincrc` procedure to intersect an arc of radius R3 having its center point at that of the flatter curve with the offset parallel line established under **Step 16**.

19. Queries the returned parameter `INTRSC` for having a value of 2, and if so sets `IERR` to be 11 and aborts; else accepts the returned coordinates as being those of the center point of the sharper curve (CCN2, CCE2) and continues.

20. Computes the values of δ1 and δ2 of the spiral (see Figure 9.2a) by use of Equations 9.2.8a and 9.2.8b.

21. Determines the azimuth of the normal to the temporary back tangent line by subtracting 90° multiplied by the temporary `xROTN` to account for a counterclockwise rotation, and if the said normal azimuth is checked:
    - First for being negative, and if so it is set to be the positive 360° complement, and
    - Then it is checked for being greater then 360°, and if so 360° are subtracted.

22. Calls the `icforce` procedure to determine the azimuth AZ3 from the center point of the flatter curve towards the center point of the sharper curve (see Figure 9.5e).

23. Calls the `icstrdln` procedure to computes the angle ε using the temporary `xROTN`, and then computes the angle of rotation γ again using the temporary `xROTN` to account for a counterclockwise rotation.

24. Applies the angle of rotation γ multiplied by `xROTN` to the azimuths preceding those of the CS and SC and calls the `icptl` procedure to computes the coordinates of the CS and SC.

25. Calls the `icprjct` procedure to project the coordinates of the center point of the sharper curve (CCN2, CCE2) as computed under **Step 19** on the temporary ahead tangent line to compute the coordinates of the PT.

26. Calls the `icintrs` procedure to intersect the two temporary tangent lines and compute the coordinates of the PI.

27. Computes the modified value of the central angle:
    - `ALPHA1` of the flatter circular curve by adding the rotation angle γ computed under **Step 23**.
    - `ALPHA2` of the sharper curve by subtracting from `DELTA` the values of the spiral angle θ_S and the central angle of the flatter curve `ALPHA1`.

28. Computes the modified value of the arc from the PC to the CS (`ArcL`) by multiplying `ALPHA2` by `RAD2`, and the terminates.

Spiral transcription from a deed.

When transcribing a parcel of land from a deed, a parcel side may be a curve offset from a main spiral of which the information usually provided includes the:

(a) *length of the spiral L_S,*
(b) *radius of the circular curve R,*
(c) *spiral angle θ_S and*
(d) *offset(s) from the main spiral.*

Also the k and p distances could be specified in the deed.

The position of the TS, SC, CS and ST may be defined by description of the adjacent line and, or circular curve courses of the parcel of land.

With this information and use of the equations presented in this chapter the main spiral and offset side may be laid out and the transcription continued. The main issue in writing a procedure for this task is its logic and not the mathematics.

For INTRSC to have a value of 2 indicates a possible input error not being detected by the built-in checking criteria.

The procedure `icCase3spr1` is called by the `icCase3spiral` procedure twice to compute the coordinates of the control points of the curves using the temporary parameters established in **Step 9**. This has been done to provide for the spiral to be near the back tangent line regardless of which tangent line has been specified first in order to maintain a forward specification of tangents.

```
Sub icCase3spr1 (xPCN, xPCE, xAZ1, RAD1, RAD2,
                 xROTN, ALPHA1, _
                 THETA, SPRL, STL, STS, xAZ2, _
                 PIN, PIE, PCN, PCE, CSN, CSE,
                 SCN, SCE, PTN, PTE, _
                 xCCN1, xCCE1, xCSN, xCSE,
                 xSCN, xSCE,
                 xCCN2, xCCE2, xPTN, xPTE)
```

The input parameters (see Figure 9.5e) to this procedure are the:

xPCN, xPCE	The north and east temporary coordinates of the PC.
xAZ1	The north azimuth (radians) of the temporary back tangent pointing from the PC to the PI.
RAD1	The radius of the flatter circular curve in feet (meters) should be positive.
RAD2	The radius of the sharper circular curve in feet (meters) should be positive.
xROTN	The reverse of ROTN.
ALPHA1	The central angle of the sharper circular curve.
THETA	The spiral angle θ_S as computed by the icComspr procedure.
SPRL	The length of the spiral in feet (meters) that transitions from the flatter towards the sharper circular curve.
STL	The long tangent line of the spiral as computed by the icComspr procedure.
STS	The short tangent line of the spiral as computed by the icComspr procedure.
xAZ2	The north azimuth (radians) of the temporary ahead tangent pointing from the PI to the PT.

The returned parameters (see Figure 9.5e) from this procedure are the:

xCCN1, xCCE1	The north and east coordinates of the temporary center point of the flatter curve.
xCSN, xCSE	The north and east coordinates of the temporary CS.
xSCN, xSCE	The north and east coordinates of the temporary SC.
xCCN2, xCCE2	The north and east coordinates of the temporary center point of the sharper curve.
xPTN, xPTE	The north and east coordinates of the temporary PT.

Upon invocation the procedure:

1. Calls the icptl procedure to set the xCCN1, xCCE1 coordinates traversing from xPCN, xPCE along xAZ1 with a zero plus and a RAD1 offset multiplied by xROTN.
2. Sets the azimuth AZcs from CC1 to the CS by subtracting 90° from xAZ1 and adding ALPHA1 both multiplied by xROTN, and then calls the icptl procedure to establish

the xCSN, xCSE coordinates traversing from xCCN1, xCCE1 along AZcs with a RAD1 plus and a zero offset.
3. Calls the icptl procedure to establish the coordinates xPISN, xPISE of the PI of the spiral (see Figure 9.2a) traversing from xCSN, xCSE along AZcs with a zero plus and an STL offset.
4. Sets the azimuth AZsc from xPISN, xPISE to the SC by adding 90° and THETA both multiplied by xROTN, and then calls the icptl procedure to establish the xSCN, xSCE coordinates traversing from xPISN, xPISE along AZcs with an STS plus and a zero offset.
5. Calls the icptl procedure to establish the coordinates xCCN2, xCCE2 traversing from xSCN, xSCE along AZsc with a zero plus and a RAD2 multiplied by xROTN offset.
6. Calls the icptl procedure to establish the coordinates xPTN, xPTE traversing from xCCN2, xCCE2 along xAZ2 with a zero plus and a RAD2 multiplied by xROTN offset, and then returns to the calling program.

9.6   Spiral Intersections

The three subsections presented below address the intersection of a spiral with a straight line, a circular curve and with another main or offset spiral. In each of these three intersections the spiral to be intersected may be a main spiral, or an offset spiral having a constant offset. When intersecting an offset spiral having a nonuniform offset throughout the length of the main spiral, the offset should be constant in the vicinity of where the intersection is to be located. However, if this is not the case, and an intersection must be made with an offset spiral having a variable offset distance from the main spiral, the offset spiral then should be constructed as discussed in the preceding sections, thus generating a polyline which may then be intersected with a line, a circular curve, or with another main or offset spiral that has been generated as a polyline.

Three individual tool commands could be created in the calling program to treat each one of these intersections individually, or there could be a single tool command to address all three spiral intersections. If so, the calling program will need to select the two features to be intersected, and then decipher the type of features that have been selected and accordingly call the appropriate of the procedures presented below. This is an individual programmer's or system developer's preference. Each of these two choices has its own merits and drawbacks.

Do not confuse the distance A in Figure 9.1a with point A (the start of the spiral).

9.6.1   Spiral Intersection with a Straight Line

Figure 9.6.1a represents a main spiral and two offset spirals one on each side. Two straight lines, a solid blue and a dashed blue, intersect these spirals as shown. The slope of these lines with respect to the line tangent to the spiral at the spiral's point of beginning (point A) is identified as $S = tan(\alpha)$. The slope S maintains the sign of the trigonometric function of the said angle according to the Cartesian quadrant. Hence, the slope shown is positive for the solid line and negative for the dashed line. Had the direction of these lines been reversed 180°, the slope would remain the same. In the said figure let X and Y represent the spiral coordinates of either the point of intersection of the straight line with the main spiral, or the projection on the main spiral of the point of intersection of the intersecting line with the offset spiral (represented by X_O and Y_O in the said figure). At each of these two intersections there is an instantaneous radial line forming a θ and θ_O angle, respectively. The subscripts in the distances X and Y and in the angle θ are used only for graphic distinction and not in the equations below. So in the said figure it may be shown that

$$X = A + \left(W \cos\theta - Y\right)\left(\frac{N}{S}\right) - W \sin\theta \quad (9.6.1a)$$

where

$N = +1$ for a clockwise curving spiral as shown in the said figure,
$N = -1$ for a counterclockwise curving spiral,
$W < 0$ offset spiral inside the main spiral,

$W = 0$ there is no offset spiral,
$W < 0$ offset spiral outside the main spiral,
$A =$ the distance along the tangent line from the start of the spiral to its intersection with the line to be intersected with the spiral.

Also the absolute values of θ, X, and Y are used to ensure proper signs since N denotes rotational direction. Thus, there is no differentiation between main and offset spirals since a main spiral may be considered to be an offset spiral with a zero offset distance. Equation 9.6.1a may now be expressed as a function of L, the distance from the beginning of the spiral to the point of projection of the intersection on the main spiral as

$$f(L) = X + \left(\frac{N}{S}\right)Y + W \sin\theta - \left(\frac{N}{S}\right)W \cos\theta - A = 0$$
$$(9.6.1b)$$

The variables θ, Y and X may be determined in terms of L by Equations 9.1.4a and 9.1.15c.

The method of solving for such an intersection is that of successive approximations. An approximate L is assumed, the error of the approximation is found and used to obtain a better approximate value of L. This process is repeated until the error is small enough to be considered negligible. Limits for this error may be set up by the individual user according to personal desires. Tests by the author have shown that a solution within a tenth of a second (0.1 s) of the instantaneous theta (θ) may be obtained within three trials depending on the first approximation of L.

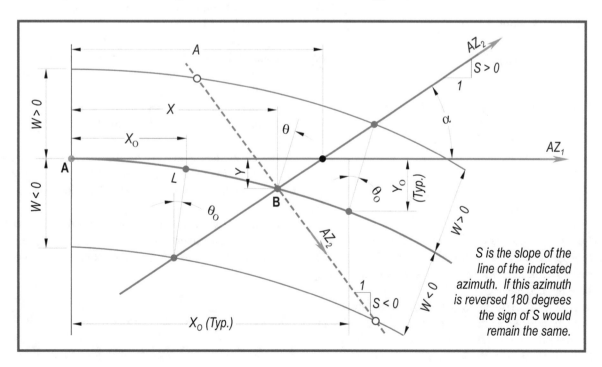

Figure 9.6.1a Intersection of a Line with a Main or Offset Spiral

If $L = \xi$ is the solution to Equation 9.6.1b, and $L = x$ is an approximation, hen

$$x' = x - \frac{f(x)}{f'(x)} = x - K \qquad (9.6.1c)$$

where x' is the new approximation and K is the error of the approximation (the function of Equation 9.20 divided by its first derivative). This may be written as

$$K = \frac{f(x)}{f'(x)} \qquad (9.6.1d)$$

and

$$L_{new} = \left| L_{assumed} - K \right| \qquad (9.6.1e)$$

When the first approximation of L is rather unreasonable and the actual point of intersection is near the beginning of the spiral, the value of K may be greater than that of the first approximation of L. Because of this the absolute value is taken in Equation 9.6.1e to ensure consistency with θ, X, and Y, which are always positive. This could be taken care of in other ways as well depending on the ingenuity of the programmer. When K does become large the only affect it has on the solution is the increase of iterations.

The value of $f'(L)$ is found as:

$$f'(L) = \frac{d}{dL}(X) + \frac{N}{S}\frac{d}{dL}(Y) + W\cos\frac{d}{dL}(\theta) + \frac{WN}{S}\sin\theta\frac{d}{dL}(\theta)$$

$$(9.6.1f)$$

in which from Equation 9.1.4b

$$\frac{d}{dL}(\theta) = \frac{D_1(L_S - L) + D_2 L}{100 L_S} \qquad (9.6.1g)$$

and, from Equations 9.1.10a and 9.1.10b, respectively

$$\frac{d}{dL}(Y) = \theta - \frac{\theta^3}{3!} + \frac{\theta^5}{5!} - \frac{\theta^7}{7!} \qquad (9.6.1h)$$

and

$$\frac{d}{dL}(X) = 1 - \frac{\theta^2}{2!} + \frac{\theta^4}{4!} - \frac{\theta^6}{6!} \qquad (9.6.1i)$$

where θ is expressed by Equation 9.1.4b. The basic concepts of this intersection have appeared in Appendix G of T.F. Hickerson's book titled *Route Surveys and Design*, and they have been applied herein.

The `icSpiralLine` procedure presented below may be employed to intersect a main or offset spiral with a straight line. Before proceeding with the description of this procedure let have a look at Figure 9.6.1b which is the left portion

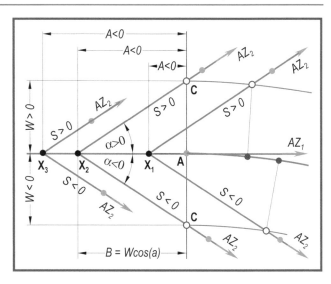

Figure 9.6.1b Intersection of a Line with a Spiral Near the Spiral's Start

of the main and offset spirals of Figure 9.6.1a. In this figure six lines intersect the tangent to the spiral at point X_1, X_2, and X_3 a distance A back from the start point A of the spiral. Also shown in this figure are two offset spirals, one inside and the other outside the main spiral and offset therefrom a distance W. Due to their position with respect to each other and the rotation of the spirals, the line intersecting the tangent at:

- X_1 and having a positive slope S intersects only the outside offset spiral, and that with the negative slope intersects only the inside spiral.
- X_2 and having a positive slope S passes through the endpoint of the outside offset spiral, and that with the negative slope passes through the endpoint of the inside spiral.
- X1 and X2 do not intersect the main spiral.
- X_3 do not intersect any of the spirals.

Before attempting to solve for the intersections the procedure checks for these conditions and returns the pertinent coordinates, or an error message if no intersection has been found.

In the last two referenced figures point A represented the start of a spiral providing a transition from a line onto a circular curve. In that case the above commentary would still hold true.

```
Sub icSpiralLine (PTNN, PTNE, AZ1, STAT, RAD1,
                  RAD2, SPRL, W, ROTN, _
                  STAT2, NORTH1, EAST1, NORTH2,
                  EAST2, AZTWO, IERR)
```

The input parameters to this procedure are the:

PTNN	The north coordinate of the start point of the main spiral.
PTNE	The east coordinate of the start point of the main spiral.
AZ1	The north azimuth (radians) of the tangent line to the main spiral at point A.
STAT	The station value, without the plus sign, of the main spiral's start point.
RAD1	The radius of the curve back of the main spiral (back curve), which should be positive or zero if it is a straight line.
RAD2	The degree of curvature in radians of the curve ahead of the main spiral (forward curve), which should be positive and it cannot be zero.
SPRL	The length of the main spiral in feet (meters).
ROTN	1 Denotes a clockwise alignment rotation of the curves.
	−1 Denotes a counterclockwise alignment rotation of the curves.
W	The uniform offset distance in feet (meters) of the offset spiral from the main spiral in the vicinity of the intersection.
	W>0 Offset spiral on the outside of the main spiral.
	W<0 Offset spiral on the inside of the main spiral.
	W=0 Intersection with the main spiral.
PTMN	The north coordinate of a point on the straight line to be intersected with the above specified spiral.
PTME	The east coordinate of a point on the straight line to be intersected with the above specified spiral.
AZ2	The north azimuth (radians) of the straight line to be intersected with the above specified spiral.
CNTlimit	The maximum number of iterations to be performed.

The returned parameters from this procedure are the:

STAT	The station value along the main spiral, without the plus sign, of the point of intersection of the straight line with the main or offset spiral.
NORTH1	The north coordinate of the point of intersection of the straight line with the main spiral, or of the point on the main spiral opposite the intersection point with the offset spiral.
EAST21	The east coordinate of the point of intersection of the straight line with the main spiral, or of the point on the main spiral opposite the intersection point with the offset spiral.
NORTH2	The north coordinate of the point of intersection of the straight line with the main or offset spiral. If W=0 then NORTH2=NORTH1.
EAST2	The east coordinate of the point of intersection of the straight line with the main or offset spiral. If W=0 then EAST2=EAST1.
AZTWO	The north azimuth in radians normal to the main spiral at NORTH1, EAST21 pointing towards the PI
IERR	An indicator the value of which denotes:
	0 An intersection has been computed.
	1 Negative or zero SPRL value (abort).
	2 Negative RAD1 or RAD2 value (abort).
	3 Zero RAD2 value (abort).
	4 RAD1>0 and RAD2>=RAD1 value (abort).

5 ROTN is neither +1 nor −1 (abort).

6 The line intersects the spiral tangent back of the TS, or ahead of the ST (abort).

7 No intersection, limit of iterations has been exceeded (abort).

8 The line intersects the main spiral at the or ahead of the SC, or at the or back of the CS (non fatal).

Upon invocation the procedure initializes the parameters to be returned, and then:

1. Queries SPRL for being negative or equal to zero within a tolerance of 0.005 feet (1.524 mm), if so sets IERR to be 1 and aborts; else continues.
2. Queries RAD1 and RAD2 for being negative, and if so sets IERR to be 2 and aborts; else continues.
3. Queries RAD2 for being zero (a spiral provides a transition from a line or curve to a sharper curve), and if so sets IERR to be 3 and aborts; else continues.
4. Queries if RAD1 is not zero and RAD2 is equal to or greater than RAD1, and if so sets IERR to be 4 and aborts; else continues.
5. Queries ROTN for being 1 or −1, and if not sets IERR to 5 and aborts; else continues.
6. Queries RAD1 for being equal to zero within a tolerance of 0.005 feet (1.524 mm), if so sets DC1 (the degree of curvature) to be zero; else computes the degree of curvature of the back curve as being equal to $(100/\text{RAD1})$ $(180/\pi)$, and the degree of curvature of the ahead curve as being equal to $(100/\text{RAD12})(180/\pi)$.
7. Calls the icintrs procedure to intersect the given straight (AZ2) line with the tangent line to the main spiral at the said spiral's start point (AZ1), and find the distance **A** and the angle of intersection α (see Figure 9.6.1a).
8. Queries the distance from the start of the spiral to the point of intersection A from the above step for being negative, and if so continues with the next step; else branches to **Step 10**.
9. Computes the distance **B** in Figure 9.6.1b and then
   (a) If the absolute value of **A** is greater than the absolute value of **B**, sets IERR to be 6, and aborts; else continues.
   (b) If the absolute value of **A** is equal to the absolute value of **B**, calls the icptl procedure to compute the coordinates of the endpoint of the offset spiral and sets them to those of the intersection, and sets the values of the other parameters to be returned accordingly.
10. Computes the slope of the intersecting line S in terms of the cotangent of the angle of intersection α from **Step 7** and taking into account the rotation of the spiral (ROTN), and then assumes as the first approximation of L one half of the length of the given main spiral (SPRL).

11. Commences a loop of a series of steps to compute a new approximation of L. These steps are performed until the value $K \leq 0.00001$, or until the limit of iterations (CNTlimit) has been reached, whichever is encountered first.

  (a) Calls the icDeriv procedure to compute the:
    • Spiral angle θ_S, the X and Y distances along and off the tangent to the spiral of the overall spiral, and
    • Instantaneous angle θ and the differentials dX and dY that corresponds to the last defined approximation of L.

  (b) Increments the counter of iterations and then compares it with the upper limit CNTlimit which if it has been reached sets IERR to be 7 and aborts; else continues.

  (c) Computes the error of approximation K as per Equation 9.6.1d in which the numerator is defined by Equation 9.6.1b, and the denominator is defined by Equation 9.6.1f which employs Equations 9.6.1g, 9.6.1h, and 9.6.1i.

  (d) Computes the new L using Equation 9.6.1e.

12. Calls the icCeg8y procedure to compute the main spiral data with the last L value as the final distance of the point of intersection from the spiral's point of beginning, and then calls the icptl procedure to compute the coordinates of the intersection with the offset spiral and terminate.

9.6.2   Spiral Intersection with a Circular Arc

The icSpiralArc procedure presented below may be employed to intersect a main or offset spiral with a circular arc. Figure 9.6.2a displays a given main spiral with an offset spiral on each side. Also displayed in this figure is the spiral tangent at point A, and a circular curve of radius R intersecting the main and offset spirals. The said point A could be a TS or CS point.

Much like with the intersection of a line with an offset spiral of Figure 9.6.1b it is possible for a circular arc to intersect only the inside or outside offset spiral if it is located near the spiral's start point A. As for the number of intersections between a spiral and a circular curve, theoretically a circular curve may intersect a spiral in more then one point depending on the position and size of these two curves. However, it is improbable that in actual practice a circular curve will intersect a spiral at more than one point. Such a case would occur if the radius of the circle is rather small, or if it is large and the spiral is extended indefinitely. Thus, only one intersection is to be considered herein. The methodology for solving for this intersection is based in relating the given elements to a local coordinate system in which its origin is at the said point A and its positive X axis direction is along the spiral tangent pointing towards the PI. Displayed in Figure 9.6.2b are the four possible positions that the center point of the circular curve may occupy with respect to the said local coordinate system.

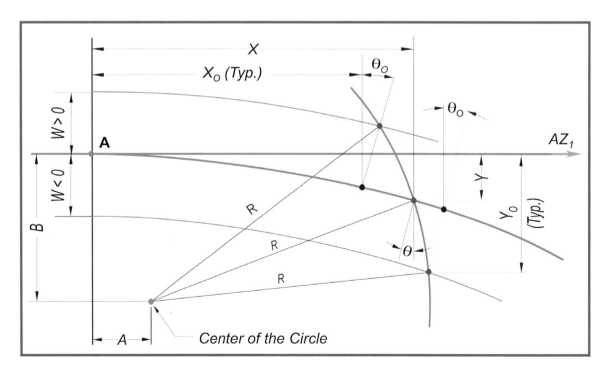

Figure 9.6.2a Intersection Of A Spiral With A Circular Arc

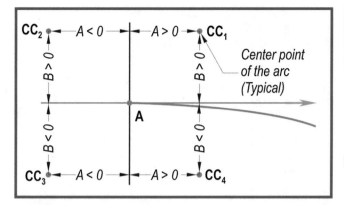

Figure 9.6.2b Relationship of the Arc's Center to the Tangent and to the Spiral's Point of Beginning

The reader's attention is drawn to the sign convention established by the A and B dimensions of Figure 9.6.2b (plus and offset distances of the center point of the intersecting curve from the tangent line to the spiral).

> *Since the* icSpiralArc *procedure considers only one intersection between a spiral and a circular curve, in special cases the spiral and circular curve once created may be treated as polylines and then intersected with each other.*

Using the Pythagorean theorem in Figure 9.6.2a, and replacing the radius R with R_C yields

$$(B - NY + NW \cos\theta)^2 + (A - X - W \sin\theta)^2 - R_C^2 = 0$$

(9.6.2a)

in which once more $N = \pm 1$ and denotes the direction of the spiral curvature (clockwise is positive and counterclockwise is negative). It is easily proved that Equation 9.6.2a will hold true for all positions of the circular origin for either a clockwise or counterclockwise spiral. Expanding the two polynomials to the second power, Equation 9.26 becomes

$$B + (NY)^2 + (NW \cos\theta)^2$$
$$+ 2B(-NY + NW \cos\theta) +$$
$$- 2NY(NW \cos\theta) + A^2 + X^2$$
$$+ (W \sin\theta)^2 + 2A(-X - W \sin\theta)$$
$$+ 2X(W \sin\theta) - R_C^2 = 0$$

(9.6.2b)

> *Do not confuse the distance A in Figure 9.6.2a with point A (the start of the spiral).*

Bearing in mind that $N^2 = 1$ and that

$$(NW \cos\theta)^2 + (W \sin\theta)^2 = W^2$$

(9.6.2c)

letting

$$C = A^2 + B^2 - R_C^2 + W^2$$

(9.6.2d)

and expressing θ, Y and X in terms of L as done previously

$$f(L) = Y^3 + X^2 - 2BNY - 2AX + 2BNW \cos\theta - 2AW \sin\theta$$
$$- 2YW \cos\theta + 2XW \sin\theta + C = 0$$

(9.6.2e)

As with the intersection of a spiral with a line an approximate L is assumed and the iteration process proceeds with Equation 9.6.1d in which $f(L)$ is given by Equation 9.6.2e. Thus

$$f'(L) = 2Y \frac{d}{dL}(Y) + 2X \frac{d}{dL}(X)$$
$$- 2BN \frac{d}{dL}(Y) - 2A \frac{d}{dL}(X)$$
$$- 2BNW \sin\theta \frac{d}{dL}(\theta) - 2AW \cos\theta \frac{d}{dL}(\theta)$$
$$+ 2YW \sin\theta \frac{d}{dL}(\theta) - 2W \cos\theta \frac{d}{dL}(\theta)$$
$$+ 2XW \cos\theta \frac{d}{dL}(\theta) + 2W \sin\theta \frac{d}{dL}(\theta)$$

(9.6.2f)

which by collecting terms and factoring becomes

$$f'(L) = 2(Y - BN - w\cos\theta)\frac{d}{dL}(Y)$$
$$+ 2(X - A + w\sin\theta)\frac{d}{dL}(X)$$
$$+ W_2 \left[\begin{array}{c} (X - A)\cos\theta \\ + (Y - BN)\sin\theta \end{array} \right] \frac{d}{dL}(\theta)$$

(9.6.2g)

in which θ is found from Equation 9.1.4b and $f'\theta$, $f'(Y)$, and $f'(X)$ from Equations 9.6.1g, 9.6.1h, and 9.6.1i, respectively.

With the values of $f(L)$ and $f'(L)$ known, the iterative process of the preceding subsection is used to find the value of $K = f(L)/f'(L)$ with Equation 9.6.1d, and a new value of L by use of Equation 9.6.1e, and the above process is repeated until K is small enough.

From an historic perspective, the values of A, B, and R_C in Equation 9.6.2d may be rather large numbers (R_C for a 30 min curve is 11,459.1559 feet). In the early days of electronic computation with 16 bit computers, the squaring of the terms of the said equation resulted in the loss of significant digits when computing C. For example, the square of the said radius was recorded as 131,312,250.000 instead of the actual 131,312,251.649. The loss of the ninth significant digit was not serious, but it had to be kept in mind.

Before proceeding with the description of this procedure let have a look at Figure 9.6.2c which depicts the vicinity of the spiral in the area depicted in Figure 9.6.1a. In its left portion the figure displays a main spiral and two offset spirals one on each side all in a clockwise rotation, while the right portion displays the same in a counterclockwise rotation. In each of these two portions the four potential locations of the center points of an intersecting circle are displays much like in Figure 9.6.2b. A line of length D is drawn from each of these center points to the start of the main spiral and to each of the corresponding offset spiral endpoints. A table at each quadrant summarizes the A and B plus and offset components of the circle center points for each figure portion and underneath four tables identify under which conditions an intersection would not occur between a main or offset spiral and a circular curve. In general, if the center point of the circular curve is located to the:

- Right of point A (A>0) and the radius of the said curve is larger than the distance D, then there is no intersection with the main or offset spiral to which D is measured.
- Left of point A (A<0) and the radius of the said curve is smaller than the distance D, then there is no intersection with the main or offset spiral to which D is measured.
- Left or to the right of point A (A<0 or A>0) and the radius of said curve is equal to the distance D, then the endpoint of the main or offset spiral is the intersection point.

```
Sub icSpiralArc (PTNN, PTNE, AZ1, STAT, RAD1
          , RAD12, SPRL, ROTN, W, _
          PTMN, PTME, R, CNTlimit, _
          STAT2, NORTH1, EAST1, NORTH2,
          EAST2, AZTWO, IERR)
```

The input parameters to this procedure are the:

PTNN, PTNE	The north/east coordinates of the start point of the main spiral.
AZ1	The north azimuth (radians) of the spiral tangent pointing towards the PI.
STAT	The station value in feet (meters), without the plus sign, of the main spiral's start point.
RAD1	The radius in feet (meters) of the back curve of the main spiral, which should be positive or zero if it is a straight line.
RAD2	The radius in feet (meters) of the ahead curve of the main spiral, which should be positive and it cannot be zero, nor greater than RAD1.
SPRL	The length of the main spiral in feet (meters).
ROTN	1 Denotes a clockwise alignment rotation of the curves. −1 Denotes a counterclockwise alignment rotation of the curves.
W	The uniform offset distance in feet (meters) of the offset spiral from the main spiral in the vicinity of the intersection. W>0 Offset spiral on the outside of the main spiral. W<0 Offset spiral on the inside of the main spiral. W=0 Intersection with the main spiral.
PTMN, PTME	The north/east coordinates of the center point of the circular arc to be intersected with the main or offset spiral.
R	The radius in feet (meters) of the circular arc to be intersected with the main or offset spiral.
CNTlimit	The maximum number of iterations to be performed.

The returned parameters from this procedure are the:

STAT2	The main spiral station, without the plus sign, of the intersection point.
NORT1, EAST1	The north/east coordinates on the main spiral opposite the point of intersection of the circular arc with the main or offset spiral.
NORT2, EAST2	The north/east coordinates of the point of intersection which is the same as NORT1, EAST1 if W=0.
AZTWO	The north azimuth (radians) normal to the main spiral at NORTH1, EAST1 pointing towards the PI
IERR	An indicator the value of which denotes: 0 An intersection has been computed. 1 Negative or zero SPRL value (abort). 2 Negative RAD1 or RAD2 or R value (abort). 3 Zero RAD2 or R value (abort). 4 RAD1>0 and RAD2>=RAD1 value (abort). 5 ROTN is neither +1 nor −1 (abort). 6 Spiral tangent intersected back of the TS, or ahead of the ST (abort). 7 No intersection, limit of iterations has been exceeded (abort). 8 The line intersects the main spiral at the or ahead of the SC, or at the or back of the CS (non fatal).

Upon invocation the procedure initializes the parameters to be returned, and then:

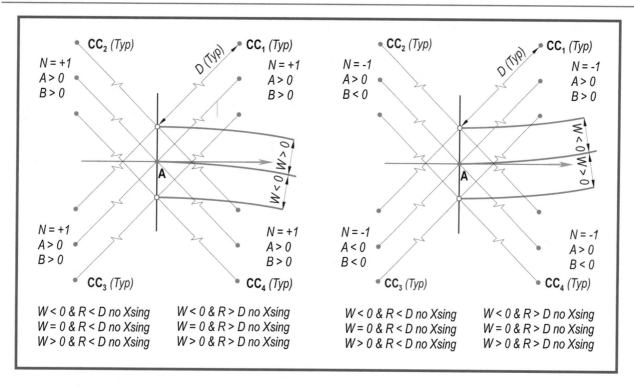

Figure 9.6.2c Intersection of a Circular Curve with a Spiral Near the Spiral's Start

1. Queries SPRL for being negative or equal to zero within a tolerance of 0.005 feet (1.524 mm), and if so sets IERR to be 1 and aborts; else continues.
2. Queries RAD1, RAD2, and R for being negative, and if so sets IERR to be 2 and aborts; else continues.
3. Queries RAD2 for being zero (a spiral provides a transition from a line or curve to a sharper curve), and if so sets IERR to be 3 and aborts; else continues.
4. Queries if RAD1 is not zero and RAD2 is equal to or greater than RAD1, and if so sets IERR to be 4 and aborts; else continues.
5. Queries ROTN for being 1 or −1, and if not sets IERR to 5 and aborts; else continues.
6. Queries RAD1 for being equal to zero within a tolerance of 0.005 feet (1.524 mm), and if so sets DC1 (the degree of curvature) to be zero; else computes the degree of curvature of the back curve as being equal to (100/RAD1)(180/π), and the degree of curvature of the ahead curve as being equal to (100/RAD2)(180/π).
7. Calls the icprjct procedure to project the center point of the circular arc on the tangent line to the main spiral at its start point to find the distances *A* and *B* according to the sign convention of Figure 9.6.2b, as well as the distance D from the center point of the circular arc to the start point A of the main spiral.

8. Queries the spiral offset distance W and if it is not equal to zero proceeds with the next step; else branches to **Step 11**.
9. Calls the icptl procedure to traverse from point A along AZ1 with a plus distance of zero and an offset distance minus W times ROTN (for a clockwise spiral a positive offset is to the left, and the reverse for a counterclockwise spiral) to locate the endpoint of the main or offset spiral.
10. Calls the icforce procedure to compute the distance D from the center point of the circular arc to the endpoint of the offset spiral opposite its start point A.
11. Queries the plus distance A as computed in **Step 7** for being positive or negative, and the radius R for being smaller or greater than the distance D computed in **Step 7** or **Step 10** to determine whether any of the criteria of Figure 9.6.2c have been met, or not, and
    - If any of the criteria are met signifying that there is no intersection between the specified spiral and the circular curve, IERR is set to be 6 and aborts.
    - If the said criteria are not met the returned parameters are set to those of the endpoint of the main and/or offset spiral, and to the back azimuth and start station values, and exits.

12. Assumes as the first approximation of L one half of the length of the given main spiral (SPRL).
13. Commences a loop of a series of steps to compute a new approximation of L. These steps are performed until the value $K \leq 0.00001$, or until the limit of iterations (CNTlimit) has been reached, whichever is encountered first.
    (a) Calls the icDeriv procedure to compute the:
        • Spiral angle θ_S, the X and Y distances along and off the tangent to the spiral of the overall spiral, and
        • Instantaneous angle θ and the differentials dX and dY that corresponds to the last defined approximation of L.
    (b) Increments the counter of iterations and checks the upper limit CNTlimit which if it has been reached sets IERR to be 7 and aborts; else continues.
    (c) Computes the error of approximation K as per Equation 9.6.1d in which the numerator is defined by Equation 9.6.2e, and the denominator is defined by Equation 9.6.2g which employs Equations 9.6.1g, 9.6.1h, and 9.6.1i.
    (d) Computes the new L using Equation 9.6.1e.
14. Calls the icCeg8y procedure to compute the main spiral data with the last L value as the final distance of the point of intersection from the spiral's point of beginning, and then calls the icptl procedure to compute the coordinates of the intersection with the offset spiral and terminates.

9.6.3 Spiral Intersection with a Another Spiral

Figure 9.6.3a displays two main spirals, *Spiral₁* and *Spiral₂*, each with an offset spiral on each side. The subscripts 1 and 2 differentiate between identical elements of the two spirals. The following sign conventions are used in this subsection.

• The tangent lines to the main spirals at their point of beginning are referred to by their north azimuths AZ_1 and AZ_2.
• The angle Δ is defined as the angle formed by AZ_2 pivoted about the point of intersection of the said two tangent lines and rotated until it coincides with AZ_1.
• The angle α is defined as the angle formed by AZ_1 pivoted about its point of tangency and rotated until it coincides with the line connecting the points of beginning of the two spirals (dimension Z).
• A positive Δ or α angle denote a clockwise rotation of *Spiral₂* with respect to *Spiral₁*, and a negative δ or α angle denote a counterclockwise rotation.
• The constants N_1 and N_2 denote the rotation of their respective spirals with a value of +1 denoting a clockwise spiral rotation and a −1 a counterclockwise rotation.

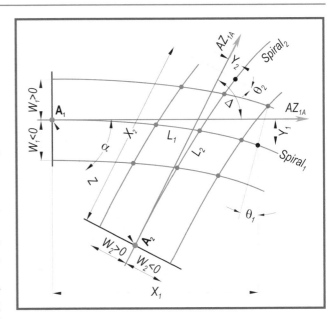

Figure 9.6.3a Intersection of a Spiral with Another Spiral

• Points A_1 and A_2 denote the start point of their respective spirals and either may be a TS or CS point.
• The distances L_1 and L_2 denote the distance along the main spiral from point A_1 or A_2 to the point of intersection with the other main spiral or with the point of projection of an offset spiral on the main spiral.

To solve for the intersection of two spirals there are two unknowns L_1 and L_2. Thus, two equations need be formulated to solve for these two parameters and they are:

• The algebraic sum of all projections on AZ_1 should equal zero, and
• The algebraic sum of all projections on a line normal to AZ_2 should also be equal to zero.

Nine possible intersections are shown in Figure 9.6.3a. The formulas derived below apply to all nine and to any other such intersection regardless of the position of the spirals and of their rotational direction (angles Δ and α may very in the range $-180° <$ angle $< 180°$). Of the possible intersections between the two spirals of the said figure the one chosen for demonstration is that of the offset spiral of *Spiral₁* having a positive W offset that intersects the offset spiral of *Spiral₂* having a negative W offset and each one curving clockwise. Figure 9.6.3b is a blowup of Figure 9.6.3a in the vicinity of said chosen intersection. The two equations for these intersections are:

$$\left(K_1 \frac{\partial}{\partial L_1} + K_2 \frac{\partial}{\partial L_2} \right) f(X_1, X_2) = f(X_1, X_2) \qquad (9.6.3a1)$$

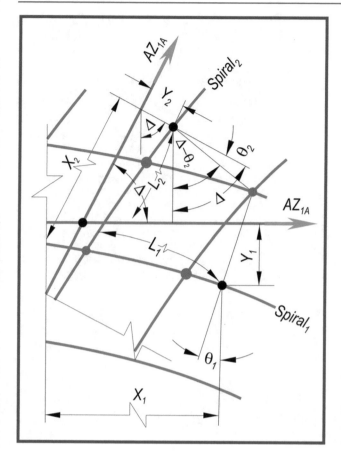

Figure 9.6.3b Blowup of Figure 9.6.3a

and

$$\left(K_1 \frac{\partial}{\partial L_1} + K_2 \frac{\partial}{\partial L_2} \right) f\left(X_1, X_2\right) = f\left(X_1, X_2\right) \qquad (9.6.3a2)$$

in which θ_1, Y_1, and X_1 may be expressed in terms of L_1, and θ_2, Y_2, and X_2 may be expressed in terms of L_2 from Equations 9.1.4b, 9.1.15a, and 9.1.15b. The method of solution of this type of intersection is similar to the two previous types with the exception that there are two equations, each being a function of the variables L_1 and L_2. If $L_1 = \xi_1$ and $L_2 = \xi_2$ are the solutions to Equations 9.6.3a1 and 9.6.3a2, and $L_1 = X_1$ and $L_2 = X_2$ are approximations, then from Equation 9.6.1e

$$X_1{}' = X_1 - K_1 \qquad \text{and} \qquad X_2{}' = X_2 - K_2$$

in which $X_1{}'$ and $X_2{}'$ are the new approximations, and K_1 and K_2 are the errors of approximation. Since Equation 9.6.1e has proved that $K = f(X)/f\,'(X)$, then by partial derivatives this new function of the two variables K_1 and K_2 may be expressed as

$$\frac{\partial}{\partial L_1} f\left(L_1, L_2\right) = \frac{d}{dL_1}\left(X_1\right) + W_1 \cos\theta_1 \frac{d}{dL_1}\left(\theta_1\right) = P_1 \quad (9.6.3b1)$$

and

$$\frac{\partial}{\partial L_1} f\left(L_1, L_2\right) = N_1 \frac{d}{dL_1}\left(Y_1\right) + N_1 W_1 \sin\theta_1 \frac{d}{dL_1}\left(\theta_1\right) = P_2$$

$$(9.6.3b2)$$

with Equations 9.6.3b1 and 9.6.3b2 applying to Equations 9.6.3a1 and 9.6.3a2, respectively. The partial derivatives with respect to L_1 for each of the two simultaneous equations are

$$f\left(L_1, L_2\right) = X_1 + W_1 \sin\theta_1 + N_2 W_2 \sin\left(\delta - N_2\theta_2\right) - N_2 Y_2 \sin\delta$$
$$- X_2 \cos\delta - Z \cos\alpha = 0 = G_1$$

$$(9.6.3c1)$$

and

$$f\left(L_1, L_2\right) = N_1 Y_1 - N_1 W_1 \cos\theta_1$$
$$+ N_2 W_2 \cos\left(\delta - N_2\theta_2\right) - N_2 Y_2 \cos\delta$$
$$+ X_2 \sin\delta - Z \sin\alpha = 0 = G_2$$

$$(9.6.3c2)$$

Remember that for spirals rotating:

- *Clockwise $N = 1$*
- *Counterclockwise $N = -1$*

The partial derivatives with respect to L_2 for each of the two simultaneous equations are

$$\frac{\partial}{\partial L_2} f\left(L_1, L_2\right) = -W_2 \cos\left(\delta - N_2\theta_2\right)\frac{d}{dL_2}\left(\theta_2\right)$$
$$- N_2 \sin\delta \frac{d}{dL_2}\left(Y_2\right) - \cos\delta \frac{d}{dL_2}\left(X_2\right) = Q_1$$

$$(9.6.3d1)$$

and

$$\frac{\partial}{\partial L_2} f\left(L_1, L_2\right) = +W_2 \sin\left(\delta - N_2\theta_2\right)\frac{d}{dL_2}\left(\theta_2\right)$$
$$- N_2 \cos\delta \frac{d}{dL_2}\left(Y_2\right) + \sin\delta \frac{d}{dL_2}\left(X_2\right) = Q_2$$

$$(9.6.3d2)$$

Hence, Equations 9.6.3b1 and 9.6.3b2 may be rewritten as

$$K_1 P_1 + K_2 Q_1 = G_1 \qquad (9.6.3e1)$$

and

$$K_1 P_2 + K_2 Q_2 = G_2 \qquad (9.6.3e2)$$

which when solved simultaneously yield

$$K_1 = \frac{Q_1 G_2 - Q_2 G_1}{Q_1 P_2 - P_1 Q_2} \qquad (9.6.3f1)$$

and

$$K_2 = \frac{P_2 G_1 - P_1 G_2}{Q_1 P_2 - P_1 Q_2} \qquad (9.6.3f2)$$

The `icSpiralSpiral` procedure presented below may be employed to intersect a main or offset spiral with another main or offset spiral.

```
Sub icSpiralSpiral (PT1N,  PT1E,  AZ1,  STA1, _
                    Rbck1, Rhad1, SPRL1, ROTN1, _
                    W1, _
                    PT2N,  PT2E,  AZ2,  STA2, _
                    Rbck2, Rhad2, SPRL2, ROTN1, _
                    W2, _
                    CNTlimit, _
                    STAx1, Nx1, Ex1, AZTWOx1, _
                    STAx2, Nx2, Ex2, AZTWOx2, _
                    NORT2, EAST2, IERR)
```

The input parameters to this procedure are the:

PT1N, PT1E	The north/east coordinates of the start point of the main Spiral$_1$.
AZ1	The north azimuth in radians of the spiral tangent of Spiral$_1$ pointing towards its PI.
STA1	The station value in feet (meters) without the plus sign of the start point of the main Spiral$_1$.
Rbck1	The radius in feet (meters) of the back curve of the main Spiral$_1$, which should be positive or zero if it is a straight line.
Rhad1	The radius in feet (meters) of the ahead curve of the main Spiral$_1$, which should be positive and it cannot be zero, nor greater than `Rbck1`.
SPRL1	The length in feet (meters) of the main Spiral$_1$ in feet (meters).
ROTN1	An indicator, the value of which if it is 1 denotes a clockwise rotation, and if it is −1 denotes a counterclockwise rotation for the main Spiral$_1$.
W1	The offset distance in feet (meters) from the main Spiral$_1$, if an offset spiral is to be intersected with another spiral. A positive value denotes an offset spiral on the outside, and a negative value denotes an offset spiral on the inside of the said main spiral. A zero denotes the said main spiral.
PT2N, PT2E	The north/east coordinates of the start point of the main Spiral$_2$.
AZ2	The north azimuth in radians of the spiral tangent of Spiral$_2$ pointing towards its PI.
STA2	The station value in feet (meters) without the plus sign of the start point of the main Spiral$_2$.

Rbck2	The radius in feet (meters) of the back curve of the main Spiral$_2$, which should be positive or zero if it is a straight line.
Rhad2	The radius in feet (meters) of the ahead curve of the main Spiral$_2$, which should be positive and it cannot be zero, nor greater than `Rbck2`.
SPRL2	The length in feet (meters) of the main Spiral$_2$ in feet (meters).
ROTN2	An indicator, the value of which if it is 1 denotes a clockwise rotation, and if it is −1 denotes a counterclockwise rotation for the main Spiral$_2$.
W2	The offset distance in feet (meters) from the main Spiral$_2$, if an offset spiral is to be intersected with another spiral. A positive value denotes an offset spiral on the outside, and a negative value denotes an offset spiral on the inside of the said main spiral. A zero denotes the said main spiral.
CNTlimit	The maximum number of iterations to be performed until an intersection is attained.

The preceding two spiral inter-section procedures provide specific testing for whether an intersection may exist or not near the beginning of a spiral. The `icSpiralSpiral` procedure does not provide for such a specific test. However, if an intersection does not exist an error message is displayed.

The returned parameters from this procedure are the:

STAx1	The station value in feet (meters) without the plus sign along the main Spiral$_1$ of the point of intersection.
Nx1, Ex1	The north/east coordinates on the main Spiral$_1$ which could be the intersection point or the point opposite the point of intersection.
AZTWOx1	The north azimuth (radians) normal to the main Spiral$_1$ at the intersection point.
STAx2	The station value in feet (meters) without the plus sign along the main Spiral$_2$ of the point of intersection.
Nx2, Ex2	The north/east coordinates on the main Spiral$_2$ which could be the intersection point or the point opposite the point of intersection.
AZTWOx2	The north azimuth (radians) normal to the main Spiral$_2$ at the intersection point.
NORT2, EAST2	The north/east coordinates on the intersection point. Note that if: W1=0 & W2<>0 (NORT2, EAST2)=(Nx1, Ex1) W2=0 & W1<>0 (NORT2, EAST2)=(Nx2, Ex2) W1=0 & W2=0 (NORT2, EAST2)=(Nx1, Ex1)=(Nx2, Ex2) W1<>0 & W2<>0 (NORT2, EAST2)<>(Nx1, Ex1)<>(Nx2, Ex2)
IERR	An indicator the value of which denotes the following:
0	An intersection has been computed.
1	Negative or zero `SPRL1` and/or `SPRL2` value (abort).

2 Negative value of any one of `Rbck1`, `Rhad1`, `Rbck2` or `Rhad2` (abort).

3 Zero `Rhad1` and/or `Rhad2` value (abort).

4 `Rbck1`>0 and `Rhad1`>=`Rbck1` and/or `Rbck2`>0 and `Rhad2`>=`Rbck2` value (abort).

5 `ROTN1` and/or `ROTN2` are neither +1 nor −1 (abort).

6 No intersection, limit of iterations has been exceeded (abort).

7 The intersection occurs at or ahead of the SC, or at or back of the CS of Spiral$_1$ (non fatal).

8 The intersection occurs at or ahead of the SC, or at or back of the CS of Spiral$_2$ (non fatal).

Upon invocation the procedure initializes the parameters to be returned, and then:

1. Queries `SPRL1` and `SPRL2` for being negative or equal to zero within a tolerance of 0.005 feet (1.524 mm), and if either or both are so sets `IERR` to be 1 and aborts; else continues.

2. Queries `Rbck1`, `Rahd1`, `Rbck2`, and `Rahd2` for being negative, and if either or both are so sets `IERR` to be 2 and aborts; else continues.

3. Queries `Rahd1` and `Rahd2` for being zero (a spiral provides a transition from a line or curve to a sharper curve), and if either or both are so sets `IERR` to be 3 and aborts; else continues.

4. Queries if:
   - `Rbck1` is not zero and `Rahd1` is equal to or greater than `Rbck1`, and
   - `Rbck2` is not zero and `Rahd2` is equal to or greater than `Rbck2`, and
   if either or both are so sets `IERR` to be 4 and aborts; else continues.

5. Queries `ROTN1` and `ROTN2` for being 1 or −1, and if either or both are not so sets `IERR` to be 5 and aborts; else continues.

6. Queries `Rbck1` and `Rbck2` for being equal to zero within a tolerance of 0.005 feet (1.524 mm), and if either or both are so sets the degree of curvature of the back curve `DCbck1` and/or `DCbck2` to be zero; else whichever of these two degrees of curvature have not been set to zero it is set to be equal to $(100/\text{Rbck1})(180/\pi)$ and/or $(100/\text{Rbck2})(180/\pi)$.

7. Sets the degree of curvature of both ahead curves as being equal to $(100/\text{Rahd1})(180/\pi)$ and $(100/\text{Rahd2})(180/\pi)$.

8. Calls the `icintrs` procedure to intersect the tangent line at the start point of Spiral$_1$ with that of Spiral$_2$ to find the angle Δ, which may be positive or negative.

9. Calls the `icforce` procedure to determine the distance Z between the start points A$_1$ and A$_2$ of the two spirals of Figure 9.6.3a.

10. Determines the angle α of Figure 9.6.3a, which angle may be either positive or negative, as the difference between the two said tangents, and sets the sine and cosine of the above computed angle Δ in distinct variable arguments.

11. Assumes as the first approximation of L_1 and L_2 as one half of the corresponding length of the given first and second main spirals (`SPRL` and `SPRL2`).

12. Initializes the limit of convergence E to 0.00001, the K_1 and K_2 factors of Equations 9.6.3b1 and 9.6.3b2, respectively to 999999#, and the counter of the number of iterations to zero.

13. Commences a loop of a series of steps to compute a new approximation of L_1 and L_2. These steps are performed until the value $K_1 \leq 0.00001$ and of $K_2 \leq 0.00001$, or until the limit of iterations (`CNTlimit`) has been reached, whichever is encountered first.
    (a) Calls the `icDeriv` procedure to compute for each of the two intersecting main spirals the:
       - Spiral angle θ_S, the X and Y distances along and off the tangent to the spiral of the overall spiral, and
       - Instantaneous angle θ and the differentials dX and dY that corresponds to the last defined approximation of L
       using Equations 9.1.4b, 9.1.15a, and 9.1.15b, respectively.
    (b) Computes the values of G_1 and G_2 of the two intersections to be intersected using Equations 9.6.3a1, and 9.6.3a2, respectively.
    (c) Computes the values of $f'(\theta_n)$, $f'(Y_n)$, and $f'(X_n)$, where the subscript n is 1 or 2 denoting the respective spiral, using Equations 9.6.1g, 9.6.1h, and 9.6.1i, respectively.
    (d) Computes the values of P_1 and P_2 of the partial derivative Equations 9.6.3c1 and 9.6.3c2.
    (e) Computes the values of Q_1 and Q_2 from Equations 9.6.3d1 and 9.6.3d2.
    (f) Computes the errors of approximation K_1 and K_2 as per Equations 9.6.3f1 and 9.6.3f2.
    (g) Computes the new estimates of the distances L_1 and L_2 using Equation 9.6.1e.
    (h) Checks the number of iterations that have been performed up until now against the upper limit specified in `CNTlimit` and if it has been reached sets `IERR` to be 6 and exits the loop. In this case the last computed values of the distances L_1 and L_2 are used to continue with the next step and complete the process.

    Note that the results that are being returned in this case may not be quite correct. Thus, in this case the calling program should test for the error message of `IERR`=6 and act accordingly.

14. Calls the `icCeg8y` procedure to compute the data of the main Spiral₁ using the last computed L_1 value as the final distance from the spiral's point of beginning to the point of intersection, or its projection on the main spiral if an offset spiral is involved.

15. Compares the last computed L_1 value with that of the overall spiral length of the main Spiral₁ and if it is equal to or greater than `SPRL1` sets `IERR` to be 7 (non fatal error message) and continues.

16. Repeats the last two steps for the main Spiral₂ using the last computed L_2 value which is compared with `SPRL2` for setting `IERR` to be 8 (non fatal error message) if need be and continues.

17. Calls the `icptl` procedure to compute the coordinates of the intersection point using the data of Spiral₁ and terminates.

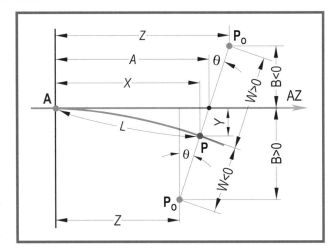

Figure 9.7a Point Projection On A Spiral

9.7   Point Projection on a Spiral

One of the most common problems in highway design is that of referencing a point to a line with a plus and an offset. Referencing points to a base line is a simple matter as in most cases a base line is composed of straight lines (tangents) and perhaps sometimes of circular arcs. What about though a retaining wall located opposite a spiral center line, or a bridge wholly or partially on a spiral whose points have to be referenced to the said spiral? Such projections are a little more intricate than those on straight lines and circular arcs.

In general the projection of a point on a base line may be thought of as an intersection of the said line with a line normal or radial to it and passing through the said point, the offset point being defined by a pair of coordinates. When the reference line is straight, the direction of the normal line is found by adding or subtracting 90° to the reference line direction; when the reference line is circular the direction of the radial line may be found by forcing between the offset point and the center point of the arc, which center point must be known in order that the reference line be established. The latter applies to a spiral reference line as well as shown in Figure 9.7a with the exception that the instantaneous radius and its origin are not known.

To project a point on a spiral, the offset point P_O in the said figure is first projected on the tangent line to the spiral at its point of beginning A while point P denotes its point of projection on the spiral. The offset distance B of point P_O from the said tangent line is negative when P_O is on the left of the said tangent line looking towards the PI, and positive when on the right side. However, the offset distance W distance of the said point from the spiral is positive when on the outside of the spiral and negative when on the inside. If the distance Z is (a) negative then there is no projection point,

(b) zero then point A is the projection point, and if it is (c) positive then there is a distinct projection point which may be located beyond the end of the spiral. The sign of the offset distance W is dependent on the rotation of the spiral which is denoted with $N = +1$ for a clockwise spiral and with a $N = -1$ for a counterclockwise spiral With reference to Figure 9.7a the distances A and X may be expressed as

$$A = Z + NB\tan\theta \qquad (9.7a)$$

and

$$X = A - Y\tan\theta \qquad (9.7b)$$

Expressing θ, Y and X in terms of L, substituting Equation 9.7a for A in Equation 9.7b, and rearranging terms yields

$$f(L) = X + (Y - NB)\tan\theta - Z = 0 \qquad (9.7c)$$

Again using the previously defined method of iteration by approximating L

$$f'(L) = \frac{d}{dL}(X) + (Y - NB)\sec^2\theta\frac{d}{dL}(\theta) + \tan\theta\frac{d}{dL}(Y) \qquad (9.7d)$$

in which $f'(\theta), f'(Y)$ and $f'(X)$ are found from Equations 9.6.1g, 9.6.1h, and 9.6.1i.

The `icSpiralPrjct` procedure presented below may be employed to project an offset point on a main spiral.

```
Sub icSpiralPrjct (PTN,   PTE,   AZ,   STA,   RAD1,
                   RAD2,   SPRL,   ROTN,   NORT2,
                   EAST2, _
                   CNTlimit, _
                   NORT1,   EAST1,   STA1,   OFF,
                   AZONE,   IERR)
```

The input parameters to this procedure are the:

PTN, PTE	The north/east coordinates of the start point of the main spiral.
AZ	The north azimuth in radians of the spiral tangent pointing towards the PI.
STA	The station value, without the plus sign, of the main spiral's start point.
RAD1	The radius in feet (meters) of the back curve of the main spiral, which should be positive or zero if it is a straight line.
RAD2	The radius in feet (meters) of the ahead curve of the main spiral, which should be positive and it cannot be zero, nor greater than RAD1.
SPRL	The length of the main spiral in feet (meters).
ROTN	An indicator, the value of which if it is a positive one (1) indicates a clockwise rotation, and if it is a negative one (−1) indicates a counterclockwise rotation for the main spiral.
NORT2, EAST2	The north/east coordinates of the point to be projected on the main spiral.
CNTlimit	The maximum number of iterations to be performed.

The returned parameters from this procedure are the:

NORT1, EAST1	The north/east coordinates of the point of projection on the main spiral.
STA1	The station value, without the plus sign, of the said point of projection.
OFF	The offset distance of NORT2, EAST2 from the main spiral (positive to the outside and negative to the inside of the spiral).
AZONE	The north azimuth in radians of the tangent line to the spiral at the said point of projection pointing towards the PI.
IERR	An indicator the value of which denotes:
	0 A projection has been computed.
	1 Negative or zero SPRL value (abort).
	2 Negative RAD1 and/or RAD2 value (abort).
	3 Zero RAD2 value (abort).
	4 RAD1 > 0 and RAD2 >= RAD1 value (abort).
	5 ROTN is neither +1 nor −1 (abort).
	6 No projection, the point to be projected lies back of the start point of the spiral (abort).
	7 No projection, limit of iterations has been exceeded (abort).
	8 The projection occurs at or ahead of the SC, or at or back of the CS of the spiral (non fatal).

Upon invocation the procedure initializes the parameters to be returned, and then:

1. Queries SPRL for being negative or equal to zero within a tolerance of 0.005 feet (1.524 mm), and if so sets IERR to be 1 and aborts; else continues.

2. Queries RAD1 and RAD2 for either being negative, and if so sets IERR to be 2 and aborts; else continues.

3. Queries RAD2 for being zero (a spiral provides a transition from a line or curve to a sharper curve), and if so sets IERR to be 3 and aborts; else continues.

4. Queries if RAD1 is not zero and RAD2 is equal to or greater than RAD1, and if so sets IERR to be 4 and aborts; else continues.

5. Queries ROTN for being 1 or −1, and if not sets IERR to 5 and aborts; else continues.

6. Queries RAD1 for being equal to zero within a tolerance of 0.005 feet (1.524 mm), and if so sets DC1 (the degree of curvature) to be zero; else computes the degree of curvature of the back curve as being equal to (100/RAD1) (180/π), and the degree of curvature of the ahead curve as being equal to (100/RAD2)(180/π).

7. Defines the constant of convergence to be E = 0.0005, the iterations counter to be CNT = 0 and the convergence criterion to be ckTest = 999999#.

8. Calls the icprjct procedure to project the offset point PO of Figures 9.7a and 9.7b on the tangent line to the spiral at point A to find the distances Z and B according to the sign convention of Figure 9.7b.

9. Queries the plus distance Z for being:
   - Zero within a tolerance of 0.005 feet (1.524 mm) sets the coordinates of the projection point to be

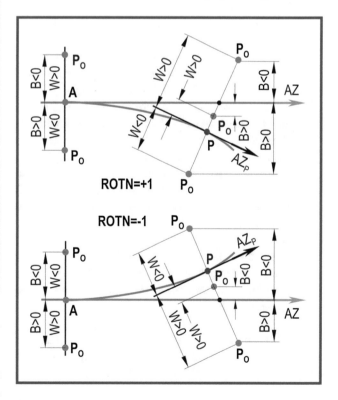

Figure 9.7b Sign Convention of Point Projection On A Spiral

those of point A, the normal azimuth to be that of the tangent line, and the offset to be W = −B(N), and then exits.

- Negative sets IERR to be 6 and aborts.
- Positive continues.

10. Assumes as the first approximation of L one half of the length of the main spirals (`SPRL`).
11. Commences a loop of a series of steps to compute a new approximation of L. These steps are performed until the values of ckTest ≤ E and K ≤ 0.00001, or until the limit of iterations (`CNTlimit`) has been reached, whichever is encountered first.
    (a) Calls the `icDeriv` procedure to compute the:
        - Spiral angle θ_S, the X and Y distances along and off the tangent to the spiral of the overall spiral, and the
        - Instantaneous angle θ and the differentials dX and dY that corresponds to the last defined approximation of L
          using Equations 9.1.4b, 9.1.15a, and 9.1.15b, respectively.
    (b) Determines the value of $f(L)$ from Equation 9.7c.
    (c) Computes the value of $f'(L)$ from Equation 9.7d.
    (d) Uses the above computed values of $f(L)$ and $f'(L)$ to find the value of $K = f(L)/f'(L)$ with Equation 9.6.1d.
    (e) Computes the values of L from Equation 9.6.1d.
    (f) Checks the number of iterations against the upper limit of `CNTlimit` and if it has been reached sets IERR to be 7 and exits the loop, in which case the last value of L is used to continue with the results not being quite correct (the calling program should test for IERR to be 7 and act accordingly).
12. Calls the `icCeg8y` procedure to compute the main spiral data for the last computed final distance L from the spiral's point of beginning to the point of projection.
13. Calls the `icptl` procedure to project the offset point P_O on the normal line (instantaneous tangent) to the instantaneous radial line at the projection point to find the offset distance W, and then queries ROTN for being positive (clockwise spiral) and if so reverses the sign of W; else continues.
14. Compares the last computed L value with that of the overall spiral length of the spiral and if it is equal to or greater than `SPRL` sets IERR to be 8 (non fatal error message) and terminates.

The projection of an offset point on an offset spiral has not been discussed here as the instantaneous angle θ for such a spiral is identical with that of the main spiral.

9.8   Tangent to a Spiral from an External Point

The construction problem of a tangent from an offset point (given point) to a main or offset spiral is not much different from that of projecting a point on a spiral. Figure 9.8a illustrates that the ability to construct a tangent line to a spiral (within its limits between points A and B) is dependent on the position of the said point with respect to the spiral and to the line tangent to the spiral at its beginning point A (spiral tangent). It is noted that in the following three statements the word spiral implies either a main or an offset spiral. If the offset point from which the new tangent is to pass is located:

1. **Inside the spiral** as point P_X then a tangent line cannot be constructed.
2. **Outside the spiral and beyond the spiral tangent** as point P_1, then only one tangent line to the spiral can be constructed pointing towards the end of the spiral (point B) and is referred to herein as the *single forward tangent*. The method of addressing this construction problem is similar to that employed in the preceding construction problems.
3. **Between the spiral and the spiral tangent** as point P_2, then there are two possible tangent lines, one pointing towards the end of the spiral (point B) and referred to herein as the *forward tangent*, and the other pointing towards the start of the spiral (point A) and referred to herein as the *backward tangent*. If point P_2 is near point B the forward point of tangency may lie beyond the spiral's limits. The method of addressing this construction problem is different from that employed in the preceding construction problems for the reasons stated later on.

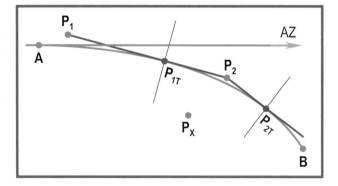

Figure 9.8a   Tangent to a Spiral from an External Point

As for the construction of a tangent line at a specific point on the spiral, the point of tangency may be created by any of the preceding procedures and the instantaneous radial azimuth can be used to create the tangent line.

A point of caution regards the condition in which the given point from which a forward tangent is to be created is located so that the tangency point may lie beyond the end point of the spiral. The subject procedure issues a warning message to that effect, but it does return the coordinates of such a point. Regarding the alignment of the spiral beyond its ending point reference is made to the last section of this chapter.

9.8.1   The Case of the Single Forward Tangent Line

Case (a) of Figure 9.8b displays a clockwise main spiral (N=1) with tangent lines drawn to it from three external points, P_1, P_3, and P_5. Since the forward tangent from point P_5 is similar to that for point P_3 it is not shown. Cases (b) and (c) of the same figure display a counterclockwise main spiral (N=−1) each with an offset spiral. A tangent line is drawn in Case (b) to an inside (negative W offset) spiral from an external point P_1, while a tangent line is drawn in Case (c) to an outside (positive W offset) spiral from an external point P_1.

In each of these cases the external point is projected on the spiral tangent, and the distances Z and B are found according to preestablished sign conventions (see side-bar for Figure 9.8b). Assuming a local coordinate system with origin at point A and a positive X direction along the spiral tangent the distance from point A to the intersection point of the spiral tangent with the tangent from point P_1 is

$$A = Z - N\frac{B}{\tan\theta} = Z - BN\cot\theta \quad (9.8a)$$

Equations 9.8a, 9.8b, 9.8c, 9.8d1, 9.8d2, and 9.8d3 may be used for any forward tangent line but in the subject procedure they are used only for the single forward tangent line.

Note that the said equations are sensitive to the seed or initial assumption of the location of the point of tangency.

Note the signs of the variables B and N in each of the cases in Figure 9.8b. The longitudinal local coordinate of the required tangency point in Case (a), or of the point of projection on the main spiral of the point of tangency at an offset spiral in Cases (b) and (c) is

$$X = A + (Y - W\cos\theta)\cot\theta - W\sin\theta \quad (9.8b)$$

with both X and Y being positive. With reference to point P_3 at this time we are only addressing the forward tangent line. Knowing that θ, Y and X may be expressed in terms of L, substituting of Equation 9.8a in Equation 9.8b, and rearranging terms yields

$$f(L) = X - (Y - BN)\cot\theta + W\sin\theta - Z = 0 \quad (9.8c)$$

Using once more the previous method of iteration by approximating L results in

Summary of the sign conventions that are used:

(a) *X and Y are measured along the spiral tangent from point A (the TS or CS) to the projection of the point of tangency P2 on the said tangent and they are positive.*

(b) *Z and B denote the plus and offset distances of the external point P1 or P3 with respect to point A (the TS or CS) along the spiral tangent, and they are signed as indicated in the three parts of Figure 9.8b.*

(c) *Distance A is measured along the spiral tangent from point A (the TS or CS) to the intersection of the said tangent with the tangent being constructed and it is positive.*

(d) *The instantaneous angle θ of the main spiral is at the point of projection of the point of tangency P2 or P4 on the main spiral and it is positive.*

$$f'(L) = \frac{d}{dL}(X) + (Y - NB)\csc^2\theta\frac{d}{dL}(\theta) - \cot\theta\frac{d}{dL}(Y) + W\cos\theta\frac{d}{dL}(\theta) \quad (9.8d1)$$

or

$$f'(L) = \frac{d}{dL}(X) - \cot\theta\frac{d}{dL}(Y) + \left[(Y - NB)\csc^2\theta + W\cos\theta\right]\frac{d}{dL}(\theta) \quad (9.8d2)$$

or

$$f'(L) = \frac{d}{dL}(X) - \cot\theta\frac{d}{dL}(Y) + \left[\frac{(Y - NB)}{\sin^2\theta} + W\cos\theta\right]\frac{d}{dL}(\theta) \quad (9.8d3)$$

Figure 9.8b Examples of a
Forward Tangent Line to a
Spiral from an External Point

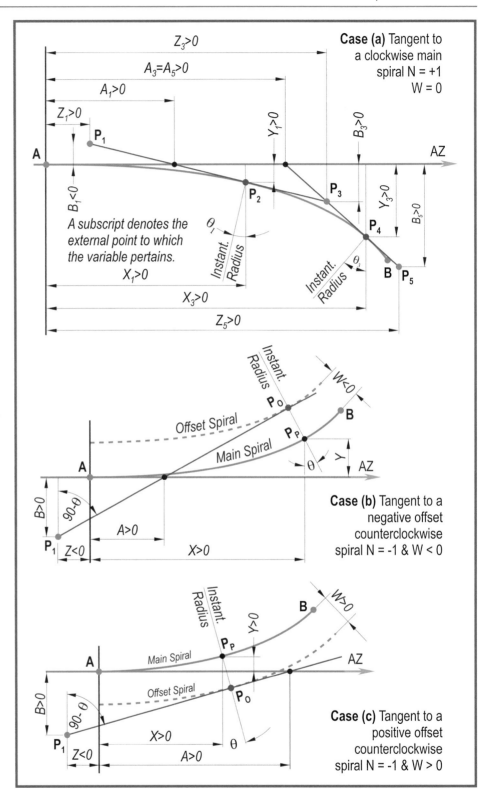

9.8.2   The Case of the Backward and Forward Tangent Lines

The method of solution presented in the preceding subsection is sensitive to the initial assumption of the location of the point of tangency. That is, depending on the location of the given point with respect to its proximity to the start or end of the spiral, and on its offset distance from the spiral the resulting tangent line might be the backward tangent and not the forward tangent if the given point is located between the spiral and the spiral tangent. Therefore the methodology used for the last said case of the location of the given point is

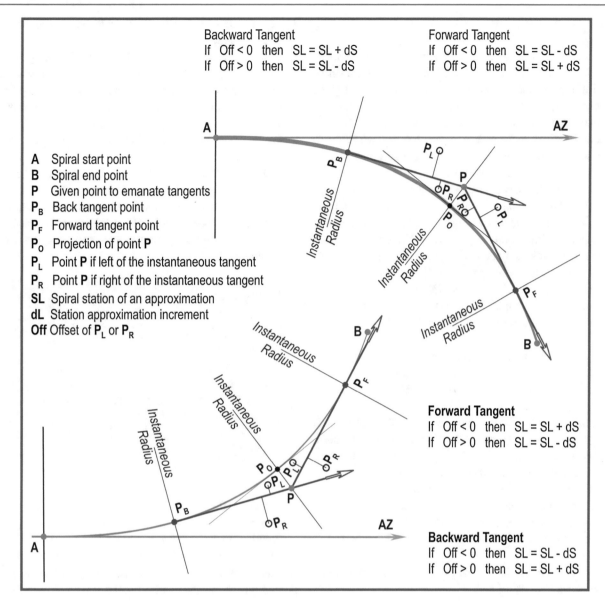

Figure 9.8c   Examples of a Backward Tangent Line to a Spiral from an External Point

different. This different methodology may also be used for the single forward tangent line. However, for demonstration purposes it is not so used in the subject procedure.

   Shown in Figure 9.8c is clockwise spiral in the top part and a counterclockwise spiral in the bottom part of the figure. Each of these spirals could be a main spiral or an offset spiral. The overall spiral tangent lines of these two spirals and the given point P are also shown. The general process of the solution is as follows:

1. Point P is projected on the main spiral to find the point of projection (station value and coordinates) and the instantaneous tangent to the spiral at the said point.

2. Point P_O is located to the left or to the right of the main spiral in the case of an offset spiral. In the case of a main spiral P_O is the said point of projection.

3. Point P_O is projected on the said instantaneous tangent line and the sign of the offset distance is queried. Depending on the said sign, the station value is adjusted backwards or forwards as indicated by the equation in the said figure by some predetermined value.

4. Point P_O is now positioned at the new adjusted station and **Step 3** is repeated. If the offset distance becomes equal to or less than some predetermined allowable distance and the iteration process terminates. If not, and the offset sign the:
   • Same as that of the preceding offset **Step 4** is repeated;
   • Opposite to that of the preceding offset the station adjustment in **Step 3** is halved prior to repeating **Step 4**.

   The above process is repeated until it is terminated as indicated in Step 4, or until a predetermined number of iterations has been reached.

9.8.3   The Procedure

The `icSpiralTang` procedure may be employed to determine the coordinates of the point of tangency of (a) a single forward tangent line to a main or offset spiral from a given point utilizing iteratively Equations 9.8c and 9.8d2, or of (b) a backward and a forward tangent lines using the iterative process described in the preceding subsection. Two issues to be addressed are the existence of a tangency point and the seed approximation of the said point.

Since there can be two, one, or no tangency points depending on the spiral and position of the given point a check must be made to determine if any tangent can be constructed, and if so whether a backward tangent is possible or not. To check for these conditions reference is made to Figure 9.8d in which a clockwise main spiral ($N = 1$) is shown on the left side and a counterclockwise spiral ($N = -1$) is shown on the right side, each having an inside and an outside offset spiral. For each set of spirals there are nine points labeled:

- P_{I1}, P_{I2}, and P_{I3} pertaining to the inside offset spiral,
- P_{M1}, P_{M2}, and P_{M3} pertaining to the main spiral, and
- P_{O1}, P_{O2}, and P_{O3} pertaining to the outside offset spiral.

Of these nine points the ones with the subscript I1, M1, and O1 are located on the inside of their respective spiral, those with the subscript I2, M2, and O2 are located between their respective spiral and its tangent line and those with the subscript I3, M3, and O3 are located on the outside of the said tangent line. The said point is projected on the main spiral to find the offset distance H from which the spiral offset W is subtracted ($H = H - W$). If the so modified H is:

- Negative denotes that the point is inside the spiral and a tangent cannot be created.

- Zero denotes that the point lies on the spiral and the point is the tangency point.
- Positive denotes that the point is outside the spiral and there can be only a forward tangent, or both a forward and a backward tangent. To determine which, the said point is projected on the spiral tangent line of the main or offset spiral to find the offset distance B as indicated in the said figure. Then the product of B times H times N is tested and if it is found to be:
  - Positive, it denotes that the point lies between the main or offset spiral and its spiral tangent and hence there can be both a forward and a backward tangent line.
  - Negative, it denotes that the point lies on the outside of the said spiral tangent and there can be no backward tangent line.

Note that in the `icSpiralTang` *procedure*

- *The maximum number of iterations to be performed is an input parameter.*
- *The initial position of the tangency point is hard coded as the*
  - *Half point of the spiral length for the single forward tangent.*
  - *The point of projection of the given point on the spiral.*

Regarding the initial or seed approximation of the point of tangency, in Figure 9.8b the spiral is very exaggerated in the Y direction. In real life the curvature of the spiral to the left of point P_3 is very slight and increases to the right. Thus, if the said point is rather close to the spiral, the two tangency point might be located nearby, but if this point is closer to the

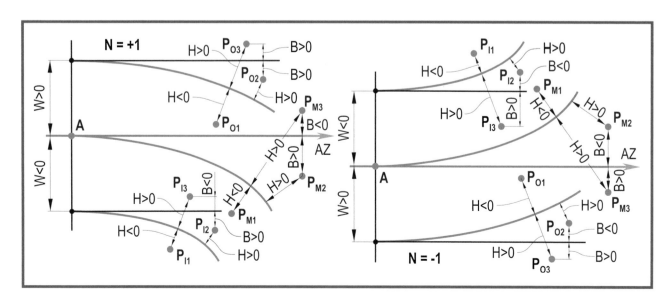

Figure 9.8d   Determination of Which if Any Tangents Can Be Constructed

spiral tangent line than to the spiral the backward tangency point may be located near the beginning of the spiral. Because of the flatness of the spiral the convergence of the solution for the backward tangent is very sensitive and becomes very slow. In most all of the other spiral procedures in this publication the initial assumption of the point is question is the half point of the overall spiral length. This is fine for the location of the forward point of tangency, but it may or it may not be so. If the half point is assumed as the initial position of the solution it is possible that instead of locating the backward point of tangency that of the forward tangency is produced. Thus, instead of assuming the half point of the overall spiral length its quarter point might be more appropriate. An approximation of her should be avoided as it results in a division by zero against which a check is not made. This is issue of seed approximation pertains only to the single forward tangent line. It is possible for the process used for the backward and forward tangent lines could also be used for the single forward tangent.

```
Sub icSpiralTang (PTN, PTE, AZ, STA, RAD1,
                  RAD2, SPRL, ROTN, W, PN, PE, _
                  CNTlimit,_
                  NORT1, EAST1, STA1, AZT1,
                  NORT2, EAST2, STA2, AZT2, IERR)
```

The input parameters to this procedure are the:

PTN, PTE	The north/east coordinates of the start point of the main spiral.
AZ	The north azimuth (radians) of the tangent line to the main spiral at its start point.
STA	The station value, without the plus sign, of the main spiral's start point.
RAD1	The radius in feet (meters) of the curve back of the main spiral which should be zero for a straight line.
RAD2	The degree of curvature in radians of the ahead curve of the main spiral which should not be zero and should smaller than RAD1 if it is not zero.
SPRL	The length in feet of the main spiral.
ROTN	An indicator, the value of which if it is a positive one (1) indicates a clockwise rotation, and if it is a negative one (−1) indicates a counterclockwise rotation for the main spiral.
W	The offset distance in feet (meters) from the main spiral, if the tangent line to be constructed is that to an offset spiral. A positive value denotes an offset spiral on the outside of the said main spiral, and a negative value denotes an offset spiral on the inside of the main spiral. A zero denotes the main spiral.
PN, PE	The north/east coordinates of the given point from which the tangent line or lines are to emanate.
CNTlimit	The maximum number of iterations to be performed.

The returned parameters from this procedure are the:

NORT1	The north coordinate of the forward tangency point, if any, or zero.
EAST1	The east coordinate of the forward tangency point, if any, or zero.
STAT1	The main spiral station value, without the plus sign, of the said tangency point.
AZT1	The forward pointing north azimuth of the forward tangent line.
NORT2	The north coordinate of the backward tangency point, if any, or zero.
EAST2	The east coordinate of the backward tangency point, if any, or zero.
STAT2	The main spiral station value, without the plus sign, of the said tangency point.
AZT2	The forward pointing north azimuth of the forward tangent line.
IERR	An indicator the value of which denotes:

0	One or two tangencies have been computed.
1	Negative or zero SPRL value (abort).
2	Negative RAD1 and/or RAD2 value (abort).
3	Zero RAD2 value (abort).
4	RAD1>0 and RAD2>=RAD1 value (abort).
5	ROTN1 and/or ROTN2 is neither +1 nor −1 (abort).
6	No tangents because the offset point cannot be projected on the main spiral. Check the input data (abort).
7	No tangents because the offset point is inside the spiral (abort).
8	There is a forward but not a backward tangent (non fatal).
9	The given point lies on the spiral and has been made the point of tangency (non fatal).
10	The given point lies on the spiral tangent and the spiral start point has been made the point of tangency (non fatal).
11	The limit of iterations for the single forward tangent line has been exceeded and the last computed tangency point has been used (non fatal). There is no backward tangent.
12	The forward tangency point is at or ahead of the SC, or at or back of the CS of the spiral (non fatal).

Upon invocation the procedure initializes the parameters to be returned, assumes that a backward tangent exists ($backTan = 1$) and then:

Since most computer compilers do not have subroutines for the cosecant and cotangent trigonometric functions, use the following substitutions in the equations:
$$csc^2\theta = 1/sin^2\theta$$
$$cot\,\theta = 1/tan\,\theta$$
or
$$cot\,\theta = cos\theta/sin\theta$$

1. Queries `SPRL` for being negative or equal to zero within a tolerance of 0.005 feet (1.524 mm), and if so sets `IERR` to be 1 and aborts; else continues.

2. Queries `RAD1` and `RAD2` for either being negative, and if so sets `IERR` to be 2 and aborts; else continues.

3. Queries `RAD2` for being zero (a spiral provides a transition from a line or curve to a sharper curve), and if so sets `IERR` to be 3 and aborts; else continues.

4. Queries if `RAD1` is not zero and `RAD2` is equal to or greater than `RAD1`, and if so sets `IERR` to be 4 and aborts; else continues.

5. Queries `ROTN` for being 1 or −1, and if not so sets `IERR` to be 5 and aborts; else continues.

6. Queries `RAD1` for being equal to zero within a tolerance of 0.005 feet (1.524 mm), and if so sets `DC1` (the degree of curvature) to be zero; else computes the degree of curvature of the back curve as being equal to $(100/\text{RAD1})$ $(180/\pi)$, and the degree of curvature of the ahead curve as being equal to $(100/\text{RAD2})(180/\pi)$.

7. Calls the `icSpiralPrjct` procedure to project the given offset point on the spiral to compute the point's offset distance H from the spiral.

8. Queries the returned error message to ascertain if a problem has been encountered and if so sets `IERR` to be 6 and aborts; else subtracts `W` from H and continues.

9. Queries the adjusted offset distance H for being zero within a tolerance of 0.005 feet (1.524 mm) and if:
   - $H = 0$ (the point lies on the spiral) accepts the point as the tangency, sets the parameters to be returned to be those of the given point and `IERR` to be 9, and terminates.
   - $H < 0$ (the point is inside the spiral) sets `IERR` to be 7 and aborts.
   - $H > 0$ continues.

10. Calls the `icptl` procedure to establish a temporary point opposite the start of the main spiral to define the start of the tangent line to the offset spiral.

11. Calls the `icprjct` procedure to project the given offset point on the offset spiral's tangent line to determine the Z and B distances of Figures 9.8b and 9.8c.

12. Queries the absolute value of the offset B for being equal to or less than 0.005 feet (1.524 mm) and if so sets the start point of the spiral as the tangency point and `IERR` to be 10 and terminates; else continues.

13. Queries the product of Z times B times `ROTN` for being negative, and if so sets `IERR` to be 8 and the indicator *backTan* to be 0, and in either case continues.

14. Queries *backTan* for being 0, and if so continues; else branches to **Step 20**.

15. Assumes as the first approximation of L as one half of the spiral length (`SPRL`).

    The procedures now solves for the tangency point of the single forward tangent.

16. Commences a loop of a series of steps to compute a new approximation of L. These steps are performed until the values of ckTest $\leq E$ and $K \leq 0.00001$, or until the limit of iterations (`CNTlimit`) has been reached, whichever is encountered first.

    (a) Calls the `icDeriv` procedure to compute the:
    - Spiral angle θ_S, the X and Y distances along and off the tangent to the spiral of the overall spiral, and the
    - Instantaneous angle θ and the differentials dX and dY that corresponds to the last defined approximation of L

    using Equations 9.1.4b, 9.1.15a, and 9.1.15b, respectively.

    (b) Determines the value of $f(L)$ from Equation 9.8c.

    (c) Computes the value of $f'(L)$ from Equation 9.8d3.

    (d) Uses the above computed values of $f(L)$ and $f'(L)$ to find the value of $K = f(L)/f'(L)$ with Equation 9.6.1d.

    (e) Computes the values of L from Equation 9.6.1d.

    (f) Checks the number of iterations against the upper limit of `CNTlimit` and if it has been reached sets `IERR` to be 11 and exits the loop, in which case the last value of L is used to continue with the results not being quite correct (the calling program should test for `IERR` to be 11 and act accordingly).

17. Calls the `icCeg8y` procedure to compute the main spiral data for the last computed final distance L from the spiral's point of beginning to the point of projection of the tangency point on the main spiral.

18. Calls the `icprjct` procedure to traverse from the main spiral point along the instantaneous tangent azimuth a zero plus distance and an offset distance of W to find the coordinates of the tangency point on the offset spiral in case an offset spiral is involved.

19. Compares the last computed L value with that of the spiral length and if it is equal to or greater than `SPRL` sets `IERR` to be 12 (non fatal error). In either case terminates.

20. Queries the backward tangent indicator *backTan* the value of which would be 1, and proceeds to compute the backward and then the forward tangent lines.

21. Sets the initial station SL of the tangency point to be that determined in **Step 7**.

22. Initializes the number of iterations CNT to zero, sets the incremental station adjustment dSL to be 20 feet (meters), and sets the previous offset sign LastOff to be minus 1.

23. Commences the following series of iterative steps to compute tangency point of the backward tangent line for a maximum of `CNTlimit` iterations:

    (a) Adjusts the initial station value to be SL = SL + dSL * LastOff.

    (b) Calls the `icCeg86` procedure to find the coordinates of the point on the main spiral at the last computed

SL station and the azimuth of the instantaneous tangent line at the said station.

(c) Calls the `icptl` procedure to find the coordinates of the corresponding point on the offset spiral in case an offset spiral exists.

(d) Calls the `icprjct` procedure to find the offset distance of the given point from the instantaneous tangent line to the main or offset spiral at the last SL station.

(e) Queries the offset distance for being equal to zero within a tolerance of 0.005 feet (1.524 mm), and if so accepts it as the tangency point of the backward tangent line and branches to **Step 24**.

(f) Queries the ROTN for having a clockwise or counterclockwise spiral, and accordingly compares the sign of the current offset with that of the offset at the previous station increasing or decreasing the station by the specified station increment as per the equations of Figure 9.8c, halving the increment if there is a change in sign.

(g) Increases the counter of iterations by one and returns to **Step 23(a)** to repeat the iterative process.

24. Repeats **Steps 21**, **22**, and **24** to compute the forward tangent line with the following changes.
   - In **Step 22** sets the previous offset sign LastOff to be plus 1.
   - In **Step 23(e)** branches to Step cc when the tangency point of the forward tangent has been found.
   - In **Step 23(f)** the process is the same but the equation criteria are different again as indicate in Figure 9.8c.

25. Queries the value of the station of the forward tangency point for being equal to or greater than the spiral angle within a tolerance of 0.005 feet (1.524 mm), and if so sets IERR to be 12. In either case terminates.

9.9 Four Other Spirals

Usually the immediate thought about a spiral is that of Archimedes. In addition to this there are several others developed by various mathematicians of which the most known are those of Fermat, the reciprocal also known as the hyperbolic spiral, and the Lituus spiral which are variations of the Archimedes spiral. These spirals and variations thereof have found a home in numerous applications, but not in roadway design. Shown in Figure 9.9a are samples of the said spirals together with their basic equations in which R is the ever increasing radius depending on the values of the constant A and the ever increasing angle θ.

The Fermat spiral has been named after the Frenchman Pierre de Fermat (1601–1665), a professional lawyer with a great interest in mathematics. A contemporary of Rene Descartes (see Chap. 1 of this publication) he is also credited

Pierre de Fermat (1601–1665)

with the discovery of analytic geometry along with Descartes, but since the latter's presentation was more formal and systematic than that of Fermat it is his name that is commonly associated with it.

The Lituus spiral is the inverse of the Fermat spiral and is so named after the Latin word lituus which refers to a hooked staff as used at that time by shepherds and later used by bishops of the Roman Catholic Church. This term as assigned to the spiral by Roger Cotes Maclaurin in a collection of various papers titled *Harmonia Mensurarum* in 1722.

The `icSpiralSpecial` procedure presented below may be employed to generate one of the above four spirals. It is necessary here to note that the spirals generated by this procedure have been customized for a specific application, that of crop dusting where it is desired to have an airplane fly in ever expanding circles such as those found in the Archimedes's spiral and meeting certain requirements that affect the value of the constant A in the aforesaid spiral equations as discussed below in the operational steps.

In reviewing Figure 9.9a we see that a spiral starts at a point and then keeps curving about it with an ever increasing radius forming a series of rings (bands or loops) until it stops at some limiting point. The Archimedes and Fermat spirals do not deviate from the said pattern but the other two could deviate towards an asymptotic end. To be defined a spiral requires:

1. A center point or point of beginning (point **A** in Figure 9.9a).

2. A uniform angular increment to control the increase in θ and thus control the smoothness of the spiral graphic to be generated. In essence this value controls the number of points per each single ring of the spiral. Thus, the number of points per ring is equal to 360° divided by the said uniform angular increment. For example, if the said increment is 18 then 20 point would be created for each ring; and if it is 36 there would be 10 points per ring.

3. The distance from the center of the spiral (the said point **A**) to the end of the spiral (point **C** of the same figure) referred to as the spiral radius which is not the same as the R in the spiral equations of Figure 9.9a. The R of the said

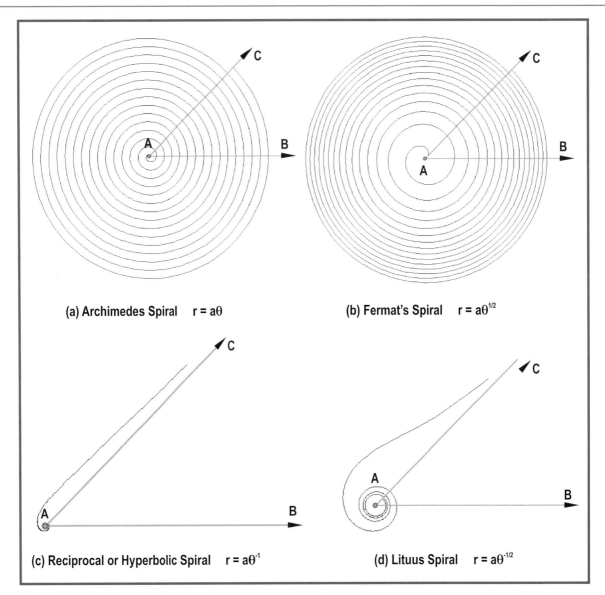

(a) Archimedes Spiral   r = aθ

(b) Fermat's Spiral   r = aθ$^{1/2}$

(c) Reciprocal or Hyperbolic Spiral   r = aθ^{-1}

(d) Lituus Spiral   r = aθ$^{-1/2}$

Figure 9.9a   Four Non-Highway Transition Spirals and Their Basic Equations

equations is computed by the procedure as described in the operational steps.

4. The band width or distance between the spiral loops to define the number of spiral rings. This band width is uniform between rings only for the Archimedes spiral.
5. The starting spiral angle defined as follows:
   • For the Archimedes and Fermat spirals it is the Cartesian direction if the line AC. In Figure 2, that is this the angle BAC.
   • For the Reciprocal and Lituus spirals it is the Cartesian direction from the center of the spiral to the start point of the innermost ring. For these spirals, the Cartesian direction from the center of the spiral (the said point **A**) to the end point of the outermost ring will be equal to the starting spiral angle plus the increment angle. For example, if the increment angle is 30° and

the starting angle is 10°, the direction from the center of the spiral to the end point of the outermost ring will be 40°.

It should be noted that since angular increment is uniform and thus the number of points per ring is constant, and since the total length of each ring is increasing as its distance from the center point increases, the spacing between points also increases, thus affecting somewhat the smoothness of the graphic.

```
Sub icSpiralSpecial (PTN, PTE, AngDeg, TheRad,
                     BandWidth, StrANG, Spiral-
                     Type, _
                     TheList, nLoops, Npoints,
                     IERR)
```

The input parameters to this procedure are the:

PTN, PTE	The north/east coordinates of the center point of the spiral.
AngDeg	The uniform angular increment in degrees (see Note 1 below).
TheRad	The distance in feet (meters) from the center point to the endpoint of the spiral (see Notes 2, 3, and 4 below).
BandWidth	The distance in feet (meters) between the spiral rings for the Archimedes spiral, and an approximate ring spacing for the other spirals to control the number of rings (see Notes 2, 3, and 4 below).
StrANG	The start angle of the spiral to orient the start chord of the Archimedes and Fermat spirals and to position the first point of the other spirals.
SpiralType	The type of spiral which should be: *Archimedes*, *Fermat*, *Lituus*, or *Reciprocal*.

NOTES

1. AngDeg cannot be zero but can be negative in which case the resultant spiral is a mirror image of the spiral with a positive AngDeg about the:
   - X-axis for the Archimedes, Fermat, and Lituus spirals, and about the
   - Y-axis for the reciprocal spiral.
2. TheRad and BandWidth cannot be zero but can be negative in which case both must be negative.
3. If both TheRad and BandWidth are negative the resultant spiral is a mirror image of the spiral of positive values about the:
   - Y-axis for the Archimedes and Fermat spirals, and about the
   - X-axis for the Lituus spiral, and about
   - Both axes for the reciprocal spiral.
4. The absolute value of TheRad must be greater than the absolute value of the BandWidth.

The returned parameters from this procedure are the:

TheList	The dynamic array of the north and east coordinates of the points comprising the polyline that represents the spiral stored sequentially (twice the number of points).
nLoops	The number of rings or loops in the spiral.
Npoints	The number of coordinates in TheList array (twice the number of points).
IERR	An indicator the value of which denotes:
	0 A spiral has been generated.
	1 The AngDeg has a zero value (abort).
	2 The TheRad has a zero value (abort).
	3 The BandWidth has a zero value (abort).
	4 The absolute value of TheRad is not greater than that of BandWidth (abort).
	5 One of the TheRad and BandWidth is negative but the other is not (abort).

Upon invocation the procedure assumes that a spiral will be generated and then:

1. Queries AngDeg for being equal to zero within a tolerance of 0.005 feet (1.524 mm), and if so sets IERR to be 1 and aborts; else continues.
2. Queries TheRad for being equal to zero within a tolerance of 0.005 feet (1.524 mm), and if so sets IERR to be 2 and aborts; else continues.
3. Queries BandWidth for being equal to zero within a tolerance of 0.005 feet (1.524 mm), and if so sets IERR to be 3 and aborts; else continues.
4. Queries the absolute values of TheRad and BandWidth and if TheRad is not greater than BandWidth sets IERR to be 4 and aborts; else continues.
5. Queries the TheRad and BandWidth and if one negative and the other is not sets IERR to be 5 and aborts; else continues.
6. Computes the number of rings spiral *nLoops* by dividing TheRad by BandWidth.
7. Computes the number of points per ring by dividing 360° by AngDeg and then multiplies this ratio by *nLoops* to get the total number of points.
8. Calls the iccegrad procedure to convert the angular increment AngDeg to radians and save it in the parameter *theAng*, and also set this value to be the initial value of *theAng* for the loop that computes the various points of the polyline to be created.
9. Computes the radial width of the overall spiral without applying a factor to determine a unit spiral width as follows for the type of spiral to be created:

   (a) For the Archimedes spiral

   $$\begin{aligned} R_U &= N_P \alpha \\ W_U &= \cos(R_U) R_U \end{aligned} \qquad (9.9a)$$

   (b) For the Fermat spiral

   $$\begin{aligned} R_U &= \sqrt{N_P |\alpha|} \\ W_U &= \cos(N_P a) R_U \end{aligned} \qquad (9.9b)$$

   (c) For the Lituus spiral

   $$\begin{aligned} R_U &= \frac{1}{\sqrt{2|\alpha|}} \\ D_X &= \cos(N_P a) R_U \\ D_Y &= \sin(N_P a) R_U \\ W_U &= \sqrt{DX^2 + DY^2} \end{aligned} \qquad (9.9c)$$

(d) For the reciprocal spiral

$$R_U = \frac{1}{2\alpha}$$
$$D_X = \cos\left(N_P a\right) R_U$$
$$D_Y = \sin\left(N_P a\right) R_U \qquad (9.9d)$$
$$W_U = \sqrt{DX^2 + DY^2}$$

where

R_U is an equivalent radius,
N_P is the number of points in the spiral,
α is the uniform angular increment `AngDeg` in radians,
D_X is the X-axis component of the unit spiral width,
D_Y is the Y-axis component of the unit spiral width, and
W_U is the unit spiral width.

10. (a) Saves in `TheList` dynamic array the center point (`PTN, PTE`) as the start point of the spiral and then sets the start point index (*iStrt*) of the said array (to be 2 for the Archimedes and Fermat spirals only).

(b) Sets the said start index (*iStrt*) to 1 for the Lituus and reciprocal spirals.

11. Commences a loop of repetitive steps from *iStrt* to N_P as indicated below to develop the polyline which is to represent the spiral:

(a) Instantiates the spiral angle for the point to be computed for the spiral by increasing *theAng* of **Step 8** by *theAng* of the same step.

(b) Instantiates the spiral radius for the type of spiral as follows:

• Archimedes spiral

$$R = \gamma \qquad (9.9e)$$

• Fermat spiral

$$R = \sqrt{|\gamma|} \qquad (9.9f)$$

• Reciprocal spiral

$$R = \frac{1}{\gamma} \qquad (9.9g)$$

• Lituus spiral

$$R = \frac{1}{\sqrt{|\gamma|}} \qquad (9.9h)$$

(c) Instantiates the Cartesian direction of the first spiral point or that of the previous point by increasing the start angle by the value of the angle of **Step 11(a)** for the Archimedes and Fermat spiral

$$\delta = \gamma + \varepsilon \qquad (9.9i1)$$

where ε is the start angle `StrANG`

(d) Modify the said angle for the reciprocal and Lituus spiral

$$\delta = \gamma + \varepsilon - \alpha \qquad (9.9i2)$$

(e) Computes the Cartesian coordinates of the point as

$$X = P_E + \left[R\cos\left(\delta\right)\left(\frac{R_S}{W_U}\right)\right] \qquad (9.9j1)$$

and

$$Y = P_N + \left[R\sin\left(\delta\right)\left(\frac{R_S}{W_U}\right)\right] \qquad (9.9j2)$$

(f) Saves the above Cartesian coordinates as north and east coordinates in the `TheList` dynamic array and terminates.

Regarding the saving of the spiral coordinates it is noted that the number of coordinates to be saved is twice the number of point N_P that have been created because the north and east coordinates are saved sequentially in the same array. Thus, the array index number for the

• North coordinate (X) of a point is equal to twice the index number of the corresponding point minus one, and for the
• East coordinate (Y) is equal to twice the index number of the corresponding point.

9.10   The Highway and the Archimedes Spirals

Presented below is a partial discussion for the sake of curiosity between the transition spiral as defined in this chapter, which we can refer to it as a highway spiral and the Archimedes spiral. For this let us consider a highway spiral which is to provide a transition from a straight line inclined at a north azimuth of 30° to a circular curve of a radius of 2000 feet (609.60 m) starting at the TS point on the said line as indicated in Figure 9.9b1 and terminating at the SC which is also referred to point C. The said highway spiral has the θ spiral angle and instantaneous radial azimuth at the SC as indicated is the said figure. With this information the center point A of the circular curve of the said radius is located as shown.

Next, using the said point A as the center point of an Archimedes spiral with a radius (`TheRad`) of 2000, a ring spacing (`BandWidth`) of 100 and a start angle (`StrANG`) of 121° 21′ 07.6″ which is the Cartesian azimuth from point A to the SC point we get the Archimedes spiral shown in the said figure. Since the two spirals meet at the SC and have the common instantaneous radius and direction thereof they are compound curves or tangent to each other. If `TheRad` had been 2400 and if `BandWidth` and `StrANG` remained the

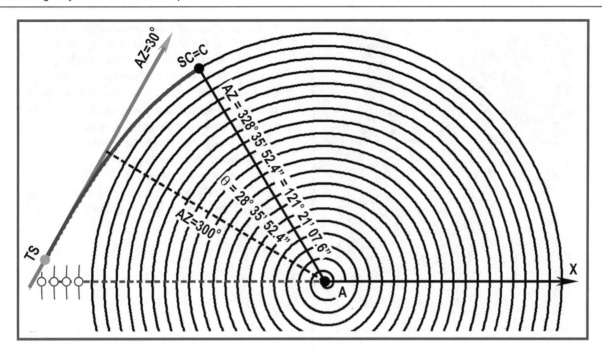

Figure 9.9b1   Highway Spiral and Archimedes Spiral

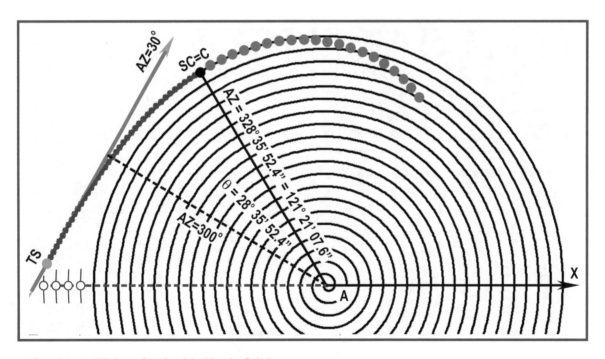

Figure 9.9b2   Extended Highway Spiral and Archimedes Spiral

same then there would be four additional rings passing through the red hollow points on the left side of the Archimedes spiral as well as through point C.

If we maintain the highway spiral as is and if we extended it by introducing points beyond the SC, at a uniform interval (station increment) along the spiral from the TS we see that the highway spiral begins to cross the rings of the Archimedes spiral as shown in Figure 9.9b2. Next let us consider extending

the highway spiral beyond the previous endpoint by introducing additional points at the same station increment as shown in Figure 9.9b3 and in which the following should be noted:

1. Shown in this figure are:
   - The various rings of the Archimedes spiral which because of the scale of the graph they are displayed as a large solid and filled circle:

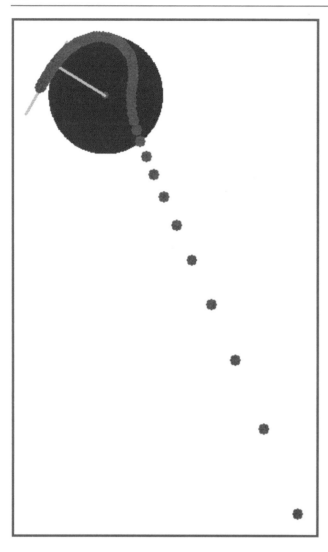

- The points along the extended highway spiral; and
- The center points corresponding to the abovesaid points. That is points 2000 meters along the instantaneous radial azimuth of each spiral point.

2. Even though the points to be generated are specified to be at a uniform station increment, the resultant point coordinates are such that the spacing between points as their corresponding station value increases.

3. The highway spiral as it is being extended further and further, after a certain point it changes rotation, from clockwise to counterclockwise and vice versa, eventually approaching asymptotically a straight line like the reciprocal spiral.

Figure 9.9b3   Further Extension of a Highway Spiral and Archimedes Spiral

In the preceding chapters we discuss various specific constructions such as intersections, tangencies, projections and the like involving lines, circular curves, and spirals. In civil engineering, however, there are certain configurations or assemblies of geometric elements that would require the invocation of several of the procedures that have been presented in the preceding chapters complete. Some of these configurations include but are not limited to (a) the treatment of dead end streets resulting in the use of cul-de-sacs or hammerheads, (b) the bypassing of an obstacle to a horizontal alignment, (c) the intersection of street ribbons such as traffic lanes, pavement edges, curb or gutter lines, and right-of-way lines to mention a few, and (d) the introduction of traffic islands and sight triangles.

In this chapter three procedures are presented that greatly simply the design of a street cul-de-sac, hammerhead, and bubble.

The procedures below address what are considered to be the most common versions of their intended application. From time to time special conditions may surface that may require a deviation that would warrant a modification to the result of a procedure, or possibly the construction by what might be referred to as the brute force method of using several of the previously presented procedures.

10.1 Cul-de-Sac Design

A common feature in many residential developments is the introduction of a cul-de-sac at the end of a dead end street. There are two basic cul-de-sac forms usually referred to as the standard cul-de-sac (see Figure 10.1a) and the offset cul-de-sac (see Figures 10.1b and 10.1c). Each one of the said three figures is composed of the main station line and its two offset right-of-way lines, the cul-de-sac bubble and its center point, and the optional fillet curves. The difference between the said two cul-de-sac forms is the offsetting of the bubble to the left, or to the right of the station line. The last two

figures illustrate an offset cul-de-sac to the right of the station line, but it could just as well have being located to the left of the said line. Note the difference between these two figures. Furthermore, note that the station line leading into the cul-de-sac could be a straight line as shown in the said three figures, or a curved line as is to be seen in figures to follow.

One common modification to either of the said standard or offset cul-de-sac types is the one that enlarges the radius of the cul-de-sac bubble, and introduces a circular island of a small radius about the center point of the bubble as illustrated by the solid blue circle in Figure 10.1d, thus causing the traffic to travel around the island in one direction only. In certain other cases, the shape of the island instead of being circular as just said, it may be composed of two semicircles of either equal or unequal radii that are connected with two tangent lines and thus form an oblong island, as illustrated by the dashed blue line in the said figure. In the latter case it is possible that the line connecting the two center points of the oblong island may slide along the center point of the cul-de-sac bubble.

For such a modified cul-de-sac design case, a generic automated geometric design of the cul-de-sac that includes a circular or oblong traffic island is possible. However, since the frequency of such occurrence is rather small, such a design is considered of rather academic nature and it is not addressed herein. When such a design is encountered, the end user may employ the tool associated with the subject procedure to create the outside shape of the cul-de-sac, and then employ the tools associated with the appropriate geometric procedures that have been presented earlier in this publication to insert the said island.

In each of the above-referenced cul-de-sac figures, the cul-de-sac is shown to be located at the end point of a straight street alignment as indicated in Figure 10.1e. However, this is not always the case. As indicated in Figure 10.1f, a cul-de-sac may also be introduced at the end point of a circular street alignment. The geometry can be complicated even further by having a short tangent line between the end point of

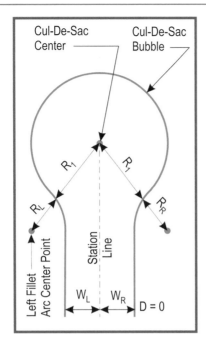

Figure 10.1a Standard Cul-de-sac

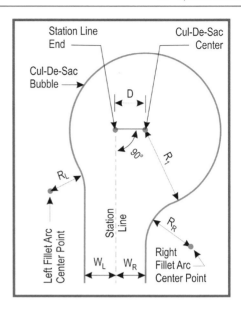

Figure 10.1c Offset Cul-de-sac (Two Returns)

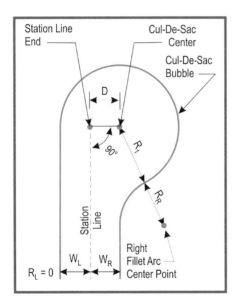

Figure 10.1b Offset Cul-de-sac (One Return)

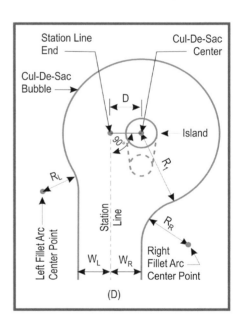

Figure 10.1d Cul-de-sac with an Island

a circular arc and the end point of the cul-de-sac so that one or both of the fillet arcs are tangent to the arc of the cul-de-sac bubble and to the arc of the right of way line which is offset from the street center or station line.

In each of the above-referenced figures, the cul-de-sac bubble is shown to be to the right of the station line. This need not be the so, and furthermore, the bubble may be on the opposite side of the curvature of the station line. In the last said two figures, it is noted that features appearing in (**a**) blue indicate input parameters to the procedure, (**b**) red indicate parameters that are returned by the procedure, and (**c**)

black indicate parameters that are computed by the procedure for internal use only. It should be noted that since generally the street right-or-way follows the shape of the street pavement, the outline of the cul-de-sac could be that of the street pavement, gutter, or curb line, or of the street right-of-way. Thus, the term "lane" is used symbolically to imply any of the abovesaid cul-de-sac outlines.

```
Sub icCulDeSac  (EN, EE, WL, WR, R1, RR, RL, A,
                 CCN, CCE, PTN, PTE, R2, _
                 CN, CE, P1N, P1E, P2N, P2E,
                 P3N, P3E, P4N, P4E, P5N, P5E, _
                 P6N, P6E, IERR)
```

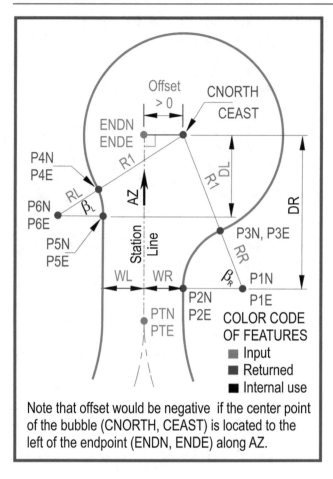

Note that offset would be negative if the center point of the bubble (CNORTH, CEAST) is located to the left of the endpoint (ENDN, ENDE) along AZ.

Figure 10.1e   Cul-de-sac at End of Line

The input parameters to this procedure are the:

EN, EE	The north and east coordinates of the end point of the street alignment.
WL, WR	The width of the left and right lanes of the street (see Note 1 below).
R1	The radius of the cul-de-sac bubble (see Note 2 below).
RR, RL	The radius of the right and left fillet arc (see Note 3 below).
A	The offset distance from the endpoint of the station line to the center point of the cul-de-sac bubble (see Note 4 below).
CCN, CCE	The north and east coordinates of the center point of the circular arc nearest to the end point of the street alignment (see Notes 5 and 6 below).
PTN, PTE	The north and east coordinates of the PT point of the circular arc nearest to the end point of the street alignment (see Note 5 below).
R2	The radius of the circular arc nearest to the endpoint of the station line:

- A positive value denotes a clockwise rotation.
- A negative value denotes a counterclockwise rotation.
- A zero value denotes a straight line.

NOTES

1. Both WL or WR should be positive. If either is negative, it is changed to positive. If one is zero, it implies that the station line is along the corresponding lane edge.
2. If R1 is equal to zero, the procedure aborts. If it is negative it set to be positive.

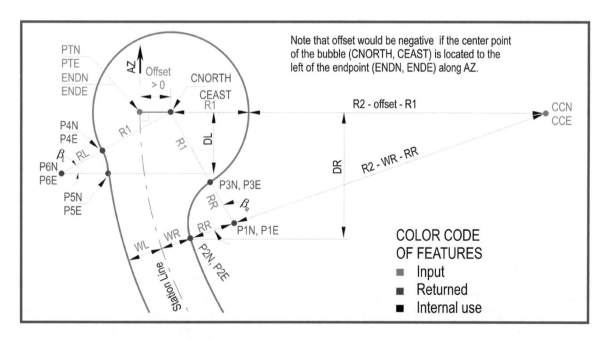

Figure 10.1f   Cul-de-sac at End of and Circular Arc

3. If either RL or RR is equal to zero, it implies that there is no fillet curve. The arc of the bubble intersects the left or right edge, respectively, of the lane at a point. If either is negative, it set to be positive.
4. The offset A is normal to a straight station line, or radial to a curved station line at the station line's endpoint (EN, EE). It is positive if the center point of the bubble is to be to the right of the station line, and negative if to the left with the station line's progression being towards the cul-de-sac bubble. This parameter is referred to as the off-set in the source code of the procedure.
5. It the PTN, PTE point is located so that either or both fillet arcs are tangent to the curve bubble arc and to the corresponding lane edge, then the coordinates of the center point (CCN, CCE) of the nearest curve of the station line to the end point (EN, EE) must be introduced. If that is not the case, that is the fillet curves are to be tangent to the arc of the bubble and to a straight lane edge, the coordinates CCN and CCE could have a zero, or any numeric value.
6. If the cul-de-sac bubble is to be located at the endpoint of a curved station line, then the coordinates (PTN, PTE) should be the same as those of (EN, EE).

The returned parameters from this procedure are the:

CN, CE	The north and east coordinate of the center of the cul-de-sac bubble.
P1N, P1E	The north and east coordinate of the center of the right fillet curve.
P2N, P2E	The north and east coordinate of the point of tangency of the right fillet curve with the right offset line to the street alignment.
P3N, P3E	The north and east coordinate of the point of tangency of the right fillet curve with the cul-de-sac bubble.
P4N, P4E	The north and east coordinate of the point of tangency of the left fillet curve with the cul-de-sac bubble.
P5N, P5E	The north and east coordinate of the point of tangency of the left fillet curve with the left offset line to the street alignment.
P6N, P6E	The north and east coordinate of the center of the left fillet curve.
IERR	An error indicator the value of which denotes the error message presented in Table 10.1. Not all messages are fatal.

Upon invocation, the procedure initializes the parameters to be returned, and then:

1. Queries the value of R1 for being negative or zero. If it is:
   - Negative sets IERR to be 4, changes R1 to be positive, and continues.
   - Zero, sets IERR to be 5, and aborts.

Table 10.1 Error Messages for the icCulDeSac Procedure

IERR	Implied Error Message	Message Type
0	No error has been encountered	No errors
1	Right Offset too large. Offset redefined	Warning
2	Left Offset too large. Offset redefined	Warning
3	Left/Right Offsets too large. Offsets redefined	Warning
4	Negative R1 changed to be positive	Warning
5	R1 is zero. Abort	Fatal
6	Negative RL changed to be positive	Warning
7	Negative RR changed to be positive	Warning
8	Cul-de-sac radius too small. Abort	Fatal
9	Can not compute left side curve return. Abort	Fatal
10	Can not compute right side curve return. Abort	Fatal
11	Negative WL changed to be positive	Warning
12	Negative WR changed to be positive	Warning

2. Queries the value of RL and RR for being negative, and if so sets IERR to be 6 or 7, and continues.
3. Calls the icforce procedure to determine the distance B (DIST in the procedure's source code) and the direction AZ from the PT point (PTN, PTE) to the end point of the cul-de-sac (EN, EE). It is assumed that point (PTN, PTE) is located back of point (E, EE) along the cul-de-sac station line.
4. Queries the value of B for being equal to zero within a tolerance of 0.00001 feet (0.003048 mm), and if it is it implies that the cul-de-sac endpoint is located at the end of a curved alignment (see Figure 10.1f), in which case the procedure computes AZ as being equal to the azimuth from (CCN, CCE) to (PTN, PTE) plus 90° if R2 (the radius of the alignment curve) is positive (clockwise), or minus 90° if R2 is negative (counterclockwise). That is, AZ is now the direction of the tangent line to the curve at the end point of the cul-de-sac.

For the statements below, reference is made to Figures 10.1g through 10.1m which display different potential cul-de-sac configurations. These configurations are divided into the following three groups depending on the magnitude of the magnitude of the length B. If B is:

- **Group 1** The length B is long enough so that all points of tangency of the fillet curves lie within the extent of the straight station line (see Figure 10.1g).
- **Group 2** The length B is not long enough so that the points of tangency of the fillet curves (P3N, P3E) and (P4N, P4E) lie within the extent of the straight line portion of the station line, and the points of tangency of the fillet curves (P2N, P2E) and (P5N, P5E) lie within the extent of the curved portion of the station line (see Figures 10.1h, 10.1i and 10.1j).

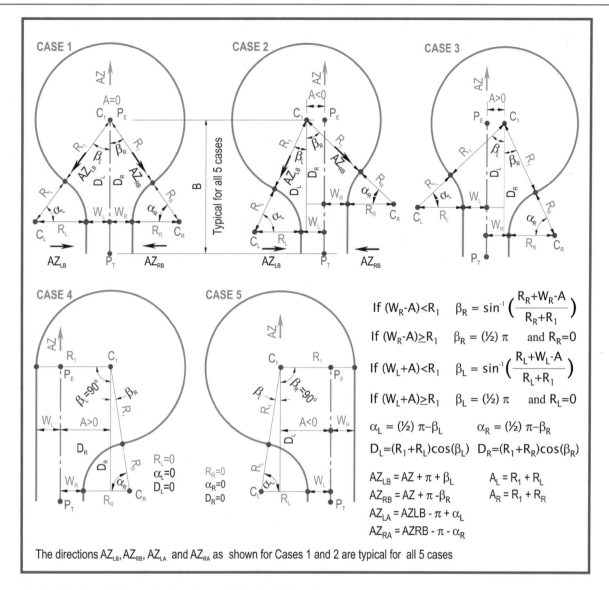

Figure 10.1g   Potential Cul-de-sac Configurations at End of a Straight Alignment

- **Group 3**   The length B is zero so that all points of tangency of the fillet curves lie within the extent of the curved station line (see Figures 10.1k, 10.1l, and 10.1m).

  These configurations are generic with the offset distance A being zero, negative or positive, and R2 being positive or negative. Slightly different configurations may occur depending on the magnitude of A with respect to WL and WR. It can be shown that the equations in the said figures hold true for all configurations.

5. Compares the value of WR and WL against the value R1, and if either WR or WL is larger then R1, the procedure sets IERR to 8, errString to it corresponding message of Table 10.1, and aborts returning to the calling program.

6. Determines whether the offset A is too large, or not to construct the cul-de-sac. If it is, the offset is redefined so as to force the right or the left edge of the corresponding lane to be tangent to the bubble's curve, thus yielding a right or left fillet radius of zero value. This is done in a two pass loop as follows:

Pass 1:   Checks to determine whether an offset adjustment is necessary by comparing first WR − A against R1 for the right side, and then WL + A against R1 for the left side of the cul-de-sac. At this time, the angles βR and βL are computed by use of the equations shown in Figure 10.1g. If the offset A is found to be too large either for the right or left side, it is modified as stated above and in Figure 10.1g, and the appropriate warning message is activated.

Figure 10.1h Cul-de-sac
Partially at Straight and Partially
at Curved Alignment (Cases 1–2)

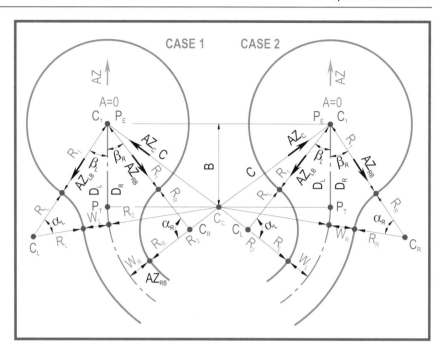

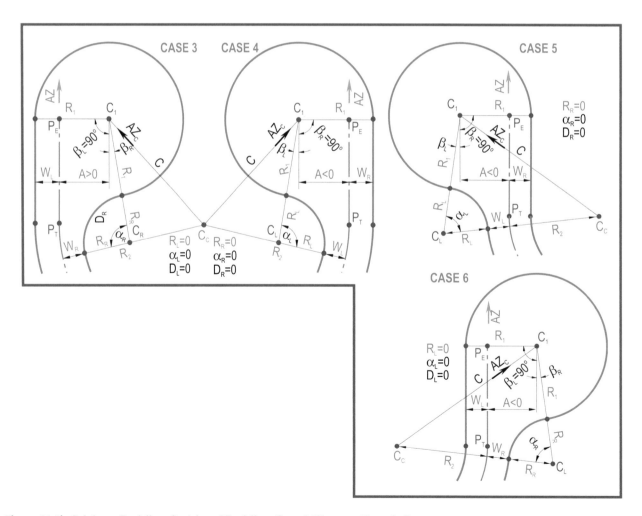

Figure 10.1i Cul-de-sac Partially at Straight and Partially at Curved Alignment (Cases 3–6)

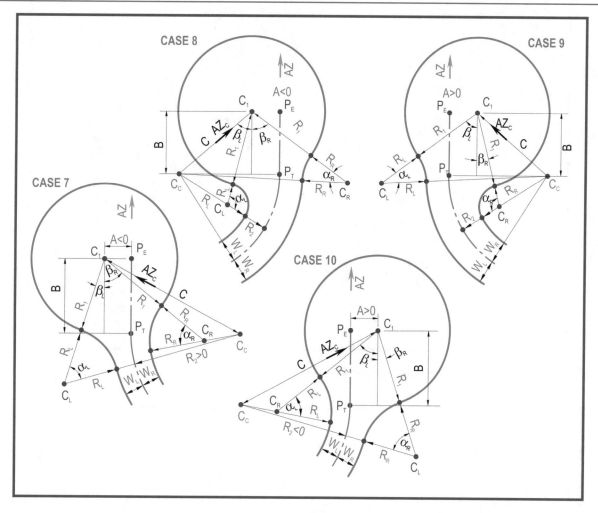

Figure 10.1j   Cul-de-sac Partially at Straight and Partially at Curved Alignment (Cases 7–10)

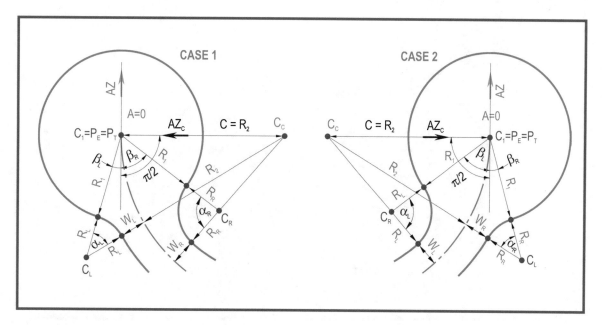

Figure 10.1k   Potential Cul-de-sac Configurations at End of a Curved Alignment

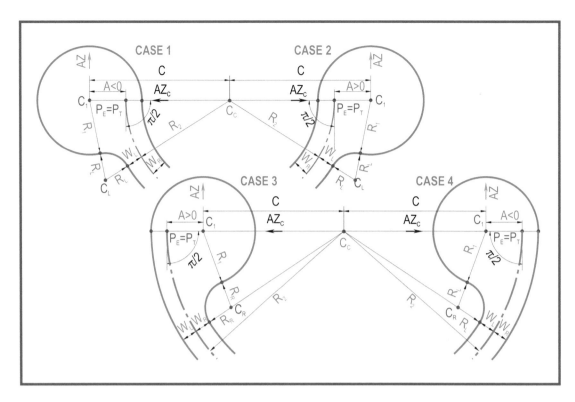

Figure 10.1l Potential Cul-de-sac Configurations at End of a Curved Alignment

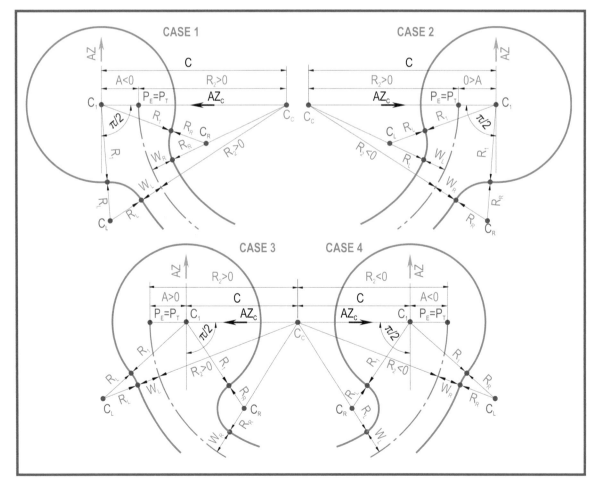

Figure 10.1m Potential Cul-de-sac Configurations at End of a Curved Alignment

Pass 2: Repeats the above tests and computation with the revised offset only if the said offset was modified during the first pass. If the said offset was not modified, this pass is not performed.

7. Computes the values of the angles αR and αL and of the distances DR and D_L by use of the equations shown in Figure 10.1g.

8. Calls the icptl procedure to determine the north and east coordinates of the center point of the cul-de-sac bubble (CNORTH, CEAST).

9. Sets the values of A_L and AR as being the sums of the radius of the cul-de-sac bubble and the radius of the left and right fillets, respectively.

10. Determines the following azimuths as indicated by the equations in Figure 10.1g:
    - AZLB and AZRB pointing from the center point of the cul-de-sac bubble towards the center point of the left and right fillets, respectively, and
    - AZLA and AZRA pointing from the center point of the left and right fillets, respectively, towards the street station or center line.

11. Calls the icptl procedure to determine the coordinates of the cul-de-sac control points (P1N, P1E) through (P6N, P6E), see Figure 10.1e, assuming that distance B is long enough so that the said control points lie along the extent of a straight station line (these points are to be modified as indicated below if a curved station line is present).

12. The procedure determines, first for the left fillet arc and then for the right fillet arc, whether points (P1N, P1E) and/or (P5N, P5E), as computed above, are located along a straight station line, or along a curved one. This is done by comparing the distance B against the distance DL for the left side and then against the distance DR for the right side of the cul-de-sac.

Thus, if distance B is:

(a) Equal to or greater than D_L, the computed points (P4N, P4E) through (P6N, P6E) are assumed as correct, and the procedure proceeds to check the right side.

(b) Less than D_L , the procedure proceeds to compute new coordinates for points (P4N, P4E) through (P6N, P6E), as indicated below.

(c) Equal to or greater than DR, the computed points (P1N, P1E) through (P3N, P3E) are assumed as being correct, and the procedure terminates.

(d) Less than DR, the procedure proceeds to compute new coordinates for points (P1N, P1E) through (P3N, P3E), as indicated below, and then the procedure terminates.

Both of the following two steps call the iccrccrc procedure of Chap. 6 to intersect two circles. It is important to note that in order to return the proper of the two potential intersections, point (CCN, CCE) must precede point (CNORTH, CEAST) as input arguments when the iccrccrc procedure is called.

To locate the control points of the fillet curves, the procedure intersects two circles to determine the center points of the left (P6N, P6E) and of the right (P1N, P1E) fillet curves. Of these circles:

- The first circle has its center point at the center point of the station line's curve, and a radius equal to the algebraic sum of R2 + WL + RL for the left side, and equal to the algebraic sum of R2 − WR − RR for the right side of the station line.
- The second circle has its center point at the center point of the cul-de-sac bubble, and a radius equal to R1 + RL for both the left and the right side of the station line.

Once the said fillet center points have been computed, the procedure:

(a) Calls the icforce procedure to determine the azimuth from point (CCN, CCE) to (P6N, P6E), and then calls the icptl procedure to determine point (P5N, P5E) from point (P6N, P6E) with a plus distance of RL.

(b) Calls the icforce procedure to determine the azimuth from point (P6N, P6E) to (CN, CE), and then calls the icptl procedure to determine point (P4N, P4E) from point (P6N, P6E) with a plus distance of RL.

(c) Calls the icforce procedure to determine the azimuth from point (CCN, CCE) to (P1N, P1E), and then calls the icptl procedure to determine point (P2N, P2E) from point (P1N, P1E) with a plus distance of RR.

(d) Calls the icforce procedure to determine the azimuth from point (CCN, CCE) to (CN, CE), and then calls the icptl procedure to determine point (P3N, P3E) from point (P1N, P1E) with a plus distance of RR.

This terminates the subject procedure. It is now up to the calling program to display the cul-de-sac that has been generated, and if necessary to introduce a traffic island.

10.2 Hammerhead Design

An alternate to the cul-de-sac turnaround at a dead end street is that of the hammerhead, or T as it might be called, as shown in Figure 10.2a. A vehicle enters the hammerhead and makes a left or right turn, backs up, and then reverses its turn

Figure 10.2a Typical
Hammerhead Configuration

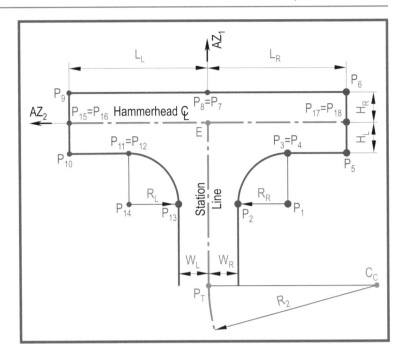

Figure 10.2b Inclined
Hammerhead Configuration

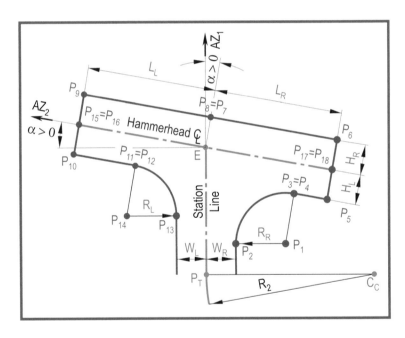

to exit. This design is usually encountered in special conditions to better utilize the land area at hand. A hammerhead is
composed of the street's station line, the hammerhead center
line, the two optional fillet curves and the widths of the two
travel lanes of the street and of the hammerhead. Although
one may argue that a hammerhead has but one lane, the subject procedure assumes that the hammerhead's center line
divides the hammerhead into two theoretical lanes.

Although the said figure displays the most common hammerhead design in which the alignment of the hammer-

head's center line is normal to the street's station line at its
endpoint, there can be several variations to it two of which
might be the one with the hammerhead's center line not
being normal to the street's station line as shown in
Figure 10.2b, and the other with of a curve in the hammerhead's center line as shown in Figure 10.2c. In the last said
case it is conceivable that the curve of the hammerhead
instead of being concave, as it shown in the said figure, to be
convex. That is the center point of the curve to lie back of
the station line's endpoint and not ahead from it along the

Figure 10.2c Curved
Hammerhead Configuration

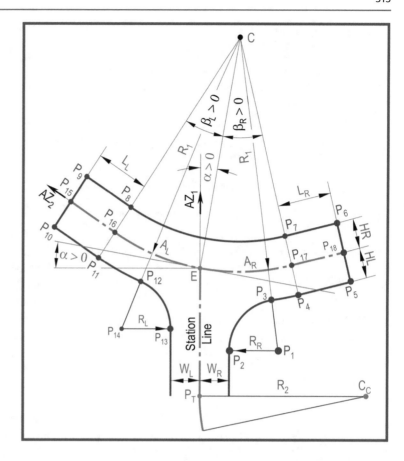

station line's direction. Although this curved condition could easily be accommodated by the procedure, it is not felt that it would produce a functional design, and therefore it is not considered.

It is noted that in each of these three hammerhead configurations the street station line may have a circular course much like the one described under the cul-de-sac. That is the curve's PT point may be located:

- Back of the fillet curves with the last course of the street station line being a "long" tangent;
- Within the extent of one or of both fillet curves and the street station line having a "short" tangent line as its last course; or
- At the end point of the street's station line.

For the case of Figure 10.2c there are various ways to define the curved portion of the hammerhead. However, the subject procedure assumes the following as being known:

- The angle α, the deflection angle of the instantaneous tangent to the hammerhead curve at the end point (E) of the station line with the normal to the station line at its end point is known.
- The arc lengths A_L and A_R at each side of the said point E.
- The length of the two tangent stubs, L_L and L_R, to each and of the said arc.

Much like the cul-de-sac procedure described in the preceding section, this procedure does not introduce any new geometric applications. It merely contains a series of steps that call previously described geometric procedures.

```
Sub icHammerHead  (EN, EE, WL, WR, ANG, R1,
                   AR, AL, LR, LL, HR, HL, _
                   RR, RL, CCN, CCE, PTN, PTE,
                   R2, _
                   CN, CE, PN, PE, IERR)
```

The input parameters to this procedure are the:

EN, EE	The north and east coordinates of the end point of the street station line, and of the point about which the hammerhead center line is pivoted (see Note 5 below).
WL, WR	The width of the left and right lanes of the street towards the direction of the hammerhead (see Notes 1 and 2 below).
ANG	The angle in radians that the "hammerhead line" (see below) makes with the:
	(a) Normal to the last course of the station line at the (EN, EE) point if the said last course is a straight line.
	(b) The radial line of the last course of the station line at the (EN, EE) point if the said last course is a circular curve.
	The "hammerhead line" is the:
	(c) Center line of a straight hammerhead.

(d)   Tangent line to the center line arc of the hammerhead at the (EN, EE) point of a curved hammerhead.

The value of ANG is positive for a counterclockwise rotation and negative for a clockwise rotation.

R1   The radius of a curved hammerhead as indicated in Figure 10.2c, or zero denoting a straight hammerhead as indicated in Figures 10.2a and 10.2b.

AL, AR   The length of the hammerhead arcs to the left and to the right of the end point of the station line (EN, EE) as shown in Figure 10.2c, or zero if the hammerhead is straight as shown in Figures 10.2a and 10.2b (see Note 2 below).

LL, LR   The length of the tangent stubs to the left and to the right of the hammerhead arcs (AL, AR), or the straight length of the hammerhead to the left and to the right of the end point of the station line (EN, EE). Neither one may be zero (see Note 2 below).

HL, HR   The width of the left and right lanes of the hammerhead with the left lane being towards the street side and the right lane being on the opposite side (see Note 2 below).

RR, RL   The radius of the right and left fillet arc (see Note 3 below).

CCN, CCE   The north and east coordinates of the center point of the circular arc of the street station line nearest to the end point (EN, EE) of the street station line (see Note 5 below).

PTN, PTE   The north and east coordinates of the PT point of the circular arc nearest to the end point of the street station line (see Note 4 below).

R2   The radius of the circular arc nearest to the endpoint of the station line (see Note 4 below) noting that:
- A positive value denotes a clockwise rotation.
- A negative value denotes a counterclockwise rotation.
- A zero value denotes a straight line and that this curve is of no use.

In Figures 10.2a, 10.2b, and 10.2c the value of R2 is of no use. However, it is used in the three cases of Figure 10.2d. In all the said figures R2 is positive. A negative R2 would place the corresponding curve on the opposite side of the street station line.

The returned parameters from this procedure are the:

CN, CE   The north and east coordinate of the center point of the hammerhead arc, if the hammerhead is curved; else 99999.0 is returned for each coordinate if the hammerhead is straight.

PN, PE   The arrays that contain the north and east coordinates of the points of tangency identified as points P1 through P18 in Figures 10.2a, 10.2b, and 10.2c. It is noted that:
- Whenever the hammerhead is straight as indicated in Figures 10.2a and 10.2b, certain points share the same pair of coordinates.
- The sequential position of a coordinate in the respective array corresponds to the number of the point indicated in the said three figures.

IERR   An error indicator the value of which denotes the error message presented in Table 10.2a. All messages are fatal.

Table 10.2a   Error messages for the icHammerHead procedure

IERR	Implied Error Message	Message Type
0	No error has been encountered	No errors
1	Zero station line radius. Abort	Fatal
2	Improper station line curve. Abort	Fatal
3	Left fillet cannot be set. Abort	Fatal
4	Right fillet cannot be set. Abort	Fatal
5	Right stub LR cannot be zero. Abort	Fatal
6	Left stub LL cannot be zero. Abort	Fatal

NOTES

1. Both WL or WR should be positive. If either is negative, it is changed to positive. If either is equal to zero, it implies that the station line is along the edge of the left or right lane. There is no checking in case either of these two conditions is desired.

2. The values of the parameters WL, WR, R1, AR, AL, LR, LL, HR, HL, RR, and RL should be positive. They are all forced to be positive at the start of the operation and they are reverted to their original value at the end of the operation. A zero value is permitted for all except for LR and LL.

3. If either RL or RR has a zero value, it implies that there is no fillet curve. That is in such cases the left or right lane edge, respectively, of the street intersects with the left lane edge of the hammerhead, and points P1, P2, and P3 are the same and points P12, P13, and P14 are also the same.

4. Regarding the last course of the street station line note that if the said course is a:
- Straight line and long enough so that all of the end points of the fillet curves of radius RR and RL project upon the said line, then the radius R2 and the center point CCN, CCE are not used and can have any value or zero.
- Straight line and only long enough so that only some or none of the end points of the fillet curves of radius RR and RL project upon the said line, then the radius R2 and the center point CCN, CCE are used and should have the appropriate values.
- Circular curve then the radius R2 and the center point CCN, CCE should have the appropriate values, and the points PTN, PTE and EN, EE should be the same.

Upon invocation, the procedure initializes the parameters to be returned, and then:

1. Preserves the values of the input parameters that should be positive, and thereafter:
- Changes any of the that are negative to be positive, and
- Checks LR and LL for being equal to zero within a tolerance of 0.005 feet (1.524 mm). If either is equal to zero the value of IERR is set to be 5 or 6, respectively.

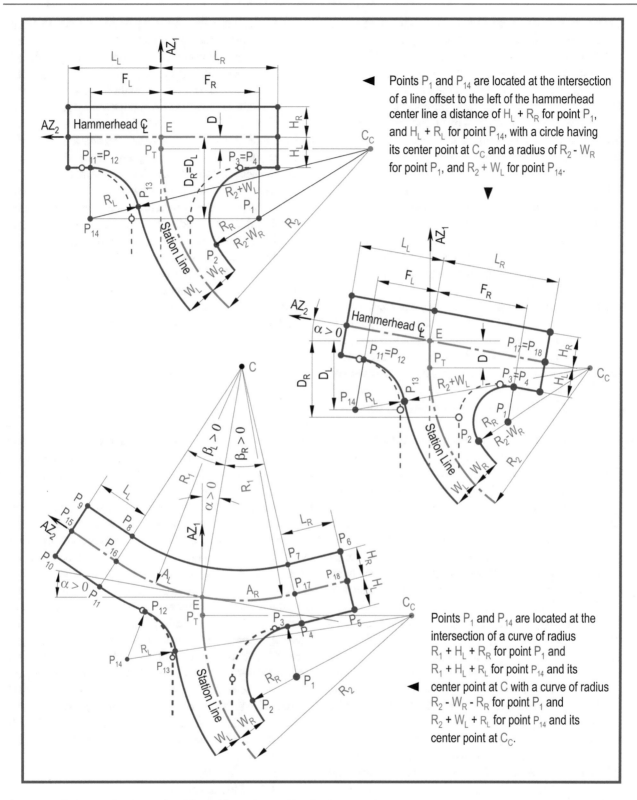

Points P_1 and P_{14} are located at the intersection of a line offset to the left of the hammerhead center line a distance of $H_L + R_R$ for point P_1, and $H_L + R_L$ for point P_{14}, with a circle having its center point at C_C and a radius of $R_2 - W_R$ for point P_1, and $R_2 + W_L$ for point P_{14}.

Points P_1 and P_{14} are located at the intersection of a curve of radius $R_1 + H_L + R_R$ for point P_1 and $R_1 + H_L + R_L$ for point P_{14} and its center point at C with a curve of radius $R_2 - W_R - R_R$ for point P_1 and $R_2 + W_L + R_L$ for point P_{14} and its center point at C_C.

Figure 10.2d Curved Hammerhead and Curved Street Station Line Configurations

2. Calls the `icforce` procedure to determine the distance `Dist1` and the direction `AZ1` from the PT point (`PTN`, `PTE`) to the end point of the street station line (`EN`, `EE`). It is assumed that the point (`PTN`, `PTE`) is located back of the endpoint (`EN`, `EE`) along the said station line.

3. Queries the value of `Dist1` for being equal to zero within a tolerance of 0.005 feet (1.524 mm), and if it is it implies that the street station line endpoint is located at the end of a curved alignment (see Figure 10.2c), in which case the procedure:
   (a) Queries the value of `R2` for being equal to zero within a tolerance of 0.005 feet (1.524 mm), and if so sets `IERR` to be 1 and aborts; else continues.
   (b) Calls the `icforce` procedure to determine the distance `Dist2` and the direction `AZradial` from the C_C point (`CCN`, `CCE`) to the end point of the street station line (`EN`, `EE`).
   (c) Compares the absolute value of the input value of `R2` with the computed distance `Dist2` for being equal to zero within a tolerance of 0.005 feet (1.524 mm), and if not sets `IERR` to be 2 and aborts; else continues.
   (d) Sets the azimuth of the tangent line of the street station line to be normal to the computed azimuth `AZradial` by adding or subtracting 90° depending on the sign of `R2`.

4. Queries the value of `Dist1` for being greater than zero within a tolerance of 0.005 feet (1.524 mm), and if so:
   (a) Sets `AZ2` to be equal to `AZ1` minus half π plus `ANG` which could be positive or negative, and then continues with Step 6; else
   (b) Sets `AZ2` to be equal to `AZradial` plus `ANG`, and if `R2` is negative reverses the direction `AZ2`.
       `AZ2` is the direction of the center line of a straight hammerhead, or of the tangent line to a curved hammerhead at the end point of the street station line (`EN`, `EE`).

5. Queries the value of `R1` for being zero within a tolerance of 0.005 feet (1.524 mm), and:
   (a) If it is zero (straight hammerhead), computes the coordinates of the points P5 through P10, and P15 through P18 along the hammerhead, excluding the fillet arcs (see Figures 10.2a and 10.2b) by starting from the endpoint of the street station line (`EN`, `EE`) and calling the `icptl` procedure to traverse along `AZ2` with the appropriate plus and offset distances.
   (b) If it is not zero (curved hammerhead), the procedure:
      • Computes the direction `AZcc` of the radial line of the curved hammerhead from the endpoint of the street station line (`EN`, `EE`) towards the center point (`CN`, `CE`) of the hammerhead arc.
      • Computes the coordinates of the said center point (`CN`, `CE`).
      • Reverses the said radial azimuth.

• Computes the central angles and directions from the hammerhead center point (`CN`, `CE`) towards the endpoints of the curved arcs corresponding to the arcs `AL` and `AR`, and then calls the `icptl` procedure to compute the coordinates of the said arc endpoints.
• Calls the `icptl` procedure to compute the coordinates of the hammerhead stubs as per Figure 10.2c.

6. Computes the coordinates of the fillet arcs as indicated by the conditions and rules (program routines A through P) of the decision Table 10.2b. The procedure queries:
   (a) The radii `R1` and `R2` for having a zero or nonzero values which results in four major condition groups.
   (b) The fillet curve radii `RL` and `RR` in each said group for having a zero or nonzero value, thus again resulting in four condition subgroups for each said group, for a total of 16 referred to as Routine A through Routine P.

7. Restores the input values of the parameters that should have been positive in case any of them had been negative and were forced to be positive, and the terminates.

10.3   Bubble Design

At times in order to maximize, or to best utilize as some designers might say, a parcel of land which is being subdivided into residential lots a so called street bubble, bulb or cluster court may be introduced into the design of a street alignment. Some samples of such bubbles are shown in Figure 10.3a. Many additional bubble designs may be generated depending on the overall design layout of the land development project. Such bubbles may be classified into two general groups, corner bubbles as those labeled C2 and E1 in the said figure, and in-line bubbles as the remainder bubbles of the said figure. Depending on the magnitude of the radius of street alignment, bubble C1 could be considered a corner bubble if the said radius is relatively small. The `icBubble` procedure presented below enables the designer to create the bubbles shown on the said figure except the one labeled D1 for the reason identified later on in this section. This is not because of impossibility or of any major difficulty but basically for the simplification of the input data.

Before addressing the `icBubble` procedure the reader should note the following regarding the bubbles of Figure 10.3a:

• The C2 bubble has two circular fillet arcs at its entrance, while the B1 fillet has a similar fillet at the right side, and no fillet at its left side, and all other bubbles have straight line fillets. All these conditions can be accommodated by the subject procedure.
• Some of the bubbles have islands and some do not. The `icBubble` procedure creates only the bubble itself and

Table 10.2b Hammerhead Fillet Arcs

	Given	Solution Conditions															
		1	2	3	4	5	6	7	8	9	10	11	12	13	14	15	16
Hammer Head	Straight	×	×	×	×	×	×	×	×								
	Curved									×	×	×	×	×	×	×	×
Station Line	Straight	×	×	×	×					×	×	×	×				
	Curved					×	×	×	×					×	×	×	×
Left Fillet	RL=0	×				×				×				×			
	RL<>0		×				×				×				×		
Right Fillet	RR=0			×				×				×				×	
	RR<>0				×				×				×				×
Program Routines for Fillet Points	P1			C	D			G	H			K	L			O	P
	P2			C	D			G	H			K	L			O	P
	P3			C	D			G	H			K	L			O	P
	P12	A	B			E	F			I	J			M	N		
	P13	A	B			E	F			I	J			M	N		
	P14	A	B			E	F			I	J			M	N		

	Fillet	Intersect
A	No left fillet	The left edge of the street station line with the left edge of the hammerhead to set the coordinates of points P12, P13, and P14
B	Left fillet	A line parallel to the station line with a line parallel to the hammerhead and both the said lines passing through the center point of the left fillet curve to set the coordinates of points P12, P13, and P14
C	No right fillet	The right edge of the street station line with the left edge of the hammerhead to set the coordinates of points P1, P2, and P3
D	Right fillet	A line parallel to the station line with a line parallel to the hammerhead and both said lines passing through the center point of the right fillet curve to set the coordinates of points P12, P13, and P14
E	No left fillet	The left edge of the street station line with the left edge of the hammerhead to set the coordinates of points P12, P13, and P14
F	Left fillet	A curve with its center point at (CCN, CCE) with a line parallel to the hammerhead and both line and curve passing through the center point of the right fillet curve to set the coordinates of points P12, P13, and P14
G	No right fillet	The right edge of the street station line with the left edge of the hammerhead to set the coordinates of points P1, P2, and P3
H	Right fillet	A curve with its center point at (CCN, CCE) with a line parallel to the hammerhead and both line and curve passing through the center point of the right fillet curve to set the coordinates of points P12, P13, and P14
I	No left fillet	The left edge of the street station line with the left edge of the hammerhead to set the coordinates of points P12, P13, and P14
J	Left fillet	A curve with its center point at (CNORTH, CEAST) with a line parallel to the station line and both the said line and curve passing through the center point of the left fillet curve to set the coordinates of points P12, P13, and P14
K	No right fillet	The right edge of the street station line with the left edge of the hammerhead to set the coordinates of points P12, P13, and P14
L	Right fillet	A curve with its center point at (CNORTH, CEAST) with a line parallel to the station line and both said line and curve passing through the center point of the left fillet curve to set the coordinates of points P12, P13, and P14
M	No left fillet	The left edge of the street station line with the left edge of the hammerhead to set the coordinates of points P12, P13, and P14
N	Left fillet	A curve with its center point at (CCN, CCE) with a curve with its center point at (CNORTH, CEAST) and both curves passing through the center point of the right fillet curve to set the coordinates of points P12, P13, and P14
O	No right	The right edge of the street station line with the left edge of the hammerhead to set the coordinates of points P1, P2, and P3
P	Right fillet	A curve with its center point at (CCN, CCE) with a curve with its center point at (CNORTH, CEAST) and both curves passing through the center point of the right fillet curve to set the coordinates of points P1, P2, and P3

not its associated island, if there is one. Such islands may be introduced after the creation of the bubble by a variety of the other procedures presented in this publication.

- Some of the bubbles are oblong, that is they are composed of the main bubble arc and a tangent line on either or both sides, while others are composed of only the bubble arc. The fillet arcs or lines may be present in either case. What

Figure 10.3a   Some Typical
Street Bubbles

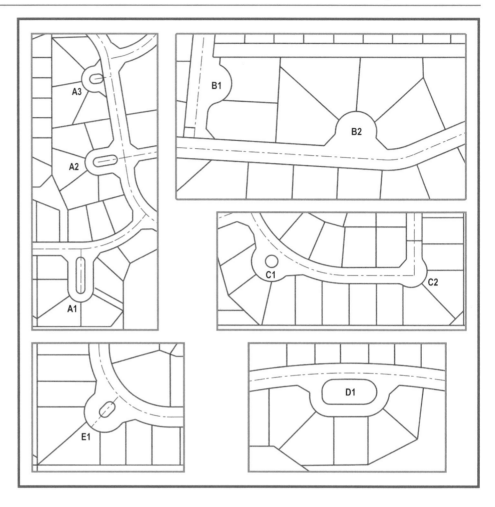

makes the bubble at D1 incompatible with the subject procedure is that the bubble is not composed of a single bubble arc but of two bubble arcs separated by a common tangent line.
- The street alignment may be composed of a straight line (bubble B1), two intersecting lines (bubble C2), a circular curve (bubbles C1, D1, and E1), a tangent line and a curve (bubbles A1, A2, A3, and B2), or two compound or reverse curves.
- The street alignment curves of bubbles A1, A2, A3, B2, C1, D1, and E1 are convex, which is the most common case for a bubble. However, they could be concave as that of bubble D1, provided it is not incompatible for the reason stated above.

In the `icBubble` procedure a bubble is composed of five basic elements, three of which are pairs of similar type as presented in the four generic bubble cases in Figure 10.3b. These elements are:

1. The pair of **street elements** referred to as *street element one* and *street element two*. These two elements define the street alignment with street element one being the start of the alignment and street element two being the end, and which are connected at a common point refereed to as point P_A. It is noted that:
   - The street alignment refers to the right-of-way side of the street to which the bubble is attached and not to the street's center line or any other station line.
   - Either street elements or both can be a straight line or circular arc, and they need not be tangent to each other, although in most cases except in corner lots they are. That is, a bubble may straddle two street elements (reference is made to a statement of caution under the fillet arcs below).
   - If the street alignment is composed of only one element, then this element should be broken in two elements preferably at the point of intersection of the street alignment with the bubble's center line (see below).
2. The **bubble's center line** is a straight line which starts at a point on one of the two street elements referred to as point P_B, and proceeds towards the end of the bubble, referred to as point P_{10}, generally along a direction radial or normal to the said street element. Note that:
   - The direction of the bubble's center line could deviate from the abovesaid normality.

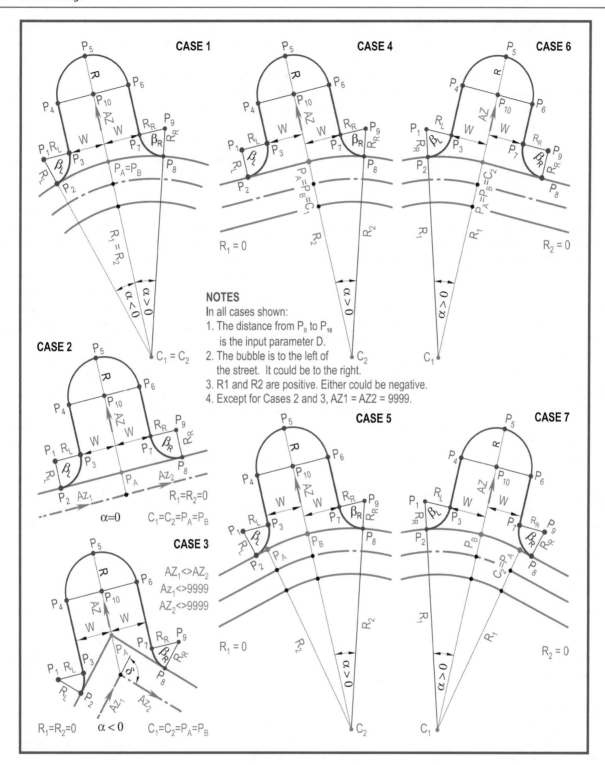

Figure 10.3b   The Geometry of the Basic Street Bubble Types 1 through 7

- The bubble's center line cannot be curved, and it may be directed to either side of a street alignment. That is the bubble may be to the left or to the right of the street alignment along the progression of the street alignment.

- Points P_A and P_B could coincide.
- Point P_{10} could be located on the opposite side of the alignment from point P_B.
3. The **bubble's arc** the center point of which is located at the end point of the bubble's center line.

4. The pair of the **bubble's sides** referred to as the *left side* and the *right side* which are two straight lines tangent to the bubble's arc. Note that:

   - The left side of the bubble must always interact with street element one, and the right side must always interact with street element two. If both sides are to interact with the same street element, then the said element should be broken into the required two street elements. This definition of left and right side is dependent on the interaction of a side with a street element regardless of the side of the street alignment on which the bubble is located.
   - Depending on the use of the optional fillets (see below) either of both of the said sides may be absent.

5. The pair of **optional fillets** either of which can be either a straight line chamfer or circular arc tangent to a bubble side and a street element. Note the following words of caution:

   - If the bubble arc is rather close to a street element so that the two intersect, and the radius of the bubble's arc is large and the fillets are also large, it is possible that the fillet will be applied between the bubble's arc and the street element without a bubble side.
   - If the common point between the two street elements is located somewhere within the extent influenced by a fillet, the end product may not be as desired. In such case a manual intervention with one of the other procedures presented in this publication may be necessary.

Figures 10.3b and 10.3c display the 13 basic bubble type conditions that can be generated with the `icBubble` procedure. In each of these conditions the bubble is shown to be on the left of the alignment, and the curved street elements are all shown to be clockwise arcs. However, this is not always the real life condition. The total number of potential conditions that could be encountered by the subject procedure is dependent on the:

(a) Street alignment for which the said basic conditions yield thirteen (13) variations;

(b) Bubble sides, left and right, each of which can interact with a street element differently from the other, thus yielding two (2) variations;

(c) Fillet type which can be a line, an arc, or absent, thus yielding three (3) variations;

(d) Fillet position between a bubble side and a street element each of which can be a line or an arc, thus yielding four (4) variations;

(e) Bubble's location with respect to the street alignment, left or right, thus yielding two (2) variations, and

(f) Points P_A and P_B position with respect to each other, that is whether they coincide or one is back or ahead of the other, thus yielding three (3) variations.

Thus, the number of case variations that could be encountered is $13 \times 2 \times 3 \times 4 \times 2 \times 3 = 1872$ excluding the variations pertaining to the words of caution regarding the fillets under paragraph 5 above. To accommodate these potential encounters the following sign conventions and point annotations have been defined:

- The directions of the two street elements should point along alignment's progression if they are line features, as well as the rotation of the said elements if they are curved features.

- A positive radius of a street element denotes a clockwise rotation, and a negative radius denotes a counterclockwise rotation as the street alignment progresses from street element one towards street element two.

- The radii of the two fillets and of the bubble arc are specified as positive values but internally in the procedure they are assigned the abovesaid sign convention and they are returned as such to the user without overriding the input positive values.

- Point P_1 refers to the center point of the left fillet, point P_2 refers to the point of tangency of the left fillet with street element one, and point P_3 refers to the point of tangency of the left fillet with the left side of the bubble. If the left fillet is a line then points P_1 and P_2 are the same and the said three points represent the intersection of the left fillet with street element one and the left side of the bubble. If there is no left fillet, all the said points are the same and denote the intersection of the bubble's left side with street element one.

- Point P_4 refers to the point of tangency of the bubble's left side with its arc. If there is no left side then points P_4 and P_3 are the same, and if there is no left side and no fillet (the bubble arc and street element one intersect) then the first said four points are the same.

- Point P_5 refers to the point along the arc of the bubble at the extension of the center line of the bubble.

- Points P_6, P_7, P_8, and P_9 correspond to said points P_4, P_3, P_2, and P_1, respectively, for the fight side of the bubble replacing street element one with street element two.

The above-stated sign conventions and point identifications are indicated in Figures 10.3b, 10.3c, 10.3d, 10.3e, and 10.3f.

A special notice should be paid to Figure 10.3f which in addition to the said sign convention and point notation presents four special bubble conditions. These pertain to the instances in which the arc of the bubble intersects either or both of the street elements. In these instances depending on the size of the bubble's arc and the size of the fillets it is possible for a fillet to be introduced between:

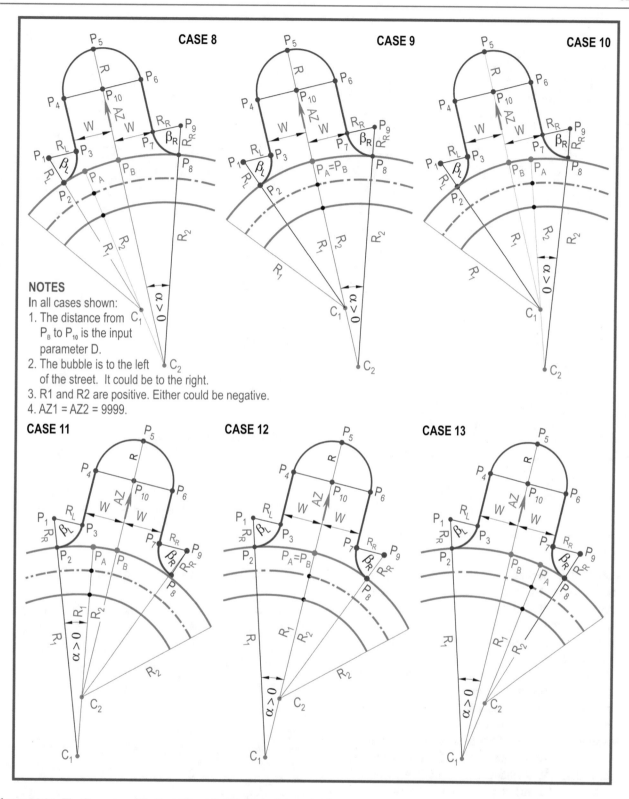

Figure 10.3c The Geometry of the Basic Street Bubble Types 8 through 13

- A bubble line side and its corresponding street element as indicated for the bubbles on the left side of the street alignment in parts (A), (B), and (C) and for the left side of the bubble in part (D) of Figure 10.3f, or
- The bubble's arc and the street element as indicated for the bubbles on the right side of the street alignment in

each of the four parts of Figure 10.3f, and for the right side of the bubble in part (D) of said figure.

Although this possibility exists, the reader is alerted that the `icBubble` procedure in it current form introduces fillets only between the bubble's arc and a street ele-

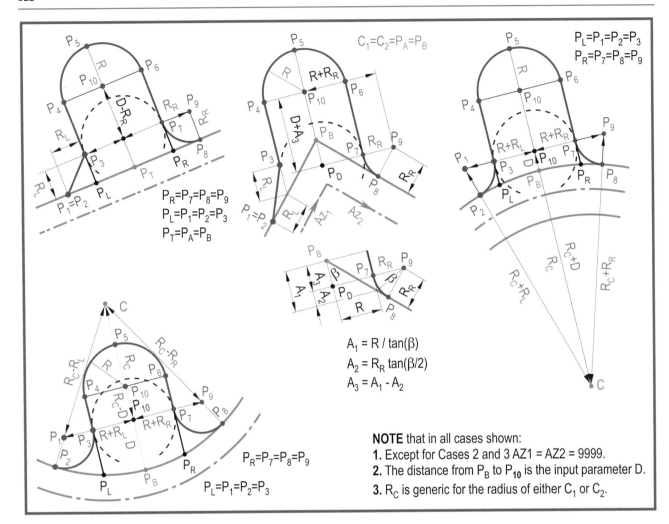

Figure 10.3d Additional Street Bubble Conventions

$$A_1 = R / \tan(\beta)$$
$$A_2 = R_R \tan(\beta/2)$$
$$A_3 = A_1 - A_2$$

NOTE that in all cases shown:
1. Except for Cases 2 and 3 AZ1 = AZ2 = 9999.
2. The distance from P_B to P_{10} is the input parameter D.
3. R_C is generic for the radius of either C_1 or C_2.

ment. If it is desired to introduce a small straight side between the arc of the bubble and the street element when the two intersect, then the bubble that is generated by the subject procedure may be modified by the user by use of any of the other procedures that have been presented in this publication.

Although line fillets are not shown in Figures 10.3f, they could be inserted following the same sign convention, point annotation and restrictions as those of an arc fillet. If fillets are not to be introduced, the points to be generated are those of the intersection of the bubble's arc with the street element (the small black dots in Figure 10.3f). Again manual modifications as stated above may be used to intersect a bubble's small straight side with a street element.

```
Sub icBubble  (PAN, PAE, PBN, PBE, C1N, C1E, _
               R1, C2N, C2E, R2, AZ, R, D, _
               FLTL, RL, FLTR, RR, AZ1, AZ2, SIDE, _
               PN, PE, Rnew, RLnew, RRnew, _
               IERR, TheCASE)
```

The input parameters to this procedure are the:

PAN, PAE — The north and east coordinates of the common point between the first and second street elements. It could be:
- The point of intersection of two street elements either of which could be a straight line or a circular arc.
- The PC point if the first street element is a line and the second street element is a curve.
- The PT point if the first street element is a curve and the second street element is a line.

PBN, PBE — The north and east coordinates of the start point of the bubble's center line, and which point could be located on either the first or second street element.

C1N, C1E — The north and east coordinates of the center point (C_1) of the first street element, if it is a curve, or the same as the (PAN, PAE) point if it is a line.

R1 — The radius of the first street element if it is a curve, or zero if it is a line (see Note 1).

C2N, C2E — The north and east coordinates of the center point (C_2) of the second street element, if it is a curve, or the same as the (PAN, PAE) point if it is a line.

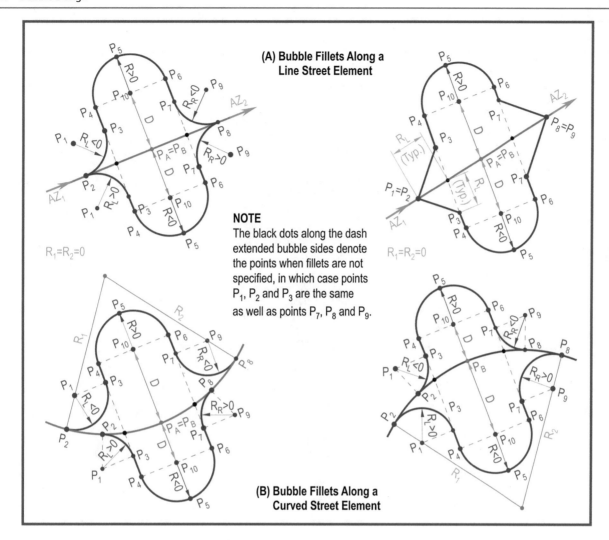

Figure 10.3e Sign Convention for Fillet Radii Applied Internally and Returned to the Calling Program

R2	The radius of the second street element if it is a curve, or zero if it is a line (see Note 1).
AZ	The azimuth in radians of the bubble's center line from the (PBN, PBE) point towards the center point of the bubble's arc.
R	The radius of the bubble's arc and half the bubble's width (see Note 2 below).
D	The distance from the (PBN, PBE) point to the center point of the bubble's arc (P_{10}). It could be zero or negative.
FLTL	The type of fillet to be introduced at the left side of the entrance to the bubble.
	C or c Denotes that an arc fillet is to be introduced.
	L or l Denotes that a line fillet is to be introduced.
	N or n Denotes that a fillet is not to be introduced. The bubble's side to intersect the first street element in which case $P_1=P_2=P_3$ would be the same.
	Any other value aborts the procedure. A lower case entry is changed to upper case.
RL	• The fillet's radius if FLTL contains C (see Note 2 below).

	• The length along the bubble's side and along the first street element measured from the intersection of the said side with the said street element if FLTL contains L.
	• A zero does not create a fillet (it is assumed that FLTL should have been N).
FLTR	The type of fillet to be introduced at the right side of the entrance to the bubble.
	A or a Denotes that an arc fillet is to be introduced.
	L or l Denotes that a line fillet is to be introduced.
	N or n Denotes that a fillet is not to be introduced. The bubble's side to intersect the second street element in which case $P_7=P_8=P_9$ would be the same.
	Any other value aborts the procedure. A lower case entry is changed to upper case.
RR	• The fillet's radius if FLTR contains C (see Note 2 below).
	• The length along the bubble's side and along the second street element measured from the intersection of the said side with the said street element if FLTR contains L.

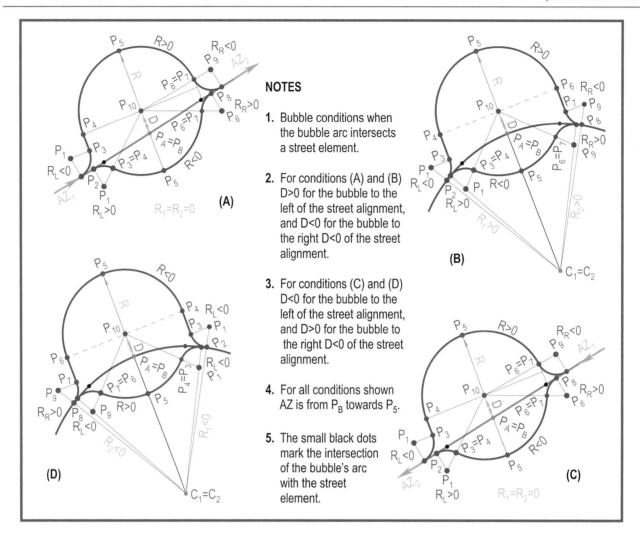

Figure 10.3f Sign Convention for Fillet Radii Applied Internally and Returned to the Calling Program

NOTES

1. Bubble conditions when the bubble arc intersects a street element.

2. For conditions (A) and (B) D>0 for the bubble to the left of the street alignment, and D<0 for the bubble to the right D<0 of the street alignment.

3. For conditions (C) and (D) D<0 for the bubble to the left of the street alignment, and D>0 for the bubble to the right D<0 of the street alignment.

4. For all conditions shown AZ is from P_B towards P_5.

5. The small black dots mark the intersection of the bubble's arc with the street element.

- A zero does not create a fillet (it is assumed that FLTL contains N).

AZ1 — The azimuth in radians of the first street element for Case 1 (see Note 3 below).

AZ2 — The azimuth in radians of the second street element for Case 1 (see Note 3 below).

SIDE — The side of the street, with respect to the progression of the two street elements on which the bubble is to be located.

    L or l   Denotes that the bubble is on the left side of the street.

    R or r   Denotes that the bubble is on the right side of the street.

Any other value aborts the procedure. A lower case entry is changed to upper case.

The returned parameters from this procedure are the:

PN(i), PE(i) — The arrays, dimensioned to 10, that contain the north and east coordinates of the ten points of tangency and center points of the bubble and optional fillets.

Rnew — The input value of the bubble's radius (R) designated as positive for a clockwise rotation and negative for a counterclockwise rotation.

RLnew — The input value of the radius of the left fillet (RL) designated as positive for a clockwise rotation and negative for a counterclockwise rotation. The same for a line fillet or for no fillet.

RLnew — The input value of the radius of the right fillet (RR) designated as positive for a clockwise rotation and negative for a counterclockwise rotation. The same for a line fillet or for no fillet.

IERR — A numeric error flag denoting a warning, or command abortion, i.e.:

0=No input errors have been detected.
1=Fatal error. Improper bubble radius R.
2=Fatal error. Improper left fillet type FLTL.
3=Fatal error. Improper right fillet type FLTR.
4=Fatal error. Improper street side SIDE.
5=Warning. Zero left fillet radius. No fillet.
6=Warning. Zero right fillet radius. No fillet.
7=Warning. R1=R2 within tolerance (set R2=R1).
8=Fatal error. No intersection with left side.

9 = Fatal error. No intersection with right side.

10 = Fatal error. Cannot create left side fillet.

11 = Fatal error. Cannot create right side fillet.

12 = Fatal error. Right arc fillet cannot be set.

7 = Warning. Points 3 and 4 are the same, and points 6 and 7 are the same.

TheCASE The number of the four cases that has been encountered.

NOTES

1. If R1 and R2 are of the same sign and equal to each other in absolute value within a tolerance of 0.005 feet (1.524 mm), R2 is set to be equal to R1. Also, a positive R1 or R2 denotes a clockwise curve rotation, and a negative one denotes a counterclockwise curve rotation with respect to the progression of the two street elements.

2. The radii R, R1 and R2 are expected to be positive. If they are negative, they are forced to be positive for internal program purposes. The input value is not changed.

3. AZ1 and/or AZ2 must have appropriate values for Cases 1, 2 and 3, and AZ1 must point towards street element two, and AZ2 must point away from street element one.. For case 4 these parameters are not used. If they are located along the same line but oppose each other in direction within a tolerance, AZ2 is reversed.

4. Regardless of which side of the street the bubble is located, the bubble's left side is the one that interacts with street element one, and the bubble's right side is the one that interacts with street element two.

5. Although certain input parameters may be changed for computational purposes by the program as indicated in the input data and in these notes, their original values are retained and returned.

Upon invocation, the procedure initializes to zero the parameters of the coordinates to be returned sets IERR to zero, and then:

1. Preserves the values of the input parameters that are altered by the procedure for internal purposes, and thereafter:
   - Checks whether R1 and R2 are of the same sign and equal to each other within a tolerance of 0.005 feet (1.524 mm), and if so they are set to be zero, and if so sets R2 equal to R1 and IERR to be 7;
   - Checks all radii for being equal to zero within a tolerance of 0.005 feet (1.524 mm), and if so they are set to be zero;
   - Changes R, RL and RR to be positive if they are negative;
   - Compares AZ2 with AZ1 to determine whether they are along the same line but opposite in direction

within a tolerance of 0.0000048 radians (0.00 s), and if so sets AZ2 to be the same as AZ1.
   - Changes the values of FLTL, FLTR and SIDE to be in upper case if they are in lower case font;
   - Sets the values of Rnew to be positive, and RLnew and RRnew to be negative if the bubble is to the left of the street alignment, and the opposite if the bubble is to the right of the street alignment;
   - Checks R, FLTL, FLTR and SIDE for potential fatal errors, and if any are found sets the appropriate error code to IERR, resets the input values that have been changed, and aborts; and then
   - Checks RL and RR for being equal to zero and sets IERR to be 5 or 6 as the case may be and continues.

2. Defines a distance D1 to be equal to twice the larger of the fillet radii RL and RR plus three times the radius of the bubble's arc R, then queries R1 and R2 for being equal to zero to determine the case (TheCASE) 1, 2, 3 or 4 that has been encountered, and depending on the case encountered calls the icptl or the icarcoff procedure to establish a point (StartN, StarE) a distance D1 back of point (PAN, PAE) along street element one, and a point (EndN, EndE) a distance D1 ahead of point (PAN, PAE) along street element two.

   As of this version of the icBubble procedure the parameter TheCASE is used by the said procedure in its internal operations, but the said distance and the start and end points are not being used. They are reserved for potential expansion of the said procedure.

3. Calls the icptl procedure to establish the coordinates of the center point of the bubble [PN(10), PE(10)], and the coordinates of the bubble points [PN(4), PE(4)], [PN(5), PE(5)], and [PN(6), PE(6)] by traversing along AZ from point (PBN, PBE) a distance D and R taking into consideration whether the bubble is to the left or to the right of the street alignment.

   At this point the reader should note that the icBubble procedure has two associate procedures, the icBubble1 and the icBubble2, which are called to perform certain operations for the left and for the right sides and fillets of the bubble.

4. Sets the parameter STRT = 1 to denote street element one and calls the icBubble1 procedure to intersect the left side of the bubble with street element one, and the arc of the bubble with the same street element.
   - If the left side of the street, regardless whether it is to be used or not, does not intersect street element one, the input values that have been changed are restored, IERR is set to be 8, and the procedure aborts.
   - If a valid intersection is found the point of intersection (PX1n, PX1e) is preserved

5. Sets the parameter STRT = 2 to denote street element two and calls the icBubble1 procedure to intersect

the night side of the bubble with street element two, and the arc of the bubble with the same street element.

- If the right side of the street, regardless whether it is to be used or not, does not intersect street element two, the input values that have been changed are restored, IERR is set to be 8, and the procedure aborts.
- If a valid intersection is found the point of intersection (PX2n, PX2e) is preserved

6. Queries FLTL for having a value of N, and if so sets the values of points [(PN(1), PE(1)], [(PN(2), PE(2)], and [(PN(3), PE(3)] to be those of point (PX1n, PX1e).

7. Queries FLTR for having a value of N, and if so sets the values of points [(PN(7), PE(7)], [(PN(8), PE(8)], and [(PN(9), PE(9)] to be those of point (PX2n, PX2e).

8. Queries FLTL and FLTR for both having a value of N, and if so terminates the command, that is the bubble is not to have any fillets. Else it continues.

9. Resets the parameter STRT=1 to denote street element one and calls the icBubble2 procedure to introduce a line or arc fillet between the left side of the bubble and street element one.

- If IERR is returned with a value of 12 it is an indication that there is no line side to the bubble and point [(PN(4), PE(4)] is set to be the same as point [(PN(3), PE(3)],
- If IERR is returned with a value other than 0 or 12 it is an indication that a fillet cannot be introduced, in which case the input values that have been changed are restored and the procedure aborts.
- Else it continues.

10. Resets the parameter STRT=2 to denote street element two and calls the icBubble2 procedure to introduce a line or arc fillet between the right side of the bubble and street element two.

- If IERR is returned with a value of 12 it is an indication that there is no line side to the bubble and point [(PN(6), PE(6)] is set to be the same as point [(PN(7), PE(7)],
- Else the input values that have been changed are restored and the procedure terminates.

```
Sub icBubble1 (PAN, PAE, CCN, CCE, AZS, RS, _
               PTN, PTE, CBN, CBE, AZB, _
               RB, SIDE, TheCASE, STRT, _
               PXN, PXE, DIST, INTRSC, PYN, PYE, _
               INTRSX)
```

The input parameters to this procedure are the:

PAN, PAE	The same coordinates as those of icBubble.
CCN, CCE	The same coordinates as those of icBubble for C1N, C1E or C2N, C2E.
AZS	The same as the azimuth of icBubble for AZ1 or AZ2.

RS	The same radius of street element one or two as that of icBubble.
PTN, PTE	The coordinates of the point on the left or right side of the bubble [(PN(4), PE(4)] or [(PN(6), PE(6)], respectively.
CBN, CBE	The coordinates of the center point of the bubble's arc [(PN(10), PE(10)].
AZB	The same azimuth of the bubble's center line as that of icBubble.
RB	The same radius of the bubble as that of icBubble.
SIDE	The same side of the street alignment that the bubble is positioned as that of icBubble.
TheCASE	The case 1, 2, 3 or 4 as determined by icBubble.
STRT	1 to denote street element one and 2 to denote street element two.

The returned parameters from this procedure are the:

PXN, PXE	The coordinates of the point of intersection of the left or right bubble side with street element one or two, respectively.
DIST	The line or arc distance from (PAN, PAE) to from (PXN, PXE) with positive being along the progression of the street alignment.
INTRSC	The numeric indicator of the intersection of the bubble side with the street element that has been encountered. 1 = The bubble's side intersects the street element. 2 = The bubble's side does not intersect the street element.
PYN, PYE	The coordinates of the point of intersection of the bubble's arc with street element.
INTRSX	The numeric indicator of the intersection of the arc of the bubble with the street element that has been encountered. 1 = The bubble's arc intersects the street element or is tangent thereto. 2 = The bubble's arc does not intersect the street element.

Upon invocation, the procedure initializes to zero the parameters to be returned, and then:

1. Preserves the values of the input parameters that are altered by the procedure for internal purposes, and thereafter:

2. Queries TheCASE and the STRT parameters and if they imply that two lines are to be intersected:

   2a. Calls the icintrs procedure to intersect the street element defined by the point (PAN, PAE) and the azimuth AZS with the bubble side defined by the point (PTN, PTE) and azimuth AZB to determine the coordinates of point (PXN, PXE).

   2b. Queries the value of the returned parameter GANG.
   - If GANG <> −9999.0 (an intersection has been found) sets INTRSC to 1, calls the icprjct procedure to project the said intersection point on the

said street element to find the distance DIST as defined above, and continues with Step 2c. DIST is computed for use with the start and end points of Step 2 of the icBubble procedure and reserved for future expansion of the said procedure.

- If GANG=−9999.0 (the two lines are parallel), sets INTRSC to 2 and aborts.

2c. Queries the value of the STRT parameter.

- If it is 1 (left side of the bubble) calls the iclincrc procedure to intersect the line street element defined by the point (PAN, PAE) and an azimuth opposing that of AZS with the bubble arc having its center point at (CBN, CBE) and a radius of RB and return the intersection point (PYN, PYE).

- If it is 2 (right side of the bubble) calls the iclincrc procedure as per above but without reversing the AZS azimuth.

3. Queries TheCASE and the STRT parameters and if they imply that the bubble side is to intersect a curved street element:

3a. Calls the iclincrc procedure to intersect the bubble's side defined by the point (PTN, PTE) and the azimuth AZB with the street element arc defined by its center point (CCN, CCE) and radius RS to determine the coordinates of point (PXN, PXE). It is noted that:

- For a clockwise street element and for the bubble on its right side, and for a counterclockwise street element and for the bubble on its left side the azimuth AZB is reversed, while

- For a counterclockwise street element and for the bubble on its left side, and for a clockwise street element and for the bubble on its right side the azimuth AZB is not reversed.

For either case the intersection procedure returns the azimuth AZZ from the center point of the street element towards point the intersection point.

3b. Queries the value of INTRSC and if it is 2 (an intersection has been found)

- Calls the icstrdln procedure to determine the angle ANG from AZS towards AZZ taking into account the possibility of crossing the north direction. It is noted that for the left side of the bubble or street element one ANG is negative and for the right side of the bubble ANG is positive.

- Computes the arc length DIST as defined above from point (PAN, PAE) to the point of intersection (PXN, PXE) by multiplying ANG by RS. DIST is being computed to be used as per Step 2b above.

3c. Calls the iccrccrc procedure to intersect the street element arc defined by its center point (CCN, CCE)

and radius RS with the bubble's arc defined by its center point at (CBN, CBE) and a radius of RB and return the intersection point (PYN, PYE). It is noted that:

- For a counterclockwise street element one and for clockwise street element two the radius RB is set to be negative, while

- For a clockwise street element one and for a counterclockwise street element two the radius RB is set to be positive.

4. The procedure terminates and returns to the icBubble procedure.

Note that the choices that are made in the use of the iclincrc and iccrccrc procedures are in order to select the appropriate of the two potential intersetions.

```
Sub icBubble2  (PAN, PAE, CCN, CCE, AZS, RS,
                PTN, PTE, CBN, CBE, AZB, _
                RB, PXN, PXE, PYN, PYW, SIDE,
                TheCASE, STRT, RF, FLT, _
                INTRSX, _
                PX1n, PX1e, PX2n, PX2e, PX3n,
                PX3e, IERR)
```

The input parameters to this procedure are the:

PAN, PAE	The same coordinates as those of icBubble.
CCN, CCE	The same coordinates as those of icBubble for C1N, C1E or C2N, C2E.
AZS	The same as the azimuth of icBubble for AZ1 or AZ2.
RS	The same radius of street element one or two as that of icBubble.
PTN, PTE	The coordinates of the point on the left or right side of the bubble [(PN(4), PE(4)) or [(PN(6), PE(6)), respectively.
CBN, CBE	The coordinates of the center point of the bubble's arc [(PN(10), PE(10)).
AZB	The same azimuth of the bubble's center line as that of icBubble.
RB	The same radius of the bubble as that of icBubble.
PXN, PXE	The same coordinates as those returned by icBubble1.
PYN, PYE	The same coordinates as those returned by icBubble1.
SIDE	The same side of the street alignment that the bubble is positioned as that of icBubble.
TheCASE	The case 1, 2, 3 or 4 as determined by icBubble.
STRT	1 to denote street element one and 2 to denote street element two.
RF	The same fillet radius as that of icBubble for RL or RR.

FLT   The same fillet type as that of icBubble for FLTL
                   or FLTR.
INTRSX   The numeric indicator the value of which denotes
                   the following:
                   1 or 3   The bubble arc intersects with or is tangent
                            to the street element.
                   2   The bubble arc does not intersect with the street
                            element.

The returned parameters from this procedure are the:

PX1n, PX1e   The coordinates of the center point of the fillet.
PX2n, PX2e   The coordinates of the fillet on street element one
                   or two.
PX3n, PX3e   The coordinates of the fillet on left or right side
                   of the bubble.
IERR   The same numeric indicator as that of
                   icBubble.

Upon invocation, the procedure:

1. Queries FLT for having a value of "A" (arc fillet) or "L"
   (line fillet), and INTRSX for having a value 2 (a fillet
   between the street element and a line bubble side). If
   not, branches to Step 6, and if so continues. In continuing below note that there are four potential conditions
   depending on whether the fillet is an arc or line fillet, and
   on whether the street element is a line or an arc.
2. Queries FLT for having a value of "A" and RS for being
   zero (line street element). If not, branches to Step 3, and
   if so (arc fillet between two straight lines) continues.
   2a. Locates the center point of the fillet arc by intersect-
       ing a line offset from the street element a distance
       OFF1 with a line offset from the center line of the
       bubble a distance OFF2. Note that:
       • OFF1 = RF
       • For the left side of the bubble (STRT = 1)
         OFF2 = RF − RB, and
       • For the right side of the bubble (STRT = 2)
         OFF2 = −RF + RB.
   2b. Calls the icptl procedure to establish a temporary
       point on the said two offset lines by traversing for-
       ward a zero distance with an OFF1 and OFF2
       offset.
   2c. Calls the icintrs procedure to intersect the two
       offset lines and determine the center point of the fil-
       let arc (PX1n, PX1e).
   2d. Calls the icprjct procedure to project the said
       center point on the street element line to locate point
       (PX2n, PX2e), and on the side of the bubble to
       located point (PX3n, PX3e), and returns to the
       icBubble procedure.
3. Queries FLT for having a value of "A" and RS for being
   nonzero (curved street element). If not, branches to Step

4, and if so (arc fillet between a line and an arc)
continues.
3a. Locates the center point of the fillet arc by inter-
    secting a curve colinear to the street element and
    offset thereof a distance OFF1 with a line offset
    from the center line of the bubble a distance OFF2.
    Note that:
    • OFF1 = RS − RF
    • For the left side of the bubble (STRT = 1)
      OFF2 = RF − RB, and
    • For the right side of the bubble (STRT = 2)
      OFF2 = −RF + RB.
3b. Calls the icptl procedure to establish a temporary
    point on the said offset line by traversing forward a
    zero distance with an OFF2 offset.
3c. Calls the iclincrc procedure to intersect the said
    colinear curve with the said offset line and deter-
    mine the center point of the fillet arc (PX1n,
    PX1e). Note that in order to retrieve the proper of
    the two intersections
    • For a clockwise street element reverses the azi-
      muth of the bubble center line when the bubble is
      to the right of the street element, and does not
      reverse it when the bubble is to the left of the said
      element.
    • For a counterclockwise street element reverses
      the azimuth of the bubble center line when the
      bubble is to the left of the street element, and
      does not reverse it when the bubble is to the right
      of the said element.
3d. Queries the indicator INTRSC for being 2 (no inter-
    section) and if so sets IERR to being 10 when
    STRT = 1, and to 11 when STRT = 2.
3e. Calls the icprjcrc procedure to project the said
    center point on the street element curve to locate
    point (PX2n, PX2e).
3f. Calls the icprjct procedure to project the said
    center point on the side of the bubble to locate point
    (PX3n, PX3e), and returns to the icBubble
    procedure.
4. Queries FLT for having a value of "L" and RS for being
   zero (line street element). If not, branches to Step 5, and
   if so (line fillet between two straight lines) continues.
   4a. Sets the distance OFF1 to be:
       • OFF1 = RF for the left side of the bubble when
         the bubble is to the left of the street element, and
         for the right side of the bubble when the bubble
         is to the right of the street element.
       • OFF1 = −RF for the opposite conditions.
   4b. Calls the icptl procedure to traverse from the
       intersection point of the street element with the bub-
       ble side (PXN, PXE):

- Along AZS an OFF1 distance with a zero offset to establish the line fillet point (PX2n, PX2e) on the street element line.
- Along AZB an |RF| distance with a zero offset to establish the line fillet point (PX3n, PX3e) on the bubble side line.

4c. Sets the coordinates of the fillet center point (PX1n, PX1e) to be those of point (PX2n, PX2e), and returns to the icBubble procedure.

5. Queries FLT for containing "L" and RS for being nonzero (curved street element). If not, branches to Step 6, and if so (arc fillet between a line and an arc) continues.

5a. Sets the distance OFF1 as per Step 4a.

5b. Calls the icforce procedure to determine the azimuth from the center point of the street element curve towards the point of intersection of the said element with the side of the bubble.

5c. Calls the icarcoff procedure to traverse along the street element arc an arc length of OFF1 and zero offset to located point (PX2n, PX2e) and then calls the icptl procedure to traverse from the intersection point of the street element with the side of the bubble
- A distance of −RF if the bubble is to the left of the street element, or
- A distance of RF if the bubble is to the right of the street element to locate the point (PX3n, PX3e) on the side of the bubble.

5d. Sets the coordinates of the fillet center point (PX1n, PX1e) to be those of point (PX2n, PX2e), and returns to the icBubble procedure.

6. Queries FLT for having a value of "A" (arc fillet) or "L" (line fillet), and INTRSX for not having a value 2 (a fillet between the street element and the bubble's arc). If not, terminates and returns to the icBubble procedure, and if so continues.

7. Queries FLT for having a value of "A" and RS for being zero (line street element). If not, branches to Step 8, and if so (arc fillet between a line street element and an arc offset from the bubble's arc) continues.

7a. Locates the center point of the fillet arc by intersecting a line offset from the street element a distance OFF1 = RF with a circle colinear to the arc of the bubble and offset therefrom a distance OFF2 = |RF| + |RB|.

7b. Calls the iclincrc procedure to intersect the said offset line with the said colinear curve to establish the center point of the fillet arc (PX1n, PX1e).

7c. Queries the indicator INTRSC for being 2 (no intersection) and if so sets IERR to being 10 when STRT = 1, and to 11 when STRT = 2.

7d. Calls the icprjct procedure to project the said center point on the street element line to locate point (PX2n, PX2e), and on the side of the bubble to

located point (PX3n, PX3e), and returns to the icBubble procedure.

8. Queries FLT for having a value of "A" and RS for being nonzero (curved street element). If not, branches to Step 9, and if so (arc fillet between two arcs) continues.

8a. Locates the center point of the fillet arc by intersecting a curve colinear to the street element and offset therefrom a distance OFF1 with a curve colinear with the arc of the bubble and offset therefrom a distance OFF2. Note that:
- The distance OFF1 = |RS − RF|
- The distance OFF2 is first set to be equal to |RS| + |RF| and then for the left side of the bubble it is changed to OFF2 = −OFF2.

8b. Calls the iccrvcrc procedure to intersect the said two colinear curves to determine the center point of the fillet arc (PX1n, PX1e).

8c. Queries the indicator INTRSC for being 2 (no intersection) and if so sets IERR to being 10 when STRT = 1, and to 11 when STRT = 2.

8d. Calls the icprjcrc procedure to project the said center point on the colinear curve to the street element curve to establish point (PX2n, PX2e), and on the colinear curve to the arc of the bubble to establish point (PX3n, PX3e), and returns to the icBubble procedure.

9. Queries FLT for having a value of "L" and RS for being zero (line street element). If not, branches to Step 10, and if so (line fillet between a line element and the arc of the bubble) continues.

9a. Sets the distances OFF1 and OFF2 to be:
- For the left side of the bubble OFF1 = −|RF| and OFF2 = |RF|, and
- For the right side of the bubble OFF1 = |RF| and OFF2 = −|RF|.

9b. Calls the icptl procedure to traverse from the intersection point of the street element with the arc of the arc of the bubble (PYN, PYE) along AZS a distance OFF1 and with a zero offset distance to establish point (PX2n, PX2e).

9c. Calls the icforce procedure find the azimuth GANG from the center point of the bubble's arc (CBN, CBE) towards the said intersection point (PYN, PYE).

9d. Calls the icarcoff procedure to traverse from the said intersection point (PYN, PYE) along AZS a distance RB with a zero offset distance to establish point (PX3n, PX3e).

9e. Sets the coordinates of the fillet center point (PX1n, PX1e) to be those of point (PX2n, PX2e), and returns to the icBubble procedure.

10. Queries FLT for having a value of "L" and RS for being nonzero (curved street element). If not, returns to the icBubble procedure, and if so (arc fillet between two arcs) continues.

10a. Sets the distance OFF1 as per Step 9a.

10b. Calls the `icforce` procedure to determine the azimuth GANG from the center point of the street element curve towards the point of intersection of the said element with the arc of the bubble point (`PYN, PYE`).

10c. Calls the `icarcoff` procedure to traverse along the street element arc an arc length of OFF1 and zero offset to located point (`PX2n, PX2e`).

10d. Calls the `icforce` procedure to determine the azimuth GANG from the center point of the bubble's arc towards the said intersection point (`PYN, PYE`).

10e. Calls the `icarcoff` procedure to traverse along the bubble's arc an arc length of OFF2 and zero offset to located point (`PX3n, PX3e`).

10f. Sets the coordinates of the fillet center point (`PX1n, PX1e`) to be those of point (`PX2n, PX2e`), and returns to the `icBubble procedure`.

Vertical Parabolic Curves

The preceding chapters address various geometric constructions concerning the creation and positioning of features on a horizontal plane, the said plane being defined by a set of X (east) and Y (north) coordinates. In civil engineering work most all of the geometric work is being carried out in the first quadrant of the said coordinate system, but it really could be carried out in any quadrant provided appropriate negative coordinates are employed. Even though this coordinate system assumes a flat earth for the position of various features (roads, buildings, dams, bridges, tunnels, and other features), the real earth within the confines of this plane is not flat. Thus, provision must be made to account for a third dimension Z (elevation). Thus, in this chapter we are to present certain computing procedures that address the vertical positioning of routes along a horizontal alignment.

It is stated earlier in this publication that travel along a horizontal alignment is made along a series of straight lines (tangents) and circular curves (curves) with or without intermediate spirals (spirals). As a vehicle travels along such a path, it has to ascend and descend terrain hills and valleys. In order to attain a smooth and comfortable travel along the vertical plane of the said horizontal alignment, the vertical alignment is composed of a series of straight lines (grades) and parabolic curves (curves). The vertical alignment is located on a coordinate system in which the X axis (station line) follows the path of the horizontal alignment, and the Z axis (elevation) is normal to the plane of the horizontal alignment along the horizontal alignment (see Figure 11.0a). The measure along the station line is that of the horizontal alignment stationing, and the measure along the Z axis is the elevation value from an assumed datum elevation.

In civil engineering work, much like with the horizontal alignment, the control points of a vertical alignment are the PVC, the PVI, and the PVT (see Figure 11.0b), of which:

- The PVC is the point at which the back grade line is tangent to the parabolic curve,
- The PVI is the point of intersection of the back and forward grade lines, and

- The PVT is the point at which the parabolic curve is tangent to the forward grade line.

In reference to the last said figure, it is noted that:

- In this chapter the Y and Z axes are used interchangeably to imply the same axis.
- The back and forward grade lines (tangents) are positive when pointing towards the positive direction of the Y axis (upwards), and negative when pointing towards the negative direction of the Y axis (downwards). It is possible for both grade lines to be either positive, or negative.
- The length of the vertical curve is measured along the projection of the curve on the station line (X axis), and not along the curve itself.
- There are two types of vertical curves, the symmetrical vertical curve in which the PVI is located midway along the said length as indicated in Figure 11.0b, and the unsymmetrical vertical curve as indicated in Figure 11.0c, in which the PVI is located off the midpoint of the curve's length.
- The symmetrical vertical curve is the standard or most common vertical curve encountered in civil engineering work.
- An unsymmetrical (occasionally referred to as unequal length) vertical curve may be encountered from time to time in order to fit certain conditions. In these cases, an unsymmetrical vertical curve may be broken down into two symmetrical vertical curves that are compounded at an in-between point. That is, they are tangent to the same grade line at a specific point. Thomas F. Hickerson in his book *Route Location and Design*, McGraw-Hill Book Company, 1959 presents a discussion in treating such a vertical curve as a single parabolic curve. However, since such curves are encountered in special cases, the designer may be afforded an easier control in treating a special condition with two compound vertical curves by varying the location and elevation of the point of compounding and the common grade.

Figure 11.0a The Vertical
Alignment Plane

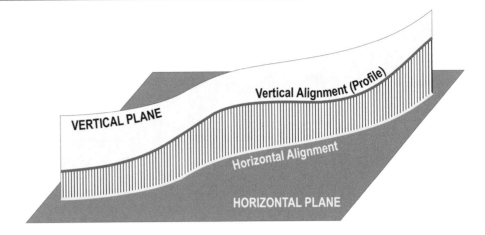

Figure 11.0b The Control
Points of a Vertical Curve

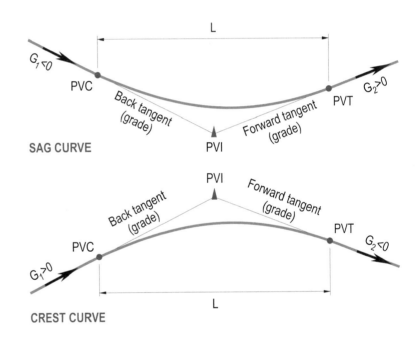

Figure 11.0c Unsymmetrical
Vertical Curve

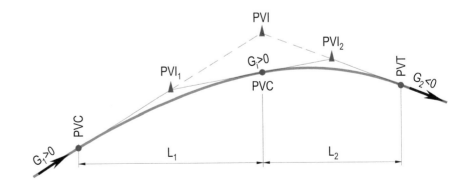

After a general discussion of the general parabolic curve, the procedures presented in this chapter are divided into three groups, those that:

- Define the control and other points along a vertical curve when the key parameters of a vertical curve (grades and lengths) are known.
- Fit a grade line or a vertical curve through a series of field observed station and elevation points.
- Perform certain construction operation involving vertical curves and grade lines such as tangencies, intersections and curve generation.

11.1   The Parabola

When intersecting a circular cone with a plane surface, the intersection generates a curve referred as a conic section. Depending on the orientation of the intersecting plane surface with respect to the base of the cone, there are four types of conic sections that may generate the conic curves (or just plain conics) as indicated in Figure 11.1a. Thus, the generated curve is:

- A circle if the intersecting plane is parallel to the base of the cone.
- An ellipse if the intersecting plane is inclined with respect to the base of the cone, and does not intersect the base of the cone within the confines of the cone.
- A parabola if the intersecting plane is inclined with respect to the base of the cone, and in addition it intersects the base of the cone within the confines of the cone.
- A hyperbola if the intersecting plane is normal to the base of the cone, and in addition it intersects the base of the cone within the confines of the cone.

In the cases of the parabola and hyperbola, the line of intersection of the intersecting plane with the base of the cone is not a part of the generated curve. Furthermore, circles and ellipses may be generated by intersecting a cylinder with a plane surface.

11.1.1   The Conic Curve

There are different methodologies of addressing the conic curves, one of which (see Figure 11.1b) defines a conic curve as the locus of points having a constant ratio (eccentricity) of the distance from a fixed point (focus) to the normal distance from a fixed line (directrix). The value of this constant ratio defines whether the conic curve is a circle, ellipse, parabola, or hyperbola. In the said figure the directrix is along the X axis and the focus point **F** along the Y axis at $x = 0$. However, the directrix could very well be along the Y axis and the focus along the X axis so that the conic curve could be rotated 90°. Figure 11.1b is used herein because it represents the cases encountered in civil engineering work.

In the said figure, let us consider any point **T** along the conic curve, for which curve the eccentricity is expressed as

$$e = \frac{T_F}{T_D} \quad \text{or} \quad T_F = e\left(T_D\right) \tag{11.1a}$$

where

T_F is referred to as the focal radius, and
T_D is referred to as the directrix offset.

For a conic curve, there are an infinite number of focal radii and directrix offsets, which may be expressed in terms of Cartesian coordinates as

$$T_F = \sqrt{\left(y - P\right)^2 + x^2} \quad \text{and} \quad T_D = y \tag{11.1b}$$

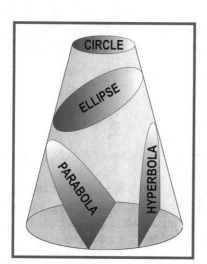

Figure 11.1a   The Conic Sections

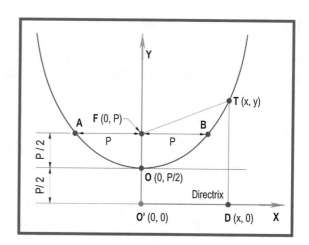

Figure 11.1b   The Conic Curve

By substituting Equations 11.1b in Equation 11.1a and squaring both sides of Equation 11.1a we have

$$\sqrt{(y-P)^2 + x^2} = e.y$$
$$y^2 + P^2 - 2Py + x^2 - e^2 y^2 = 0 \qquad (11.1c)$$

or

$$y^2\left(1-e^2\right) - 2Py + x^2 + P^2 = 0 \qquad (11.1d)$$

with Equation 11.1d being the general equation form of a conic curve.

If we now let x = 0 we can obtain the Y axis intercepts of the conic curve as

$$y^2\left(1-e^2\right) - 2Py + P^2 = 0$$
$$\left[y(1-e)-P\right]\left[y(1+e)-P\right] = 0 \qquad (11.1e)$$
$$y_1 = \frac{P}{1-e} \quad \text{and} \quad y_2 = \frac{P}{1+e}$$

and if we then let y = 0 we can obtain the X axis intercepts of the conic curve as

$$x2 = -P2$$
$$x = \pm P\sqrt{-1} \quad \text{or} \quad x = \pm Pi \qquad (11.1f)$$

which means that for the orientation of the conic curve as per Figure 11.1b the X axis intercepts are imaginary, and nor real. In reviewing Equation 11.1e we see that if:

- **e** = 0, we have a circle with the directrix being located at infinity from its center point.
- **e** < 1, we have two real Y axis intercepts, and the curve is an ellipse.
- **e** = 1, we have a parabola with one real Y axis intercept, and one approaching infinity.
- **e** > 1, we have one hyperbola with a positive real Y axis intercept, and another with a negative real Y axis intercept.

11.1.2 The Parabolic Curve

Now, let us turn our attention to the general conic Equation 11.1d and substitute 1 for the value of eccentricity **e** in order to obtain the general form of the parabolic curve as

$$y^2\left(1-1^2\right) - 2Py + x^2 + P^2 = 0$$
$$-2Py + x^2 + P^2 = 0 \qquad (11.1g)$$
$$x^2 = 2Py - P^2 \quad \text{or} \quad x^2 = 2P\left(y - \frac{1}{2}P\right)$$

where x and y are measured with respect to the origin of the X and Y axes at point O' of Figure 11.1b. If we translate the origin of these axes to point O of the said figure as to

coincide with the vertex of the parabolic curve, then the corresponding measures would be

$$x = x_n$$

and

$$y = y_n + \frac{1}{2}P$$

where x_n and y_n are the measures with respect to the origin of the X and Y axes at point O of Figure 11.1b. Thus, Equation 11.1g becomes

$$x_n{}^2 = 2P\left(y_n + \frac{1}{2}P - \frac{1}{2}P\right) = 2Py_n \qquad (11.1h)$$

If for the sake of simplicity we now drop off the subscripts from x_n and y^n, and solve for y instead of x, we get

$$y = \frac{1}{2P}x^2 \qquad (11.1i)$$

In Figure 11.1b let C represent the focal radius to any point **T**, so that C equals the length T_F, and it may be represented in terms of x and y coordinates using the Pythagorean theorem as

$$C^2 = \left(y - \frac{1}{2}P\right)^2 + x^2$$
$$C^2 = y^2 - Py + \frac{1}{4}P^2 + x^2 \qquad (11.1j)$$

Substituting Equation 11.1h, without the subscripts, in Equation 11.1i we get

$$C^2 = y^2 - Py + \frac{1}{4}P^2 + 2Py = y^2 + Py + \frac{1}{4}P^2$$
$$= \left(y + \frac{1}{2}P\right)^2 \qquad (11.1k)$$
$$C = y + \frac{1}{2}P$$

We now may notice that the value of C as expressed in Equation 11.1k is equal to the lengths TF and TD with y being measured from the X axis at O and not O'. Thus,

$$C = T_F = T_D$$

or

$$\frac{T_F}{T_D} = 1 \qquad (11.1l)$$

which satisfies the definition of the conic curve and that of the parabolic curve.

In the above equations of this subsection, the origin of the coordinates has been at the vertex of the parabolic curve. In order to make the parabolic Equation 11.1i more generic, we can translate the X and Y axes so that

$$y = Y - h$$

and

$$x = X - k$$

Thus, the said equation now becomes

$$Y - h = \frac{(X-K)^2}{2P} = \frac{X^2 - 2kX + k^2}{2P} = \frac{X^2}{2P} - \frac{kX}{P} + \frac{k^2}{2P}$$

$$Y = \left(h + \frac{k^2}{2P}\right) - \left(\frac{k}{P}\right)X + \left(\frac{1}{2P}\right)X^2 \quad (11.1m)$$

$$Y = A_0 + A_1 X + A_2 X^2$$

where

$$A_0 = h + \frac{k^2}{2P} \quad (11.1n1)$$

$$A_1 = -\frac{k}{P} \quad (11.1n2)$$

$$A_2 = -\frac{1}{2P} \quad (11.1n3)$$

If we solve Equation 11.1n3 for P we get

$$P = \frac{1}{2A_2} \quad (11.1o1)$$

If we substitute the value of P of Equation 11.1o1 for that in Equation 11.1n2 and then we solve for k we get

$$A_1 = -2A_2 k$$

$$k = \frac{-A_1}{2A_2} \quad (11.1o2)$$

If we substitute the value of P and k of Equations 11.1o1 and 11.1o2, respectively, for those in Equation 11.1n1 and then we solve for h we get

$$A_o = h + \left(\frac{-A_1}{2A_2}\right)^2 (4A_2) = h + \frac{A_1^2}{4A_2}$$

$$h = A_o - \frac{A_1^2}{4A_2} \quad (11.1o3)$$

If we now review the three Equations of each of 11.1n and 11.1o, we can summarize that

1a. If $A_2>0$, hence $P>0$, then the parabola is concave up (sag curve).
1b. If $A_2<0$, hence $P<0$, then the parabola is concave down (crest curve).
1c. If $A_2=0$, hence $P=0$, then a straight line exists in which A_1 is the slope of the line and A_0 is the Y axis intercept.
2a. If both A_1 and A_2 are either positive, or negative, hence $P>0$ and $k<0$, then the vertex of the parabola is to the left of the Y axis.

2b. If one of A_1 and A_2 is positive and the other is negative, or the reverse, hence $P<0$ and $k>0$, then the vertex of the parabola is to the right of the Y axis.
2c. If $A_1=0$, hence $k=0$, then vertex of the parabola lies on the Y axis, and the parabola is concave up or down depending on the sign of A_2.
3a. If $A_0>0$, then the focus point of the parabola lies above the X axis.
3b. If $A_0<0$, then the focus point of the parabola lies below the X axis.
3c. If $A_0=0$, then the focus point of the parabola lies on the X axis.

A graphic representation of the above comments, excepting those for which the vertex or the focus point of a parabola may line on either axis, is shown in Figure 11.1c.

11.1.3 The Vertical Curve

Now let us return to civil engineering work. When working with a parabolic curve, the civil engineer is concerned only with a portion of the said curve in which the grade lines (slopes) at the endpoints of the desired portion are considered to be known as indicted in Figure 11.1d. Note that at this time we are not concerned about symmetrical and unsymmetrical vertical curves. In reviewing this figure note the following:

• The vertical curve shown is a sag curve (concave up). It could very well have been shown as a crest curve (concave down).
• The engineer has no knowledge of the location of the focal point of the parabola, nor of its eccentricity.
• The vertical curve begins at the PVC point, and terminates at the PVT point along the parabola.
• The slope of the curve at the PVC is G_1, and the slope of the curve at the PVT is G_2.
• D_G is the algebraic difference of the said two slopes defined as $D_G = G_2 - G_1$.
• The tangent lines of the said two slopes intersect at the PVI point which is located midway along the L distance from the PVC towards the PVT (for symmetric vertical curves).
• The vertex point of the parabola V may be located anywhere between the PVC and the PVT points, or it may be located outside the limits of the vertical curve if both grades are either positive, or negative. At this point, the slope is equal to zero. When the vertex point is located within the limits of the vertical curve, it is referred to as the turning point of the vertical curve. It is also referred as the low point for a sag curve, or the high point for a crest curve.

Figure 11.1c Potential Position of Parabolas in a Cartesian Coordinate System

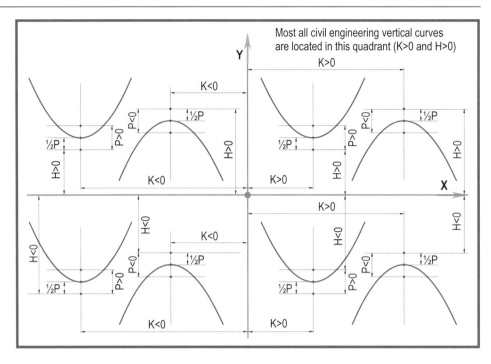

Figure 11.1d The Vertical Curve (sag curve shown)

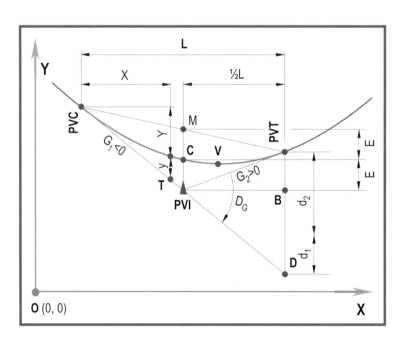

The slope at any point along the parabola is defined by the first derivative of the parabolic cure Equation 11.1m as

$$\frac{dY}{dX} = A_1 + 2A_2 X \qquad (11.1p)$$

and the rate of change of slope of the parabolic curve is defined by its second derivative as

$$\frac{d^2Y}{dX^2} = 2A_2 \qquad (11.1q)$$

which implies that the rate of change in elevation is constant.

In Figure 11.1d, the origin of the X–Y coordinate system is shown in the low left corner more or less resembling an actual case in which the elevation datum and the beginning of the station line are away and back from the vicinity of the vertical curve under consideration. However, let us consider for a moment that the origin of the coordinate system is translated to the PVC point with the X and Y measures as indicated in the said figure, and y being measured from the back tangent line extended forward towards the curve.

The rate of change R in Equation 11.1r4 is similar to the K factor referred to in the AASHTO and other publications and used with certain tables in vertical curve selection. Refer to Sect. 11.1.7 of this chapter.

From Equation 11.1p,

$$at \quad X = 0 \quad the\ slope\ is\ G_1, \quad or \quad A_1 = G_1 \quad (11.1r1)$$

and at $X = L$ the slope is G_2, or

$$G_2 = A_1 + 2A_2L = G_1 + 2A_2L \quad (11.1r2)$$

$$or \quad G_2 = G_1 + 2A_2L \quad (11.1r3)$$

$$or \quad A_2 = (G_2 - G_1)/2L \quad (11.1r4)$$

and if we substitute the value A_2 from Equation 11.1r3 in Equation 11.1q we can express the rate of change in elevation as

$$\frac{d^2Y}{dX^2} = R = \frac{G_2 - G_1}{L} \quad (11.1r5)$$

If in Equation 11.1m, repeated below, we substitute the values of $A1$ and A_2 from Equations 11.1r1 and 11.1r3, respectively, we get

$$Y = A_0 + A_1X + A_2X^2$$

$$Y = A_0 + G_1X + \left(\frac{G_2 - G_1}{2L}\right)X^2 \quad (11.1s1)$$

Since at $X=0$, or at the PVC, $Y=A_0$, and thus $Y=0$

$$Y = G_1X + \left(\frac{G_2 - G_1}{2L}\right)X^2 \quad (11.1s2)$$

The measure y from the back tangent to the curve may be found from the ratio

$$G_1 = \frac{Y + y}{X} \quad (11.1t1)$$

or

$$Y = G_1 - y \quad (11.1t2)$$

If we now substitute the value of Y from Equation 11.1t2 in Equation 11.1s2 and we then solve for y, we obtain

$$y = -\left(\frac{G_2 - G_1}{2L}\right)X^2 \quad (11.1t3)$$

At times it is desired to measure y in the opposite direction from the measure of Y, and replace X with x. Hence, Equation 11.1t3 becomes

$$y = -\left(\frac{G_2 - G_1}{2L}\right)x^2 \quad (11.1t4)$$

At the location of the PVI, $y=E$ and $x=l/2$, thus

$$E = \left(\frac{G_2 - G_1}{2L}\right)\left(\frac{L}{2}\right)^2 = \left(\frac{G_2 - G_1}{8}\right)L \quad (11.1u1)$$

From Figure 11.1d we see that

$$d_2 + d_1 = \frac{L}{2}(G_2 - G_1) \quad (11.1u2)$$

and that

$$\frac{2E}{L/2} = \frac{d_2 + d_2}{L}$$

$$4E = d_2 + d_2 = \left(\frac{G_2 - G_1}{2}\right)L \quad (11.1u3)$$

$$E = \left(\frac{G_2 - G_1}{8}\right)L$$

so that Equations 11.1u1 and 11.1u3 are the same.

Based on the above, the shape of the civil engineering vertical curve is dependent firstly upon the slopes of the back and forward grade lines G_1 and G_2, and secondly upon the length of the curve L. Shown in Figure 11.1e are the six potential shapes of a vertical curve. In this figure, the red line work indicates a unit of the vertical alignment as treated by the various procedures that are presented later on in this chapter, and the blue line work indicates the start portion of the subsequent vertical alignment unit. Thus:

- Point PVL is the start point of the first vertical alignment unit, or the PVT point of the previous vertical alignment unit for all other units.
- Point PVC is the start point of a vertical curve of a vertical alignment unit.
- Point PVT is the end point of a vertical curve of a vertical alignment unit, and the PVL point of the next alignment unit.
- The first PVL point could be the first PVC point, and the PVT point of a vertical curve could be the PVC point of the following vertical curve.
- The vertical alignment terminates at the endpoint of the blue line at the end of the last vertical alignment unit. This point could be a repetition of the last PVT point.

With these in mind, the elevation E_x of any point between a PVL and a PVC is

$$E_x = E_a + G_1x_1 \quad (11.1v1)$$

where

x_1 is the horizontal distance of the said any point from point PVL, and

E_a is the elevation of the PVL point,

Figure 11.1e The Six Potential Shapes of a Vertical Curve

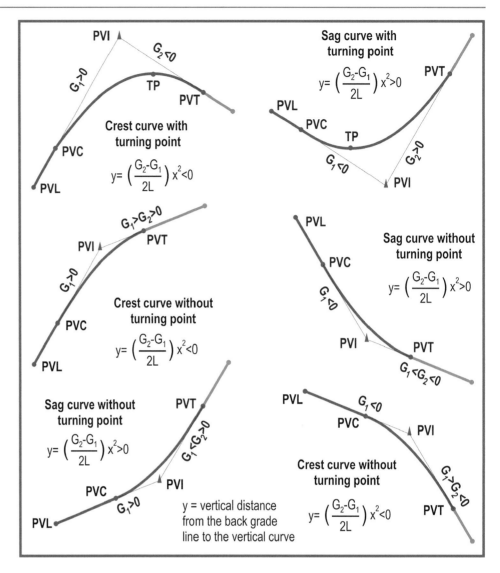

y = vertical distance from the back grade line to the vertical curve

and the elevation E_x of any point within the limits of a vertical curve from PVC to PVT, by use of Equation 11.1t4, is

$$E_x = E_b + G_1 x_2 + \left(\frac{G_2 - G_1}{2L}\right) x_2^2 \qquad (11.1v2)$$

Where

x_2 is the horizontal distance of the said any point from point PVC, and
E_b is the elevation of the PVC point.

The right most term in Equation 11.1v2 may be represented by the lower case y letter to represent the vertical distance from the back grade line to the vertical curve, so that

$$y = \left(\frac{G_2 - G_1}{2L}\right) x_2^2 \qquad (11.1v3)$$

or also at times y is expressed as

$$y = \frac{1}{2} R x_2^2 \qquad (11.1v4)$$

where

$$R = \frac{G_2 - G_1}{L} \qquad (11.1v5)$$

denotes the rate of change of grade, and corresponds to the focal distance P of the parabola.

In reviewing the magnitude of the grades of the tangent lines with respect to zero (whether they are positive or negative) and with respect to each other, Equation 11.1v2 holds true for all cases of Figure 11.1e, since the signs of G_1 and G_2 will dictate the sign of the y distance.

11.1.4 Turning Point of a Vertical Curve

In Figure 11.1e it is noted that a vertical curve may have a turning point (TP), high or low point, only when G_1 and G_2 are opposite in sign. The location of the turning point is then at the point at which the slope of the tangent line to the curve has a value of zero. Thus, differentiating Equation 11.1v2 and setting it equal to zero we obtain

Refer to Equations 11.3k8 and 11.3k9 in Sect. 11.3.2 for locating the station and elevation of the high or low point of a vertical curve using the parabolic curve equation $f(x) = A_0 + A_1 X + A_2 X^2$.

$$\frac{dE_x}{dx} = G_1 + \left(\frac{G_2 - G_1}{L}\right) x_2 = 0 \qquad (11.1w1)$$

or

$$x_2 = \frac{G_1}{\left(\frac{G_2 - G_1}{L}\right)} = +\frac{G_1 L}{G_1 - G_2} \qquad (11.1w2)$$

11.1.5 The Parabolic Length of a Vertical Curve

It has been previously stated that the length of a vertical alignment for the determination of elevations is measured along the alignment's projection on the x axis, or station line. However, at times it is desired to know the exact, or almost as close to the exact as possible, the actual length along vertical alignment itself. Reasons for such a length include the:

- Determination of quantities of the various components of the pavement structure, exclusive of earthwork quantities which are determined by different means.
- Assessment of pavement conditions along a roadway by various new technologies, one of which is the that of ground radar penetration (GPR) and in which conditions are observed along the surface of the pavement as a vehicle travels along the roadway. In addition to GPR, there are certain other survey operations that are obtained by traveling along the pavement and thus measuring length along the actual pavement and not along its projection on the pavement's station line.

To compute the actual length of the vertical alignment along its straight line segments is an easy task. One may generate a straight line equation, or simply determine the difference in elevation between two stations, and then use the Pythagorean theorem to compute the length of surface, the hypotenuse. To find the length of a segment of the vertical alignment along a vertical curve, the programmer may employ integral calculus to determine the length of an infinitesimal length along the vertical curve using Equation 11.1v4 in which the subscript of x may be dropped. Thus, from integral calculus the infinitesimal length of curve may be expressed as

$$\frac{ds}{dx} = \sqrt{dx^2 + dy^2} \qquad (11.1x1)$$

or

$$ds = \sqrt{1 + \left(\frac{dy}{dx}\right)^2}\, dx \qquad (11.1x2)$$

We may now integrate between two limits say $S1$ and $S2$, each representing a station value to solve for the distance s as follows:

$$s = \int_{S1}^{S2} \sqrt{1 + \left(\frac{dy}{dx}\right)^2}\, dx \qquad (11.1x3)$$

If we substitute the derivative of the said Equation 11.1v4 in Equation 11.1x3, we get

$$s = \int_{S1}^{S2} \sqrt{1 + (Rx)^2}\, dx \qquad (11.1x4)$$

where R is defined by Equation 11.1v5. To solve Equation 11.1x4 we can let $k = (Rx)^2$ to change Equation 11.1x5.

$$s = \int_{S1}^{S2} (1 + k)^{1/2}\, dx \qquad (11.1x5)$$

The mathematical solution of Equation 11.1x5 is in essence an approximation of an infinitesimal increment of x. That is, the arc length of the parabola is the summation of very small chords. As seen in the description of the `icVcurve` procedure, curve elevations can be computed at any desired station interval. Thus, the smaller the interval, the smoother the curve's appearance and the more precise its arc length. As an example consider an 800 foot long curve +10 % of back grade and 15 % forward grade. Thus, for a station increment of

•	50 feet, the arc length is	806.306868
•	25	806.307106
•	10	806.307173
•	1 foot, the arc length is	806.307185

Thus, for all intents and purposes a 25 foot station increment will yield a precise enough length of parabolic arc.

11.1.6 Offset Vertical Curves

Quite often it becomes necessary to create an offset vertical curve such that every point along the offset vertical curve has an elevation which is lower, or higher than the corresponding point of the main or base vertical curve by the same elevation differential. For examples consider the profile of a:

- Curb or gutter line of a city street which is to be located at a constant cross slope from the station line alignment.
- Utility line, such as a water main that has to be located at a specific minimum depth below a surface that is being defined by a vertical curve that denotes the ground surface.

- Theoretical (imaginary) profile above a roadway surface below which profile an aerial cable or structure cannot attain.

In such cases, any of the procedures presented in the remaining of this chapter may be used provided that the tangent grade lines are lowered, or raised by the same increment without a change of the grade, and the length of the base and offset vertical curves is the same.

11.1.7 Stopping Sight Distance

The stopping sight distance is the distance traveled by a vehicle from the time its driver sees an obstacle and stops the vehicle without impacting the obstacle. This distance is composed of three time segments: (**a**) the time that it takes the driver to become aware of the obstacle, (**b**) the time that it takes the driver to react and apply the brakes, and (**c**) the time it takes the vehicle to come to a full stop position after the application of the brakes. The first two depend solely on the driver and they are jointly referred to as the *break reaction time*, while the third is dependent upon the speed of travel, the vehicle type and condition, the weather conditions and the pavement type and condition as they affect the coefficient of friction, and the geometry of the roadway upon which the travel occurs as it affects visibility due to a horizontal and/or vertical alignment curvature referred to a the *breaking time*. Along a horizontal or vertical straightway there is no visibility issue other than that of the earth's curvature which for all intents and purposes can be overlooked. The geometric issues due to horizontal curves are different and much simpler than those due to vertical curves.

Since the intent of this publication is not to delve into the aspects of highway safety and appropriate design, we will address only the aspects of the geometry as they relate to the stopping sight distance, and particularly those that pertain to the geometry of the vertical curvature of the roadway alignment.

As of the date of this publication the AASHTO recommended values of h_1 and h_2 are:

$h_1 = 3.5\ feet\ (1.07\ m)$
$h_2 = 0.5\ feet\ (0.15\ m)\ for\ stopping$
$h_2 = 4.25\ feet\ (1.3\ m)\ for\ passing.$

The stopping sight distance along a vertical curve is dependent upon whether the vertical curve is a crest curve as in curves A, B, or F of Figure 11.1e, or a sag curve as in curves C, D, or E of the same figure. Displayed in Figures 11.1f and 11.1g are samples of crest curves and in Figures 11.1h and 11.1i are samples of sag curves representing the conditions: (**a**) when the stopping sight distance **S** is less than the length **L** of the vertical curve, and (**b**) when the said **S** is greater than the said **L**. As seen in these figures:

1. For a crest curve the controlling criteria are the height h_1 of the driver's eye above the pavement, the height h_2 of the obstacle to be seen above the pavement, and the geometry of the curve defined by the **K** factor which is the inverse of the rate of change **R** as defined by Equation 11.1r4, or $K = L/A$ where $A = |G_2 - G_1|$ the difference in forward and back grades.
2. For a sag curve the controlling criteria are the headlight range of the traveling vehicle defined by the height h_1 of the vehicle's headlights above the pavement, the angle β (one half the angle of the headlight beam provided the headlights are properly adjusted to the vehicle), and the geometry of the curve again expressed by the said factor **K**.

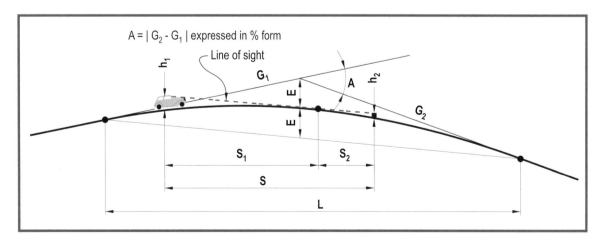

Figure 11.1f Stopping Sight Distance on a Crest Vertical Curve with S < L

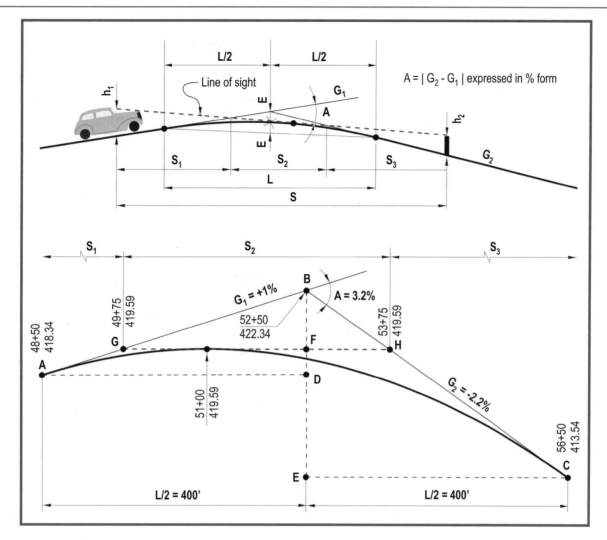

Figure 11.1g   Stopping Sight Distance on a Crest Vertical Curve with S > L

3. The line of sight is the line connecting the driver's eye with the top of the obstacle and for crest curves it is tangent at some point to the curve. For a sag curve the line of sight is the line joining the midpoint of the headlight with the point on intersection of the top ray of the headlight beam with the curve.

4. Generally h_1 is greater than h_2, and h_1 is greater for trucks then for a cars. Since the line of sight is tangent to the curve, the shorter h_1 yields the shorter stopping sight distance and is therefore the more conservative one.

5. All other symbols in the said figures are the same as described previously for a vertical curve.

The algebraic difference A of $G_2 - G_1$ is positive for sag curves and negative for crest curves, but in the sight distance formulas the absolute value of A is used.

The so called K factor is equal to the length of the vertical curve L divided by A as defined above.

AASHTO and other governmental agencies publish tables and graphs that relate the speed of travel and driver reaction time studies to the said **K** factor. In the design of a new roadway the designer having knowledge of the two grade lines and the speed of travel with aide of these tables and charts selects the required stopping sight distance **S** and by the use of the formulas presented below computes, or by graph or table lookup selects the length **L** of the vertical curve. However, for an existing roadway the vertical curve length is fairly much a fixed entity. Therefore the need exists to determine the allowable speed by computing **S** if **L** is known, and then determining the allowable speed by use of the said tables or charts. Presented below are the formulas for computing both **S** and **L** given the other.

11.1.7.1   General Commentary

It is necessary to point out that the contents of this section pertain to individual vertical curves of a vertical alignment and not to vertical alignments with compound, reversed and/

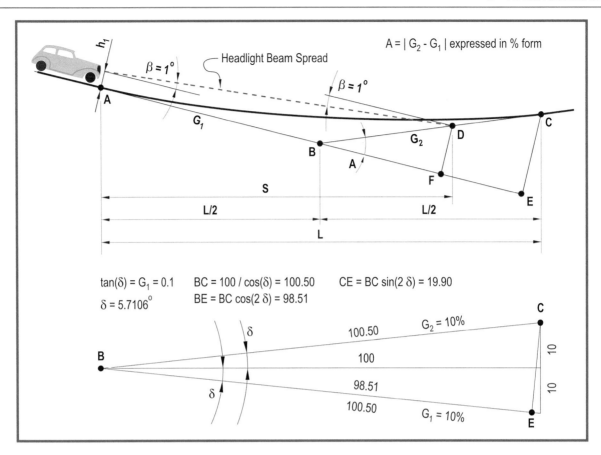

Figure 11.1h   Stopping Sight Distance on a Sag Vertical Curve with S < L

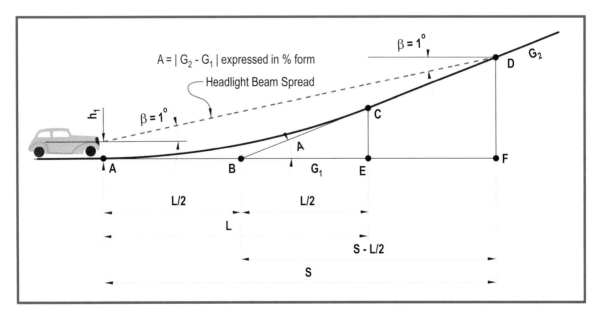

Figure 11.1i   Stopping Sight Distance on a Sag Vertical Curve with S > L

or broken back vertical curves. This is because in such cases if the stopping sight distance S of a vertical curve is greater than its length L, as is to be discussed below, it could encroach on the vertical curve ahead.

11.1.7.2 Crest Curves with $S<L$ (See Figure 11.1f)

From Equation 11.1t3 we see that the offset distance from the tangent line to the parabolic curve varies as the square of the distance from the point of tangency. This parabolic curve property is used in the equations of this section. Thus, using E from Equation 11.1u3:

$$\frac{h_1}{E} = \frac{S_1^2}{(L/2)^2} = \frac{4S_1^2}{L^2} \qquad (11.1y1)$$

and similarly

$$\frac{h_2}{E} = \frac{S_2^2}{(L/2)^2} = \frac{4S_2^2}{L^2} \qquad (11.1y2)$$

or

$$S_1 = \sqrt{\frac{L^2 h_1}{4E}} \quad \text{and} \quad S_2 = \sqrt{\frac{L^2 h_2}{4E}} \qquad (11.1y3)$$

If in Equation 11.1u3 we express the grades G_1 and G_2 in percent form and let $G_2 - G_1$ be A, then $E = AL/800$. So with S being equal to the sum of S_1 and S_2 in Equation 11.1y3 we get

$$S = \sqrt{\frac{800h_1 L^2}{4AL}} + \sqrt{\frac{800h_2 L^2}{4AL}} = \sqrt{\frac{200h_1 L}{A}} + \sqrt{\frac{200h_2 L}{A}}$$
$$S = \left(\sqrt{200h_1}\right) + \sqrt{200h_2}\sqrt{\frac{L}{A}} \qquad (11.1y4)$$

Solving the above equation for L and calling it L_M for minimum vertical curve length we get

$$L_M = \frac{AS^2}{100\left(\sqrt{2h_1} + \sqrt{2h_2}\right)^2} \qquad (11.1y5)$$

which is the standard AASHTO equation for the stopping sight distance when $S<L$.

For a value of 3.5 for h_1 and 0.5 for h_2 the denominator of Equation 11.1y5 becomes 1329 to the nearest whole number.

11.1.7.3 Crest Curves with $S>L$ (See Figure 11.1g)

For crest curves when the stopping sight distance is greater than the length of the vertical curve reference is made to Figure 11.1g which is composed of two parts. In the upper part a vertical curve and its two tangent grade lines are shown with the said tangents extending beyond the curve limits to display the total length of the stopping sight distance. At the lower portion of this figure the vertical curve portion of the top part has been enlarged and exaggerated for clarity purposes to justify a comment to be made shortly. In these two parts of the said figure note the following:

1. Whereas in the top part the vertical alignment is fairly much to the same horizontal and vertical "scale" (more or less), in the lower part the vertical curve is that of an 800 foot long vertical curve with the indicated grade lines of +1 % and −2.2 %, and for which the vertical scale is three times larger than the horizontal scale in relative terms and neither scale is to any specific graphic units.
2. Whereas in the top part the vehicle is located so that the line of sight is tangent to the curve and inclined in a downward orientation, in the lower part the said vehicle is located so that the line of sight is horizontal and tangent to the curve at its high point. This has been done to simplify the discussion the comment made below. It is noted that the line of sight and hence the stopping sight distance is composed of three parts, S_1, S_2 and S_3.
3. One of the properties of a parabolic vertical curve is that the line of sight, the tangent line at a point of the curve, intersects the back and forward tangent grade lines G_1 and G_2 at the end points A and C of the said curve at points G and H, respectively, with the horizontal distance between G and H being equal to one half the length L of the vertical curve between A and C. That is, the said segment S_2 of the stopping sight distance shown in the top part of the figure is equal to one half the length of the vertical curve L.
4. In the lower part of Figure 11.1g the triangles BFG and BDA are similar, as are the triangles BFH and BEC. Hence

$$\frac{BF}{BD} = \frac{GF}{AD} \quad \text{and} \quad \frac{BF}{BE} = \frac{FH}{EC} \qquad (11.1y6)$$

Solving for GF and FH we get

$$GF = (BF)\frac{(AD)}{(BD)} \quad \text{and} \quad FH = (BF)\frac{(EC)}{(BE)} \qquad (11.1y7)$$

Noting that (a) $S_2 = GF + FH$, (b) $L/2 = AD = EC$ and that the two fractions in Equation 11.1y7 represent the inverse of the grades G_1 and G_2, we have

$$S_2 = (BF)\frac{(AD)}{(BD)} + + (BF)\left(\frac{EC}{BE}\right)$$

$$S_2 = \frac{(BF)}{(G_1)} + \left(\frac{BF}{G_2}\right) \qquad (11.1\text{y}8)$$

From Figure 11.1g we see that $G_1=0.01$, $G_2=-0.022$ and $BF=422.34-419.59=2.75$. If we substitute these values in Equation 11.1y8 we have $S_2=(2.75/0.01)+(2.75/0.022)=400.00$ which is equal to one half the 800 curve length. Note that for this test we did not use the percent form of the grades, and that $FH=AG$ and that $GF=EC$.

Now back to the top part of Figure 11.1g in which $S=S_1+S_2+S_3$, or

$$S = 100\left(\frac{h_1}{G_1}\right) + \frac{L}{2} + 100\left(\frac{h_2}{G_2}\right) \qquad (11.1\text{y}9)$$

Noting that the rate of change between grades G_2 and G is equal and opposite to that between G_1 and G we can set the derivative of Equation 11.1y9 to zero to minimize the distance S. Thus

$$\frac{dS}{dG} = \frac{h_1}{(G_1)^2} - \frac{h_2}{(G_2)^2} = 0 \qquad (11.1\text{y}10)$$

or

$$G_2 = G_1\sqrt{\frac{h_2}{h_1}} \qquad (11.1\text{y}11)$$

Now consider the parameter A as a scalar measure or as an angle composed of the sum of the angle represented by the slopes G_1 and G_2 so that

$$A_{scalar} = G_1 + G_2 = G_1 + G_1\sqrt{\frac{h_2}{h_1}} = G_1\left(\frac{\sqrt{h_2}+\sqrt{h_1}}{\sqrt{h_1}}\right) \qquad (11.1\text{y}12)$$

Solving Equation 11.1y12 for G_1, and then substituting this value of G_1 in Equation 11.1y11 we have

$$G_1 = A\left(\frac{\sqrt{h_1}}{\sqrt{h_2}+\sqrt{h_1}}\right) \quad \text{and} \quad G_2 = A\left(\frac{\sqrt{h_2}}{\sqrt{h_2}+\sqrt{h_1}}\right) \qquad (11.1\text{y}13)$$

If we now substitute the values of G_1 and G_2 from the Equations 11.1y13 in Equation 11.1y9 and we factor the common parameters in the right two terms we have

$$S = \frac{L}{2} + \frac{100}{A}\left[\frac{h_1\left(\sqrt{h_1}+\sqrt{h_2}\right)}{\sqrt{h_1}} + \frac{h_2\left(\sqrt{h_1}+\sqrt{h_2}\right)}{\sqrt{h_2}}\right] \qquad (11.1\text{y}14)$$

By multiplying the left and right terms within the brackets with the denominators of the other of the said two terms to create a common denominator and expanding the terms within the parentheses and combining terms we get

The design of vertical curves must also consider the comfort of the occupants of the vehicle. This is more important in sag curves, and generally is not considered in crest curves. The comfort is dependent upon the speed of travel and the sharpness of the curve. The sharper the curve, the greater the centrifugal force that causes discomfort. This aspect of design is not considered a goal of this publication.

$$S = \frac{L}{2} + \frac{100}{A}\left[h_1 + \frac{2h_1 h_2}{\sqrt{h_1}\sqrt{h_2}} + h_2\right]$$

$$S = \frac{L}{2} + \frac{100}{A}\left[h_1 + 2\sqrt{h_1}\sqrt{h_2} + h_2\right] \qquad (11.1\text{y}15)$$

$$S = \frac{L}{2} + \frac{100}{A}\left(\sqrt{h_1} + \sqrt{h_2}\right)^2$$

If we now substitute the abovesaid value of A for the difference in grades in Equation 11.1y4 and we solve for L by squaring both sides of the said equation, which L we can now call L_M for minimum vertical curve length we get

$$L_M = 2S - \frac{200\left(\sqrt{h_1} + \sqrt{h_2}\right)^2}{A} \qquad (11.1\text{y}16)$$

which is the standard AASHTO equation for the stopping sight distance when $S<L$.

11.1.7.4   Sag Curves with $S<L$ (See Figure 11.1h)

The top part of this figure displays a sag vertical curve with a vehicle entering the curve at point A, the start of the curve, its headlights located a distance of h_1 above the pavement and having a spread angle of 1° intercepting the vertical curve at point C. Much like the crest curves the ratio A/100 represents the vertical difference per foot of length along the vertical curve. Thus

$$CE = \frac{A}{100}\frac{L}{2} \qquad (11.1\text{y}17)$$

Before continuing let us look at the lower part of the said figure in which the right half of the top part has been enlarged to true scale, and in which $G_1=-10\%$, $G_2=+10\%$ and $L=200$. Note that the 10 % grades are a bit on the high side for a highway. The object of this enlargement of the sag curve is to show that the difference between the curve as shown on the upper part and a curve with $G_1=0\%$ and $G_2=20\%$ would be rather negligible especially since the final selection of the vertical curve length would also be based in part on certain empirical observations. Thus, the equations presented below assume that the back tangent line is located along the station line at which the length of the vertical curve and the sight distance are measured.

Since as said previously from Equation 11.1t3 we see that the offset distance from the tangent line to the curve varies as the square of the distance from the point of tangency, and noting that the triangles DFB and CEB are similar we have

$$DF = \left(\frac{S}{L}\right)^2 CE = \left(\frac{S}{L}\right)^2 \frac{A}{100}\frac{L}{2}$$

$$= \left(\frac{S}{L}\right)^2 \frac{AL}{200} \qquad (11.1y18)$$

From the same figure we also see that

$$DF = h_1 + S\tan(\beta) = h_1 + S\tan(1) = h_1 + 0.0175S \quad (11.1y19)$$

If we next equate 11.1y18 and 11.1y19 and reduce terms we get

$$\frac{S^2 A}{200L} = h_1 + 0.0175S \ or \ AS^2 = 200Lh_1 + 3.5SL \quad (11.1y20)$$

If we solve Equation 11.1y20 for L we get the standard AASHTO equation

$$L = \frac{AS^2}{200h_1 + 3.5S} \qquad (11.1y21)$$

in which the height of the headlights h_1 is 2 feet (610 mm). The stopping distance for a given curve length may be determined by solving Equation 11.1y20 for S so that

$$AS^2 - 3.5LS = 200h_1L$$

$$S^2 - \frac{3.5L}{A}S = \frac{200h_1L}{A} \qquad (11.1y22)$$

We can now solve Equation 11.1y22 for S by completing the square on the left side, or by moving the right side of the said equation to the left and using the quadratic equation solution formula, both methods of which are shown below, first by completing the square

$$S^2 - \frac{3.5L}{A}S + \left(\frac{1.75L}{A}\right)^2 = \frac{200h_1L}{A} + \left(\frac{1.75L}{A}\right)^2$$

$$\left(S - \frac{1.75L}{A}\right)^2 = \frac{200h_1L}{A} + \left(\frac{1.75L}{A}\right)^2 \qquad (11.1y23)$$

$$S = 1.75\frac{L}{A} \pm \sqrt{200h_1\frac{L}{A} + 3.0625\left(\frac{L}{A}\right)^2}$$

and next by the quadratic equation solution formula

$$S^2 - \frac{3.5L}{A}S - \frac{200h_1L}{A} = 0$$

$$S = \frac{3.5L \pm \sqrt{\left(\frac{3.5L}{A}S\right)^2 + 4\left(\frac{200h_1L}{A}\right)}}{2} \qquad (11.1y24)$$

$$S = 1.75\frac{L}{A} \pm \sqrt{3.0625\left(\frac{L}{A}\right)^2 + 200h_1\frac{L}{A}}$$

both methods yielding the same result of two roots for S. Of the two solutions, the one with the negative radical can be disregarded since the resulting root of the equation would be negative and hence not realistic.

The above equations are based on a β angle of 1° as indicated in Equation 11.1y19 and in which the value of $200 \ tan(\beta) = 3.4910$ has been rounded up to 3.5. If it is desired to employ an angle value for β other than 1°, then the constants in Equations 11.1y21 through 11.1y24 should be changed as shown below:

• Replace	3.5	with	$200 \ tan(\beta)$
•	1.75		$100 \ tan(\beta)$
•	3.0625		$[100 \ tan(\beta)]^2$

11.1.7.5   Sag Curves with $S > L$ (See Figure 11.1i)

Figure 11.1i is similar to the upper part of Figure 11.1h with two noted differences: (a) the headlight beam spread intercepts the forward tangent grade line beyond the end of the vertical curve (point C), and (b) the entire configuration has been rotated counterclockwise about point B as argued in the side bar of the preceding subsection. There is still a slight imprecision regarding the distances CE and DF but as discussed in the previous subsection it is rather negligible. Thus, from the triangles DFB we have

$$DF = \frac{A}{100}\left(S - \frac{L}{2}\right) \quad and \quad DF = h_1 + S\tan(\beta) \quad (11.1y25)$$

By equating the two sides

$$h_1 + S\tan(\beta) = \frac{A}{100}\left(S - \frac{L}{2}\right) = \frac{A}{100}S - \frac{A}{100}\frac{L}{2} \quad (11.1y26)$$

Solving for L we get

$$L = \frac{200}{A}\frac{A}{100}S - \frac{200}{A}h_1 - \frac{200}{A}S\tan(\beta)$$

$$L = 2S - \frac{200h_1 + 200S\tan(\beta)}{A} \qquad (11.1y27)$$

If we now let $h_1 = 2$ feet (610 mm) and β = 1 we get the standard AASHTO equation for the required curve length for a specified stopping sight distance.

$$L = 2S - \frac{400 + 3.5S}{A} \qquad (11.1y28)$$

The stopping distance for a given curve length may be determined by solving Equation 11.1y27 for S so that

$$L = 2S - \frac{200h_1}{A} - \frac{200}{A}S\tan(\beta) \qquad (11.1y29)$$

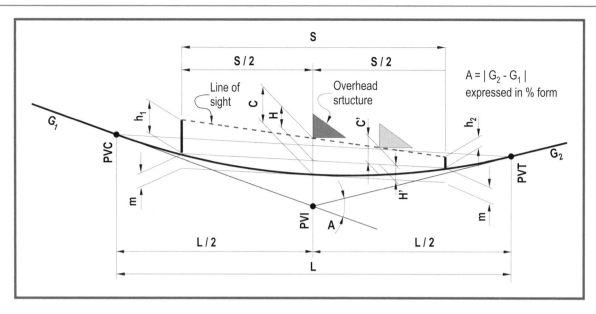

Figure 11.1j Stopping Sight Distance at an Underpass with S<L

$$LA = 2AS - 200h_1 - 200S \tan(\beta)$$
$$LA + 200h_1 = S\left[2A - 200 \tan(\beta)\right] \qquad (11.1y30)$$

or

$$S = \frac{LA + 200h_1}{2A - 200 \tan(\beta)} \qquad (11.1y31)$$

If we now let $h_1 = 2$ feet (610 mm) and $\beta = 1$ we get the equation for the stopping distance for a given curve length as

$$S = \frac{LA + 400}{2A - 3.5} \qquad (11.1y32)$$

11.1.7.6 Underpass at Sag Curves with S<L (See Figure 11.1j)

Figure 11.1j portrays the condition in which the bridge fascia beam interferes with the line of sight. If there is no bridge, the night conditions become the controlling factor in which case it is the headlights that control the distance. However, a bridge affects the driver's line of vision day and night. Thus, the heights h_1 and h_2 used for the crest curves must be taken in consideration. Regarding h_1 it was previously stated that it pertains to the driver of a passenger vehicle and not to that of a truck driver who seats higher and thus has a longer line of vision. However, due to this higher line of vision the chances are that a bridge would be more critical for a truck driver than for the driver of a passenger vehicle. Thus, a higher h_1 value should be used.

When a roadway traverses under another roadway carried by a bridge it is possible that the underside of the bridge might affect stopping sight distance. This is of no concern along crest curves as the bridge would be much higher than the top of the obstacle. However, this is not the case in sag curves.

In Figures 11.1j and 11.1k the PVI the overhead structure (shown in purple) is located directly under the critical clearance point, and C denotes the minimum clearance between the said structure and the curve. This is done to simplify the relationships presented below. If the overpass structure is located off the location of the PVI (shown in tan color with C' and H' in Figure 11.1j) the following relationships should be valid provided the structure is not located more than 200 feet (60 meters) from the PVI. Regarding this figure note that:

- The minimum stopping sight distance S or ling of sight occurs when the line of sight is bisected at the location of the minimum clearance line regardless of the size of the heights h_1 and h_2, so that $H = (h1 + h2)/2$.
- The line joining the locations where h_1 and h_2 meet the curve is parallel to the long chord of the vertical curve, that is, the line joining the PVC and the PVT.
- The line tangent to the curve at the location of the PVI is also parallel to the said long chord, and it is offset from

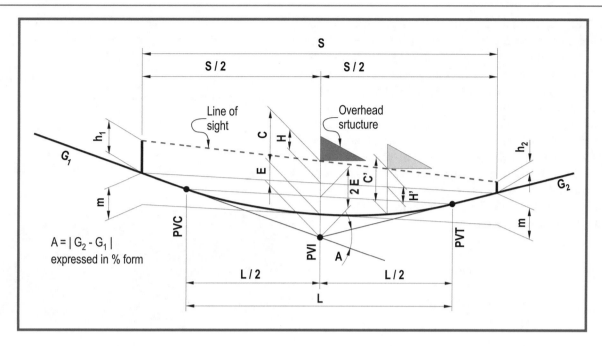

Figure 11.1k   Stopping Sight Distance at an Underpass with S>L

the line between h_1 and h_2 a distance $m = C - H$. If we now look at this line a bit closer we see that it quite the same as sight line for a crest curve when $S < L$ (see Sect. 11.1.7.2) in an upside form, with each of h_1 and h_2 equal to m and with $S_1 = S_2$. Thus, Equation 11.1y4 may be expressed as

$$S = \left(\sqrt{200m} + \sqrt{200m}\right)\sqrt{\frac{L}{A}} = 2\sqrt{m}\sqrt{\frac{200L}{A}}$$

$$S = \sqrt{\frac{800Lm}{A}} = \sqrt{\frac{800L}{A}\left(C - \left(\frac{h_1 + h_2}{2}\right)\right)} \qquad (11.1y33)$$

Solving Equation 11.1y33 for L we get

$$L = \frac{AS^2}{800\left(C - \left(\frac{h_1 + h_2}{2}\right)\right)} \qquad (11.1y34)$$

11.1.7.7   Underpass at Sag Curves with S>L (See Figure 11.1k)

The introductory comments made in the preceding subsection regarding Figure 11.1j hold true for this case. We could use the inverse reference to the crest curve situation with m replacing each of h_1 and h_2 as done in the preceding case, but instead we will use similar triangles. Thus

$$\frac{S/2}{E+m} = \frac{L/2}{2E}$$

$$S = \frac{LE + Lm}{2E} \qquad (11.1y35)$$

and substituting Equation 11.1u3 for E the above equation becomes

$$S = \frac{L}{2} + \frac{Lm}{2E} = \frac{L}{2} + \frac{Lm}{2}\frac{8}{LA} = \frac{L}{2} + \frac{4m}{A} \qquad (11.1y36)$$

If we now substitute the expression for m and expressing the grades (that is, A) in percent from the equation becomes

$$S = \frac{L}{2} + \frac{400}{A}\left(C - \left(\frac{h_1 - h_2}{2}\right)\right) \qquad (11.1y37)$$

We can next solve the above equation for L to get

$$L_M = 2S - \frac{800\left(C - \frac{h_1 - h_2}{2}\right)}{A} \qquad (11.1y38)$$

which is the standard AASHTO equation for the stopping sight distance when $S > L$.

11.1.7.8   Safe Stopping Distance

It was stated at the start of this section that the stopping distance is composed of the sum of the distances traveled during the break reaction time D_R and breaking time D_B. Of these two distance the break reaction distance may be defined as $D_R = VT$, where V is the speed of travel in miles per hour (kilometers per hour), T is the break reaction time, and D_R is in feet (meters).

$$D_R = 5280\frac{feet}{mile}\frac{1}{3600}\frac{hour}{seconds}V\frac{miles}{hour}T \ seconds$$

$$D_R = 147VT feet or VT 0.278 meters \qquad (11.1y39)$$

The breaking distance D_B may be computed from the principles of mechanics that state that the force (F) of the vehicle while traveling is equal to its mass (m) times the deceleration (a), or $F=ma$ or $F=Wa/g$ where W is the weight and g is the gravitational acceleration of 32.2 feet per second square (9.81 m/s²). Also during the deceleration of the vehicle the force may be defined as $F=Wf$ where f is the coefficient of friction which is assumed to be constant during the deceleration process. By equating the two force equations we get $a=fg$. Again from mechanics we know that the distance traveled from a velocity of v feet per second to rest is equal to $v^2/2a$. Thus, the breaking distance on a relatively level grade may be expressed as

$$D_B = \frac{V^2}{2a}$$

$$D_B = \frac{1}{2}\frac{V^2}{a}\left(\frac{miles}{hour}\right)^2\left(\frac{5280\,feet}{mile}\right)^2$$
$$\left(\frac{hour}{3600\ \sec onds}\right)^2\left(\frac{\sec ond^2}{feet}\right) \qquad (11.1y40)$$

$$D_B = 1.0756\frac{V^2}{a}\,feet \quad or \quad D_B = 0.039\frac{V^2}{a}\,meters$$

At times, such as in the case of a sloped pavement (see below) it may be desired to express the breading distance in terms of the coefficient of friction f rather than in terms of deceleration a in which case the breaking distance is

$$D_B = \frac{V^2}{2a} = \frac{V^2}{fg} = 1.0756\frac{V^2}{f\,32.2} = \frac{V^2}{30f}\,feet \quad or$$
$$D_B = \frac{V^2}{254f}\,meters \qquad (11.1y41)$$

Thus, the total stopping distance along a relatively level ground is

$$D = 1.47VT + 1.0756\frac{V^2}{a}\,feet or$$
$$D = 0.278VT + 0.039\frac{V^2}{a}\,meters \qquad (11.1y42)$$

On an uphill roadway grade gravity comes to the aide of the vehicle during the deceleration process but has the opposite effect on a downhill roadway grade. Thus, in Equation 11.1y41 the deceleration a has to be modified by the ratio of the said deceleration a over the gravitational acceleration g and which ratio is added to the algebraic value of the grade line G' which is the previously used grade G in percent form divided by 100. Thus, the breaking distance on a sloped grade line may be expressed as

$$D = 1.47VT + \frac{V^2}{30\left(\left(\frac{a}{32.2}\right)\pm G'\right)}\,feetor$$
$$D = 0.278VT + \frac{V^2}{254\left(\left(\frac{a}{9.81}\right)\pm G'\right)}\,meters \qquad (11.1y43)$$

which is shorter than that of a level ground for an uphill grade (G'>0) and longer for a downhill grade (G'<0).

As of the date of this publication the AASHTO recommended values for reaction break time T and deceleration rate a are:
$T=2.5$ s
$a=11.2$ feet/s² (3.4 m/s²)

It was stated earlier in this section that the procedures that are presented below assume a given vertical curve length and determine the stopping sight distance S which can then be compared with that computed by the appropriate of the above presented equations for the total breaking distance D. Alternatively the programmer could first compute the said distance D using the equations of this subsection and then use it as the value of S of the stopping sight distance equations presented in the preceding subsections and modify the said procedures to solve for L rather than for S. This is left as an exercise for the reader or user of the procedures that are provided.

In the equations of this subsection the parameters V, T and a have been used. Of these three, the velocity V is a given parameter of known or theoretical value. The values of the reaction time T and the acceleration a are empirical observations of various tests and experiments.

11.1.8 Passing Sight Distance

In the discussion of the stopping sight distance use was made of the driver's eye distance h_1 and of the obstacle's distance h_2 both above the pavement. In addition to this stopping distance it is also desirable to know the passing sight distance; that is the minimum distance for one vehicle to safely pass another vehicle without impacting any another vehicle. For this maneuver the passing distance is not as stringent as the stopping distance because the obstacle to be seen is another vehicle that has a larger h_2 distance. Thus, the equations presented in the preceding section for the stopping sight distances may be employed with the only difference being that the h_2 distance may be set to be the same as the h_1 distance or even slightly higher.

11.2   Elevations Along a Vertical Alignment

11.2.1   Station Points and Elevations

Much like for the horizontal alignment the generation of a vertical alignment is a product of trial and decision to accept or not an assumed design. There are various means to generate a trial geometry all of which require the knowledge of the back and forward slopes and the length of the vertical curve. The avVcurve procedure enables the programmer to generate a series of points along the vertical curve at a specified station increment to:

- Generate a neat column based text report that contains the various stations and elevations together with other pertinent curve information identified below.
- Create two collections, one for stations and another for curve elevations at a relatively small station interval to display a smooth graphic curve.

For the first function the station increment would usually be say 50 feet or 20 m, while for the second function the station increment would be say three to four feet or one meter, or even smaller depending on the desired degree of smoothness of the graphic curve. For this reference is made to the notes below regarding the use of the station increment parameter.

This procedure addresses one vertical curve at a time and should be incorporated within the overall alignment design process.

To define a vertical curve the avVcurve procedure assumes knowledge of its curve length and of the station and elevation of three points, the PVI, one point back of the PVI and another ahead of it. The last two points could be the preceding and the following PVI's, or the starting and ending points of the vertical alignment, respectively. Thus, an overall file could be created containing such points and respective vertical curve lengths, and can be processed by calling repeatedly the avVcurve procedure. Following the description of this procedure are certain other procedures that could be used to supplement or modify its results for reprocessing.

```
Sub avVcurve    (Bsta, Belv, PVIsta, PVIelv,
                 Fsta, Felv, VL, StaInc, _ STA,
                 ElevT, ElevC, TPsta, TPelv,
                 Bgrd, Fgrd, MO, VCLNG, Kfactor,
                 IERR)
```

The input parameters to this procedure are the:

Bsta	The station value of a point back of the PVI. It could be the preceding PVI, or the start point of the alignment, or any point of which the elevation is known.
Belv	The elevation of the BackSta point.
PVIsta	The station value of the PVI of the vertical curve being processed.
PVIelv	The elevation of the PVI of the vertical curve being processed.
Fsta	The station value of a point ahead of the PVI. It could be the following PVI, or the end point of the alignment, or any point of which the elevation is known.
Felv	The elevation of the FrwdSta point.
VL	The length of the vertical curve along the station line in feet (meters).
StaInc	The station increment at which stations and elevations are to be generated along the vertical curve (see notes below).

All input and returned station values are without the plus (+) sign.

The returned parameters from this procedure are the:

STA	The collection that contains the station value of each profile point.
ElevT	The collection that contains the elevation on the back tangent line of each station.
ElevC	The collection that contains the elevation on the curve of each station.
TPsta	The station value of the turning point (see notes below).
TPelv	The elevation of the turning point (see notes below).
Bgrd	The back grade G_1 in percent form.
Fgrd	The forward grade G_2 in percent form.
MO	The middle ordinate (y distance) at the specified PVI station.
VCLNG	The length of the vertical curve computed as the sum of the curve chords between station increments (length of the parabolic arc when StaInc is small enough)
Kfactor	The AASHTO K factor which is equal to absolute value of the length of the vertical curve L divided by the difference in grades in percent form to the nearest whole number.
IERR	An error indicator the value of which denotes the following:
	0 No errors have been encountered.
	1 Zero VL within a tolerance of 0.005 feet (1.524 mm), or negative.
	2 Zero StaInc within a tolerance of 0.005 feet (1.524 mm), or negative.
	3 The back and forward grades have been computed to be the same within a tolerance 0.000001 (0.0001 %).

If it is desired to have the program return the station and elevation of the turning point when such point is located off the limits of the vertical curve the programmer may remove the If Then statement that checks the sign of the two grade lines.

NOTES

1. The first station and elevation in the STA and ELV collections is that of the PVC point of the vertical curve being processed, while the last point of the said collections is the station and elevation of the PVT point of the said curve.
2. A positive grade ascends, and a negative grade descends in a station order.
3. If the back and forward grades are both ascending (positive) or descending (negative), then the turning point is beyond the limits of the vertical curve and the TPsta and TPelv station and elevation are each set to be zero.
4. If the PVI station (PVIsta) and the vertical curve length (VL) are such that the PVC station [STA(1)] is computed to be not a whole or half station, then the procedure computes the first full or half station and thereafter begins to increase the stations by the station increment (StaInc). That is, if the PVI station and the length of curve are such that the PVC station is 15+36.25 and StaInc is 50 in the U. S. Customary units system, the next station to be returned would be that of 15+50 and not 15+86.25. Similarly in the SI units if the PVC station is 185+215.48 and StaInc is 20 the next station to be returned would be that of 185+220 and not 185+235.48. The procedure assumes that the station increment StaInc would not have any decimal digits and would be a multiple of 10 not exceeding 990. This assumption is valid only for the preparation of a report by stations. Actually StaInc could be of any value.

Upon invocation, the procedure initializes the parameters to be returned and then

1. Checks the value of the vertical curve (VL) and station increment (StaInc) for being equal to zero or negative, and if so sets IERR to be 1 or 2, respectively, and aborts; else it continues.
2. Computes the values of the back grade (G1) from the back point towards the PVI point, and the forwards grade (G2) from the PVI towards the forward point, and then compares the said two grades for being equal to each other within a tolerance of 0.000001, and if so sets IERR to be 3 and aborts; else it continues.
3. Computes the middle ordinate at the PVI station by means of Equation 11.1u3.

4. Computes the station and elevation of the PVC and PVT by assuming a symmetrical curve. Thus, the PVC station is set to be the PVI station minus half the length of the vertical curve, and the PVT station is set to be the PVI station plus the said half length. For each of these two stations the corresponding tangent and curve elevations are computed by traversing along the back and forward grade lines, respectively. Of these two points only the PVC station and elevation are saved in the corresponding arrays to be returned (STA, ElevT, and ElevC) at this time.
5. Computes the station and elevation of the turning point (TPsta, TPelv) of the vertical curve if the two grades are of opposite sign; else the said station and elevation are assigned the value of zero.
6. Computes the station adjustment X as indicated by the two statements below
   iVal = PVCsta/StaInc
   X = PVCsta − (iVal * StaInc)
   If X is:
   - Greater than zero (denotes that a station adjustment is necessary):
     – Computes the next full or half station and the back tangent and curve elevations;
     – Saves them in their corresponding arrays (STA, ElevT, and ElevC);
     – Sets the parameters StaX, YT, and YC to be those of the first full or half station and thus represent the station and elevation of the preceding point for subsequent computations;
     – Computes the difference in elevation DY between the curve elevation at the PVC and that of the StaX;
     – Sets the previous curve elevation to be that of YC; and
     – Initiates the curved length VCLNG of the vertical curve to be returned as the square root of the sum of the squares of the difference in elevation and difference in station values.
   - Equal to zero (denotes that the PVC is at a full or half station and that no adjustment is necessary) sets the parameters StaX and YP to be those of the PVC, and initiates the value of VCLNG to be zero.
7. Computes the station and elevation of all points up to and excluding the PVT by the station increment (StaInc), saves the computed values in the corresponding arrays to be returned (STA, ElevT, and ElevC), and increases the value of VCLNG by the corresponding length between two successive station values.
8. Saves previously computed values of the station and elevation of the PVT in the corresponding arrays to be returned (STA, ElevT, and ElevC), and increases the value of VCLNG by the corresponding length between the last two station values.

9. Changes the two grades G1 and G2 into percent form and saves them in the corresponding `Bgrd` and `Fgrd` parameters to be returned, computes the AASHTO K factor, and terminates the procedure.

11.2.2 Location of the Turning Point

The `avVcHighLow` procedure computes the station and elevation of the turning (high or low) point of a vertical curve given the station and elevation of the PVI, the back and forward grades and the length of the vertical curve. Whereas the preceding procedure as part of its overall function computes the location of the turning point, it does so only if the said point is located within the limits of the vertical curve. The subject procedure computes the location of the said point regardless of whether it is located within or outside the limits of the curve. In addition, the required input is slightly different as is stated below. However, the method of computation of the said point is the same.

Regarding the case in which the turning point is located outside the limits of the vertical curve length reference is made to cases (B), (C), (E), and (F) of Figure 11.1e in which the said point is located outside the curve. The turning point would be located back of the curve if the forward grade (G_2) is greater in absolute value (sharper) than that of the forward grade (G_1), and it would be located ahead of the curve if the forward grade (G_2) is less in absolute value (flatter) than that of the forward grade (G_1). Thus, in cases (C) and (F) the turning point is back of the curve, and in cases (B) and (E) the turning point is ahead of the curve.

> *All input and returned station values are without the plus (+) sign.*

```
Sub avVcHighLow (PVIsta, PVIelv, Bgrd, Fgrd,
                 VL, _ TPsta, TPelv, IERR)
```

The input parameters to this procedure are the:

PVIsta	The station value of the PVI point of the vertical curve.
PVIelv	The elevation of the PVI point of the vertical curve.
Bgrd	The back grade in percent form.
Fgrd	The forward grade in percent form.
VL	The length of the vertical curve in feet (meters).

The returned parameters from this procedure are the:

TPsta	The station value of the turning point of the vertical curve.
TPelv	The elevation of the turning point of the vertical curve.
IERR	An error indicator the value of which denotes the following:
	0 No errors have been encountered.
	1 Zero VL within a tolerance of 0.005 feet (1.524 mm), or negative.
	2 The back and forward grades have been computed to be the same within a tolerance 0.000001 (0.0001 %).

Upon invocation, the procedure initializes the parameters to be returned and then

1. Checks the value of the vertical curve (VL) for being equal to zero or negative, and if so sets IERR to be 1 and aborts; else it continues.
2. Converts the back and forward grades from the percent form into a non-percent form, and then compares the said two grades for being equal to each other within a tolerance of 0.000001, and if so sets IERR to be 2 and aborts; else it continues.
3. Computes the station and elevation of the turning point by use of Equations 11.1w2 and 11.1v2, respectively, and then terminates.

11.2.3 Sight Distance on a Vertical Curve

The `avVcSightDist` procedure computes the stopping, or passing sight distance of a vertical curve given the back and forward grades, the length of the vertical curve, and the heights h_1 and h_2 above the curve representing the height of the driver's eye, and that of the obstacle, respectively, as well as the angle of the headlight beam. The value of $h2$ determines whether the returned sight distance is for stopping or passing.

The computation of the said distance is as described in Sects. 11.1.7 and 11.1.8 of this chapter. Since there are two potential conditions for each of the sag and crest type curves which are dependent upon the length of the stopping or passing distance with respect to the length of the vertical curve, the procedure returns three distances, one on the assumption that the said distance is less than the curve's length, the other on the assumption that it is greater, and a distance selected from the said two distance as stated under the operational steps below.

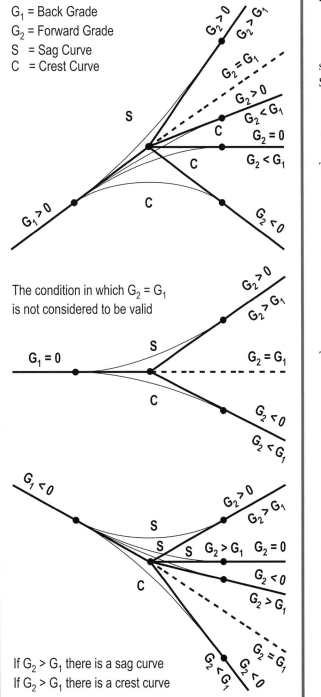

Figure 11.2 Sag and Crest Vertical Curve Conditions

- The two grades cannot be the same as that would create a straight grade line.

It is noted that the subject procedure does not consider the safe passing distance at an underpass as addressed in Sect. 11.1.7.8 if this chapter.

```
SubavVcSightDist  (G1,G2,VL,H1,H2,BETA, _ SPlong,
                    SPshort, SPdist, CvType, K, IERR)
```

The input parameters to this procedure are the:

G1, G2   The back and forward grades in percent form.
VL   The length of the vertical curve in feet (meters).
H1   The height of the driver's eye, or of the headlights above the pavement in feet (meters). It should be positive.
H2   The height of the obstacle, or of the headlights of an oncoming vehicle above the pavement in feet (meters). It should be positive.
BETA   The angle in degrees that the headlight beam makes with the horizontal. It should be positive.

The returned parameters from this procedure are the:

SPlong   The vertical curve stopping or passing distance assuming it is longer than the vertical curve.
SPshort   The vertical curve stopping or passing distance assuming it is longer than the vertical curve.
SPdist   The stopping or passing distance of the vertical curve which is selected from the above two distances as stated below.
CvType   The type of the vertical curve, either CREST or SAG is returned.
K   The AASHTO K factor which is equal to absolute value of the length of the vertical curve L divided by the difference in grades in percent form to the nearest whole number.
IERR   An error indicator the value of which denotes:
- 0   No errors have been encountered.
- 1   Zero VL within a tolerance of 0.005 feet (1.524 mm), or negative.
- 2   Zero H1 or H2 within a tolerance of 0.005 feet (1.524 mm), or negative.
- 3   The back and forward grades are the same within a tolerance 0.00001 (0.001 %).
- 4   The BETA value is negative or zero.
- 5   The BETA value is too large (zero divide).

Upon invocation, the procedure initializes the parameters to be returned and then

1. Checks the value of the vertical curve length (VL) for being equal to zero within a tolerance of 0.005 feet (2.524 mm) or negative, and if so sets IERR to be 1 and aborts; else it continues.
2. Checks the values of H1 and H2 for being equal to zero or negative, and if so sets IERR to be 2 and aborts; else it continues.

Displayed in Figure 11.2 are the relationships between the back and forward grades G_1 and G_2 from which it is seen that when G_2 is greater than G_1 the vertical curve is a sag curve and when G_2 is less than G_1 the vertical curve is a crest curve. It is noted that:

- This holds true regardless whether the two grades are of the same or opposite sign, and that

3. Compares the grades G1 and G2 for being the same within a tolerance of 0.001 % and if so sets IERR to be 3 and aborts; else it continues.

4. Compares the grades G1 and G2 against each other to determine the type of vertical curve. If G2>G1 sets CvType to be "SAG", and if G2<G1 sets CvType to be "SAG".

5. Checks the values of BETA for being equal to zero within a tolerance of 0.001°, or negative, and if so sets IERR to be 4 and aborts; else it continues.

6. Computes the rate of parabolic rate of change A and the AASHTO K factor by dividing the vertical curve length by the said parabolic change.

7. Determines whether the vertical curve is a crest or sag curve and the computes the stopping/passing distance:
   - SPshort by use of Equation 11.1y4 for a crest curve.
   - SPlong by use of Equation 11.1y15 for a crest curve.
   - SPshort by use of Equation 11.1y23 for a sag curve substituting angle BETA for the constants of the said equation as noted at the end of Sect. 11.7.1.4.
   - SPlong by use of Equation 11.1y31 for a sag curve.

8. Selects from the above two distances the SPdist to be

For a crest curve:	If SPshort is greater than VL	SPdist=SPlong
	else	SPdist=SPshort
For a sag curve:	If SPlong is greater than or equal to VL	SPdist=SPlong
	else	SPdist=SPshort

This step terminates the procedure.

11.3 Vertical Profile Fitting

The design of the vertical alignment of a new roadway is a process of repetitive trials in meeting various controls that have been defined by the engineer to meet site specific conditions. As long as such controls are maintained the engineer has a certain degree of flexibility to vary the design. However, there is one case in which the control conditions are numerous that limit the degree of the said flexibility, and that is the major reconstruction of an existing roadway beyond the case of a simple resurfacing with a new coat of asphaltic overlay. This subject is further complicated for existing roadways for which design and as-built information is lacking, and the new design is based on the field survey of the ground profile of the subject roadway. In other words, the design problem is one in which we are given a set of stations and elevations, and it is required to determine the grades of straightaways, and the vertical curve length of a curved profile so as to minimize the deviations between computed and field measured elevations.

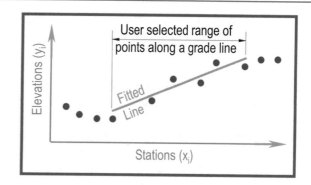

Figure 11.3a Fitting a Grade Line to a Set of Points

The solution to this problem invites the use of a least square curve fitting process. However, since the overall profile of a roadway is a series of connected straight lines and parabolic curves, profile curve fitting must be done in distinct operations. To complicate the problem, we have the requirement that the straight lines and the parabolic curves must be tangent to each other.

To assist the engineer in the profile fitting process the avFitGrade and the avFitCurve procedures are presented below. To use these procedures, the engineer displays on the computer monitor the overall set of field surveyed profile points, and then visually segregates the points into point groups that seem to constitute straight lines, and point groups that seem constitute curves. The first procedure may then be used to fit a straight line, and the second procedure may be used to fit a parabola. An alternative to the avFitCurve procedure is the avVc3points procedure that constructs a vertical curve passing through three points each defined by a station and elevation.

However, the use of these procedures by themselves alone does not solve the profile redesign problem. They just provide certain base information as to what is at hand and they constitute the "first cut" of the redesign process. The engineer may now begin certain design plays by using the construction procedures presented in the next subsection of this chapter to produce the required tangencies of the straight and parabolic sections of the vertical alignment.

11.3.1 Grade Line Fitting Through Field Observed Points

A set of x_i and y_i coordinate points is displayed in Figure 11.3a representing field survey points along a roadway, in which the x_i coordinate denotes the station value of a point and the y_i coordinate denotes the elevation of a point. Also displayed in this figure is (**a**) the range of points that have been selected to comprise the point set through which a straight line is to be fitted by the method of least squares, and (**b**) the said fitted line which is represented by the equation

$$f(y_1) = A_0 + A_1 x_1 \tag{11.3a}$$

where A_0 and A_1 are two constants or coefficients that need to be evaluated in order to obtain the best fit of the said straight line to the said set of selected points by the method of least squares. To attain this, the sum of the squares of the errors of approximation between the computed elevations and those of the observed elevations should be minimized. Thus, the error of approximation may be expressed as

$$d_i = y_i - f(x_i) = y_i - (A_0 + A_1 x_i) \tag{11.3b}$$

and the sum of these errors may then be expressed as

$$\sum_{i=1}^{n} d_i^2 = \sum_{i=1}^{n} \left[y_i - (A_0 + A_1 x_i) \right]^2 \tag{11.3c}$$

For the purpose of simplification,

- let Σ denote the full expression of the limits of summation, and let
- let **C** represent the left side of Equation 11.3c, and after that
- let us expand the square of the right side of the same equation to define **C** as

$$C = \sum \left(\begin{array}{c} y_1^2 - 2A_0 y_i + 2A_1 x_i y_i + 2A_0 A_1 x_i + \\ 2A_1^2 x_i^2 + A_0^2 \end{array} \right) \tag{11.3d}$$

In order to minimize the square of the sum of the errors, the partial derivatives of Equation 11.3d with respect to A_0 and A_1 must be set to zero, so that

$$\frac{\partial C}{\partial A_0} = \sum (-2y_i + 2A_1 x_i + 2A_0) = 0 \tag{11.3e1}$$

and

$$\frac{\partial C}{\partial A_1} = \sum (-2x_i y_i + 2A_0 x_i + 2A_1 x_i^2) = 0 \tag{11.3e2}$$

If we now divide both sides of each of Equations 11.3e1 and 11.3e2 by 2, and we rearrange the terms, we obtain

$$\sum y_i = \sum A_0 + \sum A_1 x_i = n A_0 + A_1 \sum x_i \tag{11.3f1}$$

and

$$\sum x_i y_i = \sum A_0 x_i + \sum A_1 x_i^2 = A_0 \sum x_i + A_1 \sum x_i^2 \tag{11.3f2}$$

where *n* is the number of profile point observations along a grade line.

If we now solve Equation 11.3f1 for A_0 in terms of A_1, then substitute this value for A_0 in Equation 11.3f2 and solve for A_1, and then substitute the value of A_1 back in Equation 11.3f1 we obtain the values of A_0 and $_1$ as

$$A_0 = \frac{\sum y_i \sum x_i^2 - \sum x_i \sum x_i y_i}{n \sum x_i^2 - \left(\sum x_i \right)^2} \tag{11.3g1}$$

and

$$A_1 = \frac{n \sum x_i y_i - \sum x_i \sum y_i}{n \sum x_i^2 - \left(\sum x_i \right)^2} \tag{11.3g2}$$

The values of A_0 and A_1 may now be introduced in Equation 11.3a to (a) define the equation of the grade line passing through the selected observed points, (b) compute elevations at the stations of the said points, and (c) compare elevations.

To assess the quality of the fit, the mean value *m* may be computed as

$$m = \frac{1}{n} \sum_{n=1}^{n} y_i \tag{11.3h1}$$

the standard deviation σ of the observations may be determined as

$$\sigma = \frac{1}{n} \sum_{n=1}^{n} \left[f(x_i) - m \right]^2 \tag{11.3h2}$$

and the coefficient of determination **R2** factor may be computed as

$$R^2 = 1 - \frac{\sum_1^n E_C^2}{\sum_1^n E_M^2} = 1 - \frac{\sum_1^n \left[f(x_i) - y_i \right]^2}{\sum_1^n \left[f(x_i) - m \right]^2} \tag{11.3h3}$$

where

E_C = Computed value minus observed value,
E_M = Computed value minus the mean value as computed by Equation 11.3h1, and
n = number of observation points.

In assessing the values of the fit, the closer the value of $\mathbf{R}^2$ is to 1, the better the fit. However, a visual observation of the tabular results that could be created, or of a plot of the fitted line and observed points on the computer monitor might offer an as good if not more appropriate acceptance or rejection of the fit.

The `avFitGrade` procedure constructs by the method of least squares a grade line that passes through a set of field observations of stations and elevations specified in two collections, and returns (a) the coefficients A_0 and A_1 of Equation 11.3a, (b) the elevations computed by the said equations for each given station, (c) the error of deviation between the observed and computed elevation, (d) the mean value of the observed elevations, (e) the standard deviation of the line fitting, (f) the coefficient of determination R^2, and (g) the stations and elevations of the PVC, PVI, and PVT based on certain user input controls as described below.

The XList and YList collections of the observed points should be in a station ascending order. All station values are without the plus (+) sign.

If the stations are specified in a decreasing order the slope will be returned with an opposite sign.

```
Sub avFitGrade(XList, YList, ListY, ListD,
_ A0, A1, MEAN, SIGMA, R2, IERR)
```

The input parameters to this procedure are the:

XList	The collection of the station values of the observed points (see Note 1).
YList	The collection of the elevations of the observed points (see Note 1).
ListY	The name of the collection in which the elevations to be computed by the fitting process are to be saved (see Note 2).
ListD	The name of the collection in which the errors of deviation between the observed elevations and their mean value are to be saved (see Note 2).

NOTES
1. The XList and ListY collections should be of the same size.
2. Both the ListY and ListD collections should be empty as input parameters. They are populated and returned by the procedure.

The returned parameters from this procedure are the:

ListY	The collection of the computed elevations at the stations of the observed points.
ListD	The collection of the deviations between the observed points and their mean value.
A0	The coefficient A_0 of Equation 11.3a as computed by the procedure.
A1	The coefficient A_1 of Equation 11.3a as computed by the procedure.
MEAN	The mean value of the observed elevations.
SIGMA	The standard deviation of the line fit.
R2	The coefficient of determination R^2 of the line fit.
IERR	An error indicator the value of which denotes the following:
	0 No errors have been encountered.
	1 The collections ListY and ListD are not of the same size. The procedure aborts.
	2 The collections ListY and ListD are not are not empty. The procedure aborts.
	3 The denominator of Equations 11.3g1 and 11.3g2 is zero. Most probably there are some erroneous data or the line is vertical. The procedure aborts.

4   The sum of the errors of deviation between the observed elevations and their mean has a value of zero resulting in a division by zero in computing the coefficient of determination R^2. Check the data. The procedure aborts.

If the values of the observations are too large in significant digits, then to increase precision the lowest station and elevation could be subtracted from each respective collection and that value could be added after the computation of the line fit.

It is felt that for the purpose of the subject procedure this is not warranted.

Upon invocation, the procedure makes certain that the collections that contain the observed points are of the same size, and that the two collections to be populated with the computed values are empty. If any of the tests fails the procedure sets the value of IERR to 1 or 2 and aborts, else it continues as follows:

1. Initializes to zero the summation parameters of Equations 11.3g1 and 11.3g2, and then carries out their summations.
2. Computes the common denominator of the said equations and checks its value for being equal to zero. If it is, IERR is set to be 3 and aborts; else it continues.
3. Computes the numerators and the constants A_0 and A_1 of the said equations.
4. Computes the elevations at each given station using Equation 11.3a and the corresponding error of deviation and store the results in the ListY and ListD collections.
5. Computes the mean, standard deviation, and R^2 values using Equations 11.3h1, 11.3h2, and 11.3h3, respectively.

This concludes the function of the avFitGrade procedure. The calling program may now display in tabular form the contents of the ListY and ListD collections, the mean, standard deviation, and R^2 values, and plot the fitted line in the form of points or as a single line for approval or rejection.

11.3.2 Parabola Fitting Through Field Observed Points

A set of x_i and y_i coordinate points is displayed in Figure 11.3b representing field survey points, in which the x_i coordinate denotes the station values of a point and the y_i coordinate

Figure 11.3b Fitting a Vertical Curve to a Set of Points

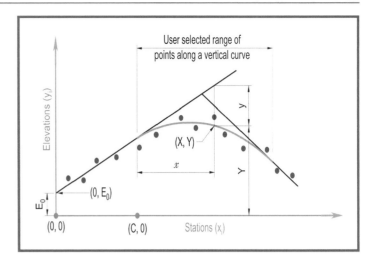

denotes the elevation of a point. Also displayed in this figure is a vertical curve that represents the parabola that best fits the observed points, and which is represented by Equation 11.1v2 rewritten in the form

$$Y = E_0 + G_1 X + y \qquad (11.3i1)$$

where

Y denotes the elevation on the curve of a point at station X,
y denotes the distance from the elevation on the back grade G_1 at station X minus the elevation of the said station on the curve,
E_0 denotes the Y axis intercept, and
G_1 denotes the back grade.

Remember that y is negative for a crest curve and positive for a sag curve. This is the reason for the $+y$ and not $-y$ in Equation 11.3i1.

If we now recall Equations 11.1v3, 11.1v4, and 11.1v5, we can rewrite Equation 11.3i1 in the form

$$Y = E_0 + G_1 X + \frac{1}{2} R x^2$$
$$Y = E_0 + G_1 X + \frac{1}{2} R (X - C)^2 \qquad (11.3i2)$$

where

C denotes the station value of the PVC of the vertical curve, and
x denotes the distance of station X from station C.

If we expand the square in Equation 11.3i2, and we rearrange terms, the equation becomes

$$Y = E_0 + G_1 X + \frac{1}{2} R X^2 + \frac{1}{2} R C^2 - RXC$$
$$Y = \left(E_0 + \frac{1}{2} R C^2 \right) + \left(G_1 - RC \right) X + \frac{1}{2} R X^2 \qquad (11.3i3)$$

and can be generalized as

$$f(X) = A_0 + A_1 X + A_2 X^2 \qquad (11.3i4)$$

where

$$A_0 = E_0 + (1/2) R C^2 \qquad (11.3i5)$$

$$A_1 = G_1 - RC \qquad (11.3i6)$$

$$A_2 = (1/2) R \qquad (11.3i7)$$

For a set of observed points with coordinates (X_i, Y_i), the application of the method of least squares to Equation 11.3i4 generates the following three equations:

$$A_0 N + A_1 \sum X_i + A_2 \sum X_i^2 = \sum Y_i \qquad (11.3j1)$$

$$A_0 \sum X_i + A_1 \sum X_i^2 + A_2 \sum X_i^3 = \sum X_i Y_i \qquad (11.3j2)$$

$$A_0 \sum X_i^2 + A_1 \sum X_i^3 + A_2 \sum X_i^4 = \sum X_i^2 Y_i \qquad (11.3j3)$$

in which the symbol Σ implies the summation of the coordinates to its right of all points in the sample set of observed point coordinates. That is, $\Sigma X_i^2 Y_i = \Sigma (X_i^2 Y_i)$. The above three equations may now be solved simultaneously by any algorithm as may be desired by the programmer. The `avFit-Curve` procedure calls the `avGaussian3x4` procedure to solve these three equations. It is a simplistic Gaussian elimination algorithm for a 3x4 matrix represented as `S(i,j)` where `i` denotes the rows and `j` denotes the columns. In this matrix

- The first row contains the number of points of observation, and the summations of Equation 11.3j1,
- The second row contains the summations of Equation 11.3j2, and
- The third row contains the summations of Equation 11.3j3.

This algorithm is specific for the conditions of the subject problem, and it does not represent a generic application of the Gaussian elimination method to the solution of simultaneous equations. To solve for the coefficients of A_0, A_1, and A_2 the procedure:

1. Divides all four coefficients of each of the Equations 3j1, 11.3j2 and 11.3j3, by the first coefficient of the said equations. This causes each coefficient of A_0 to be 1.
2. Subtracts the coefficients, as computed in Step 1, of Equation 11.3j1 from the respective coefficients from Equations 11.3j2 and 11.3j3. This causes the first coefficient of the said equations to be 0.
3. Divides all four coefficients, as computed in Step 2, of each of the Equations 11.3j2 and 11.3j3, by the second coefficient of each said equations. This causes each coefficient of A_1 to be 1.
4. Subtracts the coefficients, as computed in Step 3, of Equation 11.3j2 from the respective coefficients from Equation 11.3j3. This causes the second coefficient of the said equation to be 0.
5. Computes the value of A_2 by dividing the fourth coefficient by the third coefficient of the third equation, each as computed in Step 4.
6. Computes the value of A_1 by substituting the value of A_2 in the second as computed in Step 3, and then rearranging the terms of the equation.
7. Computes the value of A0 by substituting the value of A_1 and A_2 in the first as computed in Step 2, and then rearranging the terms of the equation.

Having determined the values of A_0, A_1, and A_2 we may now compute elevations along the fitted vertical curve at each observed station value by solving Equation 11.3i4. These computed elevations may then be compared with their observed counterparts to determine the individual deviations, and then perform an error analysis and an assessment of the curve fitting process in a manner similar to that described for the `avFitGrade` procedure in the preceding section utilizing Equations 11.3h1, 11.3h2, and 11.3h3.

Before declaring this curve fitting process as being accomplished, there is on more task to be done because the ultimate goal is to locate the PVC, PVI, and PVT of the curve. To do this the values of G_1, G_2 and L are needed. With the values of A_0, A_1, and A_2 now known we can use:

- Equation 11.3i7 to solve for R from which we see that

$$R = 2A_2 \qquad (11.3k1)$$

- Equation 11.3i6 to solve for C (the PVC station) or G_1 provided the other is assumed to be known, that is

$$C = \frac{G_1 - A_1}{R} \qquad (11.3k2)$$

- or

$$G_1 = A_1 + RC \qquad (11.3k3)$$

- Equation 11.3i5 to solve for E0, the Y axis intercept of the back grade as indicated in Figure 11.3b and being

$$E_0 = A_0 - \frac{1}{2}RC^2 \qquad (11.3k4)$$

- which may be used to determine the elevation of the PVC provided C and G_1 have been determined.
- Equation 11.1v5 may be used to solve for G_2 if G_1 and L have been assumed

$$G_2 = RL + G_1 \qquad (11.3k5)$$

- Equation 11.1v5 may also be used to solve for G_1 if G_2 and L have been assumed

$$G_1 = G_2 - RL \qquad (11.3k6)$$

- Equation 11.1v5 may again be called to solve for L if G_2 and G_1 have been assumed

$$L = \frac{G_2 - G_1}{R} \qquad (11.3k7)$$

The slope of the tangent line to the parabola at any point is represented by the first derivative of Equation 11.3i4 as

$$\frac{dY}{dX} = A_1 + 2A_2X \qquad (11.3k8)$$

from which the value of X may be expressed as

$$X = \frac{\frac{dy}{dx} - A_1}{2A_2} = \frac{G - A_1}{2A_2} \qquad (11.3k9)$$

in which G is the assumed grade, and X is the corresponding station. Thus, if either G_1 or G_2 is assumed, the returned value of X would be the station of the PVC or of the PVT, respectively, and the difference between these two station would yield the length of the vertical curve.

Shown in Table 11.1 are the five valid conditions of parameter pairs to be assumed by the user in order to complete the construction of the vertical curve. These parameters include the PVC station (referred to as C in the said table and associated equations), the back and forwards grades G_1 and G_2, respectively, and the vertical curve length L. For each valid pair specification the table displays the number of the equations to solve for the remaining unknown pairs of parameters. Regarding these parameters note the following notes:

1. There are certain pairs of parameters that are not valid because of geometric consideration. For example if the PVC station is assumed as known, the back grade G_1 cannot be known. It has to be computed by the procedure because there is only one possible grade at that point.
2. The PVI and PVT stations could be considered to fix the vertical curve, but they are not considered by the procedure. If either of these points is to be considered as a

Table 11.1 Conditions Table for Completing the Curve Fit

		1	2	3	4	5
Assumed or Given	R	11.3k1	11.3k1	11.3k1	11.3k1	11.3k1
	C	•	•			
	G1			•	•	
	G2		•		•	•
	L	•		•		•
Solve for	G1	11.3k3	11.3k3			11.3k6
	C			11.3k2	11.3k2	Note 3
	G2	11.3k5		11.3k5		
	L		11.3k7		11.3k7	Note 3
	Eo	11.3k4	11.3k4	11.3k4	11.3k4	11.3k4

desired user specification, then Figure 11.3b and Equations 11.3i1 and 11.3k2 through 11.3k7 will have to be modified so that the said point would replace the PVC.

3. For condition 5 when G_2 and L are given, Equation 11.3k6 is used to compute the PVT station and the curve length L is subtracted therefrom to compute the PVC station.

4. For condition 5 the Y axis intercept E0 must be computed after the value of the PVC has been determined.

The `avFitCurve` procedure constructs by the method of least squares a vertical curve that passes through a set of field observations of stations and elevations passed in as two collections. This procedure utilizes the user assumption conditions as presented in Table 11.1, and the associated equations.

The `XList` *and* `YList` *collections of the observed points should be in a station ascending order. All station values are without the plus (+) sign.*

   Of the `G1, C, E0, G2, L` *input arguments, only two may have a non-zero value.*

```
Sub avFitCurve    (XList, YList, ListY, ListD,
                  PVCsta, G1, G2, L, _ A0, A1,
                  A2,  MEAN,  SIGMA,  R2,  E0,
                  Kfactor_    PVCelv,    PVIsta,
                  PVIelv, PVTsta, PVTelv, IERR)
```

The input parameters to this procedure are the:

`XList, YList`	The collections of the stations and elevations of the observed points (see Note 1).
`ListY, ListD`	The name of the collections in which the elevations and stations to be computed by the fitting process are to be saved (see Note 2).

`PVCsta`	The user assumed station at which the fitted curve is to start, or zero if this parameter is not to be considered assumed (see Note 3).
`G1, G2`	The user assumed back and forward grade to the fitted curve, or zero if this parameter is not to be considered assumed (see Note 3).
`L`	The user assumed length of the fitted curve, or zero if this parameter is not to be considered assumed (see Note 3).

If the values of the observations are too large in significant digits, then to increase precision the lowest station and elevation could be subtracted from each respective collection and that value could be added after the computation of the line fit.

   It is felt that for the purpose of the subject procedure this is not warranted.

NOTES

1. The `XList` and `ListY` collections should be of the same size.
2. Both the `ListY` and `ListD` collections should be empty as input parameters. They are populated and returned by the procedure.
3. Of the above optional parameters (`PVCsta`, `G1`, `G2`, and `L`) two must be selected as per Table 11.1 and assigned an appropriate value. The other two should have a value of zero. Their appropriate values are assigned by the procedure. The proper pairs for a valid selection identified by the parameter `CONDI` are as follows:

`CONDI` 1 `PVCsta` and `L`	`CONDI=2` `PVCsta` and `G2`	
`CONDI=3` `G1` and `L`	`CONDI 4` `G1` and `G2`	`CONDI=5` `G2` and `L`

Negative values are considered to be valid, such as a grade, but a negative station and curve length could cause a problem in most, but not in all, cases:

The returned parameters from this procedure are the:

ListY	The collection of the computed elevations at the stations of the observed points.
ListD	The collection of the deviations between the observed points and their mean value.
A0	The coefficient A_0 as computed by the procedure.
A1	The coefficient A_1 as computed by the procedure.
A2	The coefficient A_1 as computed by the procedure.
MEAN	The mean value of the curve fit.
SIGMA	The standard deviation of the curve fit.
R2	The R^2 value of the curve fit.
E0	The assumed elevation E_0, or zero if not assumed.
PVCelv	The fitted elevation at the PVC station.
PVIsta	The fitted PVI station value.
PVIelv	The fitted elevation at the PVI station (not the elevation on the curve).
PVTsta	The fitted PVC station value.
PVTelv	The fitted elevation at the PVC station.
IERR	An error indicator the value of which denotes the following:
	0 No errors have been encountered.
	1 The collections ListY and ListD are not of the same size. The procedure aborts.
	2 The collections ListY and ListD are not are not empty. The procedure aborts.
	3 The assumed or computed grades G1 and G2 are equal within a tolerance of 0.0000.1 (0.0001 %). The procedure aborts.
	4 An appropriate condition has not been specified. The procedure aborts.
	5 One of the X summation values is zero. The procedure aborts.

Upon invocation, the procedure makes certain that (a) the collections that contain the observed points are of the same size, (b) the two collections to be populated with the computed values are empty, and (c) the two grades are not equal to each other if both have been assumed as given. If any of the tests fails the indicator IERR is set to be 1, 2, or 3 depending on the failure encountered and the procedure aborts, else it continues as follows:

1. Initializes the condition parameter CONDI to zero and then queries pairs of the optional input arguments PVCsta, G1, G2, L and determines which appropriate pair has been assigned a non-zero value as previously stated. If a proper pair has not been encountered IERR is

set to 3 and the procedure aborts; else the other two parameters are initialized to zero and continues.
2. Initializes to zero the summation coefficients ΣX_i, ΣX_i^2, ΣX_i^3, ΣX_i^4, ΣY_i, $\Sigma X_i Y_i$, and $\Sigma X_i^2 Y_i$ of Equations 11.3j1, 11.3j2, and 11.3j3, and then carries out their summations.
3. Populates the array $S(i,j)$ with the said summations and calls the avGaussian3x4 procedure to compute the coefficients A0, A1, and A2 of Equations 11.3j1, 11.3j2, and 11.3j3.
4. Checks the returned indicator IERR for having a value of 1, and if so aborts; else continues.
5. Computes the elevations at each given station using Equation 11.3i4 and the corresponding error of deviation and store the results in the ListY and ListD collections.
6. Computes the mean, standard deviation, and R^2 values using Equations 11.3h1, 11.3h2, and 11.3h3, respectively.
7. Computes the value of the rate of parabolic change R by Equation 11.3k1, and then depending on the value of the parameter CONDI and the equations of Table 11.1 computes the values of the user non-specified optional input parameters.
8. Computes the value of the Y axis intercept E_0 by Equation 11.3k4. Note that for condition 5 this must be computed after the computation of the PVC.
9. Checks the final values of the G1 and G2 grades for being equal to each other, thus implying two collinear tangent lines.
10. Computes the value of the AASHTO K factor as the absolute value to the nearest whole number of the curve length L divided by the algebraic difference of the back and forward grades expressed in percent form.
11. Computes the values of the remaining arguments PVCelv, PVIsta, PVIelv, PVTsta, PVTelv to be returned, and sets IERR to be 0 denoting no detected errors.

In the subject procedure the K factor is not computed as $K = |L/A|$ but as $K = |L/(G_2 - G_1)|$, where the grades are in percent form, because the R as computed by the procedure reflects the fitted curve and not the computed curve with the final L, G_1 and G_2 values.

This concludes the function of the procedure. The calling program may now display in tabular form the contents of the ListY and ListD collections, the mean, standard deviation, and R^2 values, and plot the fitted line in the form of points or as a curve for approval or rejection.

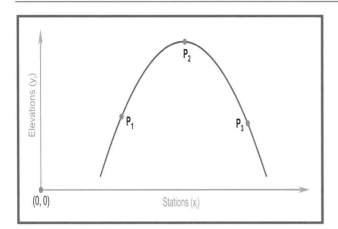

Figure 11.3c Parabolic Curve Passing Through Three Given Points

11.3.3 Vertical Curve Through Three Points

The `avVc3points` procedure enables the user to determine the constants A_0, A_1, and A_2 of a parabolic curve defined by the Equation 11.3i4 that passes through three points of known station and elevation. This procedure may be used as a prelude to the use of:

- The `avGrdTcrvFrPnt` procedure that constructs a grade line tangent to a vertical curve from a point external to the vertical curve.
- The `avGrdTcrvAtSta` procedure that determines the grade of the tangent line to a vertical curve at a point on the vertical curve defined by its station value.
- The `avGrdTcrvAtElv` procedure that determines the grade of the tangent line to a vertical curve at a point on the vertical curve defined by its elevation.
- Any procedure that needs to create a parabolic curve defined by the said Equation 11.3i4 and repeated below

$$y = A_0 + A_1 x + A_2 x^2.$$

Displayed in Figure 11.3c is a parabolic curve passing through points P_1, P_2, and P_3, each defined by a station (X) and elevation (Y) coordinate, and which three points are not collinear. The procedure checks for collinearity by inversion from P_1 to P_2 and to P_3. If the two grades are equal to each other within a tolerance of 0.0001 (0.01 %), the three points are considered to be collinear. Thus, the equation at point

$$P_1 \text{ is } \quad Y_1 = A_0 + A_1 X_1 + A_2 X_1^2 \tag{11.3m1}$$

$$P_2 \text{ is } \quad Y_2 = A_0 + A_1 X_2 + A_2 X_2^2 \tag{11.3m2}$$

$$P_3 \text{ is } \quad Y_3 = A_0 + A_1 X_3 + A_2 X_3^2 \tag{11.3m3}$$

The programmer may elect to make the 0.0001 radians tolerance as a designer input parameter, rather than a built-in control.

The above three equations could be solved by calling the `avGaussian3x4` procedure using the following data for the three matrix lines: $(1, X_1, X_1^2, Y1)$,: $(1, X_2, X_2^2, Y_2)$, $(1, X_3, X_3^2, Y_3)$. However, we solve these three equations simultaneously with determinates as follows:

$$D = \begin{vmatrix} 1 & X_1 & X_1^2 \\ 1 & X_2 & X_2^2 \\ 1 & X_3 & X_3^2 \end{vmatrix} = (1)\begin{pmatrix} X_2 & X_2^2 \\ X_3 & X_3^2 \end{pmatrix}$$
$$- (1)\begin{pmatrix} X_1 & X_1^2 \\ X_3 & X_3^2 \end{pmatrix} + (1)\begin{pmatrix} X_1 & X_1^2 \\ X_2 & X_2^2 \end{pmatrix} \tag{11.3n1}$$

$$\begin{aligned} D &= 1\left(X_2 X_3^2 - X_2^2 X_3\right) - 1\left(X_1 X_3^2 - X_1^2 X_3\right) \\ &\quad + 1\left(X_1 X_2^2 - X_1^2 X_2\right) \\ &= X_2 X_3^2 - X_2^2 X_3 - X_1 X_3^2 + X_1^2 X_3 \\ &\quad + X_1 X_2^2 - X_1^2 X_2 \\ &= X_1^2\left(X_3 - X_2\right) + X_2^2\left(X_1 - X_3\right) \\ &\quad + X_3^2\left(X_2 - X_1\right) \end{aligned} \tag{11.3n2}$$

$$A = \begin{vmatrix} Y_1 & X_1 & X_1^2 \\ Y_2 & X_2 & X_2^2 \\ Y_3 & X_3 & X_3^2 \end{vmatrix} = Y_1\begin{pmatrix} X_2 & X_2^2 \\ X_3 & X_3^2 \end{pmatrix}$$
$$- Y_2\begin{pmatrix} X_1 & X_1^2 \\ X_3 & X_3^2 \end{pmatrix} + Y_3\begin{pmatrix} X_1 & X_1^2 \\ X_2 & X_2^2 \end{pmatrix} \tag{11.3n3}$$

$$\begin{aligned} A &= Y_1\left(X_2 X_3^2 - X_2^2 X_3\right) \\ &\quad - Y_2\left(X_1 X_3^2 - X_1^2 X_3\right) + Y_3\left(X_1 X_2^2 - X_1^2 X_2\right) \end{aligned} \tag{11.3n4}$$

$$B = \begin{vmatrix} 1 & Y_1 & X_1^2 \\ 1 & Y_2 & X_2^2 \\ 1 & Y_3 & X_3^2 \end{vmatrix} = (1)\begin{pmatrix} Y_2 & X_2^2 \\ Y_3 & X_3^2 \end{pmatrix} - (1)\begin{pmatrix} Y_1 & X_1^2 \\ Y_3 & X_3^2 \end{pmatrix}$$
$$+ (1)\begin{pmatrix} Y_1 & X_1^2 \\ Y_2 & X_2^2 \end{pmatrix} \tag{11.3n5}$$

$$\begin{aligned} B &= 1\left(Y_2 X_3^2 - X_2^2 Y_3\right) - 1\left(Y_1 X_3^2 - X_1^2 Y_3\right) \\ &\quad + 1\left(Y_1 X_2^2 - X_1^2 Y_2\right) \\ &= Y_2 X_3^2 - X_2^2 Y_3 - Y_1 X_3^2 + X_1^2 Y_3 \\ &\quad + Y_1 X_2^2 - X_1^2 Y_2 \\ &= X_1^2\left(Y_3 - Y_2\right) + X_2^2\left(Y_1 - Y_3\right) \\ &\quad + X_3^2\left(Y_2 - Y_1\right) \end{aligned} \tag{11.3n6}$$

$$C = \begin{vmatrix} 1 & X_1 & Y_1 \\ 1 & X_2 & Y_2 \\ 1 & X_3 & Y_3 \end{vmatrix} = (1)\begin{pmatrix} X_2 & Y_2 \\ X_3 & Y_3 \end{pmatrix}$$
$$- (1)\begin{pmatrix} X_1 & Y_1 \\ X_3 & Y_3 \end{pmatrix} + (1)\begin{pmatrix} X_1 & Y_1 \\ X_2 & Y_2 \end{pmatrix} \tag{11.3n7}$$

$$C = 1(X_2Y_3 - X_3Y_2) - 1(X_1Y_3 - X_3Y_1) + 1(X_1Y_2 - X_2Y_1)$$
$$= X_2Y_3 - X_3Y_2 - X_1Y_3 + X_3Y_1 + X_1Y_2 - X_2Y_1 \qquad (11.3n8)$$
$$= X_1(Y_2 - Y_3) + X_2(Y_3 - Y_1) + X_3(Y_1 - Y_2)$$

$$A_o = \frac{A}{D}$$

$$A_1 = \frac{B}{D} \qquad (11.3n9)$$

$$A_2 = \frac{C}{D}$$

Alternatively we can call the `avGaussian3x4` procedure to solve Equations 11.3m1, 11.3m2, and 11.3m3.

We are now at the same point as that of the `avFitCurve` procedure at which the values of A₀, A₁, and A₂. We can call it quits and let the construction procedures of the next subsection take over, or we can make similar user assumptions as those of the `avFitCurve` procedure and summarized in Table 11.1.

SubavVc3points	(X1, Y1, X2, Y2, X3, Y3, PVCsta, G1, G2, L, _ A0, A1, A2, E0, Kfactor, _ PVCelv, PVIsta, PVIelv, PVTsta, PVTelv, IERR, IERR)

The input parameters to this procedure are the:

X1, Y1	The station and elevation of the first point through which the curve is to pass.
X2, Y2	The station and elevation of the second point through which the curve is to pass.
X3, Y3	The station and elevation of the third point through which the curve is to pass.
PVCsta	The user assumed station at which the fitted curve is to start, or zero if this parameter is not to be considered assumed (see Note 3).
G1, G2	The user assumed back and forward grades to the fitted curve, or zero if this parameter is not to be considered assumed (see Note 3).
L	The user assumed length of the fitted curve, or zero if this parameter is not to be considered assumed (see Note 3).

Reference is made to the procedures of Sect. 11.3 of this chapter regarding the determination of the constants A₀, A₁, and A₂ of the parabolic curve.

The returned parameter from this procedure is the:

A0	The coefficient A_0 as computed by the procedure.
A1	The coefficient A_1 as computed by the procedure.
A2	The coefficient A_1 as computed by the procedure.
PVCelv	The computed elevation at the PVC station.
PVIsta	The computed PVI station value.

PVIelv	The computed elevation at the PVI station (not the elevation on the curve).
PVTsta	The computed PVC station value.
PVTelv	The computed elevation at the PVC station.
IERR	An error indicator the value of which denotes the following:

0 No errors have been encountered.
3 The assumed or computed grades G1 and G2 are equal within a tolerance of 0.0000.1 (0.0001 %). The procedure aborts.
4 An appropriate condition has not been specified. The procedure aborts.
5 The denominator determinant D is zero. The three input points could be colinear. The procedure aborts.

Upon invocation, the procedure makes certain that the two grades are not equal to each other if both have been assumed as given. If any of the tests fails the indicator IERR is set to be 3 and the procedure aborts, else it continues as follows:

1. Initializes the condition parameter CONDI to zero and then queries pairs of the optional input arguments PVCsta, PVC, G2, L and determines which appropriate pair has been assigned a non-zero value as previously stated. If a proper pair has not been encountered IERR is set to 4 and the procedure aborts; else the other two parameters are initialized to zero and continues.
2. Computes the value of the determinant D by means of Equation 11.3n2.
3. Checks the absolute value of D for being equal to zero, and if it is so the value of IERR is set to be 5, and the procedure aborts.
4. Computes the value of the determinants A, B, and C by means of Equations 11.3n4, 11.3n6, and 11.3n8.
5. Computes the value of the constants A0, A1A2, and A2 by means of Equations 11.3n9.
6. Computes the value of the rate of parabolic change R by Equation 11.3k1, and then depending on the value of the parameter CONDI and the equations of Table 11.1 computes the values of the user non-specified optional input parameters.
7. Computes the value of the Y axis intercept E_0 by Equation 11.3k4. Note that for condition 5 this must be computed after the computation of the PVC.
8. Checks the final values of the G1 and G2 grades for being equal to each other, thus implying two collinear tangent lines.
9. Computes the value of the AASHTO K factor as the absolute value to the nearest whole number of the curve length L divided by the algebraic difference of the back and forward grades expressed in percent form.
10. Computes the values of the remaining arguments PVCelv, PVIsta, PVIelv, PVTsta, PVTelv that are to be returned, and sets IERR to be 0 denoting that no errors have been found.

This concludes the function of the procedure. The calling program may now display the curve for approval or rejection.

In the subject procedure the K factor is not computed as $K = |L/A|$ but as $K = |L/(G_2 - G_1)|$, where the grades are in percent form, because the R as computed by the procedure reflects the fitted curve and not the computed curve with the final L, G_1 and G_2 values.

11.4 Special Vertical Curve Constructions

In this section we present certain procedures that enable the user to (**a**) construct a grade line tangent to a vertical curve from an external point, or at a specific station or elevation of the curve, (**b**) pass a vertical curve tangent to a pair of given lines and passing through a known point, (**c**) project a point on a vertical curve, and (**d**) intersect a straight line with a vertical curve. In using these procedures, it is noted that the creation of a vertical alignment is a process of repetitive trials in meeting various controls that may have been defined by the design engineer. Thus, the end user of the computer program to be developed will need to use several of the following commands before accepting a final vertical alignment.

The construction procedures presented below assume that a vertical curve exists and is to interact with a line and/ or point. To so assume, the vertical curve may be defined by specification of one of the following two sets of parameters:

(a) The constants A_0, A_1, and A_2 of Equation 11.3i4. This could be the case in which the vertical curve has been fitted through a set of three or more points and the user wishes to modify the back and/or forward grade lines. For a general parabola it is possible for the constants A_0, and A_2 to be zero, but not for constant A_2. However, for roadway alignments the constants A_0, and A_2 would most probably not have a zero value.

(b) The terms G_1, G_2, L, C, and E_0 of Equations 11.3i2 and 11.1v5, where C and E_0 are the PVC station and elevation, respectively. This could be the case of profile realignment, or possibly bringing in the grade line from a ramp unto the vertical alignment of the main line. In this case the constants A_0, A_1, and A_2 are determined by Equations 11.3i5, 11.3i6, and 11.3i7 where R is defined by Equation 11.1v5, and E_0 is the elevation of the PVC point.

In the procedures to follow a vertical curve may be defined by either the constants A0, A1, and A2 of the parabola equation, or by the terms G1, G2, L, and PVC station and elevation. If one set of curve definition parameters has valid input value, all others must have a zero value.

11.4.1 Tangent Grade to a Curve from an External Point

The avGrdTcrvFrPnt procedure presented below enables the user to construct a grade line tangent to a vertical curve from a point external to the vertical curve as indicated in Figure 11.4a. Shown in this figure are three points external to the curve from which a tangent line to the curve is to be constructed and which points are labeled (X_{i1}, Y_{i1}) and (X_{i2}, Y_{i2}) where the subscript "i" refers to any point identification number. As seen in this figure from any such point it is possible to construct two tangent lines, one point located ahead of the given external point designated by the subscript 1, and the other located back of the said point designated by the subscript 2. The slope of these two tangent lines is identified in the said figure as g_1 and g_2 with the subscripts corresponding to the said ahead and back tangent lines. It is assumed that the vertical curve is defined by one of the two manners discussed in the preamble to this section of this chapter. For simplification purposes the subscripts have been eliminated in the equations presented below.

The slope of the tangent line to be created may be defined by the first derivative of the parabolic curve defined by Equation 11.3k8 and repeated below as

$$\frac{dY}{dX} = A_1 + 2A_2X = G = \frac{Y-E}{X-S} \quad (11.4a)$$

where S and E are the station and elevation of the given point through which the tangent line is to pass. Solving the above equation for Y, e get

$$Y = A_1(X-S) + 2A_2X(X-S) + E \quad (11.4b1)$$

or

$$Y = A_1X - A_1S + 2A_2X^2 - 2A_2SX + E \quad (11.4b2)$$

or

$$Y = 2A_2X^2 + (A_1 - 2A_2S)X + (E - A_1S) \quad (11.4b3)$$

If we now equate Equation 11.4b3 to Equation 11.3i4

$$A_0 + A_1X + A_2X^2 = 2A_2X^2 \\ + (A_1 - 2A_2S)X + (E - A_1S) \quad (11.4c1)$$

Figure 11.4a Line Tangent to Vertical Curve from an External Point

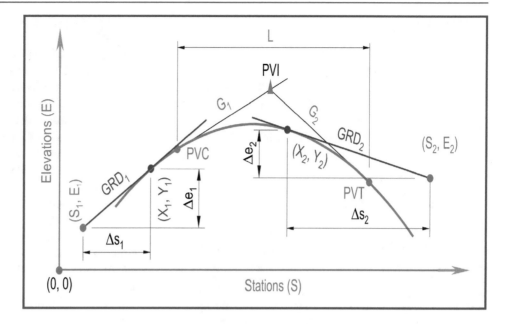

or

$$A_2X^2 - 2A_2SX + (E - A_1S - A_0) = 0 \qquad (11.4c2)$$

and if in the above quadratic equation we let $F = E - A_1S - A_0$, and then solve for X

$$X = \frac{2A_2S \pm \sqrt{4A_2^2S^2 - 4A_2F}}{2A_2}$$

$$X = S \pm \sqrt{S^2 - \frac{F}{A_2}} \qquad (11.4d1)$$

from which the station of the point of tangency would be either X_1 if the given external point is located back of the PVI, or X_2 if the given external point is located ahead of the PVI.

$$X_1 = S + \sqrt{S^2 - \frac{F}{A_2}} \qquad (11.4d2)$$

$$X_2 = S - \sqrt{S^2 - \frac{F}{A_2}} \qquad (11.4d3)$$

In Figure 11.4a, if the given external point is:

- At point (S_1, E_1), station X_2 would place the tangent grade towards the low left corner of the figure.
- At point (S_2, E_2), station X_1 would place the tangent grade towards the low right corner of the figure.
- Somewhere between the PVC and the PVT of the vertical curve, then either solution could be the desired solution. Therefore the procedure returns both solutions and leaves the selection of the desired solution up to the calling program.

Having selected the appropriate X station of the point of tangency, we can now substitute this value for X in Equation 11.3i4 and solve for the elevation Y of the point of tangency.

Now let us take a closer look at the discriminant of Equation 11.4d1, that is, the value under the square root. If this value is:

- Positive, there are two distinct solutions as defined by Equations 11.4d2 and 11.4d3,
- Zero, there are two identical solutions implying that the given point lies on the curve.
- Negative, there is no real solution, that is, the point is located within the vertical curve.

```
Sub avGrdTcrvFrPnt (A0, A1, A2, G1, G2, L,
                    PVCS, PVCE, S, E, _ X1,Y1,
                    X2,Y2, GRD1, GRD2, IERR)
```

The input parameters to this procedure are the:

A0	The A_0 constants of Equation 11.3i4, or zero.
A1	The A_1 constants of Equation 11.3i4, or zero.
A2	The A_2 constants of Equation 11.3i4, or zero.
G1	The back grade of the vertical curve (not in percent form), or zero.
G2	The forward grade of the vertical curve (not in percent form), or zero.
L	The length of the vertical curve, or zero.
PVCS	The station of the PVC point of the vertical curve without the plus (+) sign, or zero.
PVCE	The elevation of the PVC point of the vertical curve, or zero.
S	The station without the plus (+) sign of the given external point.
E	The elevation of the given external point.

NOTE that:

1. The vertical curve may be defined by specifying the constants A0, A1, and A2 of the parabola equation, or the parameters G1, G2, L, PVCS, and PVCE of the vertical curve. When specifying one set of parameters, all others must have a zero value.
2. Reference is made to the operational instructions below on how the selection of the proper curve definition is made.

The returned parameter from this procedure is the:

X1	The station of the first point of tangency.
Y1	The elevation of the first point of tangency.
X2	The station of the second point of tangency.
Y2	The elevation of the second point of tangency.
GRD1	The slope of the first tangent line from the (S, E) point to the vertical curve.
GRD2	The slope of the second tangent line from the (S, E) point to the vertical curve.
IERR	An error indicator, the value of which if it is:

    0   Denotes that an error has not been encountered.

    1   Denotes that the vertical curve has not been properly defined (A2 = 0).

    2   Denotes that the given point is within the curve, and hence there is no solution.

    3   Denotes that the vertical curve has not been properly defined (L = 0).

    4   Denotes that there is no solution (the point is most probably within the curve).

When the vertical curve is too flat, the procedure becomes rather sensitive to the degree of precision in the values of the constants A_0, A_1, and A_2. So in such cases it may be advantageous to use the vertical curve definition that computes the said constants.

Upon invocation, the procedure initializes to zero the parameters to be returned, and then:

1. Queries the values of G1, G2, L, PVCS, and PVCE for all of them having a zero value. If they all have a zero value, it is assumed that A0, A1, and A2 have been specified and continues; else it branches to Step 1(b).
   (a) Queries the value of A2 for having a zero value. If so, IERR is set to be 1 and the procedure aborts; else continues.
   (b) It is assumed that G1, G2, L, PVCS, and PVCE and queries the value of G1 and G2 for being collinear within a tolerance of 0.000001 (0.0001 %). If so,

IERR is set to be 2 and the procedure aborts; else continues.
   • Queries L for having a value of zero within a tolerance of 0.005 (1.524 mm). If so, IERR is set to be 3 and the procedure aborts; else continues.
   • the procedure computes the values of
     – R by use of Equation 11.1v5,
     – A0 by use of Equation 11.3i5,
     – A1 by use of Equation 11.3i6, and
     – A2 by use of Equation 11.3i7,
     where
     C = PVCS and
     E_0 = PVCE − PVCS (G1), the Y axis intercept of the vertical curve equation.
2. Computes the value of $F = E − A_1 S − A_0$.
3. Computes the value of the discriminant $S^2-(F/A_2)$, and if it is
   • Negative, the procedure sets IERR to be 4, and then aborts.
   • Zero, the procedure sets X1 = X2 = S and Y1 = Y2 = E and terminates (see below).
   • Positive, continues with the next step.
     When testing for a discriminant value of zero, the procedure checks for a "pure" zero and not for a zero within any tolerance. Thus, it is possible to get two points of tangency that are extremely close together.
4. Computes the:
   • Stations of the points of tangency X1 and X2 by use of Equations 11.4d2 and 11.4d3, respectively.
   • Elevations of the points of tangency Y1 and Y2 by introducing the values of X1 and X2, respectively, in Equation 11.3i4.
   • The slopes GRD1 and GRD 1 of the tangent lines, and terminates.

If the procedure terminates successfully (IERR = 0), it is up to the calling program to determine which tangency solution is the desired one. The selection would most probably involve a comparison of the stations X1 and X2 with respect to the station of the PVC.

11.4.2 Tangent Grade to a Curve at a Given Station

The avGrdTcrvAtSta procedure presented below enables the user to determine the grade of the tangent line to a vertical curve at a point on the vertical curve defined by its station value as indicated in Figure 11.4b. It is assumed that the vertical curve is defined by one of the two manners discussed in the preamble to this section of this chapter.

The slope G of the tangent line to be determined may be defined by the first derivative of the parabolic curve defined by Equation 11.3k8 and repeated below as

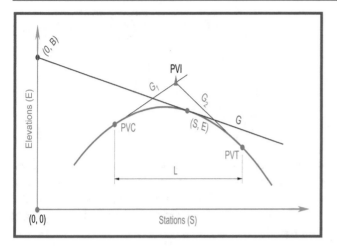

Figure 11.4b Line Tangent to a Vertical Curve at a Station

$$G = \frac{dY}{dX} = A_1 + 2A_2 X \qquad (11.4e)$$

At $X = S$

$$G = A_1 + 2A_2 S \qquad (11.4f)$$

The equation of the tangent line passing through point (S, E) may be expressed as

$$E = GS + B \qquad (11.4g1)$$

where B is the Y (elevation) axis intercept.

The elevation E at station S may be expressed by Equation 11.3i4 as

$$E = A_0 + A_1 S + A_2 S^2 \qquad (11.4g2)$$

If we equate Equation 11.4g1 to Equation 11.4g2 and solve for B, we obtain

$$B = A_0 + A_1 S + A_2 S^2 - GS$$
$$B = A_0 + (A_1 - G)S + A_2 S^2 \qquad (11.4g3)$$

Equation 11.4g3 implies that at $S = 0$ the Y axis intercept B is the constant A_0.

If we solve Equation 11.4g1 of the slope G, we have

$$G = \frac{E - B}{S} \qquad (11.4g4)$$

In Figure 11.4b we observe that $B > E$, so that $G < 0$. Had station S been located further to the left, the slope would have reversed direction and become positive. The same concept would hold true for any configuration of a vertical curve.

```
Sub avGrdTcrvAtSta    (A0, A1, A2, G1, G2, L, PVCS,
                       PVCE, S, _ G, E, B, IERR)
```

The input parameters to this procedure are the:

A0	The A_0 constants of Equation 11.3i4, or zero.
A1	The A_1 constants of Equation 11.3i4, or zero.
A2	The A_2 constants of Equation 11.3i4, or zero.
G1	The back grade of the vertical curve (not in percent form), or zero.
G2	The forward grade of the vertical curve (not in percent form), or zero.
L	The length of the vertical curve, or zero.
PVCS	The station of the PVC point of the vertical curve without the plus (+) sign, or zero.
PVCE	The elevation of the PVC point of the vertical curve, or zero.
S	The station without the plus (+) sign of the given point on the curve.

NOTE that:

1. The vertical curve may be defined by specifying the constants A0, A1, and A2 of the parabola equation, or the parameters G1, G2, L, PVCS, and PVCE of the vertical curve. When specifying one set of parameters, all others must have a zero value.
2. Reference is made to the operational instructions below on how the selection of the proper curve definition is made.
3. If the given station is located outside the confines of the vertical curve, the curve is considered to be extended to the said station.

The returned parameter from this procedure is the:

G	The grade (non-percent form) of the tangent line at the given station S of the vertical curve.
E	The elevation of the point at the given station of the curve.
IERR	An error indicator, the value of which if it is:
0	Denotes that an error has not been encountered.
1	Denotes that the vertical curve has not been properly defined (A2 = 0).
2	Denotes that the given point is within the curve, and hence there is no solution.
3	Denotes that the vertical curve has not been properly defined (L = 0).

Upon being called, this procedure initializes to zero the parameters to be returned, and then:

1. Defines the vertical curve by executing Step 1 of the operational steps of the avGrdTcrvFrPnt procedure.
2. Computes the grade G using Equation 11.4f.
3. Computes the elevation at station S using Equation 11.4g2.
4. Computes the Y axis intercept B (elevations) by use of Equation 11.4g3, and then terminates.

11.4.3 Tangent Grade to a Curve at a Given Elevation

The avGrdTcrvAtElv procedure presented below enables the user to determine the grade of the tangent line to a vertical curve at a point on the vertical curve defined by its elevation

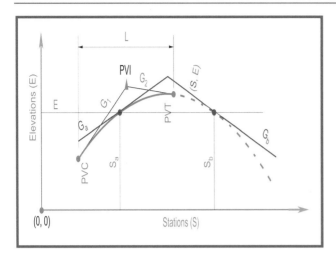

Figure 11.4c Line Tangent to a Vertical Curve at an Elevation

as indicated in Figure 11.4c. Due to the nature of the parabolic curve in its full extent a given elevation E may occur at two distinct points S_a and S_b, at which there can be the tangent lines with grades G_a and G_b, respectively. However, since a roadway vertical curve occupies only a portion of a parabola of infinite length, for any given elevation there may be two stations, or only one station within the confines of a given vertical curve.

In the said figure a vertical curve occupies a portion of the corresponding parabolic curve and is extended to the right beyond the PVT. At first glance one might say that in this case only the left of the two solutions (S_a) might be the desired one. However, the case may be made for having the solution at the right side (S_b) for either a roadway reconstruction or realignment, or possibly for an exiting ramp. Thus, the procedure returns both solutions (station and tangent line slope), and it is up to the calling program to select the desired one.

From the said figure it is evident that the given elevation cannot be above the high point of a crest curve, nor below the low point of a sag curve. The procedure checks for this condition by use of Equations 11.3k9 and 11.3i4, and flags an error message if that is the case.

It is assumed that the vertical curve is defined by one of the two manners discussed in the preamble to this section of this chapter. The equation of the parabolic curve 11.3i4 for an elevation E may be written as

$$A_2 X^2 + A_1 X + A_0 - E = 0 \tag{11.4h}$$

where X could be either S_a or S_b. Thus, solving the above quadratic equation of X we get

$$X = \frac{-A_1 \pm \sqrt{A_1^2 - 4A_2(A_0 - E)}}{2A_2} \tag{11.4i1}$$

or

$$S_a = \frac{-A_1 - \sqrt{A_1^2 - 4A_2(A_0 - E)}}{2A_2} \tag{11.4i2}$$
$$(for\ sag\ curve)$$

and

$$S_b = \frac{-A_1 + \sqrt{A_1^2 - 4A_2(A_0 - E)}}{2A_2} \quad (for\ a\ sag\ curve) \tag{11.4i3}$$

With the location of the stations corresponding to the given elevation the slope of the tangent line G_A and G_B at those stations may be computed by use Lof Equation 11.3k8 using stations S_A and S_B, and the elevations of the Y axis intercepts may be computed as

$$B_A = E - (S_A * G_A) \quad (for\ a\ sag\ or\ rest\ curve) \tag{11.4i4}$$

and

$$B_B = E + (S_B * G_B) \quad (for\ a\ sag\ curve) \tag{11.4i5}$$

In Figure 11.4c S_a is shown to be located on the left side, that is, towards the start of the vertical curve in station ascending order, and S_b is shown to be located on the right side, that is, towards the end of the vertical curve in station ascending order. **It is important to note** that in order to maintain this location sequence of S_a and S_b, the Equations 11.4i2, 11.4i3, and 11.4i5 pertain to a sag curve. For a crest curve Equations 11.4i6, 11.4i7, and 11.4i8 pertain should be used. The reason for this is that for a sag curve $A_2 > 0$ and for a crest curve $A_2 < 0$.

$$S_a = \frac{-A_1 + \sqrt{A_1^2 - 4A_2(A_0 - E)}}{2A_2} \quad (for\ a\ crest\ curve) \tag{11.4i6}$$

$$S_a = \frac{-A_1 - \sqrt{A_1^2 - 4A_2(A_0 - E)}}{2A_2} \quad (for\ a\ crest\ curve) \tag{11.4i7}$$

and

$$B_B = E - (S_B * G_B) \quad (for\ a\ sag\ curve) \tag{11.4i8}$$

```
Sub avGrdTcrvAtElv(A0,  A1,  A2,  G1,  G2,  L,
PVCS, PVCE, E,  _ SA, SB, GA, GB, BA, BB, IERR)
```
The input parameters to this procedure are the:

A0	The A_0 constants of Equation 11.3i4, or zero.
A1	The A_1 constants of Equation 11.3i4, or zero.
A2	The A_2 constants of Equation 11.3i4, or zero.
G1	The back grade of the vertical curve (not in percent form), or zero.
G2	The forward grade of the vertical curve (not in percent form), or zero.
L	The length of the vertical curve, or zero.

PVCS	The station of the PVC point of the vertical curve without the plus (+) sign, or zero.
PVCE	The elevation of the PVC point of the vertical curve, or zero.
E	The given elevation on the curve.

NOTE that:

1. The vertical curve may be defined by specifying the constants A0, A1, and A2 of the parabola equation, or the parameters G1, G2, L, PVCS, and PVCE of the vertical curve. When specifying one set of parameters, all others must have a zero value.
2. Reference is made to the operational instructions below on how the selection of the proper curve definition is made.

The returned parameter from this procedure is the:

SA	The station S_a of the curve at the given elevation E.
SB	The station S_b of the curve at the given elevation E.
GA	The G_a grade in non percent form of the tangent line at the computed station SA of the vertical curve.
GB	The G_b grade in non percent form of the tangent line at the computed station SB of the vertical curve.
BA	The Y (elevation) axis intercept corresponding to the computed station SA.
BB	The Y (elevation) axis intercept corresponding to the computed station SB.
IERR	An error indicator, the value of which if it is:
	0 Denotes that an error has not been encountered.
	1 Denotes that the vertical curve has not been properly defined (A2 = 0).
	2 Denotes that the two given grades are colinear.
	3 Denotes that the vertical curve has not been properly defined (L = 0).
	4 Denotes that the given elevation is lower than the low point of a sag curve.
	5 Denotes that the given elevation is higher than the high point of a crest curve.

Upon being called, this procedure initializes to zero the parameters to be returned and then

1. Defines the vertical curve by executing Step 1 of the operational steps of the avGrdTcrvFrPnt procedure. In addition to defining the vertical curve it determines whether the curve is a sag or crest curve.
2. Computes the station and elevation of the low point for a sag curve or the high point of crest curve by setting the slope to be zero and using Equations 11.3k8 and 11.3k9, respectively.
3. Compares the elevation of the low or high point with the given elevation for being equal to each other within a tolerance of 0.005 feet (1.524 mm), and if so sets both stations

of tangency and elevations to be those of the computed low (sag curve) or high (crest curve) point; else continues.
4. Compares the given elevation with the computed low or high point for being lower (sag curve) or higher (crest curve), and if so sets IERR to be 4 or 5, respectively and aborts; else continues.
5. Computes the stations SA and SB of the points of tangency using Equations 11.4i2 and 11.4i3 for a sag curve, or Equations 11.4i6 and 11.4i7 for a crest curve.
6. Computes the slope of the tangent line GA and GB at those station may be computed by use of Equation 11.3k8.
7. Computes the elevations of the Y axis intercepts using Equations 11.4i4 and 11.4i5 for a sag curve, and Equations 11.4i4 and 11.4i8 for a crest curve, and terminates.

11.4.4 Curve Through One Point and Tangent to Two Grade Lines

The avCrv1Pnt2Grd procedure enables the user to construct a vertical curve that passes through a given point PT and is tangent to two given grade lines as indicated in Figure 11.4d. In essence this procedure assumes that the PVI and the back and forward tangent grades G_1 and G_2 are known and it is desired to locate the PVC and PVT and determine the length of the vertical curve.

This procedure may be of use when it is desired to introduce a vertical curve to provide a certain amount of cover above an underground utility, or clearance under an overhead structure or other feature such as the low point of the catenary of an electric transmission wire, in which case the elevation and alignment station are known, as shown in the said figure.

With reference to Figure 11.4d, which resembles Figure 11.1d in an upside down way, if we divide Equation 11.1t4 by Equation 11.1u1 we have

$$\frac{y}{E} = \frac{\left(\dfrac{G_2 - G_1}{2L}\right)x^2}{\left(\dfrac{G_2 - G_1}{8}\right)L} = \frac{\left(G_2 - G_1\right)x^2}{2L}\frac{8}{\left(G_2 - G_1\right)L} \tag{11.4j1}$$

$$= \frac{4x^2}{L^2}$$

and solving for y, we get

$$y = 4E\left(\frac{x}{L}\right)^2 \tag{11.4j2}$$

where

y = the vertical offset distance from the tangent grade line to the vertical curve.

Figure 11.4d Vertical Curve Passing Through a Point and Tangent to Two Grade Lines

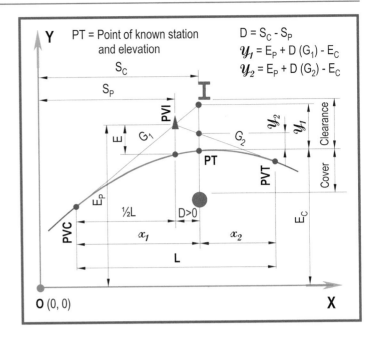

$E=$ the middle ordinate at the PVI station.

$L=$ the length of the vertical curve.

Thus, the vertical offset y varies as the square of the distance x over the curve length L times $4E$, regardless from which end of the curve the distance x is measured.

Equation 11.1u1 defines the middle ordinate E in terms of the grades G_1 and G_2 and the curve length L. For this current construction we are going to define E in terms of the station difference D, the elevation differences y_1 and y_2, the equations of which are shown in Figure 11.4d, and the vertical curve length L by working from the PVC towards the PVT, and in the reverse direction.

Regarding the values of y_1 and y_2 it should be noted that:

- Whereas with the equations previously presented the said values are positive for sag curves and negative for crest curves because they are measured from the tangent grade line to the vertical curve, with the Equations in Figure 11.4d they are the reverse because they are measured from the curve to the tangent grade line.
- Since as is to be seen below the square root of the said values is to be used, the absolute value of said values should be employed.

From the PVC towards the PVT

$$x_1 = \frac{L}{2} + D \qquad (11.4\text{k1})$$

$$y_1 = 4E\left(\frac{\frac{L}{2}+D}{L}\right)^2 \qquad (11.4\text{k2})$$

$$E = \frac{y_1}{4\left(\frac{\frac{L}{2}+D}{L}\right)^2} \qquad (11.4\text{k3})$$

From the PVT towards the PVC

$$x_2 = \frac{L}{2} - D \qquad (11.4\text{k4})$$

$$y_2 = 4E\left(\frac{\frac{L}{2}-D}{L}\right)^2 \qquad (11.4\text{k5})$$

$$E = \frac{y_2}{4\left(\frac{\frac{L}{2}+D}{L}\right)^2} \qquad (11.4\text{k6})$$

Equating Equation 11.4k3 to Equation 11.4k6 and simplifying we get

$$\frac{y_1}{\left(\dfrac{L}{2}+D\right)^2} = \frac{y_2}{\left(\dfrac{L}{2}-D\right)^2} \qquad (11.4k7)$$

or

$$\sqrt{y_1}\left(\frac{L}{2}-D\right) = \sqrt{y_2}\left(\frac{L}{2}+D\right) \qquad (11.4k8)$$

Solving the above equation for the length L of the vertical curve we get

$$L = \frac{2D\left(\sqrt{y_1}+\sqrt{y_2}\right)}{\left(\sqrt{y_1}-\sqrt{y_2}\right)} \qquad (11.4k9)$$

We can now solve for the stations and elevations of the PVC and PVT points by:

- Adding to and subtracting from the PVI station, respectively, half the value of L, and
- Multiplying the said half value by the back and forward grades G_1 and G_2, respectively, and adding them to the elevation of the PVI.

```
Sub avCrv1Pnt2Grd(PVIS, PVIE, G1, G2, PTS, PTE,
              _ PVCS, PVCE, PVTS, PVTE, L)
```

The input parameters to this procedure are the:

PVIS, PVIE	The station and elevation, respectively, of the PVI (S_P and E_P in Figure 11.4d).
G1, G2	The back and forward grades, respectively (see Figure 11.4d).
PTS, PTE	The station and elevation, respectively, of the point through which the curve is to pass (S_C and E_C in Figure 11.4d).

The returned parameter from this procedure is the:

PVCS, PVCE	The station and elevation, respectively, of the PVC point.
PVTS, PVTE	The station and elevation, respectively, of the PVT point.
L	The length of the vertical curve.
K	The AASHTO K factor.
IERR	An error indicator, the value of which if it is:
	0 Denotes that an error has not been encountered.
	1 Denotes that the two given grades are colinear.

Upon invocation this procedure initializes to zero the parameters to be returned and then:

1. Queries the values of G1 and G2 for being colinear within a tolerance of 0.000001 (0.0001 %). If so, IERR is set to be 1 and aborts; else continues.
2. Computes the distances D, y_1, and y_2 as stated above. Note that the absolute value of y_1 and y_2 is computed.
3. Computes the length of the vertical curve L by means of Equation 11.4k9.
4. Computes the station and elevation of the PVC point by traversing backwards from the PVI point a distance of $L/2$ at a G_1 grade, and of the PVT point by traversing forwards from the PVI point a distance of $L/2$ at a G_2 grade.
5. Computes the AASHTO K factor and terminates.

11.4.5   Project a Point on a Curve

With the avCrv1Pnt2Grd procedure we are able to pass a vertical curve through a point in order to maintain a certain clearance from an underground or overhead feature, with the clearance being measured in the vertical plane. However, when a clearance is of rather critical concern, such as the clearance under the bottom flanges the girders of a bridge, the normal distance from a vertical curve might be a more appropriate mensuration. Thus, with the avPrjPtCrv procedure the user is able to project a given point of known station and elevation on a given vertical curve and determine the normal offset distance of the given point from its point of projection on the curve and the station and elevation of the said point of projection. It is assumed that the vertical curve is defined by one of the two manners discussed in the preamble to this section of this chapter.

Before we proceed any further let us take a look at the three parabolas of Figure 11.4e1 in which five points external to a parabola (not located on a parabola) are projected on a parabola. In this figure it is seen that there could be one, two or three possible points of projection on a parabola depending on where the external point is located with respect to the parabola. The proof of this fact is beyond the scope of this publication, but the issue of the three projection points and the selection of the desired projection point is to be discussed later on in this subsection.

Figure 11.4e2 contains a portion of a parabola defined by Equation 11.1m repeated below and the external point (S,E). At the point of projection (X, Y) of the said external point the slope of the tangent line to the parabola is defined by the first derivative of the said curve represented by Equation 11.3k8

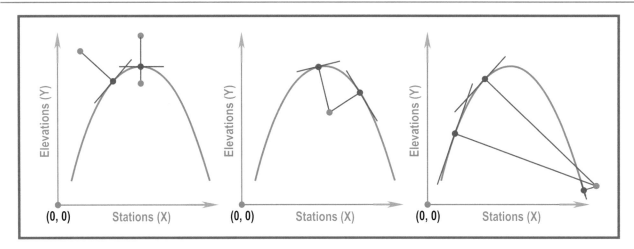

Figure 11.4e1 Point Projection on a Parabola

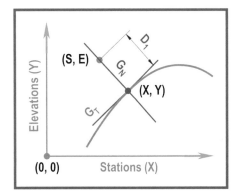

Figure 11.4e2 Point Projection on a Vertical Curve (Cubic Equation)

and repeated below, and the slope of the normal line from the said external point towards the said tangent line is the negative reciprocal of said tangent line slope expressed as:

$$f(X) = A_0 + A_1 X + A_2 X^2 = Y$$
$$\frac{dY}{dX} = A_1 + 2A_2 X = G_T \qquad (11.4m1)$$
$$G_N = \frac{-1}{A_1 + 2A_2 X}$$

From Figure 11.4e2 the slope of the normal line to the parabola from the said external point may also be expressed as

$$G_N = \frac{Y - E}{X - S} \qquad (11.4m2)$$

If we now equate the above two equations of the slope of the normal line G_N to each other and substituting the value of Y from Equation 11.4m1 into the right side of Equation 11.4m2, we get

$$\frac{A_0 + A_1 X + A_2 X^2 - E}{X - S} = \frac{-1}{A_1 + 2A_2 X} \qquad (11.4m3)$$

In Equation 11.4m3 all terms except for X are known. Thus, solving for X by first cross multiplying the two sides of Equation 11.4m3 and then collecting terms as indicated below

$$\left(A_0 + A_1 X + A_2 X^2 - E\right)\left(A_1 X - 2A_2 X\right) = S - X \qquad (11.4m4)$$

$$A_0 A_1 + A_1^2 X + A_1 A_2 X^2 - A_1 E + 2A_0 A_2 X$$
$$+ 2A_1 A_2 X^2 + 2A_2^2 X^3 - 2A_2 EX - S + X = 0 \qquad (11.4m5)$$

we get the cubic equation

$$\left(2A_2^2\right)X^3 + \left(3A_1 A_2\right)X^2$$
$$+ \left(A_1^2 + 2A_0 A_2 - 2A_2 E + 1\right)X + \left(A_0 A_1 - A_1 E - S\right) = 0 \qquad (11.4m6)$$

If we now let

$$a = 2A_2^2 \qquad\qquad b = 3A_1 A_2$$
$$c = A_1^2 + 2A_0 A_2 - 2A_2 E + 1 \quad d = A_0 A_1 - 2A_1 E - S \qquad (11.4m7)$$

and substitute a, b, c, and d in Equation 11.4m6 we get the cubic equation

$$aX^3 + bX^2 + cX + d = 0 \qquad (11.4m8)$$

The direct solution of a cubic equation could be considered to be a bit messy to deal with as it involves the use of complex or imaginary numbers that will result in one of three possible solutions or roots: (a) three real values of X, (b) three values of X at least two of which are the same, or (c) one real value of X and a pair of complex conjugate values. There are certain other methods of solution which are left to the reader's interest of research, but a common iterative method of solving a cubic and other complicated equations is referred to as the bisection method. The basic premise of this solution is that a root of the equation is flanked by a positive and a negative, or the reverse, Y value.

With this iterative method the approximate location of the desired X root must be known (assumed) and specified with a lower and a higher range limit of X (X_L and X_H, respectively) within which the desired root X is to be located. The process computes the value of Y at each range limit and at the midpoint of the range. It then compares the sign of the Y at the low limit and with that of the midpoint and halves the range by moving the low X or the high X to the midpoint until the two X values are close enough to be considered the same. To safeguard against the possibility of an erroneous assumption of low and high range limits a maximum number of iterations to be performed should be assumed. Also to be assumed is the allowable error in X values to be considered as being the same. The specific steps to carry out the bisection method are presented in the operational steps below.

Now let us consider the actual construction problem at hand, the projection of point on a roadway vertical curve. The parabola of this curve is not as pronounced as indicated in the last said two figures as it is rather flat and the general or approximate location of the desired root of the cubic equation is known or can easily be ascertained. For instance the root will not be too far away from the station of the given external point, and the lower and higher point estimates could be assumed to be 50 or 100 feet on each side of the station of the given external point. Furthermore, it is doubtful that the other two roots, even if they are real, will be of any use because in most cases will be located off the area of interest. The same argument could also be made for the rare use of a parabolic structural element requiring certain detail for a structural connection. Thus, an iterative solution does not present a problem, nor is speed an issue with the current state-of-the art of electronic computing.

```
Sub avPrjPtCrv    (A0, A1, A2, G1, G2, L, PVCS,
                   PVCE, PST, PEL, DR, Tries, _ X1,
                   Y1, X2, Y2, Snorm, Stang, I, IERR)
```

The input parameters to this procedure are the:

A0	The A_0 constants of Equation 11.3i4, or zero.
A1	The A_1 constants of Equation 11.3i4, or zero.
A2	The A_2 constants of Equation 11.3i4, or zero.
G1	The back grade of the vertical curve (not in percent form), or zero.
G2	The forward grade of the vertical curve (not in percent form), or zero.
L	The length of the vertical curve, or zero.
PVCS	The station of the PVC point of the vertical curve without the plus (+) sign, or zero.
PVCE	The elevation of the PVC point of the vertical curve, or zero.
PST	The station of the given external point to be projected.

PEL	The elevation of the given external point to be projected.
DR	One half the initial range about the PST point over which the bisection method iterations are to be carried out (50 feet should suffice).
Tries	The maximum number or bisection method iterations that should be carried out (50 trials should suffice).

The returned parameter from this procedure is the:

SPP	The station of the point of projection.
EPP	The station and elevation of point of projection farthermost from (S, E).
OFF	The offset distance of the given external point from the vertical curve.
Snorm	The slope of the normal line to the curve from the given external point.
Snorm	The slope of the tangent line to the curve at the point of projection.
I	The number of iterations to solve the equation, or the maximum number of iterations plus one if a solution has not been found.
IERR	An error indicator, the value of which if it is:
	0 Denotes that an error in the input data has not been encountered.
	1 Denotes that an error has been encountered ($A_2 = 0$).
	2 Denotes that G1 and G2 are the same within a tolerance of 0.000001 %.
	3 Denotes improper vertical curve definition (L=0).
	4 Denotes that a solution has not been found within the specified maximum number of iterations.

Upon being called, this procedure initializes to zero the argument to be returned, and then:

1. Defines the vertical curve by executing Step 1 of the operational steps of the avGrdTcrvFrPnt procedure.
2. Computes the constants a, b, c, and d of the cubic equation by use of Equation 11.4m7, and then sets the low and high stations S_L and S_H of the iteration range.
3. Caries out the iteration process of the bisection method for up to the maximum number of iteration (Tries) as follows:
   (a) Computes the station of the midpoint of the two limits $S_A = (S_L + S_H)/2$.
   (b) Computes the functions F_L and F_H of Equation 11.4m8 at S_L and S_H.
   (c) Sets $F_A = (F_L + F_H)/2$.
   (d) If $(S_L - S_H)/2$ or if $F_A = 0$ within a tolerance of 0.0009 feet (0.27432 m), then F_A is the station of the point of projection, its elevation is computed with Equation 11.1m, the iteration process terminates and branches to **Step 4**.
   (e) Else if F_L and F_H are of the same sign, then S_L is set to be SA and the iteration process from step 3 above is repeated.
   (f) Else if F_L and F_H are of the opposite sign, then S_H is set to be S_A and the iteration process from step 3 above is repeated.

4. Queries the number of iteration that have been performed, and if it exceeds the specified maximum number seta IERR to 5, beeps and aborts.
5. Calls the icforce procedure to compute the offset distance and the slope of the normal line by treating the elevation and the station as the north and east coordinates, and then converts the returned azimuth into a Cartesian angle and computes the tangent function of this angle to find the slope of the normal line.
6. Computes the slope of the tangent line at the point of projection (see Equation 11.4m1), and then terminates.

The above discussion on the bisection method has focused on the solution of the cubic Equation 11.4m8 which is a simplification of Equation 11.4m6. The same method could also be used on the basic parabola equation and its derivative (see Equation 11.4m6) without the use of the cubic equation. Just as above the location of the root is assumed as well as the maximum number of iterations and the tolerance for equality of two trial solutions. The difference is that instead of solving for X of the cubic equation, we can project the given external point on the normal to each of the two limits of the range as indicated in Figure 11.4e3. That is:

1. At each end point of the range the slope of the tangent line to the curve is computed, as is the slope of the normal line at those locations (Equation 11.4m1) as well as at the midpoint of the range. This corresponds to **Step 3(a)**.
2. The external point is projected on the said three normal lines instead of computing F_L, F_H, and F_A in Steps 3(b) and 3(c).
3. The remaining steps of the bisection method 3(d), 3(e) and 3(f) are carried out with F_L, F_H, and F_A being analogous to F_L, F_H, and F_A.

In order to compare the above two approaches a test was performed on an 800 foot long vertical curve with a back grade of −1.0 % and a −2.2 % forwards grade. For an external point of about 18 feet away from the curve, the difference in

elevation of the projection point and hence in length of the offset distance was 0.001 feet, with the solution of the cubic curve requiring 16 trials, while the approach of the offsets required 14 trials. For all intents and purposes both approaches yielded the same results. Of these two approaches the subject procedure applies the bisection method on the cubic curve. It is important to note that the location of the point of projection was extremely near the station of the external point.

11.4.6 Intersect a Vertical Curve with a Grade Line

With the avXgrdCrv procedure we are able to intersect a straight (grade) line with a vertical curve. One possible application of this procedure is in the case of a force jacking under a roadway a sewer or water pipe at a given slope along or parallel to the roadway alignment that contains a vertical curve (see Figure 11.4f) and maintaining a minimum of cover at the lowest point. The solution to this construction problem is similar to that of the avGrdTcrvAtElv procedure in which: It is assumed that the vertical curve is defined by one of the two manners discussed in the preamble to this section of this chapter.

With reference to Figure 11.4f the equation of the vertical curve 11.3i4 is repeated below as

$$Y = A_0 + A_1 X + A_2 X^2 \tag{11.4n1}$$

while the equation of the grade line may be expressed as

$$Y = GX + B \tag{11.4n2}$$

Where G is the known slope of the straight line, and B is the Y axis intercept of the said line passing through the given point (S, E). Said intercept may be expressed as

$$B = E - GS \tag{11.4n3}$$

Equations 11.4n1 and 11.4n2 may be equated to each other, with B being replaced with the right side of Equation 11.4n3 to get

$$A_0 + A_1 X + A_2 X^2 = GX + E - GS$$
$$A_2 X^2 + (A_1 - G) X + (A_0 + GS - E) = 0 \tag{11.4n4}$$

Solving the last equation for X we get

$$X = \frac{-(A_1 - G) \pm \sqrt{(A_1 - G)^2 - 4A_2 (A_0 + GS - E)}}{2A_2} \tag{11.4n5}$$

From the above equation we see that if

- $A2 = 0$, the vertical curve cannot be defined.
- The value of the square root is negative then the line and the curve do not intersect.
- The value of the square root is zero then there is only one intersection.

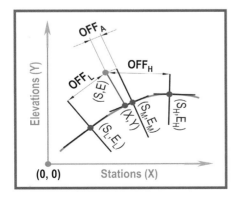

Figure 11.4e3 Point Projection on a Vertical Curve with Offsets

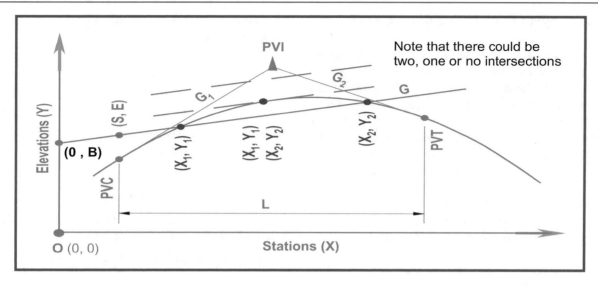

Figure 11.4f   Intersection of a Grade Line with a Vertical Curve

- The value of the square root is not zero then there are two intersections define by the following equations:

$$X_1 = \frac{-(A_1 - G) + \sqrt{(A_1 - G)^2 - 4A_2(A_0 + GS - E)}}{2A_2}$$ (11.4n6)

and

$$X_2 = \frac{-(A_1 - G) - \sqrt{(A_1 - G)^2 - 4A_2(A_0 + GS - E)}}{2A_2}$$ (11.4n7)

It is noted that for a crest curve X_1 will be smaller than X_2 in a station ascending order, and in a sag curve X_2 will be smaller than X_1.

```
Sub avXgrdCrv(A0, A1, A2, G1, G2, L, PVCS,
              PVCE, S, E, G, _
              X1, Y1, X2, Y2, IERR)
```

The input parameters to this procedure are the:

A0	The A_0 constants of Equation 11.3i4, or zero.
A1	The A_1 constants of Equation 11.3i4, or zero.
A2	The A_2 constants of Equation 11.3i4, or zero.
G1	The back grade of the vertical curve (not in percent form), or zero.
G2	The forward grade of the vertical curve (not in percent form), or zero.
L	The length of the vertical curve, or zero.
PVCS	The station of the PVC point of the vertical curve without the plus (+) sign, or zero.
PVCE	The elevation of the PVC point of the vertical curve, or zero.
S	The station of the point on the given grade line.
E	The elevation of the point on the given grade line.
G	The slope of the given grade line.

The returned parameter from this procedure is the:

X1, Y1	The station and elevation of one point of intersection.
X2, Y2	The station and elevation of the other point of intersection.
IERR	An error indicator, the value of which if it is:
	0 Denotes that an error in the input data has not been encountered.
	1 Denotes that an error has been encountered ($A_2 = 0$).
	2 Denotes that G1 and G2 are the same within a tolerance of 0.000001 %.
	3 Denotes improper vertical curve definition ($L = 0$).
	4 Denotes that there is no intersection.

Upon being called, this procedure initializes to zero the argument to be returned, and then:

1. Defines the vertical curve by executing Step 1 of the operational steps of the avGrdTcrvFrPnt procedure.
2. Sets the constants a, b, and c of the quadratic equation to represent the following terms of Equations 11.4n6 and 11.4n7:

$$a = A2 \quad b = A0 - G \quad c = A0 + GS - E$$

3. Computes the value under the square root of Equations 11.4n6 and 11.4n7, and if it is negative, sets the value of IERR to be 4, and aborts; else continues.
4. Computes the station values X1 and X2 using Equations 11.4n6 and 11.4n7, respectively.
5. Computes the elevations Y1 and Y2 at stations X1 and X2 using Equation 11.4n1, and then terminates.

11.4.7 Tangent Grade to Two Given Curves

The `avGrdTo2crvs` procedure presented below enables the user to construct a grade line tangent to two given vertical curves. We are distinguishing here between two parabolas and two vertical curves although the latter are parabolic curves. In Figure 11.4g several pairs of parabolas are shown of which only certain pairs are potential candidates to be encountered in a vertical roadway alignment. Depending on the size, orientation, and positioning of the pair of parabolas there could be:

- No common tangent line (Cases 1 through 4),
- One common tangent line (Cases 5 and 6), or
- Two common tangent lines (Cases 7 and 8).

This procedure is of use when a designer, in order to meet certain elevation controls introduces two vertical curves and then joins them with a common grade line. Since the points of tangency could be located outside the limits of the originally selected curves, the limits of these curves may need to be redefined. It is assumed that both vertical curves are defined by one of the two manners discussed in the preamble to this section of this chapter.

In Figure 11.4h there are two vertical curves (a crest curve and a sag curve) with a common tangent grade line. In this figure, the reader should be able to visualize another potential common tangent, one that is tangent to the left curve at its right side and tangent to the right curve at its left side. However, this tangent line does not represent an appropriate solution in roadway design. For the solution of this construction problem, we see from Figure 11.4h that if we let the point of tangency with the left curve be point A, and with the right curve be point B the equations of the left and right curves may be defined as

$$Y_A = A_0 + A_1 X_A + A_2 X_A^2 \qquad (11.4o1)$$

and

$$Y_B = B_0 + B_1 X_B + B_2 X_B^2 \qquad (11.4o2)$$

and the slope G of the common tangent line from point A to point B may be defined as

$$G = \frac{Y_B - Y_A}{X_B - X_A} \qquad (11.4o3)$$

If we substitute Equations 11.4o1 and 11.4o2 in Equation 11.4o3, the said slope becomes

$$G = \frac{B_0 + B_1 X_B + B_2 X_B^2 - A_0 - A_1 X_A - A_2 X_A^2}{X_B - X_A} \qquad (11.4o4)$$

Said slope may be expressed as the derivative of the curves t points A and B and defined as

$$G = \frac{dY_A}{dX_A} = A_1 + 2A_2 X_A \qquad (11.4o5)$$

and

$$G = \frac{dY_B}{dX_B} = B_1 + 2B_2 X_B \qquad (11.4o6)$$

Since the slope at points A and B should be the same Equations 11.4o5 and 11.4o6 should be equal to each other. Thus, equating these two equations we can solve for X_B to get

$$X_B = \frac{A_1 - B_1 + 2A_2 X_A}{2B_2} \qquad (11.4o7)$$

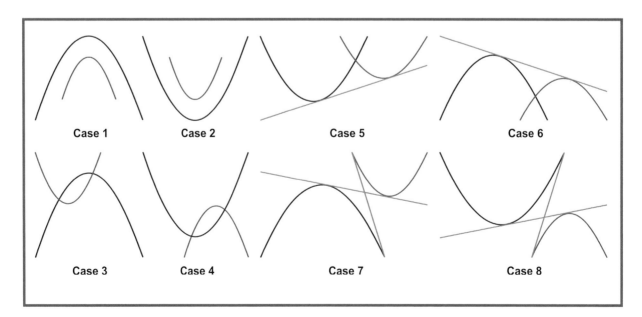

Figure 11.4g Samples of Common Tangent Lines to Two Parabolas

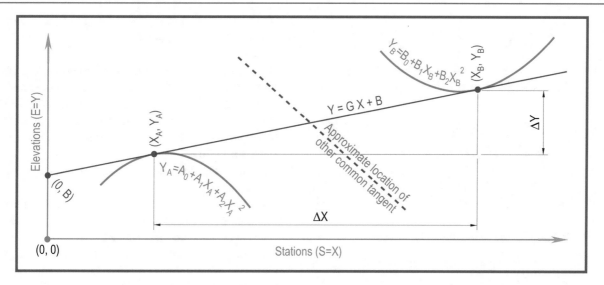

Figure 11.4h Line Tangent to Two Vertical Curves

If we now substitute Equation 11.4o7 in Equation 11.4o4, the said slope may be expressed as

$$G = \frac{B_0 + B_1\left(\dfrac{A_1 - B_1 + 2A_2 X_A}{2B_2}\right) + B_2\left(\dfrac{A_1 - B_1 + 2A_2 X_A}{2B_2}\right)^2 - A_0 - A_1 X_A - A_2 X_A^2}{\left(\dfrac{A_1 - B_1 + 2A_2 X_A}{2B_2}\right) - X_A}$$

(11.4o8)

To simplify a bit the above equation we can let

$$D = \frac{A_1 - B_1}{2B_2} \quad \text{and} \quad E = \frac{2A_2}{2B_2} = \frac{A_2}{B_2}$$ (11.4o9)

If we now substitute D and E into Equation 11.4o8, the said equation becomes

$$G = \frac{B_0 + B_1\left(D + EX_A\right) + B_2\left(D + EX_A\right)^2 - A_0 - A_1 X_A - A_2 X_A^2}{D + EX_A - X_A}$$ (11.4o10)

If we next expand the terms and then group them the last equation becomes

$$G = \frac{\left(B_0 - A_0 + B_1 D + B_2 D^2\right) + \left(B_1 E + 2B_2 DE - A_1\right)X_A + \left(B_2 E^2 - A_2\right)X_A^2}{D + (E-1)X_A}$$ (11.4o11)

We can now equate the slope of Equation 11.4o11 to the slope of Equation 11.4o5 cross multiply and solve for X_A

$$\left(B_0 - A_0 + B_1 D + B_2 D^2\right) + \left(B_1 E + 2B_2 DE - A_1\right)X_A + \left(B_2 E^2 - A_2\right)X_A^2$$
$$= \left[D + (E-1)X_A\right]\left(A_1 + 2A_2 X_A\right)$$
$$= DA_1 + \left[2DA_2 + (E-1)A_1\right]X_A + 2(E-1)A_2 X_A^2$$

(11.4o12)

If we next move all terms to the left and combine certain terms we get the quadratic equation

$$\left(B_0 - A_0 + B_1 D + B_2 D^2 - DA_1\right)$$
$$= \left[B_1 E + 2B_2 DE - A_1 - 2DA_2 - (E-1)A_1\right]X_A$$
$$= \left[B_2 E^2 - A_2 - 2(E-1)A_2\right]X_A^2 = 0$$

(11.4o13)

To simplify the last equation we can let

$$a = \left[B_2 E^2 - A_2 - 2(E-1)A_2\right]$$
$$= B_2 E^2 - (2E-1)A_2$$ (11.4o14a)

$$b = B_1 E + 2B_2 DE - A_1 - 2DA_2 - (E-1)A_1$$
$$= E\left(B_1 - A_1\right) + 2D\left(B_2 E - A_2\right)$$ (11.4o14b)

and

$$c = B_0 - A_0 + B_1 D + B_2 D^2 - DA_1$$
$$= B_0 - A_0 + D\left(B_1 + B_2 D - A_1\right)$$ (11.4o14c)

To get the quadratic equation

$$X_A = \frac{-b \pm \sqrt{b^2 - 4ac}}{2a}$$ (11.4o15)

which is applicable only if the value under the square root is not negative. If it is negative, then there is no real solution to the construction problem.

With the two values of X_A now known we can introduce them in Equation 11.4o5 to find the slopes S_1 and S_2 of the two common tangent lines The terms S_1 and S_2 have been used so as not to be confused with G_1 and G_2 used to denote the back and forward slopes of a vertical curve. The slope values S_1 and S_2 may then be introduced in Equation 11.4o6 to find the corresponding values of X_B as

$$X_{B1} = \frac{S_1 - B_1}{2B_2} \quad \text{and} \quad X_{B2} = \frac{S_2 - B_1}{2B_2} \quad (11.4o16)$$

Having found the two potential common tangent constructions it is up to the calling program to select the appropriate one.

```
Sub avGrdTo2crvs  (A0, A1, A2, G1a, G2a, La,
                   PVCSa, PVCEa, _ B0, B1, B2,
                   G1b, G2b, Lb, PVCSb, PVCEb, _
                   Xa1, Ya1, Xb1, Yb1, Xa2,
                   Ya2, Xb2, Yb2, S1, S2, IERR)
```

The input parameters to this procedure are the:

A0	The A_0 constants of Equation 11.3i4, or zero (1st curve).
A1	The A_1 constants of Equation 11.3i4, or zero (1st curve).
A2	The A_2 constants of Equation 11.3i4, or zero (1st curve).
G1a	The back grade of the vertical curve (not in percent form), or zero (1st curve).
G2a	The forward grade of the vertical curve (not in percent form), or zero (1st curve).
La	The length of the vertical curve, or zero (1st curve).
PVCSa	The station of the PVC point of the vertical curve without the plus (+) sign, or zero (1st curve).
PVCEa	The elevation of the PVC point of the vertical curve, or zero (1st curve).

B0, B1, B2, G1b, G2b, Lb, PVCSb, PVCEb similar to A0, A1, A2, G1a, G2a, La, PVCSa, PVCEa for the 2nd curve.

The returned parameter from this procedure is the:

Xa1, Ya1	The station and elevation of one tangency point on slope S1.
Xa2, Ya2	The station and elevation of one tangency point on slope S2.
Xb1, Yb1	The station and elevation of the other tangency point on slope S1.
Xb2, Yb2	The station and elevation of the other tangency point on slope S2.
IERR	An error indicator, the value of which if it is:
	0 Denotes that an error in the input data has not been encountered.
	1 Denotes that an error has been encountered in the 1st curve ($A_2 = 0$).
	2 Denotes that G1 and G2 of the 1st curve are the same within a tolerance of 0.000001 %.
	3 Denotes improper 1st vertical curve definition (L=0).
	4 Denotes that an error has been encountered in the 1st curve ($A_2 = 0$).
	5 Denotes that G1 and G2 of the 1st curve are the same within a tolerance of 0.000001 %.
	6 Denotes improper 1st vertical curve definition (L=0).
	7 Denotes that there is no intersection (negative value under the radical of Equation 11.4o15).

Upon being called, this procedure initializes to zero the argument to be returned, and then:

1. Defines the first vertical curve "A" by executing Step 1 of the operational steps of the avGrdTcrvFrPnt procedure.
2. Defines the second vertical curve "B" by executing Step 1 of the operational steps of the avGrdTcrvFrPnt procedure.
3. Computes the temporary constants D and E of Equation 11.4o9.
4. Computes the constants a, b, and c of the quadratic Equation 11.4o15.
5. Computes the value under the radical of the quadratic Equation 11.4o15 for being negative, and if it is so, sets IERR to be 7 and aborts; else continues.
6. Computes the station values of the two points of tangency on curve "A" using the Equation 11.4o15, and their corresponding elevations using Equation 11.4o1.
7. Computes the slopes of the two common tangent lines using the two station values of the above Step 6 in Equation 11.4o5.
8. Computes the station values of the two points of tangency on curve "B" using the Equation 11.4o16, and their corresponding elevations using Equation 11.4o2, and then aborts.

11.4.8 Find Elevation or Station Given Station or Elevation

At times as part of the overall design process it becomes necessary to determine the elevation on a vertical curve at a specific station, or to determine the station value of a given elevation on a vertical curve. The avVcStaELV procedure provides this opportunity. The applicable geometric for this operation is discussed often in the preceding subsections of this chapter. Thus, we proceed with the operational steps in which the vertical curve is defined by the vertical curve equation only as discussed in the preamble to this section of this chapter. However, the parabolic constants A_0, A_1, and A_2 are being computed.

```
Sub avGrdTo2crvs  (G1, G2, L, PVCS, PVCE,
                   STA, ELV, _ ELV1, STA1,
                   STA2, IERR)
```

The input parameters to this procedure are the:

G1	The back grade of the vertical curve (not in percent form).
G2	The forward grade of the vertical curve (not in percent form).
L	The length of the vertical curve.
PVCS	The station of the PVC point of the vertical curve without the plus (+) sign.
PVCE	The elevation of the PVC point of the vertical curve.

| STA | The station for which the elevation is desired, or 99999.0. |
| ELV | The elevation for which the two stations are desired, or 99999.0. |

The returned parameter from this procedure is the:

ELV1	The computed or given elevation.
STA1	The first station of elevation ELV1.
STA2	The second station of elevation ELV1.
IERR	An error indicator, the value of which if it is:
	0 Denotes that an error in the input data has not been encountered.
	1 Denotes that an error has been encountered in the input data.
	2 Denotes that G1 and G2 are the same within a tolerance of 0.000001 %.
	3 Denotes improper vertical curve definition (L=0).
	4 Denotes that STA and ELV are both unknown (99999.0).
	5 Denotes that the given elevation is lower than the low point of a sag curve.
	6 Denotes that the given elevation is higher than the high point of a crest curve.

Upon being called, this procedure initializes to zero the argument to be returned, and then:

1. Defines the vertical curve by executing Step 1 of the operational steps of the avGrdTcrvFrPnt procedure for the vertical curve and computes the values of the parabola constants A_0, A_1, and A_2.

2. Queries the values of STA and ELV for both being equal to 99999.0, and if so sets IERR to be 4 and aborts; else continues.

3. Queries the value of ELV for being equal to 99999.0, and if not branches to Step 8 below; else continues.

4. Computes the station value of the turning point by setting the first derivative of the parabolic equation to zero ($X = -A_1/2A_2$) and then compute the elevation at that station ($Y = A_0 + A_1X + A_2X_2$).

5. Compares the values of the back and forward grade lines G1 and G2, respectively, to determine whether the vertical curve is a sag or crest curve.
   (a) For a sag curve if ELV < Y, sets IERR to be 5 and aborts; else continues.
   (b) For a crest curve if ELV > Y, sets IERR to be 6 and aborts; else continues.

6. Computes the value under the radical of the quadratic formula $[A_1{}^2 - 4A_2(A_0 - ELV)]$ of the quadratic equation $[Y = (A_0 - ELV) + A_1X + A_2X_2]$ and tests it for being negative, and if it is sets IERR to be 1 and aborts; else continues.

7. Computes the values of the two stations of elevation ELV by solving the quadratic equation $X = \{-A1 \pm SQR[A_1{}^2 - 4A_2(A_0 - ELV)]\}/2A_2$.

8. Queries the value of STA for being equal to 99999.0, and if not compute the elevation by solving the parabola equation $ELV1 = A_0 + A_1STA + A_2STA^2$, sets both STA1 and STA2 to be the input value STA, and terminates.

Elliptic Features

The procedures that are addressed in this chapter are those that generate and operate on elliptically shaped features. Such features do not represent a staple curve type in civil engineering design and construction except perhaps in (a) certain structural designs to attain special architectural effects, (b) certain storm water detention or retention pond design and architectural landscape designs to produce esthetic effects, (c) hydraulic conduit (pipe) shapes where horizontal or vertical space restriction and other requirements so dictate, and (d) at times in general drafting operations in the production of engineering design and construction drawings. The ellipse is a conical section (see Figures 11.1a and 12.1a) which together with its sisters, circle, parabola, and hyperbola, draw their name from Apollonios of Perga over 2000 years ago. For those interested in etymology of words reference is made to Volume II of *A History of Mathematics* by Sir Thomas Heath page 138 of the 1981 Dover publication.

Presented below prior to addressing the procedures are the basic equations that define an ellipse and those that pertain to various areas (whole, segment, and sector of an ellipse) and perimeter arc lengths. To simplify the construction (drawing and field layout) of an ellipse and certain of its computations, various methods have been devised over the past centuries to generate curves that resemble ellipses. Some of these methods are based on the use of appropriate geometric elements of the ellipse, while others generate a figure that although it is not an exact ellipse it is a very close facsimile of an ellipse. One of these methods is referred to as the three-centered ellipse which utilizes four compound circular arcs one of which is used twice. Thus, this method is more commonly known as the four-centered ellipse and is discussed in Sect. 12.3. The Steven's method and the five-centered ellipse are modifications of the four-centered ellipse utilizing three circular arcs generate somewhat better approximations of an ellipse but are more complex to use. For more details on the above reference is made to the Internet website http://mysite.du.edu/~jcalvert/math/ellipse.htm by J.B. Calvert, created on 6 May 2002 and last revised 6 September 2005.

12.1 The Basics of the Ellipse

An ellipse (see Figure 12.1a) may be defined by the general Cartesian equation:

$$\frac{(x-X_C)^2}{A^2} + \frac{(y-Y_C)^2}{B^2} = 1 \qquad (12.1a1)$$

where

X_C and Y_C are the global coordinates of the center point of the ellipse,

x and y are the local coordinates of the ellipse with respect to its center point, and

A and B are one half the length of the major and minor axes of the ellipse.

If the origin of the coordinate system is assumed at the center point of the ellipse Equation 12.1a1 may then be simplified to its sometimes called canonical form of

$$\frac{x^2}{A^2} + \frac{y^2}{B^2} = 1 \qquad (12.1a2)$$

The above two equations may be solved for y to get respectively

$$y = Y_C \pm B\sqrt{1-\left(\frac{x-X_C}{A}\right)^2} \qquad (12.1b1)$$

and

$$y = \pm B\sqrt{1-\left(\frac{x}{A}\right)^2} \qquad (12.1b2)$$

The major and minor axes of an ellipse need not be parallel to the axes of the coordinate system. They could be inclined, rotated about the center point of the ellipse, as discussed a few pages later on in this chapter. As for the coordinate system for

© Springer International Publishing Switzerland 2016
E.C. Tonias, C.N. Tonias, *Geometric Procedures for Civil Engineers*, DOI 10.1007/978-3-319-24295-8_12

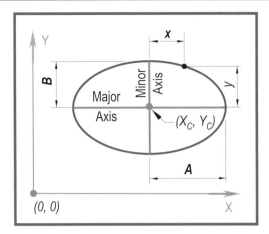

Figure 12.1a   The General Ellipse

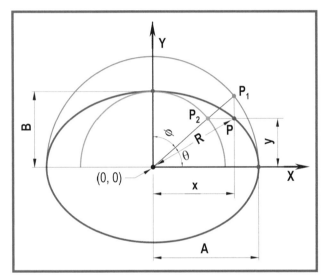

Figure 12.1b   Definition of Ellipse by Polar Coordinates

ease of computation we can forgo the global system and we can use the local one with the origin of its axes about the center point of the ellipse. Any results may always be translated to the global system after their computation.

An ellipse may also be represented in a polar coordinates system in parametric form. For this let us peruse Figure 12.1b in which two semicircles have been drawn, one with a radius of A, and the other with a radius of B representing the major and minor axes of the ellipse. A sample angles θ of arbitrary value has been drawn that intersects the two semicircles at P_1 and P_2 and the ellipse at P. From this it is seen that

$$x = A\cos(\theta) \qquad (12.1c)$$

and

$$y = B\sin(\theta) \qquad (12.1d)$$

where θ is the unit angle in the range $0 \le \theta \le 2\pi$ and A and B are the semi major and semi minor axes of the ellipse as in

Figure 12.1a. We will talk about angle ϕ later on about the area of the ellipse. The instantaneous radius R from the center point of the ellipse to point P on the ellipse may be defined by the Pythagorean theorem as

$$R = \pm\sqrt{A^2 \cos^2(\theta) + B^2 \sin^2(\theta)} \qquad (12.1e)$$

The polar coordinate form of ellipse definition proves handier to programmatically construct the ellipse than that of the Cartesian coordinate system. In either system a series of chords needs to be generated to represent the graphic of the ellipse. These chords are generated in:

- The polar coordinate system at a constantly increase uniform angle θ interval solving Equation 12.1e. In this case assumed θ interval should be such that when divided into $\pi/2$ results in a whole number of chords to assure for a point at each ellipse quarter point. The chords generated are not exactly uniform but they are quite similar.
- The Cartesian coordinate system at a uniform x interval by solving Equation 12.1b2 in which case the resulting chords are larger towards the ends of the major axis from those at the ends of the minor axis, unless a factor is introduced to equalize the chords.

Figure 12.1c displays a theoretical ellipse compared to one generated in both coordinate systems using the same number of chord segments. In either case a check must be made for a negative discriminant value under the square root of the said equations.

12.1.1 Area of an Ellipse

The area of an ellipse may be computed by determining the area of the ellipse's first quadrant and then multiplying it by four. So we can use Equation 12.1a2 and express it as

$$B^2 x^2 + A^2 y^2 = A^2 B^2 \qquad (12.1f)$$

Solving the above equation for y^2 we get

$$y^2 = \frac{A^2 B^2 - B^2 x^2}{A^2} \qquad (12.1g1)$$

or

$$y = \pm\frac{B}{A}\sqrt{A^2 - x^2} \qquad (12.1g2)$$

which is similar to Equation 12.1b2. To find the area of the first quadrant of the ellipse we can integrate the above equation from $x=0$ to $x=A$, and since we are interested in a positive area, we can drop the plus and minus sign, so that

$$Area = \int_A^B \frac{B}{A}\sqrt{A^2 - x^2}\,dx \qquad (12.1h1)$$

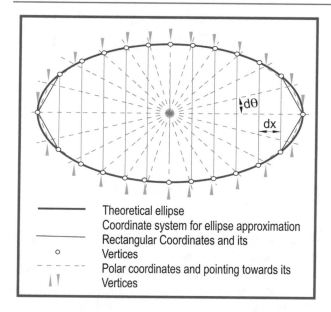

Figure 12.1c   Ellipse in Rectangular and Polar Coordinate Systems

We can evaluate this integral by replacing x with the expression $x = A\sin(\phi)$ so that x^2 is replaced with $x^2 = A^2 \sin^2(\phi)$, and dx is replaced with $dx = A\cos(\phi)d\phi$ where the angle ϕ is the complement of the angle θ in Figure 12.1b and in Equations 12.1c, 12.1d, and 12.1e. This trigonometric substitution is made to express x as a function of an angle ϕ so that at $\phi = 0$, $x = A\sin(\phi) = 0$ and at $\phi = 2\pi$, and $x = A\sin(\phi) = 1$. In this case, angle ϕ is measured from the vertical rather than from the horizontal axis. This necessitates the change of the limits of integration to be between 0 and $\pi/2$ instead of from 0 to A. So the area of the said quadrant becomes

$$Area = \frac{B}{A} \int_{0}^{\pi/2} \sqrt{A^2 - A^2 \sin^2(\theta)} \left[A\cos(\theta) \right] d\theta$$

$$Area = B \int_{0}^{\pi/2} A^2 \sqrt{1 - \sin^2(\theta)} \left[\cos(\theta) \right] d\theta$$

$$Area = AB \int_{0}^{\pi/2} \sqrt{\cos^2(\theta)} \left[\cos(\theta) \right] d\theta \qquad (12.1h2)$$

$$Area = AB \int_{0}^{\pi/2} \cos^2(\theta) d\theta$$

$$Area = AB \frac{1}{2} \left[\theta + \frac{\sin 2(\theta)}{2} \right]_{0}^{\pi/2}$$

or

$$Area = AB \frac{1}{2} \left\{ \left[\frac{\pi}{2} + \frac{\sin\left(2\frac{\pi}{2}\right)}{2} \right] - \left[0 + \frac{\sin(0)}{2} \right] \right\} \qquad (12.1h3)$$

or

$$Area = AB \frac{\pi}{4} \qquad (12.1h4)$$

which makes the area of the whole ellipse four times the area of the first quadrant, or

$$Area = AB\pi \qquad (12.1h5)$$

Note that if A = B, or if each A and B are equal to say R (a radius), that is the ellipse is a circle, the area becomes πR^2, the common equation for the area of a circle.

12.1.2 Perimeter of an Ellipse

To compute the length of the perimeter of an ellipse, let us again work with the first quadrant of an ellipse with its center at the origin of a local coordinate system as done with the area of an ellipse, but this time let us use the polar coordinate Equations 12.1c and 12.1d from which

$$\frac{dx}{d\theta} = -A\sin(\theta), \quad \frac{dy}{d\theta} = B\cos(\theta) \qquad (12.1i)$$

In calculus the length of an arc L may be expressed as

$$L = \int_{\theta_1}^{\theta_2} \sqrt{\left(\frac{dx}{d\theta}\right)^2 + \left(\frac{dy}{d\theta}\right)^2} \, d\theta \qquad (12.1j1)$$

If we now substitute Equation 12.1i in Equation 12.1j1 and we specify as the limits of integration from 0° to 90°, we get

$$L = \int_{0}^{\pi/2} \sqrt{A^2 \sin^2(\theta) + B^2 \cos^2(\theta)} \, d\theta \qquad (12.1j2)$$

Realizing that $sin^2(\theta) = 1 - cos2(\theta)$, the above equation may be expressed as

$$L = \int_{0}^{\pi/2} \sqrt{A^2 \left[1 - \cos^2(\theta) \right] + B^2 \cos^2(\theta)} \, d\theta \qquad (12.1j3)$$

If we extract A^2 out of the radical, the equation then becomes

$$L = \int_{0}^{\pi/2} A \sqrt{1 - \cos^2(\theta) + \frac{B^2}{A^2} \cos^2(\theta)} \, d\theta \qquad (12.1j4)$$

$$L = \int_{0}^{\pi/2} A \sqrt{1 - \left(\frac{A^2 - B^2}{A^2} \right) \cos^2(\theta)} \, d\theta \qquad (12.1j5)$$

In Euclidean geometry, the distance from the center of an ellipse to either of its two focal points is referred to as the eccentricity of the ellipse ε, which is expressed as

$$\varepsilon = \sqrt{\frac{A^2 - B^2}{A^2}} \qquad (12.1k)$$

If we substitute the square of the said eccentricity into Equation 12.1j5, the said equation becomes

$$L = \int_0^{\pi/2} A\sqrt{1-\varepsilon^2\cos^2(\theta)}\,d\theta \qquad (12.1\mathrm{m}1)$$

Since between 0° and 90° the sine function of an angle has the same value as the cosine function but in reverse order, and since Equation 12.1m1 represents the length of the arc of the first quadrant of the ellipse, the said equation for the entire ellipse may be expressed as

$$L = 4A \int_0^{\pi/2} \sqrt{1-\varepsilon^2\sin(\theta)}\,d\theta \qquad (12.1\mathrm{m}2)$$

Equation 12.1m2 is an elliptic integral of the second term and cannot be evaluated with elementary functions. However, it can be represented by the infinite series as

$$L = 2\pi A\left[1 - \left(\frac{1}{2}\right)^2\frac{\varepsilon^2}{1} - \left(\frac{1.3}{2.4}\right)^2\frac{\varepsilon^4}{3} - \left(\frac{1.3.5}{2.4.6}\right)^2\frac{\varepsilon^6}{5} - \cdots\right] \qquad (12.1\mathrm{n})$$

Indian stamp issued in 1962 commemorating the 75th anniversary of Ramanujan's birth

There have been several approximation made to simplify the above equation, of which perhaps the most popular is that of Srinivasa Ramanujan presented below:

$$L \approx \pi(A+B)\left(1 + \frac{3\lambda^2}{10+\sqrt{4-3\lambda^2}}\right) \qquad (12.1\mathrm{p}1)$$

where

$$\lambda = \frac{A-B}{A+B} \qquad (12.1\mathrm{p}2)$$

12.1.3 Area of a Sector or Segment of an Ellipse

When working with ellipses it becomes necessary at times to know the area and perimeter of a sector or of a segment of an ellipse. Figure 12.1d displays the conventional ellipse with its origin at the center point of the ellipse and with its major and minor semi-axes A and B. The angles θ_1 and θ_2 define the radial lines to the start and end of the chord that defines the sector or the segment in question. It is assumed that these two angles are known, or that they may be computed by some means. For example if the ellipse represents a pipe or a tank partially full, then depth of flow measured from the bottom of the pipe or tank could be used to compute the value of the said angles.

The sector and the segment that are displayed in Figure 12.1d are symmetrical about the two axes of the ellipse (the sector about the X axis and the segment about the Y axis). However, this need not be the case. For example for cases other than partially flowing pipes the chords connecting the arc endpoints could be inclined with respect to the two exes. Furthermore, either the major or the minor axis could be oriented along the X axis, or inclined thereto. Once the area of the sector has been found, the area of the segment may be found by subtracting the area of the triangle of the sector formed by the chord connecting the two

Figure 12.1d Area of sector or segment of an ellipse

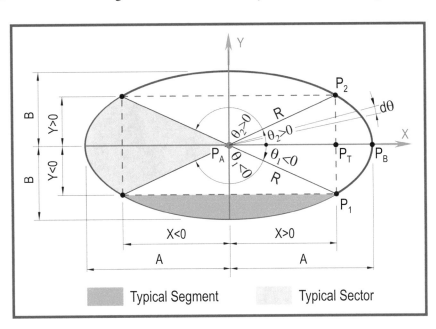

arc endpoints and radial lines from the center of the ellipse to the said endpoints. Presented below is the methodology for computing the area of the sector. A procedure for this area is not provided because of the procedure described in a subsequent subsection regarding four-centered simulated ellipse.

In Figure 12.1d the ellipse is defined by Equation 12.1a2 which is repeated below as

$$\frac{x^2}{A^2} + \frac{y^2}{B^2} = 1 \qquad (12.1q1)$$

and the projections of a radial line R on the X and Y axes may be expressed as

$$x = R\cos(\theta) \qquad (12.1q2)$$

and

$$y = R\sin(\theta) \qquad (12.1q3)$$

If we substitute the squares of x and y of the last two equations in Equation 12.1q1, we get

$$\frac{\left[R\cos(\theta)\right]^2}{A^2} + \frac{\left[R\sin(\theta)\right]^2}{B^2} = 1 \qquad (12.1q4)$$

and solving Equation 12.1q4 for R^2 we get

$$\frac{R^2\left[B\cos(\theta)\right]^2 + R^2\left[A\sin(\theta)\right]^2}{(AB)^2} = 1$$

$$\qquad (12.1q5)$$

$$R^2 = \frac{(AB)^2}{\left[B\cos(\theta)\right]^2 + \left[A\sin(\theta)\right]^2}$$

The area of the sector of the ellipse between θ_1 and θ_2 may be defined as the summation of infinitesimal elliptical triangles of a central angle $d\theta$ and it may be expressed as

$$A_{SECT} = \int_{\theta_1}^{\theta_2} \frac{1}{2}R^2 d\theta$$

$$= \frac{(AB)^2}{2} \int_{\theta_1}^{\theta_2} \frac{d\theta}{\left[B\cos(\theta)\right]^2 + \left[A\sin(\theta)\right]^2} \qquad (12.1q6)$$

Expressing the integral of Equation 12.1q6 by Equation #25 of the *Integrals Containing Sin(ax) and Cos(ax)* of *http://SOSmath.com/tables/integrals/integr21/integr21.html/* with $a = 1$ and the lower integration limit $\theta_1 = 0$ we get

$$A_{SECT} = \frac{(AB)^2}{2} \frac{1}{AB} \left\{ \tan^{-1}\left[\frac{A}{B}\tan(\theta)\right] \right\}$$

$$\qquad (12.1q7)$$

$$A_{SECT} = \frac{AB}{2} \left\{ \tan^{-1}\left[\frac{A}{B}\tan(\theta)\right] \right\}$$

The sector having the area defined by Equation 12.1q7 is the sector in the first quadrant of the ellipse. If the value of angle θ_2 is negative, then the area would be negative and the sector would be located in the fourth quadrant of the ellipse (below the previous sector). The ranges of these angles of integration limits may be $-\pi \leq \theta_1 \leq 0$ and $0 \leq \theta_2 \leq \pi$. It is noted that if the value of θ is 90° then the value between the braces would also be 90° or $\pi/2$ yielding an area of $AB\pi/4$ which is that of Equation 12.1h4.

The area as per Equation 12.1q7 assumes a low integration limit of zero. If that is not the case, then the area of the segment would found by using the said equation with a θ angle equal to the upper limit and subtracting the area computed by the same equation for a θ angle equal to the lower limit.

12.2 Ellipse Given Its Axes and Inclination Angle

The `icEllipse` procedure generates a polyline that comprises an ellipse. The ellipse is defined in polar coordinates with its center point as the origin of the said coordinate system, and with the length of its major and minor axes specified. In addition the Cartesian angle of inclination α of the major axis with respect to the X axis as indicated in Figure 12.2 is also specified, as well as the number of uniform angular segments to generate the polyline graphic of the ellipse (the greater the number of segments the smoother the curve to be generated). In addition to creating the polyline of the ellipse, the procedures returns the area and perimeter of the ellipse.

At times, alternatively to directly specifying the length of the minor axis of the ellipse it could be determined by specifying the angle of the ellipse (angle β in Figure 12.2), and then determining its length as

$$B = A\sin(\beta) \qquad (12.2)$$

where

A = one half the length of the major axis,
B = one half length of the minor axis, and
β = the angle of the ellipse defined as the angle formed by the line joining two opposite corners of a rectangle that encloses the ellipse with its sides being parallel to the major and minor axes of the ellipse.

It should be noted that in specifying the β angle:

• Its value should be in the range of $0 < \beta < \pi$.
• A value of 90° creates a circle.
• A value of 180° creates a straight line.

The major axis of the ellipse is along the X axis (due east) for an α angle of 0°. Thus, if it is desired for the major axis to be along the Y axis (due north), an α angle of 90° should be specified. There is no theoretical limit as to the value of the α angle, but in general it should be in the range of $0 < \alpha < 2\pi$.

In constructing the ellipse, the procedure assumes that the ellipse is oriented about a local X and Y polar coordinate system having its origin a the center point of the ellipse and the major axis positioned along the X axis. Once the polar coordinates of each point along the ellipse have been computed they are rotated about the center point of the ellipse an angular amount equal to the user specified α angle (see Figure 12.2).

```
Sub icEllipse   (XC, YC, ALPHA, MAJLEN, MINLEN,
                 BETA, NSEG, _
                 pl, AREA, PERIM, IERR)
```

The input parameters to this procedure include the:

XC, YC	The Cartesian coordinates of the center point of the ellipse.
ALPHA	The angle of rotation (angle α in Figure 12.2) in radians of the major axis of the ellipse with respect to the X axis (a positive angle denotes a counterclockwise rotation).
MAJLEN	The length of the major axis (2A in Figure 12.2) of the ellipse.
MINLEN	The length of the minor axis (2B in Figure 12.2) of the ellipse, or zero if BETA is to be used.
BETA	The angle β in Figure 12.2 if the length of the minor axis MINLEN is to be determined by Equation 12.2, or zero if MINLEN is to be used.
NSEG	The user desired number of angular segments to define the ellipse polyline.

The parameters that are returned from this procedure include the:

pl	A collection of NSEG+1 X and Y point coordinates that comprise the polyline of the ellipse with X and Y concatenated in series counterclockwise with the first point repeated as the endpoint at the right end of the major axis in Figure 12.2.
AREA	The area of the ellipse computed as indicated in the preceding section of this chapter, and not the area of the polygon generated by the polyline.
PERIM	The perimeter of the ellipse computed as indicated in the preceding section of this chapter, and not the length of the generated polyline.
IERR	An indicator the value of which denotes the following: 0 An error has not been encountered. 1 One or more of the MAJLEN, MINLEN, or BETA values is negative (fatal).

2   The value of BETA exceeds 180° (fatal).

3   The values of both MINLEN and BETA zero, or positive (fatal).

4   One or more of the MAJLEN, MINLEN, or BETA values is erroneous (fatal).

Upon invocation, the procedure initializes the parameters to be returned, and then:

1. Computes the number of points to define the ellipse as NSEG+1, thus repeating the start point of the ellipse which is to be at the right half of the major axis.
2. Computes the value of the unit angle θ as $\theta = 2\pi/\text{NSEG}$. Note that an NSEG value divisible by 3 creates points on the ellipse at 90°, 180°, and 270°.
3. Computes the values of the sines and cosines of ANGLE and θ for later use as

$$C_1 = Cos(\alpha) \quad S_1 = Sin(\alpha) \quad C_2 = Cos(\theta) \quad S_2 = Sin(\theta)$$

and initializes the accumulation factors as: $C_3 = 1$ and $S_3 = 0$

4. Checks for input data errors setting the appropriate IERR value, aborting the procedure if necessary, and sets the half lengths of the major and minor axes as

$$A = 0.5 \, (\boldsymbol{MAJR}) \quad \text{and} \quad B = 0.5(\boldsymbol{MINR})$$

5. Commences a computing loop from 1 up to NSEG of the following operational steps to compute the coordinates of each said point on the ellipse:
   - Computes the coordinates of a point of the assumed ellipse as

$$X = A(C_3) \quad \text{and} \quad Y = B(S_3)$$

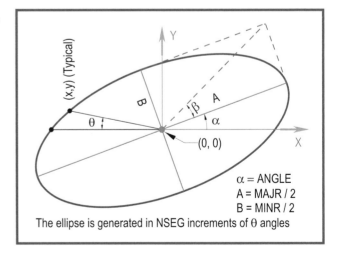

α = ANGLE
A = MAJR / 2
B = MINR / 2

The ellipse is generated in NSEG increments of θ angles

Figure 12.2   The inclined ellipse in a local coordinate system

- Computes the coordinates of the said point of the ellipse rotated by an α angle as

$$X_A = XC + X(C_1) - Y(S_1)$$
$$and \quad Y_A = YC + X(S_1) - Y(C_1)$$

- Updates the accumulation factor S_3 to $S_3 = (S_3 * C_2) + (C_3 * S_2)$

- Updates the accumulation factor C_3 to $C_3 = (C_3 * C_2) + (S_3 * S_2)$
- Note that the update of the C_3 factor must follow the update of the S_3 factor.
- Adds the rotated coordinates X_A and Y_A into a temporary collection.

6. Populates the polyline collection with the contents of the temporary collection of the rotated points.

In Figure 12.3 points G and K are shown to be located within the simulated ellipse. Depending on the dimensions H and W the said points may be located outside the ellipse.

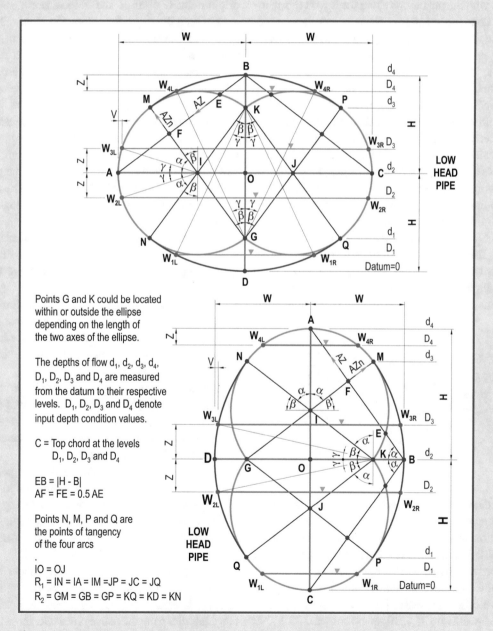

Points G and K could be located within or outside the ellipse depending on the length of the two axes of the ellipse.

The depths of flow $d_1, d_2, d_3, d_4,$ D_1, D_2, D_3 and D_4 are measured from the datum to their respective levels. D_1, D_2, D_3 and D_4 denote input depth condition values.

C = Top chord at the levels D_1, D_2, D_3 and D_4

EB = |H - B|
AF = FE = 0.5 AE

Points N, M, P and Q are the points of tangency of the four arcs

IO = OJ
R_1 = IN = IA = IM =JP = JC = JQ
R_2 = GM = GB = GP = KQ = KD = KN

Figure 12.3 The geometry of a low and a high head four-centered ellipse

12.3 Elliptic Pipes and the Four-Centered Ellipse

The most common pipe shape in hydraulic applications is the circle, but occasionally an elliptic pipe is used when horizontal or vertical spatial considerations so dictate. In such cases it becomes necessary to know the area of a partially flowing elliptic pipe, the length of the perimeter of the said pipe with which the liquid is in contact, commonly referred to as the wet perimeter, and the top width of the exposed liquid in the pipe. The ratio of the area occupied by the liquid over the wet perimeter is referred to as the hydraulic radius which is used in various hydraulic flow equations. In most cases, the elliptic pipe is laid with the major axis along the X axis (horizontally) and is commonly referred to as a low head pipe. However, there are cases in which it is the minor axis that is laid along the said X axis and is commonly referred to as a high head pipe.

The computation of the partial area of an elliptic pipe may be made as per Sect. 12.1.3, but the computation of the wet perimeter would still encounter the same difficulties as those of the whole perimeter of the ellipse. To simplify this and to provide for easier computation of other constructions, this section presents a different approach or technique to defining an ellipse. This approach had been used prior to the introduction of the computer aided drafting to draft ellipses and is referred to as a four-centered ellipse. That is, an ellipse is simulated by four compound circular arcs comprising a closed figure. This simulation technique is presented below and the following `ic4ccEllipse` procedure utilizes this approach to compute the area, wet perimeter, top width, and hydraulic radius of an elliptic pipe flowing partially or completely full.

> *The term hydraulic radius has no relationship to the radius of a circle or arc. It is the ratio of the area of the liquid in any conduit of any shape divided by the length of the inside surface of the conduit in touch with the liquid.*

The four-centered ellipse simulation is remarkably very good. For example for an elliptic pipe having a major axis of 8 units and with a minor axis of 6 units the area as computed by Equation 5.4b8 is $\pi(8/2)(6/2)=37.6991$ square units, and the perimeter as computed by Equations 5.4c11 and 5.4c12 is 22.0286 units. The corresponding area and perimeter of a four-centered ellipse represented by a closed polyline composed of 192 segments are 37.7510 square units and 22.14090 units with the four-centered ellipse being slightly bulgier

near the ends of the major axis yielding a 0.1377 % increase in area and a 0.5098 % increase in perimeter. In the case of a 300 by 200 ellipse the area and perimeter of a conical ellipse are 47,123.98 and 793.27, correspondingly, while those of the simulated ellipse are 42,286.84 and 795.97 yielding corresponding increases of 0.346 % and 0.340 % when developing the simulated ellipse by a series of chords approximately 7.5 units long, each thus resulting in 216 points along the overall perimeter. Reducing the chord length from 7.5 to 2.0 units resulted in the same unit values for the area, perimeter and corresponding increases.

Figure 12.3 displays the extent of the known (given) major and minor axes of an ellipse with its center point at the origin point **O** of an X and Y local north–south coordinate system with the major axis along the X axis. The coordinate system could very well be a Cartesian system, but the north–south system has been used in the `ic4ccEllipse` procedure to make easy use of the various geometric construction procedures have been presented in the preceding chapters of this publication. Note that in the said figure W denotes one half of the major or minor axis that is oriented along the X axis, and that H denotes one half of the major or minor axis that is oriented along the Y axis. In other words, W denotes half the width, and H denotes one half the height of the pipe for either a low head or high head pipe. The subject procedure then decides whether the pipe is a low or high head pipe.

To simulate an ellipse with four circular arcs for a low head pipe and having drawn the two axes, the person and hence the procedure with reference to Figure 12.3:

1. Draws a line from point A to point B. From point B and along the line BA point E is laid such that BE is equal to the absolute value of the difference of a half of the two axes (BE=|H−W|). Thereafter point F is laid midway between point A and point E.
2. Draws the normal line to AB passing it through point F and intersecting the major axis at point I and the minor axis at point G. Point I is the center point of the left small arc of the four arcs, and point G is the center point of the large arc of the said four arcs.
3. Repeats Step 2 for line AD to position point K, and for lines CB and DC to position points J and K, or alternatively locates the center point J by laying a distance OI from point O towards point C, and point K by laying a distance OG from point O towards point B.

For a high head pipe the above process is repeated by rotating everything 90° counterclockwise and changing left to upper to right to lower. The geometric proof of the resultant equalities from the above process as shown in Figure 12.3 are left up to the reader.

```
Sub ic4ccEllipse   (Y, X, W, H, D, LSEG, _
                    PL, Awhole, Apart, Lwhole,
                    Lpart, Chord, HydR, HiLo,
                    IERR)
```

The input parameters to this procedure include the:

Y, X	The Cartesian or north and east coordinates of the center point of the ellipse. Cartesian coordinates are treated as being north–south coordinates.
W	One half of the major or minor axis of the ellipse that is located along the X (west to east) axis of the coordinate system.
H	One half of the minor or major axis of the ellipse that is located along the Y (south to north) axis of the coordinate system.
D	The depth of flow in the pipe measured from the bottom of the pipe, may it be the low point of the minor or major axis of the ellipse.
LSEG	The length in feet (meters) of a uniform chord segment along the perimeter of the pipe to define the smoothness of the shape of the pipe.

The parameters that are returned from this procedure include the:

PL	The collection of north and east coordinates of the polyline of vertices defining the perimeter of the entire simulated ellipse.
Awhole	The area of the entire pipe.
Apart	The liquid area in the pipe.
Lwhole	The perimeter of the entire pipe.
Lpart	The wet perimeter of the pipe.
Chord	The chord length of the exposed liquid within the pipe (zero when full, or empty).
HydR	The hydraulic radius of the pipe for the given depth D of the liquid within the pipe.
HiLo	An indicator the value of which denotes the type of pipe, such that if it is:
	Hi denotes a high head elliptic pipe (H>W).
	Lo denotes a low head elliptic pipe (H<W).
	Ci denotes a circular pipe H=W within a tolerance of 0.005 feet (1.524 mm).
IERR	An indicator the value of which denotes:
	0 No errors have been encountered.
	1 No flow in the pipe (D=0).
	2 Pipe is flowing full (D=H).
	3 Pipe is surcharged (D>H).
	4 Negative depth specified (D<0); the procedure aborts.
	5 Negative or zero pipe width specified (W<0); the procedure aborts.
	6 Negative or zero pipe height specified (H<0); the procedure aborts.
	The tests for equality are made within a tolerance of 0.005 feet (1.524 mm).

Upon invocation, the procedure initializes the parameters to be returned, and then:

1. Checks the value of D for being negative, and the values of H and W for being negative or zero, and if so sets the

indicator IERR to its corresponding error value and aborts; else continues.

2. Compares H with W to determine if the simulated elliptic pipe is to be a low or high head, or a circular pipe, and sets the indicator HiLo to be Lo, Hi, or Ci, respectively. If in the process of this determination it is ascertained that HiLo is Lo or Hi, the procedure branches to **Step 3**, and if it is Ci, the procedure:

(a) Sets the radius R of the pipe to be the average of H and W and determines the area and perimeter of the entire pipe.

(b) Compares the input depth of flow D with the radius R and diameter $2R$ to determine if the pipe is empty, partially full, full, or surcharged and accordingly computes the partial area, wet perimeter, top chord, and hydraulic radius, and terminates.

In the process of computing the top chord C, wet perimeter L and area A (segment area of a circle) of a partially flowing pipe the procedures makes use of the:

- One half the central angle subtended by the top chord of the exposed liquid

$$\alpha = \cos^{-1}\left(\frac{R-D}{R}\right) = \cos^{-1}\left(1-\frac{D}{R}\right) \qquad (12.3a)$$

where

$D =$ the depth of flow of the liquid in the pipe when flowing less than half full, or the excess measured from the top of the pipe to the top chord when flowing more than half full but less than full, and $R =$ the radius of the pipe.

- The length of the top chord of the exposed liquid

$$C = 2R\sin(\alpha) \qquad (12.3b)$$

- The area of the circle segment occupied by the liquid

$$A = A_{SECTOR} - A_{TRIANGLE}$$
$$A = \frac{1}{2}R^2(2\alpha) - \frac{1}{2}C(R-D)$$
$$A = \frac{1}{2}R^2(2\alpha) - 2R\sin(\alpha)R\cos(\alpha) \qquad (12.3c)$$
$$A = \frac{1}{2}R^2\left[2\alpha - \sin(2\alpha)\right]$$

The above three equations are used in several of the steps below and in addition to relating to the bottom and top segments of a pipe, they could pertain to segments along the side of a pipe.

3. Commences the construction of the entire four-centered simulated ellipse as follows:

(a) Queries the HiLo indicator and sets the north and east coordinates of the points **A**, **B**, **C** and **D** at the ends of the major and minor axes by adding or subtracting H and W to the values of the input coordinates of the center point of the ellipse Y and X.

(b) Calls the `icforce` procedure to determine the distance and azimuth from point **B** towards point **A**.

(c) Computes the absolute value of the difference between H and W, and calls the `icptl` procedure to compute the coordinates of point **E** as being located a distance equal to the said absolute value from point **B** towards point **A**.

(d) Calls the `icptl` procedure to compute the coordinates of point **F** located midway between point **E** and point **A**.

(e) Sets the azimuth of a line normal to the line from point **B** towards point **A** in a clockwise direction by adding 90° (0.5π) which new line is to pass through point **F**.

(f) Sets the azimuth of the major axis (AZx) from point **C** towards point **A** and the azimuth of the minor axis (AZy) from point **D** towards point **B** by querying the HiLo indicator and setting:

- AZx to be due west and AZy to be due north if HiLo = "Lo".
- AZx to be due north and AZy to be due east if HiLo = "Hi".

(g) Calls the `icintrs` procedure (see Chap. 6) to intersect the said normal line through point **F** with the:

- AZx azimuth through the center point of the ellipse to set the center point **I** of the small arc, and to determine the angle α (see Figure 12.3).
- AZy azimuth through the center point of the ellipse to set the center point **G** of the large arc, and to determine the angle β (see Figure 12.3).
  Angles α and β are then reduced to be positive and less than 180° by calling the function `icangl`.

(h) Calls the `icforce` procedure to determine the distance from the center of the ellipse point **O** to point **I**, and then calls the `icptl` procedure to compute the coordinates of point **J** located back of point **O** a distance equal to the one just computed along the AZx azimuth.

(i) Calls the `icforce` procedure to determine the distance from the center of the ellipse point **O** to point **G**, and then calls the `icptl` procedure to compute the coordinates of point **K** located ahead of point **O** a distance equal to the one just computed along the AZy azimuth.

(j) Calls the `icforce` procedure to determine the radius

- R1 of the small circles from point **I** to point **A**.
- R2 of the large circles from point **G** to point **B**.

(k) Calls the `icforce` procedure to determine the azimuth between the indicated points, and then calls

the `icptl` procedure to set the indicated points by traversing a distance equal to R2:

- From point **G** to point I to set point **M**.
- From point **K** to point I to set point **N**.
- From point **G** to point J to set point **P**.
- From point **K** to point L to set point **Q**.

4. Queries the HiHo indicator and sets the flow depth d1 at the points of arc tangencies N and Q as being:

- $d1 = H - R1\ Cos(\beta)$ for a low head pipe.
- $d1 = H - R2\ Cos(\alpha)$ for a high head pipe.

5. Sets the flow depth

- d2 at the midpoint of the pipe as being $d2 = H$

- d3 at the points of arc tangencies M and P as being $d3 = 2H - d1$
- d4 at the top of the pipe as being $d4 = 2H$

6. Queries indicator HiLo to accordingly set the values of the central angles (ANGbot and ANGsid) and radii (RADbot and RADsid) of the bottom/top and left/right side arcs in order to generalize the steps below that compute the top chord, wet perimeter, and area at the above four control flow depths.

ANGbot = angB for and HiLo = "Lo"		ANGbot = angA for HiLo = "Hi"
ANGsid = angA		ANGsid = angB
RADbot = R2		RADbot = R1
RADsid = R1		RADsid = R2

7. Computes the length of the top chord, the wet perimeter, and the area at the said control depths as follow:

(a) .At level   d1

C1	use Equation 12.3b with α = ANGbot
L1	= 2 (RADbot) (ANGbot)
A1	use Equation 12.3c with α = ANGbot

(b) .At level   d2

C2	= 2W
L2	= L1 + 2 (RADsid) (ANGsid)
A2	twice the area of the side segment using Equation 12.3c with α = RADsid to account for the two side segments, plus A1, plus the area of the trapezoid between C1 and C2 and a height of d2 − d1.

(c) .At level   d3

C3	= C1
L3	= L1 + 2 (RADsid) (twice ANGsid)
A3	twice the area of the side segment using Equation 12.3c with α = 2 (RADsid) to account for the two side segments and twice the central angle, plus A1, plus the area of the rectangle between C1 and C3 and a height of d3 − d1.

(d) .At level d4

C4	= zero
L4	= L1+L3
A4	= A1+A3

8. Sets the area and wet perimeter of the whole pipe to be equal to L4 and A4.
9. Creates the polyline that represents the perimeter of the simulated ellipse as follows:
   (a) Sets the starting point of the polyline PL collection to be the north and east coordinates of point M (see Figure 12.3). The points are then to proceed in a counterclockwise rotation along the four arcs (large, small, large and small again), initialized the azimuth to be from point **I** towards point **M**.
   (b) Computes the number of points for each of the small and large arcs by:
     • Dividing LSEG by the radius of the small arcs (R1) and the radius of the large arcs (R2) to determine the equivalent uniform central angle.
     • Dividing the central angle of the small arcs (twice angA) and large arcs (twice angB) by the equivalent uniform central angle to define the number of required points.
     • Dividing the central angle of the number of required points to define the adjusted uniform central angle.
   (c) Computes the required points along the simulated ellipse for each of the number of points for an arc and for each of the four arcs by:
     • Adding to the preceding azimuth the adjusted uniform central angle.
     • Calling the icptl procedure to traverse from the arc's center point along the new accumulated azimuth a distance equal to the arc's radius to compute the coordinates of a point along the perimeter of the ellipse.
     • Introducing the said coordinates into the polyline of the perimeter.
10. Compares the specified depth D with zero and with the control depths d2 and d4 for being equal to either within a tolerance of 0.005 feet (1.254 mm) to determine if the pipe is empty, half full, full, or surcharged, and if so sets the top chord, wet perimeter, and liquid area (IERR, Chord, Lpart, and Apart, respectively) equal to the values computed under **Step 7**, and then branches to **Step 12**; else continues.
11. Compares the input depth D with each of the d1, d2, d3, and d4 depths for being equal to any within a tolerance of 0.005 feet (1.254 mm) to determine if the liquid is at any of the said levels, and if so sets said corresponding top chord, wet perimeter, and liquid area equal to the said values, and then branches to **Step 12**; else continues.

(a) Liquid level is below level d1
   • Computes half the central angle γ of the top chord at depth D using Equation 12.3a and doubling it.
   • Computes Chord using Equation 12.3b.
   • Computes Lpart by multiplying RADbot by twice the said angle γ.
   • Computes Apart of the segment using Equation 12.3c for the angle γ and RADbot.
(b) Liquid level is between level d1 and d2, or between level d3 and d3
   • Sets Z=D−H which is negative when D is below lever d2, and positive above it.
   • Computes half the central angle γ of the side arc for an excess of V (see Figure 12.3) finding the arcsine of the ratio Z over RADsid.
   • Computes the value of the said V=RADsid−|(RADsid) cos(g)|.
   • Computes Chord setting equal to twice W−V.
   • Computes the area of the trapezoid between the horizontal axis and the said top chord for a depth equal to Z (see above for its sign).
   • Computes the area of the two side segment, each one determined by Equation 12.3c for the angle γ.
   • Computes Apart as the sum of the above two areas plus the previously computed A2.
   • Computes Lpart by adding twice the product of RADsid by twice the said angle γ to the previously computed :L2.
(c) Liquid level is between level d3 and d4
   • Sets Z=2H−D.
   • Computes half the central angle γ using Equation 12.3a for the excess Z and a radius of RADbot.
   • Computes Chord using Equation 12.3b.
   • Computes Apart using Equation 12.3c.
   • Computes Lpart by subtracting from the previously computed L4 the length of the arc RADbot times the angle γ.
(d) Liquid level above level d4
   Sets the values of Chord, Lpart, and Apart to be those of the previously computed C4, L4 and A4 respectively.
12. Checks the value of Lpart for being equal to zero, and if so sets the hydraulic radius to be zero; else it sets to be the ration of Apart over Lpart.
13. Computes the theoretical area of the conic ellipse using Equation 12.1h5, and the perimeter of the said ellipse using Equations 12.1p1 and 12.1p2.
14. Computes the percent ratio of area and perimeter of the simulated over the conical ellipse, and terminates.

From a programming point of view the four-centered ellipse is a bit too cumbersome, but it provides a decent approximation for partial area and perimeter. At the end of the source code included in the accompanying CD there are six code statement lines that compute and display the percent increase in area and perimeter for a full pipe of the values computed for the simulated ellipse over those of the conic ellipse. These statements are currently commented to be inactive. They can be activated by removing the comment statement symbol.

12.4   Intersection of an Ellipse with a Line

The potential exists in certain architectural landscape and structural project for the need to intersect an ellipse with a straight line. For example, a particular large garden may require one or more elliptic layouts with various pedestrian walks intersecting, or an elliptic cylindrical high rise structure that will most probably be traversed with floor support systems that crisscross and elliptic perimeter wall. In the former example, the intersecting line may not be parallel to neither the major nor the minor axis of the ellipse, while for the latter example the intersecting lines will most probably be parallel to one or both of the said axes. For either of these examples a direct mathematical solution is relatively easy to produce, and the four-centered ellipse simulation is more than adequate. If a simulated ellipse is used, the polyline created with the `ic4ccEllipse` procedure can be intersected with a line with one of the procedures presented in Chap. 6. A mathematical solution is presented below.

Figure 12.4 displays an ellipse and a line crossing it. The ellipse is represented by equation

$$\frac{X^2}{W^2} + \frac{Y^2}{H^2} = 1 \tag{12.4a}$$

which is the same as Equation 12.1a2 and the line is represented by equation

$$Y = mX + D \tag{12.4b}$$

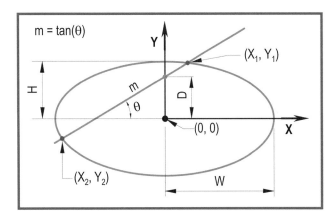

Figure 12.4 Intersection of an ellipse with a line

It is noted that whereas in Equation 12.1a2 A and B denote one half of the major and minor axes, respectively, in Equation 12.4a W and H denote one half of the axes along the X and Y axes, respectively. In this way separate equations need not be created for low and high head ellipses.

The methodology of the `icLine_Ellipse` procedure presented below for intersecting a line with an ellipse assumes that the center point of the ellipse is located at the origin point of a local coordinate system. This implies that for a site work project the calling program should transform the ellipse and line from the global coordinate system to the said local coordinates system, and after the coordinates of the points of intersection have been found, they can be transformed back to the global coordinates system of the site project. Such a transformation could most probably involve coordinate translation and rotation about the center point of the ellipse. Translation could be accomplished by the subtraction/addition of the center point's global coordinates and the `icdrot2s` procedure could be called to handle the rotation.

To determine the coordinates of the points of intersection we substitute the value of Y from Equation 12.4b into the Y of Equation 12.4a to get the equation

$$\frac{X^2}{W^2} + \frac{(mXD)^2}{H^2} = 1 \tag{12.4c}$$

which may then be expressed as

$$H^2 X^2 + W^2 (mX + D)^2 = W^2 H^2 \tag{12.4d}$$

By expanding and rearranging the above equation, it becomes

$$H^2 X^2 + W^2 (m^2 X^2 + D^2 + 2mD) - W^2 H^2 = 0$$
$$H^2 X^2 + W^2 m^2 X^2 + W^2 D^2 + 2W^2 mDX - W^2 H^2 = 0 \tag{12.4e}$$
$$\left(H^2 + W^2 m^2\right) X^2 + \left(2W^2 mD\right) X + W^2 \left(D^2 - H^2\right) = 0$$

In order to simplify the above equation we can let

$$a = H^2 + W^2 m^2 \quad b = 2W^2 mD \quad c = W^2 \left(D^2 - H^2\right) \tag{12.4f}$$

to obtain the traditional quadratic equation

$$aX^2 + bX + c = 0 \tag{12.4g}$$

for which the solution is

$$X_1 = \frac{-b + \sqrt{b^2 - 4ac}}{2a}$$

$$X_2 = \frac{-b - \sqrt{b^2 - 4ac}}{2a} \tag{12.4h}$$

The values of X_1 and X_2 may now be introduced in Equation 12.4a and solved for Y to obtain the corresponding

Y_1 and Y_2 coordinates of the points of intersection as indicated below.

$$Y = \pm\sqrt{\left(1 - \frac{X^2}{W^2}\right)H^2} \qquad (12.4i)$$

Alternatively to substituting X_1 and X_2 in Equation 12.4a, they may be substituted in Equation 12.3b which may be solved for the corresponding Y_1 and Y_2 coordinates. However, before attempting to solve for Y_1 and Y_2 it is important to note the following:

(a) If the discriminant *b2-4ac* in Equation 12.4h is:
   • Positive, then there are two valid intersections.
   • Negative, then the line and the ellipse do not intersect.
   • Zero, then there is only one solution. That is the line is tangent to the ellipse.
(b) If the line is vertical, that is it is parallel to the Y axis, then $tan(\theta)$ and hence the values of the slope m and D would be infinity. In this case consideration should be given to temporarily rotating the ellipse and the line, or have the procedure detect the condition and act accordingly as described below under the operational steps.
(c) If the slope m is zero, then the line is parallel to the X axis, $tan(\theta)$ and hence the slope m would also be zero, and D would be the distance of the line above or below from the said axis. In this case Equation 12.4e is reduce to

$$H^2 X^2 + W^2 D^2 - W^2 H^2 = 0 \qquad (12.4j)$$

from which the roots for X become

$$X = \pm \frac{W}{H}\sqrt{D^2 - H^2} \qquad (12.4k)$$

Since a value of zero slope m and intercept D does not impose a division by zero, the procedure does not check for such condition nor does it use Equation 12.4j.

```
Sub icLine_Ellipse   (CN, CE, W, H, P1N, P1E,
                      P2N, P2E, _
                      XN1, XE1, XN2, XE2, IERR)
```

The input parameters to this procedure include the:

CN, CE	The north and east coordinates of the center point of the ellipse.
W	One half of the major or minor axis of the ellipse located along the X (west to east) axis of the coordinate system.
H	One half of the minor or major axis of the ellipse located along the Y (south to north) axis of the coordinate system.
P1N, P1E	The north and east coordinates of a point on the line to be intersected.
P2N, P2E	The north and east coordinates of another point on the line to be intersected.

The parameters that are returned from this procedure include the:

X1N, X1E	The north and east coordinates of a point of intersection.
X2N, X2E	The north and east coordinates of the other point of intersection.
IERR	An indicator the value of which denotes:
	0 No errors have been encountered.
	1 Negative or zero width specified (W < 0); the procedure aborts.
	2 Negative or zero height specified (H < 0); the procedure aborts.
	3 Vertical line not intersecting the ellipse; the procedure aborts.
	4 Line not intersecting the ellipse; the procedure aborts.
	The tests for equality are made within a tolerance of 0.005 feet (1.524 mm).

Upon invocation, the procedure initializes the parameters to be returned, and then:

1. Checks the values of H and W for being negative or zero, and if so sets the indicator IERR to its corresponding error value and aborts; else continues.
2. Sets the translation adjustment dX and dY to be the coordinates of the center point of the ellipse for adjusting certain coordinates as shown later on.
3. Sets temp1 and temp2 to be the differences in north and east coordinates, respectively, between the two points that define the line for the computation of the slope.
4. Queries the value of temp2 for being equal to zero within a tolerance of 0.005 feet (1.524 mm), and if so (the line is vertical) continues; else branches to **Step 6**.
   (a) Queries the value of the east coordinate of the vertical line for being equal to that of the right side of the ellipse (the line is tangent to that side of the ellipse), and if so sets the roots XN1, XE1, XN2 and XE2 to be those of the point of tangency, and exits the procedure; else continues.
   (b) Queries the value of the east coordinate of the vertical line for being equal to that of the left side of the ellipse (the line is tangent to that side of the ellipse), and if so sets the roots XN1, XE1, XN2 and XE2 to be those of the point of tangency, and exits the procedure; else continues.
   (c) Queries the value of the east coordinate of the vertical line for being to the left (smaller) of the left side, or to the right (greater) of the right side of the ellipse, and if so sets IERR to be 3 and exits the procedure; else continues.
   (d) Computes the coordinates of the roots XN1, XE1, XN2 and XE2 by use of Equation 12.4i and terminates.
5. Computes the slope m by dividing temp1 by temp2, and the value of D with respect to the center point of the ellipse by use

of Equation 12.3b, the values of the a, b and c parameters of Equation 12.4f, and of the discriminant of Equation 12.4h.

6. Queries the value of the said discriminant for being negative, and if so sets `IERR` to be 4 and exits; or else continues.

7. Computes the values of the roots X_1 and X_2 by use of Equation 12.4h and adding thereto the values of dX and dY computed earlier, and the values of Y_1 and Y_2 by use of Equation 12.4b, and terminates.

ERRATUM TO

Geometric Procedures for Civil Engineers

Elias C. Tonias and Constantine N. Tonias

© Springer International Publishing Switzerland 2016
E.C. Tonias, C.N. Tonias, *Geometric Procedures for Civil Engineers*, DOI 10.1007/978-3-319-24295-8

DOI 10.1007/978-3-319-24295-8_13

References in this textbook to supplemental electronic material on an accompanying compact disc (CD) should direct readers to the website, http://extras.springer.com. This site provides the electronic supplements referenced throughout the book without charge. Readers will need to enter the book's print ISBN. The Publisher no longer produces compact discs.

The updated original online version for this book can be found at DOI
http://dx.doi.org/10.1007/978-3-319-24295-8

Elias C. Tonias
Tonias Engineers
Pittsford, NY, USA

Constantine N. Tonias
The CEDRA Corporation
Pittsford, NY, USA

© Springer International Publishing Switzerland 2016
E.C. Tonias, C.N. Tonias, *Geometric Procedures for Civil Engineers*, DOI 10.1007/978-3-319-24295-8_13

Appendix A: Resection Problems

At times it becomes necessary to establish the position of a point, that is, to define its coordinates, by occupying it and sighting with the transit or theodolite three points or stations of known coordinates, otherwise referred to as control points or control stations. In essence, we have to find the coordinates of a point when we know the coordinates of three other points as well as the two angles formed about the unknown point, the vertex, with lines drawn from the said vertex to each of the three known points. This problem is commonly referred to as the three-point problem, and the process of its solution is referred to as the resection method. Figure A.1 displays the five possible situations under which this problem may surface as Cases 1 through 5. Please note that these five configuration cases are not the same as the cases to be discussed later on regarding this problem.

This problem could be encountered by a surveyor when carrying out a topographic survey in an area from which three known points are visible but cannot be occupied. This was quite often encountered in the past when using the plane-table and its associate instrument the alidade. This problem could also be encountered by a ship at sea during a hydrographic survey taking bathymetric measurements (soundings). With the introduction of sub-meter accuracy of global positioning (GPS) equipment the need of resection may not be so predominant nowadays as it was a few years ago. However, even with the availability of GPS the three-point problem may raise its ugly head from time to time.

This problem was first posed in 1617 by the Dutch mathematician Willebrord Snellius (1581–1626) in his publication *Eratosthenes Batavus*. This problem and its solutions as provided by Snellius did not seem to attract much attention until 75 years later when the French mathematician Laurent Pothenot presented his solution to the French Academy. Since then several solutions have been presented by various persons some of which were tailored to the use of the plane-table and of which the US Coast Survey (later referred to as the US Coast and Geodetic Survey and now as the NOAA's Office of Coast Survey) method and the Bessel graphical solution were the most commonly used.

Over 200 years later Peter Andreas Hansen, a Danish astronomer, presented a solution to a similar resection problem in which there are two unknown points with sights at two known stations. This problem is usually referred to as the Hansen problem, or as the five element problem because there are five known parameters, the two angles at each of the two unknown point and the distance between the two known points. Typical configurations identifying the five known elements are depicted in Figure A.2.

Both of the above, the Pothenot and the Hansen problems, have been included in the 1933 book of the German mathematician Heinrich Dorrie (1873–1955) "*100 Great Problems of Elementary Mathematics*".

The contents of this appendix are based partly on a paper authored by the senior author of this book dated 1955 and prepared as fulfillment of a partial requirement for the completion of a course on route design at the Rensselaer Polytechnic Institute, Troy, New York.

For the Hansen Problem reference is made to: http://www2.washjeff.edu/users/mwoltermann/dorrie/40.pdf.

A.1 The Three-Point Problem

Of the several solutions that have been provided for the three-point problem by various mathematicians, the one presented herein is based on a modification by the senior author of a common graphic solution. This solution may also be carried out without a need for a dedicated procedure but by executing a few of the procedures that have been presented earlier in this book. Before continuing it is prudent to note that:

- In the configuration of Case 5 in Figure A.1 the three control stations A, B, and C and the unknown point P line on the same circle. In this case there are an infinite number of

© Springer International Publishing Switzerland 2016
E. C. Tonias, C. N. Tonias, *Geometric Procedures for Civil Engineers*, DOI 10.1007/978-3-319-24295-8

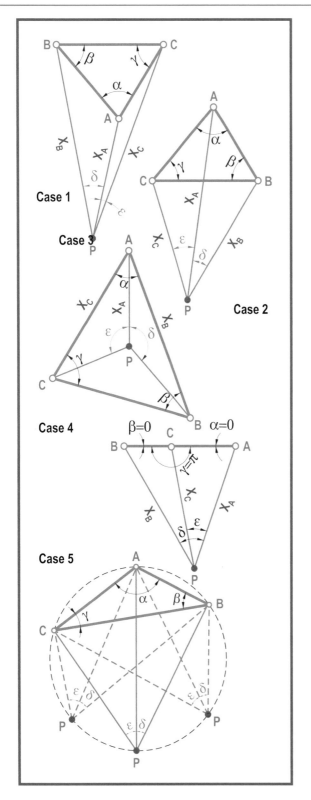

Figure A.1   Basic Configurations of the Three Point Problem

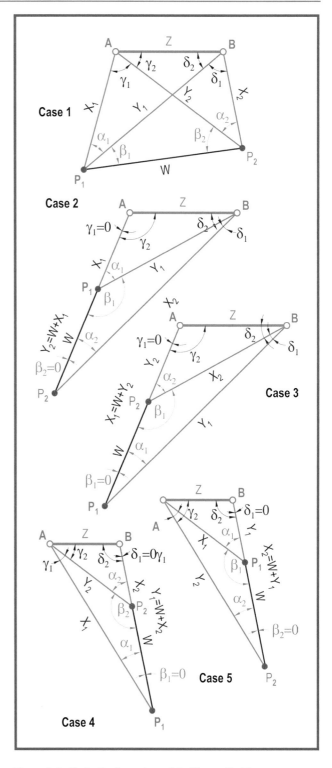

Figure A.2   Basic Configurations of the Hansen Problem

solutions because no matter where point P is to be located, angle δ will be equal to angle γ and angle ε will be equal to angle β since they respectively have their vertices on the same circle and subtend the same arc.

- Neither of the two observed angles at the unknown point P can have a value of 0° due to division by zero.
- In Pothenot's solution the three known points cannot be collinear as shown in the configuration of Case 4 in Figure A.1, again due to division by zero.

- In the procedure to be presented later on in this appendix the three known stations A_1, A_2, and A_3 may be entered in any specified sequence as long as the second point A_2 defines the reference line from the unknown point P. This is illustrated by the six cases of Figure A.3 for the configuration when the unknown point is located outside the triangle formed by the three known stations. Similar cases can be generated for the configuration in which the unknown point P is located within the triangle formed by the three known stations. These three points are rearranged by the procedure to be points A, B, and C with point B being between points A and C as shown in the generic case of Figure A.3.
- In all cases a clockwise angle is be positive and a counterclockwise angle is negative, and is measured from the reference line represented by a dot towards a line denoted with an arrowhead.

To solve for the unknown point P let assume a circle that passes points A, C, and P of the said generic case of Figure A.3 and which circle intersects line BP at point D. Since the vertices of angles α_1, α_2, δ, and ε lie on the circle's circumference and subtend the same arcs angle, then $\alpha_1 = \varepsilon$ and $\alpha_2 = \delta$. Thus, to solve for the unknown point P without a special procedure, construct at point:

- A a line at an angle δ from the line AC equal to but opposite in sign from angle α_2, and
- C a line at an angle ε from the line CA equal to but opposite in sign from angle α_1,
- Intersect these two new lines with each other to set point D, and then
- Extend line BD to meet the circle on the opposite point of the circle to set point P.

The `ic3PointProblem` procedure automates the above four steps to solve a three point problem. The procedure as it now stands returns only the coordinates of point P. However, the sides of the triangle of the three known stations and the distances and directions from point P to stations A, B, and C are being computed and could be returned if the user so desires.

Sub ic3PointProblem	(A1N, A1E, A2N, A2E, A3N, A3E, A1, A2, _ PTNP, PTEP, IERR)

The input parameters to this procedure include the:

A1N, A1E	The north/east coordinates of one of the known stations.
A2N, A2E	The north/east coordinates of the second known station being the reference point from the unknown point.
A3N, A3E	The north/east coordinates of the third known station.
A1, A2	The observed angles (radians) from the unknown point (PTNP, PTEP) to station (A2N, A2E) and towards to stations (A1N, A1E) and (A3N, A3E), respectively.

The parameters that are returned from this procedure include the:

PTNP, PTEP	The north/east coordinates of the unknown point P.
IERR	An indicator the value of which denotes the following:
	0 An error has not been encountered.
	1 Either or both A1 and A2 are zero.
	2 Points A, B, C, and P are collinear.
	3 Angle BAD is 90°.
	4 Angle BCD is 90°.
	5 Points A, B, and D are collinear.

Upon invocation, the procedure initializes the parameters to be returned, and then:

1. Checks the angles A1 and A2 for being zero within a tolerance of 0.000005 radians, and if so sets IERR to 1 and aborts; else reduces them to less than π, and continues.
2. Computes the 360° supplement ANG of the two observed angles A1 and A2.
3. Determines whether the unknown point P is located within or outside the triangle formed by the three known stations by querying the values of ANG, A1, and A2.

Point P is inside the triangle if	A1 > 0, A2 < 0 and ANG < π, or
	A1 < 0, A2 > 0 and ANG > π.
Point P is inside the triangle if	A1 < 0, A2 > 0 and ANG > π, or
	A1 > 0, A2 < 0 and ANG > π.
	A1 < 0, A2 < 0.
	A1 > 0, A2 > 0.

4. Calls the `ic3PointProblemA` procedure if point P is outside the said triangle and it calls the `ic3PointProblemB` procedure if inside. The said procedures rearrange the known stations into points A, B, and C of the generic case of Figure A.3.
5. Calls the `icforce` procedure to compute the azimuths and distances of sides AB, CB, and AC of the said triangle.
6. Computes the angles BCA and BAC depending on whether point P is inside or outside the said triangle and then uses the law of sines to compute the angles sides of the BAP and BCP triangles. As a result two values are obtained for the BP distance which are averaged.

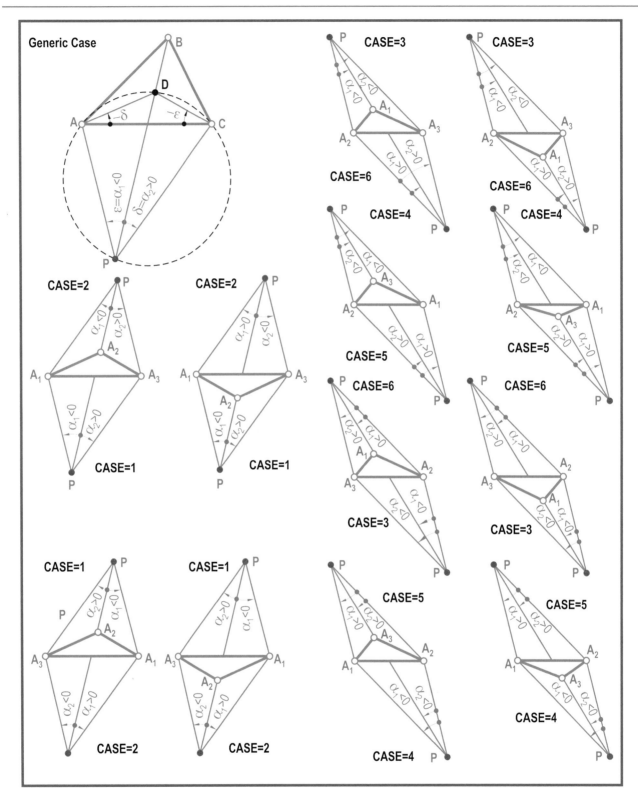

Figure A.3   Potential Configurations of Points of the Three Point Problem

7. Computes the azimuths of the sides of the said triangles and the coordinates of the unknown point P, and then terminates.

A.2   The Hansen Problem

Figure A.4 is an excerpt of Figure A.2 that illustrates four variations of the basic configuration (Case 1) of the Hansen or five element problem. The five known elements are the angles α_1, β_1, α_2, and β_2 at the two unknown points P_1 and P_2 and the distance Z since it can be computed from the known coordinates of points A and B. The lower four configurations of Figure A.2, although they may be considered as being possible, they cannot be handled by the procedure because one of the angles would be zero, thus resulting in a division by zero. Before proceeding with the solution of the problem let us establish certain conventions. In perusing Figure A.4 we see the following conventions:

- The known point B is to the right of the known point A, and that the unknown point P_2 is to the right of the unknown point P_1. That is, points P_1 and P_2 advance in the same direction as points A and B.

- The two unknown points P_1 and P_2 can be located either to the left or to the right of the two known points A and B. The lines connecting these two sets of points respectively cannot intersect each other. The indicator IND is used to denote the side of the known points to which the unknown points are to be located.

- The known angles α_1, β_1, α_2, and β_2 are measured, and considered to be positive, as indicated below regardless whether the rotation is clockwise or counterclockwise. If the rotation of the angle opposes the said indicated measure the angle should be negative.

   α_1 From the line from point P_1 to point A towards the line from point P_1 to point B.

   β_1 From the line from point P_1 to point B towards the line from point P1 to point P2.

   α_2 From the line from point P_2 to point A towards the line from point P2 to point B.

   β_2 From the line from point P_2 to point P1 towards the line from point P to point A.

   In other words, if the unknown points are to the right of the known points, a clockwise rotation is positive, and if to the left a counterclockwise rotation is positive.

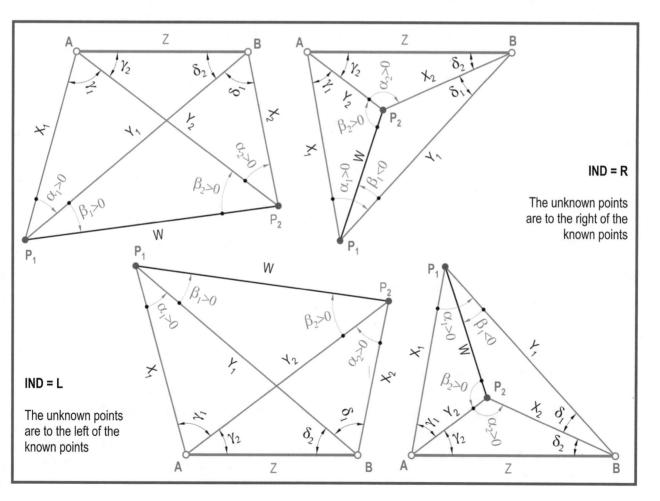

Figure A.4   The Basic Hansen Problem Configuration

From Figure A.4 it is seen that

$$\gamma_1 = \pi - \alpha_1 - \beta_1 - \beta_2 \quad\quad\quad\text{(A.1a)}$$

$$\delta_1 = \pi - \alpha_2 - \beta_1 - \beta_2 \quad\quad\quad\text{(A.1b)}$$

$$\gamma_2 = \pi - \alpha_2 - \delta_1 - \delta_2 \quad\quad\quad\text{(A.1c)}$$

$$\delta_2 = \pi - \alpha_1 - \gamma_1 - \gamma_2 \quad\quad\quad\text{(A.1d)}$$

If we add Equations A.1c and A.1d we get

$$\gamma_2 + \delta_2 = 2\pi - \alpha_1 - \alpha_2 - \delta_1 - \gamma_1 - \delta_2 - \gamma_2 \quad\text{(A.2a)}$$

By substituting Equations A.1a and A.1b in Equation A.2a and combine and reduce terms the said equation becomes

$$\gamma_2 + \delta_2 = \beta_1 + \beta_2 \quad\quad\quad\text{(A.2b)}$$

Utilizing the law of sines in the triangles A-B-P_1 and A-P_1-P_2 we see that

$$\frac{Z}{X_1} = \frac{\sin(\alpha_1)}{\sin(\delta_2)} \quad\quad\quad\text{(A.3a)}$$

and

$$\frac{X_1}{W} = \frac{\sin(\beta_2)}{\sin(\gamma_2)} \quad\text{or}\quad X_1 = W\frac{\sin(\beta_1)}{\sin(\gamma_1)} \quad\text{(A.3b)}$$

If we next substitute the value of X_1 of Equation A.3b in Equation A.3a we get

$$\frac{Z}{W} = \frac{\sin(\alpha_1)\sin(\beta_2)}{\sin(\delta_2)\sin(\gamma_1)} \quad\quad\quad\text{(A.3c)}$$

Similarly utilizing the law of sines in the triangles A-B-P_2 and B-P_2-P_1 we see that

$$\frac{Z}{X_2} = \frac{\sin(\alpha_2)}{\sin(\gamma_2)} \quad\quad\quad\text{(A.4a)}$$

and

$$\frac{X_2}{W} = \frac{\sin(\beta_2)}{\sin(\delta_2)} \quad\text{or}\quad X_2 = W\frac{\sin(\beta_1)}{\sin(\delta_1)} \quad\text{(A.4b)}$$

If we next substitute the value of X_2 of Equation A.4b in Equation A.4a we get

$$\frac{Z}{W} = \frac{\sin(\alpha_2)\sin(\beta_1)}{\sin(\gamma_2)\sin(\delta_1)} \quad\quad\quad\text{(A.4c)}$$

Equating the right sides of Equations A.3c and A.4c, and cross multiplying we get

$$\sin(\alpha_2)\sin(\beta_1)\sin(\delta_2)\sin(\gamma_1) = \sin(\alpha_1)\sin(\beta_2) \\ \sin(\gamma_2)\sin(\delta_1) \quad\text{(A.5)}$$

in which the angles g2 and d2 are unknown and all other angles are known. Thus

$$\frac{\sin(\gamma_2)}{\sin(\delta_2)} = \frac{\sin(\alpha_2)\sin(\beta_1)\sin(\gamma_1)}{\sin(\alpha_1)\sin(\beta_2)\sin(\delta_1)} = \xi \quad\text{(A.6)}$$

If we express Equation A.6 as

$$\frac{\sin(\gamma_2)}{\sin(\delta_2)} = \frac{M}{N} = \frac{\xi}{1} \quad\quad\quad\text{(A.7)}$$

we can invoke the sine-tangent theorem and apply it to Equation A.7 so that

$$\frac{\sin(\gamma_2)}{\sin(\delta_2)} = \frac{\tan\left(\dfrac{\gamma_2-\delta_2}{2}\right)}{\tan\left(\dfrac{\gamma_2+\delta_2}{2}\right)} = \frac{M-N}{N+N} = \frac{\xi-1}{\xi+1} \quad\text{(A.8a)}$$

Since the value of ξ can be found with Equation A.6 using angles α_1, α_2, β_1, β_2, γ_1, and δ_1, Equation A.8a can be written as

$$\tan\left(\frac{\gamma_2-\delta_2}{2}\right) = \frac{\xi-1}{\xi+1}\tan\left(\frac{\gamma_2+\delta_2}{2}\right) \quad\text{(A.8b)}$$

If we use Equation A.2b in Equation A.8b, the latter becomes

$$\tan\left(\frac{\gamma_2-\delta_2}{2}\right) = \frac{\xi-1}{\xi+1}\tan\left(\frac{\beta_2+\beta_2}{2}\right) \quad\text{(A.8c)}$$

which may be rewritten as

$$\frac{\gamma_2-\delta_2}{2} = \tan^{-1}\left[\frac{\xi-1}{\xi+1}\tan\left(\frac{\beta_2+\beta_2}{2}\right)\right] = K \quad\text{(A.9)}$$

from which

$$\gamma_2 = \delta_2 + 2K \quad\quad\quad\text{(A.10a)}$$

and

$$\delta_2 = \gamma_2 - 2K \qu\quad\quad\text{(A.10b)}$$

Substituting Equation A.10a in Equation A.2b, and Equation A.10b in Equation A.2b we get

$$\delta_2 + 2K + \delta_2 = \beta_1 + \beta_2 \quad\text{or}$$

$$\delta_2 = \frac{\beta_1+\beta_2-2K}{2} \quad\quad\quad\text{(A.11)}$$

and

$$\gamma_2 + \gamma_2 - 2k = \beta_1 + \beta_2 \quad\text{or}$$

$$\gamma_2 = \frac{\beta_1+\beta_2+2K}{2} \quad\quad\quad\text{(A.12)}$$

With all the angles now known we can use the law of sines to compute the lengths of the five unknown sides X_1, X_2, Y_1, Y_2, and W. Since each side belongs to two triangles it can be computed in two different ways. The procedure the sides by use of the following equations with those sides that are being computed twice being averaged.

$$X_1 = Z \frac{\sin(\delta_2)}{\sin(\alpha_1)} \qquad (A.13)$$

$$W = X_1 \frac{\sin(\gamma_2)}{\sin(\beta_1)} \qquad (A.14a)$$

and

$$W = Z \frac{\sin(\gamma_1)\sin(\delta_2)}{\sin(\beta_2)\sin(\alpha_1)} \qquad (A.14b)$$

$$X_2 = Z \frac{\sin(\gamma_2)}{\sin(\alpha_2)} \qquad (A.15)$$

$$W = X_2 \frac{\sin(\delta_1)}{\sin(\beta_1)} \qquad (A.16a)$$

and

$$W = Z \frac{\sin(\gamma_2)\sin(\delta_1)}{\sin(\alpha_2)\sin(\beta_1)} \qquad (A.16b)$$

$$Y_1 = Z \frac{\sin(\gamma_1 + \gamma_2)}{\sin(\alpha_1)} \qquad (A.17a)$$

and

$$Y_1 = W \frac{\sin(\alpha_2 + \beta_2)}{\sin(\delta_1)} \qquad (A.17b)$$

$$Y_2 = Z \frac{\sin(\delta_1 + \delta_2)}{\sin(\alpha_2)} \qquad (A.18a)$$

and

$$Y_2 = W \frac{\sin(\alpha_1 + \beta_1)}{\sin(\gamma_1)} \qquad (A.18b)$$

The `icHansenProblem` procedure described below addresses the said problem of Hansen. In perusing this procedure reference is made to Figure A.4 and to the commentary and particularly to the various conventions identified in the introduction to this section of this Appendix.

The input parameters to this procedure include the:

AN, AE	The north/east coordinates of one of the known stations.
BN, BE	The north/east coordinates of the second known station.
Alpha1	The observed angle B-P1-A in radians.

```
Sub icHansenProblem (AN, AE, BN, BE,
                     Alpha1, Beta1, Alpha2,
                     Beta2, IND, _ P1N,
                     P1E, P2N, P2E, IERR)
```

Beta1	The observed angle B-P1-P2 in radians.
Alpha2	The observed angle A-P2-B in radians.
Beta2	The observed angle A-P2-P1 in radians.
IND	An indicator denoting on which side of the from line A to B the two unknown points are located:
	R or r Denotes that the unknown points are to the right of the said line, and
	L or l Denotes that the unknown points are to the left of the said line.

The parameters that are returned from this procedure include the:

P1N, P1E	The north/east coordinates of the unknown point P_1.
P2N, P2E	The north/east coordinates of the unknown point P_2.
IERR	An indicator the value of which denotes the following:
	0 An error has not been encountered.
	1 All angles are zero.
	2 An angle greater than 180° has been encountered
	3 Improper IND has been encountered.

Upon invocation, the procedure initializes the parameters to be returned, and then:

1. Queries all angles for any one being zero or greater than 180° within a tolerance of 0.000005 radians, and if so sets IERR to 1 or 2 and aborts; else continues.
2. Queries IND for not being R, r, L, or l, and if so sets IERR to 3 and aborts; else continues.
3. Calls the `icptl` procedure to find the distance and azimuth between the two known points, and then computes the angles γ_1 and δ_1 using Equations A.1a and A.1b.
4. Calls the `icptl` procedure to find the distance and azimuth between the two known points, and then computes the angles γ_1 and δ_1 using Equations A.1a and A.1b.
5. Uses Equations A.6, A.9, A.11, and A.12 to solve for ξ, K, $\delta 2$, and $\gamma 2$, respectively.
6. Uses Equations A.13 through A.18b to solve for the sides X_1, X_2, Y_1, Y_2, and W, respectively.
7. Depending on the value of IND computes the azimuths of the said sides, and then calls the `icptl` procedure to compute the coordinates of the two unknown points.

Appendix B: Line and Curve Fitting

We have addressed in the main text of this book procedures for creating straight lines and circular arcs under various geometric conditions. However, there are times when an engineer or surveyor has a series of point coordinates that are supposed to lie along such a line or arc of which the mathematical equation is not known and it is desired to be known. Such case may be encountered in the field when locating the center line of a roadway of which design or as-built plans are not available. In this case the surveyor may observe points at a certain distance interval along the roadway by splitting the roadway midway from edge to edge of the pavement. A similar operation may be encountered in the office when such center line is being digitized from an existing map. Furthermore, certain design conditions may develop a set of points through which a line or arc is to pass. Three procedures are presented in this appendix, one to fit a line, one to fit a circular arc and one to fit a polynomial through a given set of points. A procedure has been presented in Sect. 11.3.2 of Chap. 11 for fitting a parabolic curve through a set of station and elevation points.

There are various methods for fitting such a line or arc feature through a set of points with the method of least squares being the prevalent method. This method minimizes the deviations of the observed points from the fitted feature. These deviations are the Y-axis distances of the observed points at their X-axis coordinates from the fitted feature. Although many advancements have been made in the application of least squares its utilization in this appendix is limited to its basic concept of minimizing the deviations. Weights are not being used because it is felt that during digitization all measurements are made with equal care, and that field observations are made similarly and with the same equipment. The latter statement may not be quite true if observations are made by global positioning system and the terrain changes from wooded to bare land conditions.

B.1 Line Fitting

To fit a straight line to a set of given points the `icFitLine` procedure utilizes the general line Equation B.1a as depicted in Figure B.1

$$y = mx + b \tag{B.1a}$$

where m is the slope of the fitted line and b is its Y axis intercept, and hence, any point in the given point set may then be expressed as

$$y_n = mx_n + b \tag{B.1b}$$

The given points within the set may be arranged in any sequence, and duplicate points may exist within the set. As stated previously weights may not be assigned to any of the said points; however, duplicate points act as weights and thusly do influence the construction of the fitted line. Although the fitted line may be considered as extending in either direction indefinitely, the `icFitLine` procedure commences and terminates the fitted line at the projection points of the first and last points on the given set of points on the fitted line.

In Figure B.1 the deviation D of any given point n of the set along the Y axis may be expressed as

$$D = mx_n + b - y_n \tag{B.1c}$$

and the summation of the squares of the said deviations may be defined as

$$\sum D^2 = \sum \left(mx_n + b - y_n \right)^2 \tag{B.1d}$$

or

$$\sum D^2 = \sum \left[\begin{array}{l} \left(mx_n \right)^2 + b^2 + y_n^2 + 2mx_n b - 2mx_n y_n \\ \qquad\qquad\qquad\qquad - 2by_n \end{array} \right] \tag{B.1e}$$

© Springer International Publishing Switzerland 2016
E. C. Tonias, C. N. Tonias, *Geometric Procedures for Civil Engineers*, DOI 10.1007/978-3-319-24295-8

In Equation B.1e we have two unknowns, the slope m and the Y axis intercept b. Since the method of least squares calls for the minimization of the deviations, the said equation must be set to zero. Thus, if we differentiate the said equation with respect to b, we get

$$\frac{\partial D^2}{\partial b} = 2\sum b + 2\sum m x_n - 2\sum y_n = 0$$

or

$$nb + m\sum x_n - \sum y_n = 0 \qquad (B.1f)$$

from which the Y axis intercept b becomes

$$b = \frac{\sum y_n}{n} - \frac{m\sum x_n}{n} \qquad (B.1g)$$

Similarly, if we differentiate Equation B.1e with respect to m, we get

$$\frac{\partial D^2}{\partial m} = 2m\sum x_n^2 + 2b\sum x_n - 2\sum x_n y_n = 0$$

or

$$m\sum x_n^2 + nb\sum x_n - \sum x_n y_n = 0 \qquad (B.1h)$$

If we next substitute Equation B.1g in Equation B.1h, the latter equation becomes

$$m\sum x_n^2 + n\sum x_n \frac{\sum y_n}{n} - n\sum x_n \frac{m\sum x_n}{n} - \sum x_n y_n = 0$$

or

$$m\left[\sum x_n^2 - \left(\sum x_n\right)^2\right] + \sum x_n\sum y_n - \sum x_n y_n = 0 \qquad (B.1i)$$

from which the slope of the fitted line may be expressed as

$$m = \frac{\sum x_n y_n - \sum x_n\sum y_n}{\sum x_n^2 - \left(\sum x_n\right)^2} \qquad (B.1j)$$

Having solved for the two unknowns b and m (Equations B.1g and B.1j), and to make Equation B.1j similar to Equation B.1g, let us divide each term on the right side of Equation B.1j by n to get

$$m = \frac{\dfrac{\sum x_n y_n}{n} - \dfrac{\sum x_n\sum y_n}{n}}{\dfrac{\sum x_n^2}{n} - \dfrac{\left(\sum x_n\right)^2}{n}} \qquad (B.1k)$$

and then let us set the following equalities to be substituted in Equations B.1g and B.1k

$$\overline{Y} = \frac{\sum y_n}{n} \qquad \overline{X} = \frac{\sum x_n}{n} \qquad \overline{X^2} = \frac{\sum x_n^2}{n}$$

$$\overline{XY} = \frac{\sum x_n\sum y_n}{n} \quad \overline{X}\overline{Y} = \frac{\sum x_n\sum y_n}{n} \quad \overline{X}^2 = \frac{\left(\sum x_n\right)^2}{n} \qquad (B.1m)$$

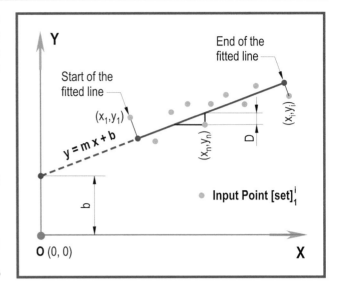

Figure B.1   Straight Line Fit Through a Set of Given Points

so that they may be expressed as

$$b = \overline{Y} - m\overline{X} \qquad (B.1n)$$

and

$$m = \frac{\overline{XY} - \overline{X}\overline{Y}}{\overline{X^2} - \overline{X}^2} \qquad (B.1o)$$

The `icFitLine` procedure automates the above process.

```
Sub icFitLine    (X1List, Y1List, _ X1,
                  Y1, X2, Y2, slope,
                  intrcpt, LENGTH, AZ,
                  IERR)
```

The input parameters to this procedure include the:

X1List	Dynamic array of the X (east) coordinates of the known points.
Y1List	Dynamic array of the Y (north) coordinates of the known points.

The parameters that are returned from this procedure include the:

X1, Y1	The X and Y (east/north) coordinates of the start point of the fitted line.
X2, Y2	The X and Y (east/north) coordinates of the end point of the fitted line.
slope	The slope of the fitted line.
intrcpt	The Y axis intercept of the fitted line.
LENGTH	The length of the fitted line.
LENGTH	The north azimuth in radians of the fitted line.
IERR	An indicator the value of which denotes the following:
	0 An error has not been encountered.
	1 The X1List and/or Y1List arrays are empty.

> 2   The X1List and Y1List arrays are not of the same size.
>
> 3   Improper input array data. One possible error condition is that in which all of the input points lie on a straight due north line.

Upon invocation, the procedure determines the limits of extent of the two arrays, extracts the number of coordinates in each of the two arrays, and then:

1. Checks whether either or both arrays are empty, or whether the said arrays are not of the same size and if so sets IERR to 1 or 2, respectively, and aborts; else continues.
2. Resets the dimensions of the two arrays, and initializes the parameters to be returned, as well as the summations.
3. Computes the summation terms of the right side of the Equations B.1m, and then computes the bar terms of the left side of the same equation.
4. Checks whether the denominator of Equations B.1n is equal to zero and if so sets IERR to 3 and aborts; else continues.
5. Computes the slope and the Y intercept of the fitted line.
6. Converts the slope to a north azimuth by:
   - Calling the icraddeg function to change the arctangent of the slope into degrees of a Cartesian rotation.
   - Calling the icdec2az function to change the Cartesian rotation in degrees into a north azimuth in degrees.
   - Calling the icdegrad function to change the azimuth in degrees into an azimuth in radians.
7. Calls the icprjct procedure to project the first and the last points of the given point set onto the new fitted line to establish the start and end points, respectively of the said line.
8. Calls the icforce procedure to compute the length and azimuth of the fitted line.

B.2   Circular Arc Fitting

The utilization of the computer in the circle fit process began since its mass introduction into the public industry in the late 1950s and early 1960s, and perhaps the first to gain popularity in this subject were P. Delogne (1972) and I. Kasa (1976). Since then numerous solutions have been developed in fitting a circle through a set of points, and comparisons of them have been conducted by many mathematicians most of whom agree that (b) there is a definite distinction between fitting a full circle and fitting a small arc (less than half of the circumference) to a set of given points. Furthermore, the application of the least squares method on minimizing the radius of fitter circle or arc provides "*a non-linear problem,*" to quote Chernov and Lesort (see side bar), "*that has no*

closed form solution. There is no direct algorithm for computing the minimum … all known algorithms are either iterative and costly or approximative by nature." However, there are certain numerical methods, Taylor's series for one, that can overcome this issue.

> *For a study of fitting circles and circular arcs to a set of points reference is made to:*
>
> *"Least squares fitting of circles and lines" by N. Chernov, C. Lesort (Submitted on 1 Jan 2003) arXiv:cs/0301001 [cs.CV] of the Cornel University Library.*
>
> *and to: "A Few Methods for Fitting Circles to Data" by Dale Urbach, Kerry Jones. IEEE Transactions on Instrumentation and Measurement, Vol. XX, No. Y 2000.*

In route alignments and similar design work a civil engineer practically rarely if ever has to be concerned with a full circle. This engineering practice is concerned with a relatively small arc length with respect to the full circumference of the corresponding circle for a given radius which may reach even thousands of feet.

When the need arises in civil engineering to fit a circular arc through a series of points, the said points are measured under rather fairly controlled conditions and not at random. Although there definitely are inherent measurement errors, they are not of great measure. Taking measurements along a pavement center line even along an old and many times repaved rural roadway it is questionable whether a measured point will be too far off its original position. Similarly, digitizing an alignment in the office from a hard copy or by heads-up digitization with a computer need not introduce major deviations unless the operator is hasty and careless in the work. At times when the need requires the establishment of a circular along a path of certain control points, "random" points may be set to locate a general location of the arc that may require a few iterations prior to finalization. Thus, the need for a very mathematically sophisticated or advanced methodology may not be considered a must use.

To fit a circular arc through a set of point pickups the icFitCircle procedure of this appendix may be used. This procedure's general premise is based on the principle that three points define one and only one circular arc (excluding of course its 360° supplement). Thus, the given points are divided into adjacent groups of three points (3-point sets) and the center point of the arc passing through each 3-point set is determined. Theoretically all such center points are to coincide. However, that seldom, if ever, happens. Therefore from these center points one is selected and the radius of the potential circular arc is determined by computing the distance from the selected center point to each of the given input point from which a radius is chosen.

This procedure assumes that all points are to be introduced in a clockwise or counterclockwise direction. That is, the points cannot advance forward, then backtrack and then go forward again. Once a backtracking is recognized, the procedure may abort, and if not the results may be flawed. Figure B.2a displays a set of observed points through which a circular arc is to pass. In this figure the following comments should be noted:

1. If there are only three given points, they are used to create the desired arc, but if there are four or more the overall rotational direction, clockwise or counterclockwise, of the points is determined by calling the `icgetdir` procedure as follows:
   - If there are four points, the first, third and fourth points are used to determine the rotation.
   - If there are more than four points, the points are divided into two groups more or less of equal number and the `icgetdir` procedure is called for each group to determine the direction of each group. If both rotations are the same, the procedure continues, else it aborts. To do so, the procedure divides the total number of points by two to create the two groups, and similarly divides each group by two to find the midpoint to be used with the `icgetdir` procedure. The division of the points in two groups is to provide for the possibility that the arc to be created may exceed 180°.

2. Duplicate and collinear points are removed.
   (a) A duplicate point is the second of two successive points that are located within a user specified distance.
   (b) A collinear point is the third of three successive points that form a deflection angle of less than 0.00014546 radians (0.00833424°=5′ 30″). It is important to note that in relatively sharp curve if three points barely exceed the said deflection limit the results may be compromised.

3. In the above point removal processes the coordinates of any remaining points are saved in a pair of dynamic arrays to preserve the coordinate arrays of the given points. Note that the first point is always retained but necessarily the last one.

4. Once any unwanted points have been removed, groups of successive 3-point sets are selected to determine the center of the circle passing through each set. If in this process :
   (a) The third point of the last 3-point group is not the last point of the whole group of input points, the last three are used to form the last 3-point group. For example, if there are:
   (b) The rotational direction of a 3-point set opposes that of the overall set of input point the said point set is disregarded.

5. Alternatively to the above methodology of 3-point set selection, all possible combination of 3 points (r) taken from a group of n points could be used. This set selection has not been employed in the procedure because it would increase immensely the number of combinations without much effect on the final result. For example, if there are 20 given points, there are 20/2=10 3-point sets. If all possible combinations are to be used then their number would be

$$cCr = \frac{n!}{(n-r)!r!} \quad \text{or} \quad 20C3 = \frac{20!}{(20-3)!3!} = 1140 \quad \text{(B.2a)}$$

6. The centroid of the above center points is computed as the mean value μ_X and μ_Y of the X and Y coordinates, respectively. This point is to be used as the center point of the sought after arc if the process below does not yield a specific center point.

7. Depending on the position of the various given points, particularly if not much care has been taken in their mensuration process, it is possible and quite probable that the center points could be scattered in the plane. Thus, certain of these center points may need to be excluded from the selection processes if they are located at the extreme ends of the center point group. To thin out the center point group the standard deviations of the X coordinates (σ_X) and of the Y coordinates (σ_Y) are computed as

$$\sigma_X = \sqrt{\frac{\Sigma(x-\mu_X)^2}{n}} \quad \text{(B.2b)}$$

and

$$\sigma_Y = \sqrt{\frac{\Sigma(y-\mu_Y)^2}{n}} \quad \text{(B.2c)}$$

where x and y denote the X and Y coordinates of a center point, μ_X and μ_Y as defined above.

Excluded from consideration are first the center points that lie outside the one half range of the standard deviation of the X coordinates ($\pm0.5\,\sigma_X$), and then those that lie outside the same range of the Y coordinates ($\pm0.5\,\sigma_Y$). The remaining points are averaged to determine the center point of the sought after arc. If for some reason all such points are excluded, the first computed mean of all center points is used.

8. The radius is computed by averaging the distances from the selected center point to each of the given points excluding duplicate and collinear points. The start and end point of the arc are computed as the points of projection of the first and last given point on the new arc. The new arc is translated so that the midpoint of its long chord is positioned at the midpoint of the line connecting the first and last given points.

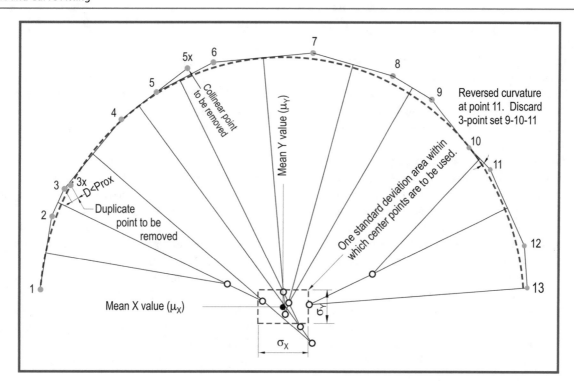

Figure B.2a   Potential Center Points of an Arc Passing Through a Set of Given Points

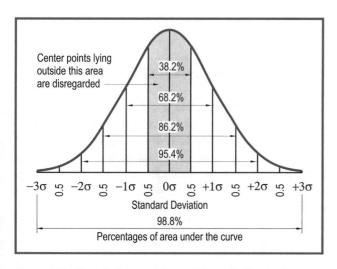

Figure B.2b   Statistical Area of Center Points to be Retained

```
Sub icFitCircle    (X1List, Y1List, Prox, _
                    X1, Y1, X2, Y2, Xc, Yc,
                    R, AZ1, AZ2, ANGLE,
                    LENGTH, IERR)
```

The input parameters to this procedure include the:

X1List	Array of the point X (east)coordinates.
Y1List	Array of the point Y (north)coordinates.
Prox	A proximity distance for duplicate point removal.

The parameters that are returned from this procedure include the:

X1, Y1	Cartesian coordinates of the start point of the new arc being the projection of the first given point on the fitted arc.
X2, Y2	Cartesian coordinates of the last point of the new arc being the projection of the last given point on the fitted arc.
Xc, Yc	Cartesian coordinates of the center point of the fitted arc.
R	Radius the fitted arc.
AZ1, AZ2	Azimuths from the center point to the start and points of the fitted arc.
ANGLE	Central angle in radians of the fitted arc.
LENGTH	Arc length of the fitted arc.
IERR	An indicator the value of which denotes the following:
	0 An error has not been encountered.
	1 The X1List and/or the Y1List are empty.
	2 The X1List and the Y1List are not of the same size
	3 The points do not advance sequentially or there is a reverse curvature.
	4 There are no potential 3-point sets.

Upon invocation, the procedure determines the size of the input arrays, checks for potential errors (empty or unequal size arrays), initializes the parameters to be returned, and then:

1. Determines and queries the number of input points and if there are:
   • Three points calling the `icgetdir` procedure to create the arc and terminates.
   • Four or more divides the points in two groups and checks their rotation as previously stated. If the two rotations are not the same, `IERR` is set to 3 and terminates; else continues.

2. Creates two temporary dynamic arrays, stores in them the X and Y coordinates of the first given point, and for each pair of successive points calls the `icforce` procedure to compute the distance between them, and then compares the said distance with `Prox`. If it is larger than `Prox`, stores the second point in the temporary arrays; if not checks the next two successive points (it is assumed that not all points are to fail this test).

3. Calls the `icforce` procedure to compute the deflection angle between three successive points as stated previously, tests for collinearity within a tolerance of 0.00014546 radians (0° 05′ 30″), and if they are not collinear the third point is saved in the temporary arrays; else the test is repeated for the next three points.

4. Selects a 3-point set as previously stated, calls the icgetdir procedure and compares the rotation of the 3-point set with that of all points. If the two rotations are not the same, the test is repeated for the next 3-point set; else calls the `icarccn3` procedure to compute the center point coordinates of the arc passing through the said 3-point set and this step is repeated until all 3-point sets have been processed.

5. Checks the number of center points that have been computed, and if there are none `IERR` is set to 4, and aborts; else continues.

6. Computes the mean values of the X and Y coordinates (μ_X and μ_Y) by averaging the corresponding coordinates and computes the standard deviations (σ_X and σ_Y) using Equations B.2b and B2c.

7. Computes the X and Y coordinate low and high range limits by subtracting and adding one half a deviation to the corresponding mean coordinate value.

8. Compares the X coordinate of each center point with the corresponding low and high coordinate limits, and saves this point in a temporary array only if it is within the said limits.

9. Checks the number of the remaining center points, and if there is none, the center point computed in **Step 6** above is used, and continues in either case.

10. Compares the Y coordinate of each of the remaining center points with the corresponding low and high coor-dinate limits, and saves this point in a temporary array only if it is within the said limits.

11. Checks the number of remaining center points, and if:
    • There is none, sets the center point computed in **Step 6** above to be the center point of the new arc.
    • There are two or more center points remaining, their coordinates are averaged to determine the final center point of the new arc.

12. Calls the `icforce` procedure to compute the radius between the center point and each given point as previously stated, and the various radii thus computed are averaged to set the radius of the new arc.

13. Calls the `icprjcrc` procedure to compute the coordinates of the start and end points, and corresponding radial direction, of the new arc by projecting thereon the first and last given points.

14. Calls the `icstrdln` procedure to compute the central angle between the said two directions and then multiplies the said angle with the radius to compute the length of the new arc.

15. Calls the `icforce` procedure to compute the arc translation distance and direction from the start point of the new arc towards the first given point.

16. Calls the `icptl` procedure to translate the two end points the center point so as to force the new arc to start at the first given point, and terminates.

B.3 Polynomial Curve Fitting

The `icFitPoly` procedure presented below fits a polynomial of any degree through a set of observation points each of which may be assigned a weight. If the degree of the desired polynomial is one, the resultant fit is a straight line, and if it is two the fit is a parabolic arc. Thus, this procedure could be employed instead of the icFitLine procedure presented above to fit a line, and instead of the avFitCurve procedure of Sect. 11.3.2 of Chap. 11. Although the procedure may easily be made to handle a polynomial of any degree, in its current form there is an arbitrarily set maximum value of 14. To increase this value of N = 14 to any higher value the array B(15,15) must be increased to B(NN, NN) where NN = N + 1. The polynomial equation being used is

$$y = a_1 + a_2x + a_3x^2 + a_4x^3 + \ldots + a_{n+1}x^n \quad \text{(B.3a)}$$

where

n denotes the degree of the desired polynomial, and a_1 through a_{n-1} represent the unknown coefficients of the equation.

For any observation point (x_j, y_j) there is a corresponding fitted y value expressed by Equation B.3a and a residual r_j defined as

$$r_j = y_j - \left(a_{1,} a_2 x_j + a_3 x_j^2 + a_4 x_j^3 + \ldots + a_{n-1} x_j^n \right) \quad \text{(B.3b)}$$

To apply the method of least squares the sum of the square of all residuals may then be expressed as

$$R^2 = \sum_{j=1}^{m} \left[y_j - \left(a_1 + a_2 x_j + a_3 x_j^2 + a_4 x_j^3 + \cdots + a_{n+1} x_j^n \right) \right]^2 \quad \text{(B.3c)}$$

where m denotes the number of observation points, and the partial derivatives of R^2 with respect to each of the unknown coefficients may be expressed as

$$\frac{\partial R^2}{\partial a_1} = -2 \sum_{j=1}^{m} \left[y_j - \left(a_1 + a_2 x_j + a_3 x_j^2 + a_4 x_j^3 + \cdots + a_{n+1} x_j^n \right) \right]^2 = 0$$

$$\frac{\partial R^2}{\partial a_2} = -2 \sum_{j=1}^{m} \left[y_j - \left(a_1 + a_2 x_j + a_3 x_j^2 + a_4 x_j^3 + \cdots + a_{n+1} x_j^n \right) \right]^2 = 0 \quad \text{(B.3d)}$$

$$\vdots$$

$$\frac{\partial R^2}{\partial a_{n+1}} = -2 \sum_{j=1}^{m} \left[y_j - \left(a_1 + a_2 x_j + a_3 x_j^2 + a_4 x_j^3 + \cdots + a_{n+1} x_j^n \right) \right]^2 = 0$$

Equation B.3d may then be represented in matrix form as indicated by Equation B.3e below and solved using the Doolittle method. As stated above each observation point may be assigned a weight w_j with a value of 1.0 retaining the observation value as is and any other value increasing or decreasing the said value. To simplify the presentation of the associated equations the parameter w_j is not included, but in the procedure it is applied to each x_j, y_j, and xy_j parameter.

$$\begin{bmatrix} m & \sum_{j=1}^{m} x_j + \cdots + \sum_{j=1}^{m} x_j^n \\ \sum_{j=1}^{m} x_j & \sum_{j=1}^{m} x_j^2 + \cdots + \sum_{j=1}^{m} x_j^{n+1} \\ \vdots & \vdots & \ddots & \vdots \\ \sum_{j=1}^{m} x_j^n & \sum_{j=1}^{m} x_j^{n+1} + \cdots + \sum_{j=1}^{m} x_j^{2n} \end{bmatrix} \begin{bmatrix} a_1 \\ a_2 \\ \vdots \\ a_{n+1} \end{bmatrix} = \begin{bmatrix} \sum_{j=1}^{m} y_j \\ \sum_{j=1}^{m} x_j y_j \\ \vdots \\ \sum_{j=1}^{m} x_j^n y_j \end{bmatrix} \quad \text{(B.3e)}$$

Once the coefficients a_1 through a_{n+1} have been computed, the procedure computes the standard error of deviation as

$$\sigma = \frac{\sum_{j=1}^{m} \left[(r_j)^2 w_j \right]}{\sum_{j=1}^{m} w_j} \quad \text{(B.3f)}$$

```
SubicFitPoly (N, X, Y, W, _ A, SEOFE,
              IERR)
```

The input parameters to this procedure include the:

N	The degree of the desired polynomial (1 through 14). The maximum polynomial degree of 14 is arbitrary, and it could increased if need be. If so, then the dimensions of the array B(15, 15) should be increased to N + 1.
X(), Y()	The Cartesian coordinate arrays of the observation points.
W()	The weight factor array of the observation points (1.0 has no effect).

The parameters that are returned from this procedure include the:

A()	The array containing the coefficients A_1 through A_{N+1} of the polynomial Equation B.3a where N is the input value N.
SEOFE	The standard error of deviation.
IERR	An indicator the value of which denotes the following:
	0 An error has not been encountered.
	1 Unequal size input arrays — abort.
	2 The X array or all arrays are empty — abort. If more than two but not all arrays are empty, then IERR is set to 1.
	3 Division by zero - abort. Check the variables DIAG or SUMWX.

Upon invocation, the procedure determines the size of the input arrays, checks for potential errors (empty or unequal size arrays) and aborts if necessary, initializes (a) the parameters to be returned, (b) the matrix counters and (c) the matrix summation parameters, and then:

1. Computes the summations of the normal Equation B.3a applying the corresponding weight.
2. Stores the summations into the forward matrix by columns and applies the Doolittle matrix method.
3. Computes the coefficients a_1 through a_{n+1} (back solution).
4. Computes the y value for each given x value by solving Equation B.3a, applying the corresponding weight, and subtracting the given y value to determine the residuals which are summarized.
5. Computes the standard deviation by use of Equation B.3f, and then terminates.

B.4   Circular Arc Fitting Revisited

Two sections above we discussed the fitting of a circular arc through a set of points. Below we will address the same issue with the additional restriction that the fitted arc must also be tangent to two given lines. This condition simulates better the mathematical recreation of an existing unknown alignment. The solution to this problem is rather simplistic and easy. In Sect. 5.1.6 of Chap. 5 we discussed the `iccrvpoc` procedure with which a circular arc was created tangent to two given lines and passing through a given point. Thus, to fit a circular arc through a set of given points tangent to two given lines the last said procedure is called for each of the points in the given set and the average of the resultant radii is computed. With this radius now known the coordinates of the PC, PT, and center point may easily be determined. The `icFitArc` procedure below performs this task.

Much like any set of observations it is possible for certain of the given points to be considered as erroneous or "outlying" and should be discarded. This is not, presently, included in the procedure below but the list of the various radii that are computed is returned for review and possible clean up of the set of given points. Alternatively, a statistical approach similar to that of the above `icFitCircle` procedure could be introduced in the `icFitArc` procedure.

```
Sub icFitArc    (P1N, P1E, AZ1, P2N, P2E,
                 AZ2, X1List, Y1List, _
                 PCN, PCE, PIN, PIE, PTN,
                 PTE, CCN, CCE, RAD, R,
                 IERR)
```

The input parameters to this procedure include the:

P1N, P1E	North/East Coordinates of a point on the back tangent line.
AZ1	North azimuth (radians) of the back tangent line pointing towards the PI.
P2N, P2E	North/East Coordinates of a point on the forward tangent line.
AZ2	North azimuth (radians) of the forward tangent line pointing ahead of the PI.
X1List()	Array of the point X (east) coordinates.
Y1List()	Array of the point Y (north) coordinates.

The parameters that are returned from this procedure include the:

PCN, PCE	North/East Coordinates of the PC of the fitted arc.
PIN, PIE	North/East Coordinates of the PI of the fitted arc.

PTN, PTE	North/East Coordinates of the PT of the fitted arc.
CCN, CCE	North/East Coordinates of the center point of the fitted arc.
RAD	Radius the fitted arc. Note that if: RAD>0 the arc rotates clockwise, and if * RAD<0 the arc rotates counterclockwise.
R()	Array of the radii corresponding to each of the points in the X1List(), Y1List() arrays. The R() array could be displayed by the calling program for review and possibly disregard outlying points.
IERR	Error indicator the value of which denotes: 0 No errors have been detected. 1 X1List() and/or Y1List() are empty. 2 X1List() and Y1List() are not of the same size. 3 Collinear tangent lines.

Upon invocation, the procedure determines the size of the input arrays, checks for potential errors (empty or unequal size arrays) and aborts if necessary setting the appropriate IERR value, initializes the parameters to be returned, and then:

1. Calls the `icintrs` procedure to intersect the two given tangent lines and determine the coordinates of the PI and the intersection angle GANG.
2. Checks the intersection angle GANG to determine whether the two tangent lines are collinear or not, within a tolerance of 0.0000048 radians (0.99 s of a degree), and if so sets IERR to 3 and aborts; else continues.
3. Initializes the accumulator sumR of the radii to be computed, and calls the `iccrvpoc` procedure for each of the given points to pass a circular arc through the point and tangent to the two given tangent lines, saves the computed radius of each such arc in the R() array, and accumulates the radii values in sumR.
4. Computes the radius RAD of the arc to be returned by dividing sumR by the number of points in the given arrays and assigns thereto the sign of the angle GANG.
5. Calls the `icsmpcrv` procedure to compute the coordinates of the PC, PT, and center point of the fitted curve, and then terminates.

Appendix C: Tangent Circle to Three Circles

About two centuries before the birth of Christ one of the great Greek geometers of the Hellenistic period, Apollonios of Perga (Απολλώνιος ο Περγαίος) circa 260–170 B.C. wrote an eight volume book titled *Conics* (*Κωνικά*) dedicated to the study of the conic sections, circle, ellipse, parabola and hyperbola. It appears that the first publication of the *Conics* may have been made by the Byzantine mathematician Eutokios (Ευτόκιος) sometime during the sixth century A.D. which was followed by various translations in Latin and Arabic. Of the eight volumes of this book the first four have been preserved in the original Greek while the next three volumes have been preserved in their Arabic translation. The eighth volume has not survived to date. In this lost volume Apollonios posed the problem of constructing a circle tangent to three given circles. According to a reference in his seventh volume Apollonios states that he "posed and solved" certain tangency (επαφαί) problems including the said tangency problem. This appears to have been confirmed by Pappus of Alexandria (Πάππος ο Αλεξανδρεύς) circa 290–350 A.D. in his seventh of eight books of *Mathematical Collection* (*Συναγωγή*).

Since that time several mathematicians, including but not limited to Adrian van Roomen, Francois Viete, Neuton and Pascal, have tackled this tangency problem and provided solutions. A solution which seems to have enthused modern day mathematicians is that of Joseph Diaz Gergonne (1771–1859) which employs circle inversion. All of these solutions are generally based on a geometric approach to the problem which can be rather complicated. Using circle inversion methodology the basic problem as posed by Apollonios has been generalized to construct a circle tangent not only to three given circles but to any to any combination of three features that may be comprised of a straight line, circle, and/or point. It should be noted that in the case of a point the new circle is not to be tangent to but passing through the point.

The solution of this problem has found home in several situations in which the location of one or more points is desirable. Perhaps the first major real world application might be during World War I in which it was used to fix the position of an artillery station to aim at three distinct enemy positions. In civil engineering there does not appear a high or even moderate demand for constructing a circle or arc thereof tangent to three given circles or arcs, but occasionally such a need may surface. To assist in such occasional cases, this appendix presents the `icCrvTan3Crv` procedure that addresses the original tangency problem as posed by Apollonios. Before proceeding with the description of this procedure let us take a look at certain configuration conditions of the subject problem.

Figure C.1 presents certain of the most common configurations of the problem in which the three given circles are colored in blue and two new tangent circles colored in red. At this time we are to concentrate on the two possible new tangent circles shown in this figure. Later on we will address the possibility of additional new tangent circles. The sample configurations in the said figure are:

- The given circles are apart from each other and arranged in two different patterns as in (a1) and (a2).
- The three given circles are arranged so as to be tangent to each other. This configuration is also referred to as the "kissing circle theorem", Soddy's Circles" after the British radio chemist Frederick Soddy's (1877–1956) "*The Kiss Precise*" poem, "Apollonian Carpet" because a series of such circles makes an appealing layout, or the "four coins problem" since the four circles could be simulated by four coins of different denominations.
- Two or three of the circles could be intersecting as in (c1) and (c2) and/or kissing as in (c3).
- Two of the given circles are apart from each other but both are inside the third given circle as in (d1).
- One of the given circles is inside another given circle and the third given circle is apart from both the said circles as in (d2), or the third given circle is inside the first given circle as in (d3) in which cases there is no new kissing circle.

E. C. Tonias, C. N. Tonias, *Geometric Procedures for Civil Engineers*, DOI 10.1007/978-3-319-24295-8

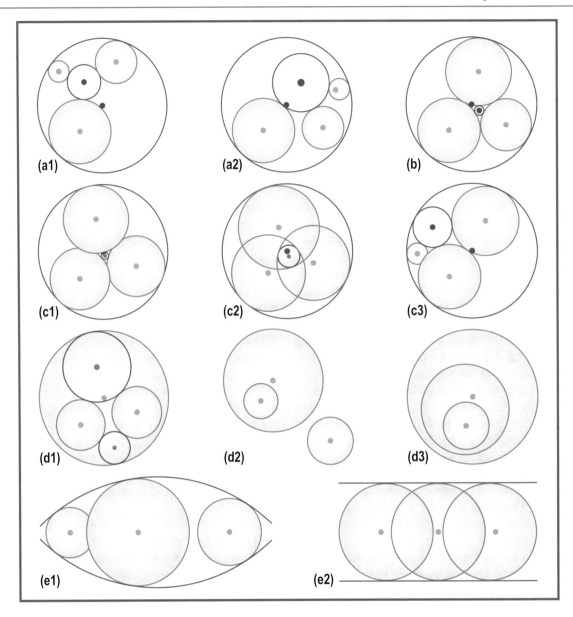

Figure C.1   Various Configurations of a Circle Being Tangent To Three Given Circles

- The center points of the given circles are collinear with the given circles having different radii as in (e1), or with all three given circles having the same radius as in (e2) in which case there is no new tangent circles, or rather there two new tangent circles each having a radius of infinite length, or two tangent straight lines.

In perusing the configurations of the three given circles it should be noted that a relatively small change in the size of a diameter or a relatively slight translation of a center point could easily transform a configuration into another. Also, regarding the relationship of a given circle to the new tangent circle it is noted the terminology below has gained familiarity among mathematicians:

- *If a given circle is external to the new tangent circle (it is located outside the new tangent circle) the new circle is external with respect to the given circle.*
- *If a given circle is internal to the new tangent circle (it is located inside the new tangent circle) the new circle is internal with respect to the given circle.*

It has been shown in the above figure that there are certain given circle configurations for which there can be no new tangent circle. Now let us take a look at Figure C.2 with specific emphasis in Figure C.1(a1), and let us note that similar possibilities may also exist for certain of the other configurations of Figure C.1. What is shown in this new figure is that there can be more than two possible new tangent lines, and

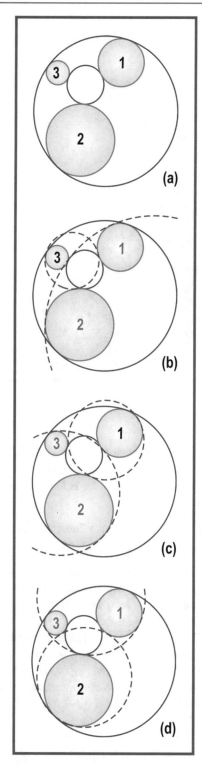

Figure C.2 The Maximum Of Eight Tangent Circles To Three Given Circles

as a matter of fact there can be up to eight possible new tangencies comprised of an odd number as well as an even number. However, it has been proven that there can not be seven new tangent circles.

Regarding the kissing circle case of Figure C.1(b) there is an interesting story involving Princess Elisabeth of Bohemia (an historic country now constituting the major part of the Czech Republic in Central Europe) and the French Mathematician Rene Descartes in 1643 A.D. Elisabeth was the daughter of Frederick V, King of Bohemia, who had lost his crown and a good part of his property thanks to the Thirty Years War between Catholics and Protestants. She was also the niece of Charles I of England (on her mother's side) who lost his head to Oliver Cromwell and his Parliamentarians. Elisabeth had a very sharp mind, she loved the Greek classics (her nickname was "The Greek"), she had taken rather fancy towards geometry and befriended (to some debatable extent) Descartes to whom she presented a solution to the kissing circles problem of the last said figure and also shown in Figure C.3. Note that in using the princess' methodology use of the center point coordinates of the given circles is not made. Also, although in Figure C.3 the circles have been numbered 1, 2, and 3 in a counterclockwise rotation, they can be numbered in the reverse direction without effect on the solution.

The princess' solution was based on the fact that since the three circles are given, the lengths of the lines joining them can be determined since the said lines lie along the common radial lines of two circles, thus being the sum of two known radii. In addition to this we know that the center point of the small new kissing circle lies within the said triangle. If radial lines are drawn from the said center point to each of the known center points three new triangles are formed and the sum of their areas should equal that of the first said triangle. Note that Elisabeth's solution is specifically addressing the smaller inner circle of Figure C.3(a). To address the outer larger circle of Figure C.3(b) slightly different geometric gymnastics must be used. In this case the center point of the new tangent circle could fall outside the abovesaid triangle, in which case instead of adding the three triangular areas, two must be added and the third area, the one on the side of the last said center point should be subtracted. Thus, knowledge of which is this triangle is necessary.

Heron's formula for finding the area of a triangle with known sides may be used to find the area of the said four triangles. In using this formula to find the area of the new triangles it is noted that each side has the length of one of the given radii plus that of the unknown new kissing circle (excepting the case of Figure C.3(b) in which the known side is subtracted). Thus, we have one equation, the sum of three

For additional reading on the princess and Descartes reference is made to J. L. Heibron's book "Geometry Civilized", Clarendon Press, Oxford (p. 208).

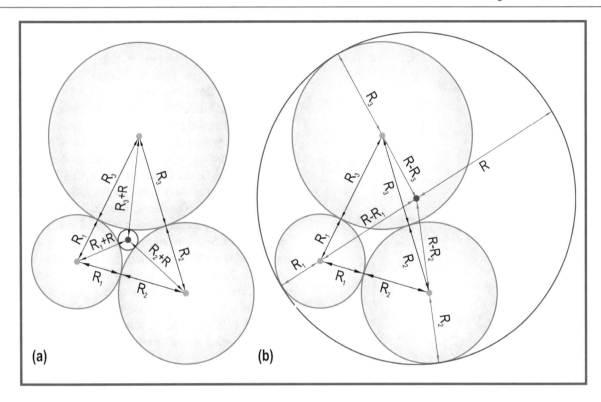

Figure C.3 The Kissing Circles Problem Using Descartes' Test Case

areas equaling a known area, and one unknown, the new
radius for which we can solve.

Heron's area formula for the triangle having as vertices
the center points of the given circles is

$$A = \sqrt{S\left[S-\left(R_1+R_2\right)\right]\left[S-\left(R_2+R_3\right)\right]\left[S-\left(R_3+R_1\right)\right]}$$

(C.a)

where A is the area of the said triangle, R_1, R_2, and R_3 are the
radii of the said triangle and S is one half the perimeter of the
triangle or $S = R_1 + R_2 + R_3$, and correspondingly the areas of
the three new triangles would be

$$A_{12} = \sqrt{S_{12}\left[S_{12}-\left(R_1+R_2\right)\right]\left[S_{12}-\left(R_2+R\right)\right]\left[S_{12}-\left(R_3+R\right)\right]}$$
$$A_{23} = \sqrt{S_{23}\left[S_{23}-\left(R_2+R_3\right)\right]\left[S_{23}-\left(R_3+R\right)\right]\left[S_{23}-\left(R_2+R\right)\right]}$$
$$A_{31} = \sqrt{S_{31}\left[S_{31}-\left(R_3+R_1\right)\right]\left[S_{31}-\left(R_1+R\right)\right]\left[S_{31}-\left(R_3+R\right)\right]}$$

(C.b)

from which

$$A = A_{12} + A_{23} + A_{31} \qquad \text{(C.c)}$$

Purportedly Elisabeth's letter to Descartes with the problem
carried out the solution of the above equations for R but it has
not survived.

Descartes' response to the princess was that her solution,
although sound, involved so many multiplications that, provided he did not make any arithmetic mistakes, it would take

him 3 months to determine the required triangle areas and
the desired radius. Instead he proceeded to present to her his
solution to the problem which leads to the equation for the
new circle's radius being

$$x^2[a^2b^2 + a^2c^2 + b^2c^2 - 2abc(a+b+c)] \qquad \text{(C.d)}$$
$$- 2x(abc)(ab+ac+bc) + (abc)^2 = 0$$

where x is the radius of the new circle and a, b, and c are the
radii of the given circles, and proceeded to solve a test case
of three circles have as radii the values of 2, 3 and 4 thus
yielding a radius 0.4333 for the small (inscribed) circle and a
radius of -7.0716 for the large (circumscribed) circle. To
locate the center points of the inscribed circle, intersect the
radii (2+0.4333), (3+0.4333 and (4+0.4333) with each other,
and to locate the circle of the circumscribed circle intersect
the absolute values of the radii |(2−7.0716)|, |(3−7.0716)|,
and |−(4-7.0716)|.

Perhaps Descartes might have overestimated his three
month period, but for the computational tools available in
1643 he might have been correct. In either case with today's
computers this estimate could have been reduced to less then
three seconds. But also, the princess' solution addressed
only one of the two solutions to the kissing circle problem,
and although Descartes' equation addresses both possible
solutions it does not address all possible cases of the
Apollonian tangency problem. Therefore we will try to solve
this problem using an algebraic methodology.

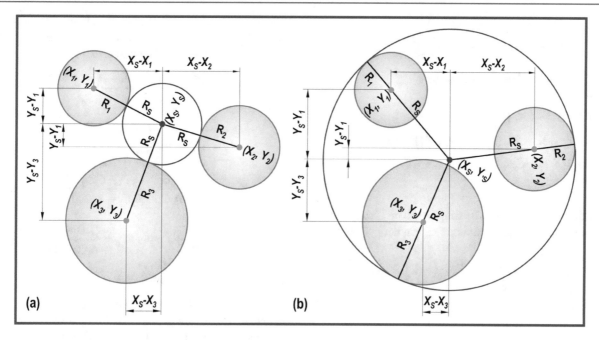

Figure C.4 Algebraic Solution of the Apollonian Tangency Problem

In Figure C.4 the three given circles 1, 2, and 3 are identified by their center point coordinates (X_1, Y_1), (X_2, Y_2), and $(X3, Y_3)$ and their radii R_1, R_2, and R_3 and the small and the large new tangent circles are similarly identified by (X_S, Y_S) and R_S. Also shown in this figure are the horizontal and vertical displacements between the center points of the three given circles and the center point of the new tangent circle. For the small tangent circle the distances between the center point of the said circle and the center point of each given circle would then be R_S+R_1, R_S+R_2, and R_S+R_3, while for the larger circle they would be R_S-R_1, R_S-R_2, and R_S-R_3. These distances may them be generalized as $R_S-S_1R_1$, $R_S-S_2R_2$, and $R_S-S_3R_3$ where S_1, S_2, and S_3 are factors of either +1 for the large tangent circle or −1 for the small tangent circle. All of the above regarding the algebraic solution could be extended to all cases of Figures C.1, C.2, and C.3. Thus, the following three equations may be used to define the above distances between the said center points.

$$\left(X_S - X_1\right)^2 + \left(Y_s - Y_1\right)^2 = \left(R_S - S_1 R_1\right)^2 \qquad \text{(C.1a)}$$

$$\left(X_S - X_2\right)^2 + \left(Y_s - Y_1\right)^2 = \left(R_S - S_2 R_2\right)^2 \qquad \text{(C.1b)}$$

$$\left(X_S - X_3\right)^2 + \left(Y_s - Y_3\right)^2 = \left(R_S - S_3 R_3\right)^2 \qquad \text{(C.1c)}$$

If we next expand the above equations to remove the squares they become

$$X_S^2 + X_1^2 - 2X_S X_1 + Y_s^2 + Y_1^2 - 2Y_S Y_1$$
$$= R_S^2 + \left(S_1 R_1\right)^2 - 2\left(S_1 R_1\right) R_S \qquad \text{(C.2a)}$$

$$X_S^2 + X_2^2 - 2X_S X_2 + Y_s^2 + Y_2^2 - 2Y_S Y_2$$
$$= R_S^2 + \left(S_2 R_2\right)^2 - 2\left(S_2 R_2\right) R_S \qquad \text{(C.2b)}$$

$$X_S^2 + X_3^2 - 2X_S X_3 + Y_s^2 + Y_3^2 - 2Y_S Y_3$$
$$= R_S^2 + \left(S_3 R_3\right)^2 - 2\left(S_3 R_3\right) R_S \qquad \text{(C.2c)}$$

In the last three equations X_S, Y_S, and R_S are of known values and their squares may be eliminated by subtracting Equation C.2b from Equation C.2a and Equation C.2c from Equation C.2b to get

$$X_1^2 - X_2^2 - 2X_S\left(X_1 - X_2\right)$$
$$+ Y_1^2 - Y_2^2 - 2Y_S\left(Y_1 - Y_2\right)$$
$$= \left(S_1 R_1\right)^2 + \left(S_2 R_S\right)^2 + 2\left(S_1 R_1 - S_2 R_2\right) R_S$$

or

$$X_1^2 - X_2^2 - 2X_S\left(X_1 - X_2\right)$$
$$+ Y_1^2 - Y_2^2 - 2Y_S\left(Y_1 - Y_2\right) - \left(S_1 R_1\right)^2 + \left(S_2 R_S\right)^2$$
$$+ 2\left(S_1 R_1 - S_2 R_2\right) R_S = 0$$

$$\text{(C.3a)}$$

$$X_2^2 - X_3^2 - 2X_S(X_2 - X_3)$$
$$+ Y_2^2 - Y_3^2 - 2Y_S(Y_2 - Y_3) = (S_2 R_2)^2$$
$$- (S_3 R_3)^2 - 2(S_2 R_2 - S_3 R_3) R_S$$

or

$$X_2^2 - X_3^2 - 2X_S(X_2 - X_3)$$
$$+ Y_2^2 - Y_3^2 - 2Y_S(Y_2 - Y_2) - (S_2 R_2)^2 + (S_3 R_3)^2$$
$$+ 2(S_2 R_2 - S_3 R_3) R_S = 0$$

$$(C.3b)$$

The last two equations can be simplified a bit if certain of the known items are grouped into one known item. Thus, let

$$A_1 = X_1^2 - X_2^2 + Y_1^2 - Y_2^2 - (S_1 R_1)^2 + (S_2 R_2)^2$$
$$B_1 = 2(X_1 - X_2)$$
$$C_1 = 2(Y_1 - Y_2)$$
$$D_1 = 2(S_1 R_1 - S_2 R_2)$$

$$(C.3c)$$

and

$$A_2 = X_2^2 - X_3^2 + Y_2^2 - Y_3^2 - (S_2 R_2)^2 + (S_3 R_3)^2$$
$$B_2 = 2(X_2 - X_3)$$
$$C_2 = 2(Y_2 - Y_3)$$
$$D_2 = 2(S_2 R_2 - S_3 R_3)$$

$$(C.3d)$$

We may next substitute Equations C.3c into Equation C.3a, and Equations C.3d into Equation C.3b to get

$$A_1 - B_1 X_S - C_1 Y_S + D_1 R_S = 0 \qquad (C.4a)$$

$$A_2 - B_2 X_S - C_2 Y_S + D_2 R_S = 0 \qquad (C.4b)$$

If we solve Equations C.4a and C.4b for X_S, equating each other and cross multiply we can then solve for Y_S in terms of R_S as indicated by the next three equations.

$$X_S = \frac{A_1 - C_1 Y_S + D_1 R_S}{B_1} = \frac{A_2 - C_2 Y_S + D_2 R_S}{B_2} \qquad (C.5a)$$

$$A_1 B_2 - B_2 C_1 Y_S + B_2 D_1 R_S$$
$$- A_2 B_1 + B_1 C_2 Y_S - B_1 D_2 R_S = 0 \qquad (C.5b)$$

$$Y_S = \frac{A_1 B_2 - A_2 B_1}{B_2 C_1 - B_1 C_2} + \frac{B_2 D_1 - B_1 D_2}{B_2 C_1 - B_1 C_2}(R_S) \qquad (C.5c)$$

Equation C.5c may simplified if we let

$$F_Y = \frac{A_1 B_2 - A_2 B_1}{B_2 C_1 - B_1 C_2} \quad \text{and} \quad G_Y = \frac{B_2 D_1 - B_1 D_2}{B_2 C_1 - B_1 C_2} \qquad (C.5d)$$

So that Y_S may now be expressed as

$$Y_S = F_Y + G_Y R_S \qquad (C.5e)$$

If we now repeat the above process solving Equations C.4a and C.4b for Y_S we can then solve for X_S in terms of R_S as indicated by the next five equations.

$$Y_S = \frac{A_1 - B_1 X_S + D_1 R_S}{C_1} = \frac{A_2 - B_2 X_S + D_2 R_S}{C_2} \qquad (C.6a)$$

$$A_1 C_2 - B_1 C_2 X_S + C_2 D_1 R_S$$
$$- A_2 C_1 + B_2 C_1 X_S - C_1 D_2 R_S = 0 \qquad (C.6b)$$

$$X_S = \frac{A_1 C_2 - A_2 C_1}{B_1 C_2 - B_2 C_1} + \frac{C_2 D_1 - C_1 D_2}{B_1 C_2 - B_2 C_1}(R_S) \qquad (C.6c)$$

$$F_X = \frac{A_1 C_2 - A_2 C_1}{B_1 C_2 - B_2 C_1} \quad \text{and} \quad G_X = \frac{C_2 D_1 - C_1 D_2}{B_1 C_2 - B_2 C_1} \qquad (C.6d)$$

$$X_S = F_X + G_X R_S \qquad (C.6e)$$

Equations C.5e and C.6e may now be substituted any of the original expanded Equations C.2a, C.2b, or C.2c to solve for R_S. We will arbitrarily select Equation C.2a to get

$$(F_X + G_X R_S)^2 + X_1^2 - 2X_1(F_X + G_X R_S) +$$
$$(F_Y + G_Y R_S)^2 + Y_1^2 - 2Y_1(F_Y + G_Y R_S) = \qquad (C.7)$$
$$R_S^2 + (S_1 R_1)^2 - 2(S_1 R_1) R_S$$

which may be expanded to become

$$F_X^2 + G_X^2 R_S^2 + 2F_X G_X R_S + X_1^2 - 2X_1 F_X - 2X_1 G_X R_S +$$
$$F_Y^2 + G_Y^2 R_S^2 + 2F_Y G_Y R_S + Y_1^2 - 2Y_1 F_Y - 2Y_1 G_Y R_S = \qquad (C.8a)$$
$$R_S^2 + (S_1 R_1)^2 - 2(S_1 R_1) R_S$$

or

$$F_X^2 + X_1^2 + F_Y^2 + Y_1^2 - (S_1 R_1)^2 - 2X_1 F_X - 2Y_1 F_Y$$
$$+ [2F_X G_X - X_1 G_X + F_Y G_Y - Y_1 G_Y + (S_1 R_1)] R_S \qquad (C.8b)$$
$$+ (G_X^2 + G_Y^2 - 1) R_S^2 = 0$$

Again letting

$$H = (G_X^2 + G_Y^2 - 1)$$
$$K = 2[F_X G_X - X_1 G_X R_S + F_Y G_Y - Y_1 G_Y + (S_1 R_1)] \qquad (C.8c)$$
$$L = F_X^2 + X_1^2 + F_Y^2 + Y_1^2 - (S_1 R_1)^2 - 2X_1 F_X - 2Y_1 F_Y$$

Equation C.8b may be expressed as the quadratic equation

$$H R_S^2 + K R_S + L = 0 \qquad (C.9)$$

from which the value of the radius of the new tangent circle R_S may be found to be

$$R_S = \frac{-K \pm \sqrt{K^2 - 4HL}}{2H} \qquad (C.10)$$

which may be substituted in Equations C.5e and C.6e to compute the center point coordinates of the new tangent

In reviewing the S factor assignment of Table C.1 and Equations C.3c, C.3d, C.5d, C.6d, and C.8c we see that the value of a factor is either +1 or −1 and hence the constants:

- *A1 and A2 do not depend on the value of a factor since the factors appear within the parentheses which are being squared.*
- *B1, C1, B2, C2, FY, and FX do not depend on the value of a factor since one is not used.*
- *D1, D2, GY, GX, H, K, and L are the only ones that depend on a factor.*

Table C.1 S_1, S_2, and S_3 Factor Assignment

Factor	1	2	3	4	5	6	7	8
S_1	=+1	=+1	=+1	=+1	=−1	=−1	=−1	=−1
S_2	=+1	=+1	=−1	=−1	=+1	=+1	=−1	=−1
S_3	=+1	=−1	=+1	=−1	=+1	=−1	=+1	=−1

circle (X_S, Y_S). Since Equation C.10 yields two values for R_S there will be two sets of the said coordinates for each of the eight conditions that result from the values assigned to the factors S_1, S_2, and S_3 summarized in Table C.1, thus yielding 16 radii with half being equal to the other half in absolute and either with the same or opposite sign.

Now let us examine the application of the above equations to the four simplistic cases (a) through (d) of Figure C.5 in which the center point coordinates and radii are stated. Associated correspondingly with these four cases are those of Table C.2 displaying the 16 radii Ra and Rb of the possible common tangent circles. In this table note the following:

- Negative radii can be disregarded. Their absolute value is the same as their positive counterparts.
- The label #DIV/0! denotes a division by zero for the corresponding value assignment of the factors S_1, S_2, and S_3 which do not produce a common tangent circle.
- The non-fatal error code of −8 denotes an imaginary root in Equation C.10.
- Duplicitous solutions are located at different table cells depending on the values of the said factors. Duplicitous radii may exist within Ra or Rb, or between Ra and Rb.
- Tables C.2a, C.2b, C.2c, and C.2d at the end of this appendix contain the computed values of the constants of Equations C.1 through C.10 from which Table C.2 has been excerpted.

The `icCrvTan3Crv` procedure automates the above process.

```
Sub icCrvTan3Crv (R1, Y1, X1, R2, Y2, X2, R3,
                  Y3, X3, _  Ra, Ya, Xa, Rb,
                  Yb, Xb, IERR)
```

The input parameters to this procedure include the:

R1	Radius of the first circle or circular arc.
Y1, X1	North/east coordinates of the first circle or circular arc.
R2	Radius of the second circle or circular arc.
Y2, X2	North/east coordinates of the second circle or circular arc.
R3	Radius of the third circle or circular arc.
Y3, X3	North/east coordinates of the third circle or circular arc. Note that as to which is the first second or third circle or circular arc is of no concern.

The parameters that are returned from this procedure include the:

Ra(8)	Array of the radii corresponding sequentially to the eight potential tangent circles and to the plus sign of the square root of the quadratic Equation C.10.
Ya(8), Xa(8)	Arrays of the north/east center point coordinates of the potential tangent circles corresponding to the Ra(8) array of radii.
Rb(8)	Array of the radii corresponding sequentially to the eight potential tangent circles and to the minus sign of the square root of the quadratic equation C.10.
Yb(8), Xb(8)	Arrays of the north/east center point coordinates of the potential tangent circles corresponding to the Rb(8) array of radii.
IERR	An indicator the value of which denotes the following: 0 Zero input radius (fatal). 1 The centers of the three given circles are collinear and they have the same size radius (fatal). 2 Division by zero in solving the quadratic Equation C.10 for the radius of the tangent circle which is set to be −9. 3 Improper input array data. One possible error condition is that in which all of the input points lie on a straight due north line (non-fatal). 4 Negative value under the square root of the quadratic Equation C.10 (imaginary radius). The radius is set to be −8 (non-fatal).

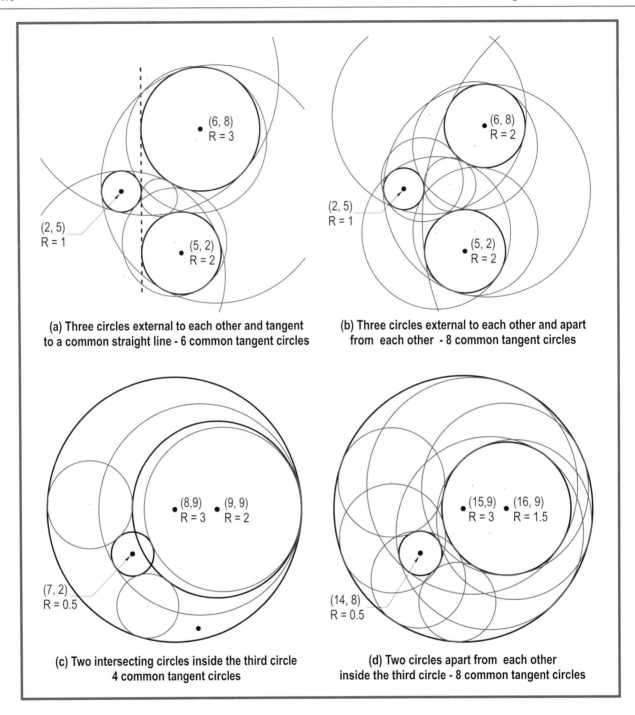

(a) Three circles external to each other and tangent
to a common straight line - 6 common tangent circles

(b) Three circles external to each other and apart
from each other - 8 common tangent circles

(c) Two intersecting circles inside the third circle
4 common tangent circles

(d) Two circles apart from each other
inside the third circle - 8 common tangent circles

Figure C.5   Certain Sample Cases of a Circle Tangent To Three Given Circles

NOTES
1. Zero and negative center point coordinates of a given circle are acceptable, but not a zero radius.
2. If the given circles are collinear within a tolerance of 0.0000048 radians (0.99 s of a degree), and if they have the same size radius within a tolerance of 0.001 feet (0.3048 mm), there is no tangent circle.

Upon invocation, the procedure initializes the variables to be returned, and then:

1. Checks if any of the given radii has a zero value, or if the three given circles or circular arcs have the same radius and their center points are collinear as stated above, and if so sets IERR to 1 or 2, respectively, and aborts; else continues.

Table C.2   Summary of radii of the four cases of Figure C.5

CASE ▶	1	2	3	4	5	6	7	8
Radii for Figure C.5(a)								
Ra	−0.917	−3.547	−2.630	#DIV/0!	#DIV/0!	−6.630	−5.105	−5.771
Rb	5.771	5.105	6.630	#DIV/0!	#DIV/0!	2.630	3.547	0.917
Radii for Figure C.5(b)								
Ra	−1.286	−3.070	−3.079	−12.180	−2.013	−4.779	−3.665	−5.102
Rb	5.102	3.655	4.779	2.013	12.180	3.079	3.070	1.286
Radii for Figure C.5(c)								
Ra	2.393	−0.729	1.861	−8	−8	−1.861	0.999	−2.393
Rb	2.393	0.999	1.861	−8	−8	−1.861	0.729	−2.393
Radii for Figure C.5(d)								
Ra	2.442	−0.894	1.937	−1.094	1.227	−1.801	1.231	−2.174
Rb	2.174	−1.231	1.801	−1.227	1.094	−1.937	0.894	−2.442

Table C.2a   Tangent Circle to Three Given Circles

CIRCLE	X	Y	R					
1	2	5	1					
2	5	2	2					
3	6	8	3					
TheCase	**1**	**2**	**3**	**4**	**5**	**6**	**7**	**8**
S1	1	1	1	1	−1	−1	−1	−1
S2	1	1	−1	−1	1	1	−1	−1
S3	1	−1	1	−1	1	−1	1	−1
			Constants—Equations C.3c, C.3d, C.5d, C.6d, C.8c and C.10					
A1	3.000	3.000	3.000	3.000	3.000	3.000	3.000	3.000
B1	−6.000	−6.000	−6.000	−6.000	−6.000	−6.000	−6.000	−6.000
C1	6.000	6.000	6.000	6.000	6.000	6.000	6.000	6.000
D1	−2.000	−2.000	6.000	6.000	−6.000	−6.000	2.000	2.000
A2	−66.000	−66.000	−66.000	−66.000	−66.000	−66.000	−66.000	−66.000
B2	−2.000	−2.000	−2.000	−2.000	−2.000	−2.000	−2.000	−2.000
C2	−12.000	−12.000	−12.000	−12.000	−12.000	−12.000	−12.000	−12.000
D2	−2.000	10.000	−10.000	2.000	−2.000	10.000	−10.000	2.000
yg	−84.000		**Denominator of Equations C.5d**					
Fy	4.786	4.786	4.786	4.786	4.786	4.786	4.786	4.786
Gy	0.095	−0.762	0.857	0.000	0.000	−0.857	0.762	−0.095
yg	84.000		**Denominator of Equations C.6d**					
Fx	4.286	4.286	4.286	4.286	4.286	4.286	4.286	4.286
Gx	0.429	−0.429	−0.143	−1.000	1.000	0.143	0.429	−0.429
H	−0.807	−0.236	−0.245	0.000	0.000	−0.245	−0.236	−0.807
K	3.918	0.367	0.980	−2.571	2.571	−0.980	−0.367	−3.918
L	4.270	4.270	4.270	4.270	4.270	4.270	4.270	4.270
K^2-4HL	29.143	4.163	5.143	6.612	6.612	5.143	4.163	29.143
Ra	−0.917	−3.547	−2.630	#DIV/0!	#DIV/0!	−6.630	−5.105	−5.771
Rb	5.771	5.105	6.630	#DIV/0!	#DIV/0!	2.630	3.547	0.917

Table C.2b Tangent Circle to Three Given Circles

CIRCLE	X	Y	R
1	2	5	1
2	5	2	2
3	6	8	2

TheCase	1	2	3	4	5	6	7	8
S1	1	1	1	1	−1	−1	−1	−1
S2	1	1	−1	−1	1	1	−1	−1
S3	1	−1	1	−1	1	−1	1	−1
			Constants—Equations C.3c, C.3d, C.5d, C.6d, C.8c and C.10					
A1	3.000	3.000	3.000	3.000	3.000	3.000	3.000	3.000
B1	−6.000	−6.000	−6.000	−6.000	−6.000	−6.000	−6.000	−6.000
C1	6.000	6.000	6.000	6.000	6.000	6.000	6.000	6.000
D1	−2.000	−2.000	6.000	6.000	−6.000	−6.000	2.000	2.000
A2	−71.000	−71.000	−71.000	−71.000	−71.000	−71.000	−71.000	−71.000
B2	−2.000	−2.000	−2.000	−2.000	−2.000	−2.000	−2.000	−2.000
C2	−12.000	−12.000	−12.000	−12.000	−12.000	−12.000	−12.000	−12.000
D2	0.000	8.000	−8.000	0.000	0.000	8.000	−8.000	0.000
yg	−84.000			**Denominator of Equations** C.5d				
Fy	5.143	5.143	5.143	5.143	5.143	5.143	5.143	5.143
Gy	−0.048	−0.619	0.714	0.143	−0.143	−0.714	0.619	0.048
yg	84.000			**Denominator of Equations** C.6d				
Fx	4.643	4.643	4.643	4.643	4.643	4.643	4.643	4.643
Gx	0.286	−0.286	−0.286	−0.857	0.857	0.286	0.286	−0.286
H	− 0.916	−0.535	−0.408	−0.245	−0.245	−0.408	−0.535	− 0.916
K	3.497	0.313	0.694	−2.490	2.490	−0.694	−0.313	−3.497
L	6.005	6.005	6.005	6.005	6.005	6.005	6.005	6.005
K^2-HL	34.231	12.952	10.286	12.082	12.082	10.286	12.952	34.231
Ra	−1.285	−3.070	−3.079	−12.180	−2.013	−4.779	−3.655	−5.102
Rb	5.102	3.655	4.779	2.013	12.180	3.079	3.070	1.285

Table C.2c Tangent Circle to Three Given Circles

CIRCLE	X	Y	R
1	9	9	2
2	7	8	0.5
3	8	9	3

The Case	1	2	3	4	5	6	7	8
S1	1	1	1	1	−1	−1	−1	−1
S2	1	1	−1	−1	1	1	−1	−1
S3	1	−1	1	−1	1	−1	1	−1
			Constants—Equations C.3c, C.3d, C.5d, C.6d, C.8c and C.10					
A1	45.250	45.250	45.250	45.250	45.250	45.250	45.250	45.250
B1	4.000	4.000	4.000	4.000	4.000	4.000	4.000	4.000
C1	2.000	2.000	2.000	2.000	2.000	2.000	2.000	2.000
D1	3.000	3.000	5.000	5.000	−5.000	−5.000	−3.000	−3.000
A2	−23.250	−23.250	−23.250	−23.250	−23.250	−23.250	−23.250	−23.250
B2	−2.000	−2.000	−2.000	−2.000	−2.000	−2.000	−2.000	−2.000
C2	−2.000	−2.000	−2.000	−2.000	−2.000	−2.000	−2.000	−2.000
D2	−5.000	7.000	−7.000	5.000	−5.000	7.000	−7.000	5.000
yg	4.000			**Denominator of Equations** C.5d				
Fy	0.625	0.625	0.625	0.625	0.625	0.625	0.625	0.625
Gy	3.500	−8.500	4.500	−7.500	7.500	−4.500	8.500	−3.500

Table C.2c (continued)

yg	−4.000			**Denominator of Equations** C.6d				
Fx	11.000	11.000	11.000	11.000	11.000	11.000	11.000	11.000
Gx	−1.000	5.000	−1.000	5.000	−5.000	1.000	−5.000	1.000
H	12.250	96.250	20.250	80.250	80.250	20.250	96.250	12.250
K	−58.625	166.375	−75.375	149.625	−149.625	75.375	−166.375	58.625
L	70.141	70.141	70.141	70.141	70.141	70.141	70.141	70.141
K^2-4HL	0.000	676.500	0.000	−127.500	−127.500	0.000	676.500	0.000
Ra	2.393	−0.729	1.861	−8.000	−8.000	−1.861	0.999	−2.393
Rb	2.393	−0.999	1.861	−8.000	−8.000	−1.861	0.729	−2.393

Table C.2d Tangent Circle to Three Given Circles

CIRCLE	X	Y	R					
1	16	9	1.5					
2	14	8	0.5					
3	15	9	3					
TheCase	**1**	**2**	**3**	**4**	**5**	**6**	**7**	**8**
S1	1	1	1	1	−1	−1	−1	−1
S2	1	1	−1	−1	1	1	−1	−1
S3	1	−1	1	−1	1	−1	1	−1
			Constants—Equations C.3c, C.3d, C.5d, C.6d, C.8c and C.10					
A1	75.000	75.000	75.000	75.000	75.000	75.000	75.000	75.000
B1	4.000	4.000	4.000	4.000	4.000	4.000	4.000	4.000
C1	2.000	2.000	2.000	2.000	2.000	2.000	2.000	2.000
D1	2.000	2.000	4.000	4.000	−4.000	−4.000	−2.000	−2.000
A2	−37.250	−37.250	−37.250	−37.250	−37.250	−37.250	−37.250	−37.250
B2	−2.000	−2.000	−2.000	−2.000	−2.000	−2.000	−2.000	−2.000
C2	−2.000	−2.000	−2.000	−2.000	−2.000	−2.000	−2.000	−2.000
D2	−5.000	7.000	−7.000	5.000	−5.000	7.000	−7.000	5.000
yg	4.000			**Denominator of Equations C.5d**				
Fy	−0.250	−0.250	−0.250	−0.250	−0.250	−0.250	−0.250	−0.250
Gy	4.000	−8.000	5.000	−7.000	7.000	−5.000	8.000	−4.000
yg	−4.000			**Denominator of Equations C.6d**				
Fx	18.875	18.875	18.875	18.875	18.875	18.875	18.875	18.875
Gx	−1.500	4.500	−1.500	4.500	−4.500	1.500	−4.500	1.500
H	17.250	83.250	26.250	68.250	68.250	26.250	83.250	17.250
K	−79.625	176.875	−98.125	158.375	−158.375	98.125	−176.875	79.625
L	91.578	91.578	91.578	91.578	91.578	91.578	91.578	91.578
K^2-4HL	21.250	789.250	12.813	81.813	81.813	12.813	789.250	21.250
Ra	2.442	−0.894	1.937	−1.094	1.227	−1.801	1.231	−2.174
Rb	2.174	−1.231	1.801	−1.227	1.094	−1.937	0.894	−2.442

2. In a loop from 1 to 8 for each of the eight potential cases of tangent circles…
   - Sets the factors S_1, S_2, and S_3 values as per Table C.1.
   - Computes the constants of Equations C.3c, C.3d, C.5d, C.6d, and C.8c.
   - Checks if the denominator of Equations C.10 is equal to zero, or negative, and if so sets IERR to be 3 or 4, respectively, and aborts; else continues.
   - Solves Equation C.10 to compute the radii Ra and Rb.
   - Solves Equations C.5e and C.6e to compute the center point coordinates.

3. Terminates. It is up to the calling program to determine the desired solution.

Acknowledgements

Listed below in their sequential order of appearance are the sources from which the photographs of various equipment, statues, and paintings, and the sketches and photographs of the vari- ous mathematicians that laid the foundation of today's mathematics have been obtained. To these sources the authors express their appreciation and thanks.

Photo	Source	Description
In Chap. 1		
Geranos02.tif	T.P. Tasios	National Polytechnic of Athens, Greece
Maps of Greece	Wikipedia	Top two Maps—Bottom map—Authors' work
Thales01.jpg	The Internet Encyclopedia of Philosophy	
		Unknown artist
Anaximander.jpg	Wikipedia	Detail of Raphael's painting *The School of Athens*, 1510–1511 depicting Anaximander leaning towards Pythagoras on his left
Pythagoras.jpg	Wikipedia	Bust of Pythagoras of Samos in the Capitoline Museums, Rome
PythagorasStatue1.jpg	http://www.bing.com/images/search?q=statue+of+pythagoras&qpvt=statue+of+pythagoras&FORM=IGRE	
		Statue in honor of Pythagoras in the island of Samos, Greece
Plato.jpg	Wikipedia	Glyptothek, Munich, Germany—Unknown artist or maker
Aristotle.jpg	Wikipedia	Wikimedia Commons—Louvre Museum
EuclidStatue.jpg	Wikipedia	Statue of Euclid in the Oxford University Museum of Natural History
Archimedes.jpg	Wikipedia	Archimedes Thoughtful by Fetti (1620)
Archimedes01.jpg	Photographs by Ken Cedeno for The New York Times	
		Archimedes' palimpsest
Archimedes02.jpg	Photographs by Ken Cedeno for The New York Times	
		Archimedes' palimpsest
Hydraulis1.jpg	Archaeological Museum of Dion, Dion, Greece	
Heron.jpg	Wikipedia	Bust of Heron from a 1688 German translation of Heron's book "*Pneumatica*"
HeronSteamEngine.jpg	Pegasus Research Consortium	
		Illustration from Heron's book "*Pneumatica*"
Hipparchos_1.jpg	University of St. Andrews, Scotland, UK	
	http://www-history.mcs.st-andrews.ac.uk/PictDisplay/Hipparchus.html	
		Unknown medieval artist's depiction of Hipparchos
Ptelemaeus.jpg	Wikipedia	A early Baroque artist's rendition of Claudius Ptolemaeus
Al-Battani.jpg	Wikipedia	Modern artist unknown
al_Biruni.jpg	Wikipedia	1973 postal stamp of the Union of Soviet Socialist Republics (USSR) honoring al_Biruni
NasirAlDin.jpg	Wikipedia	Iranian stamp honoring Nasir al-Din
Regiomontanus.jpg	Wikipédia, a enciclopédia livre (Portuguese)	
		Johannes Muller von Konigsberg
Francois_Viete.jpg	Wikipedia	Artist unknown
Diophantus.jpg	http://www.famous-mathematicians.com/diophantus/	
		Artist unknown
John_Napier.jpg	Wikipedia	Merchiston Tower, Edinburgh, Scotland
Descartes.jpg	Wikipedia	Rene Descartes. Portrait by Frans Hals, 1648
Oughtred.jpg	Wikipedia	Medieval portrait of William Oughtred by unknown artist

© Springer International Publishing Switzerland 2016

E. C. Tonias, C. N. Tonias, *Geometric Procedures for Civil Engineers*, DOI 10.1007/978-3-319-24295-8

Photo	Source	Description
Plate 1.5a	Wikipedia	Computer operators at the Lawrence Livermore National Laboratory working at the IBM 650 computer consoles, October 1956. Lawrence Livermore accepted delivery of its first IBM computer in 1954
Plate 1.5b	Wikipedia	IBM 610 computer promotional material
Plate 1.5c (Left)	Wikipedia	IBM 1620 Model 1 Level A computer with paper tape and typewriter input and output printer (the basic entry system)
Plate 1.5d1	Wikipedia	IBM 1620 computer paper tape
Plate 1.5e	Paul E. Ceruzzi	*History of Modern Computing*. Cambridge, MA: MIT Press, 1998. pp. 42–43. http://members.iinet.net.au/~dgreen/
Plate 1.5f	http://simh.trailing-edge.com/photos/lgp30.jpg	LGP 30 computer promotional material
Plate 1.5g	Wikipedia	IBM 1130 computer console as it appeared in the IBM announcement of the machine
john_backus2.jpg	Wikipedia	Unknown photographer. Year of photograph probably middle of 1960s
In Chap. 9		
Fermat.jpg	http://www-groups.dcs.st-and.ac.uk/~history/PictDisplay/Fermat.html	Portrait of Pierre de Fermat by unknown artist
In Chap. 12		
Ramanujan.jpg	http://www-history.mcs.st-andrews.ac.uk/PictDisplay/Ramanujan.html	

Also included in Chap. 1 are the following photographs and plates which are from the authors' archives: The photographs in pages 1-2, 1-3, 1-5 and 1-6, and the plates 1.4a, 1.4b, 1.4c, 1.4d, 1.4e, 1.4f, 1.5c (right photo), 1.5d2, 1.5d3, 1.5h, 1.5i, 1.5j, 1.5k, 1.5l, 1.5m, 1.5n, 1.5o, and 1.5p.

All other sketches, diagrams, flowcharts, and tables that appear in this book have been created by the authors for the purpose of this book.

Bibliography

AASHTO *"A Policy on Geometric Design of Highways and Streets"*, American Association of State Highway and Transportation Officials, Washington DC, 2001.

Aref, M. N. and Wernick, William *"Problems and Solutions In Euclidean Geometry"*, Dover Publications, Inc., Mineola, New York, 2010.

Antin, SyDavid. *"100 Great Problems of Elementary Mathematics"* a translation of the 1933 book by Henirich Dorrie, Dover Publications, Inc., 1965.

Chasen, Sylvan H. *"Geometric Principles and Procedures for Computer Graphic Applications"*, Prentice-Hall, Inc., 1978.

Cluff, P. E. M. ASCE, Jack C. *"The Tonias Algorithm"*, Journal of Surveying Engineering, Proceedings of the American Society of Civil Engineers, Volume 121, Issue No. 3, August, 1995, pp. 118–127.

Darling, Arthur Burr, et al., *"Engineering in History"*, McGraw-Hill Book Company, 1950, republished by Dover Publications, Inc., Mineola, New York, 1990, pp. 129–133.

Fenves, PhD, Steven J. and Rasdorf, P. E., William J., *"Role of ASCE in the Advancement of Computing in Civil Engineering"* Journal of Computing in Civil Engineering, October 2001.

Hartshorne, Robin, *"Geometry: Euclid and Beyond"*, Springer Science+Business Media, New York, New York, 2000.

Hastings, Jr., Cecil, *"Approximations for Digital Computers"*, Princeton University Press, 1955.

Heath, K.C.B., K.C.V.O., F.R.S., Sir Thomas L., *"A History of Greek Mathematics"*, Volumes 1 and 2, Dover Publications, Inc., 1981.

Heath, K.C.B., K.C.V.O., F.R.S., Sir Thomas L., *"The Thirteen Books of Euclid's Elements"*, Volumes 1, 2 and 3, Dover Publications, Inc., 1956.

Heilbron, J. L., *"Geometry Civilized - History, Culture, and Technique"*, Clarendon Press, Oxford, 2000.

Hickerson, Phd, M. ASCE, Thomas F., *"Route Location and Design"*, McGraw-Hill Book Company, New York, 5th Edition, 1967.

Kilgour, Frederick Gridley, et al., *"Engineering in History"*, McGraw-Hill Book Company, 1950, republished by Dover Publications, Inc., Mineola, New York, 1990, pp. 129–133.

Kirby, Richard Sheldon, et al., "Engineering in History", McGraw-Hill Book Company, 1950, republished by Dover Publications, Inc., Mineola, New York, 1990, pp. 129–133.

Maor, Eli, *"The Pythagorean Theorem - A 4000 Year History"*, Princeton University Press, 1998, Princeton, New Jersey 08540.

Maor, Eli, *"Trigonometric Delights"*, Princeton University Press, 1998, Princeton, New Jersey 08540.

McCauley, Christopher J. (Senior Editor), *"Machinery's Handbook"*, Industrial Press, Inc., New York, 27th Edition, 2004.

McGrory, P. E., Hugh (Chairman), *"A Proposal for a National Institute for Computers in Engineering"*, Civil Engineering Program Applications, 1975 - A report for the fulfillment of the requirements of National Science Foundation grant.

mlahana.de Website, *http://www.mlahanas.de/Greeks/Kythera.htm* regarding the Atikythera mechanism.

Pegasus Research Consortium, *"Antikythera Mechanism - The Ancients"*, http://www.thelivingmoon.com/forum1/index.php?topic=2592.0.

Pegasus Research Consortium, *"Heron of Alexandria - Ancient Machines"*, http://www.thelivingmoon.com/43ancients/02files/Antikythera_ Mechanism.html.

Polio, Markus Markus Vitruvius, *"Vitruvius, The Ten Books On Architecture"*, Translated by Morris Hicky Morgan, Elibron Classics Replica Edition of the Harvard University 1914 edition, Cambridge, Mass.

Reynolds, W. F., *"Manual of Triangulation Computation and Adjustment"*, U. S. Coast and Geodetic Survey, Special Publication 138, Government Printing Office, Washington, D. C. 1955.

Ryu, Myungsunn, *"Ευκλείδου Στοιχεία Γεωμετρίας"*, Greek text from the Perseus Digital Library, http//blogfile.paran.com/ BLOG_17207/200412/1104318343_elem.pdf, November, 2004.

Tonias, P. E., M. ASCE, Elias C., *"Spiral Problems For An Electronic Computer"*, Journal of the Highway Division Proceedings of the American Society of Civil Engineers, Volume 91, Issue HW2, December, 1965.

Tasios, Theodosios P., http://vimeo.com/13229017, animation of the operation of the Diolkos of Corinth.

Traite de la Construction et Principaux Usage de Instruments de Mathematique by Nicolas Bion published in France in 1709 *The Construction and Principal Uses of Mathematical Instruments*. This translation has been reprinted by *The Astragal Press* of Mendham, New Jersey in 1995.

Withington, Sidney, et al., *"Engineering in History"*, McGraw-Hill Book Company, 1950, republished by Dover Publications, Inc., Mineola, New York, 1990, pp. 129–133.

Wolf, Harold Eichholtz, *"**Introduction to Non-Euclidean Geometry**"*, The Dryden Press, 1945, republished by Dover Publications, Inc., Mineola, New York, 2012.

© Springer International Publishing Switzerland 2016
E. C. Tonias, C. N. Tonias, *Geometric Procedures for Civil Engineers*, DOI 10.1007/978-3-319-24295-8

Index

© Springer International Publishing Switzerland 2016
E.C. Tonias, C.N. Tonias, *Geometric Procedures for Civil Engineers*,
DOI 10.1007/978-3-319-24295-8

Printed in the United States
By Bookmasters